Statistical Mechanics

From Thermodynamics to the Renormalization Group

James H. Luscombe

CRC Press
Taylor & Francis Group
Boca Raton London New York

CRC Press is an imprint of the
Taylor & Francis Group, an **informa** business

First edition published 2021
by CRC Press
6000 Broken Sound Parkway NW, Suite 300, Boca Raton, FL 33487-2742

and by CRC Press
2 Park Square, Milton Park, Abingdon, Oxon, OX14 4RN

Library of Congress Cataloging-in-Publication Data

Names: Luscombe, James H., 1954- author.
Title: Statistical mechanics : from thermodynamics to the renormalization
  group / James Luscombe.
Description: First edition. | Boca Raton : CRC Press, 2021. | Includes
  bibliographical references and index.
Identifiers: LCCN 2020048146 | ISBN 9781138542976 (paperback) | ISBN
  9780367689278 (hardback) | ISBN 9781003139669 (ebook)
Subjects: LCSH: Statistical mechanics.
Classification: LCC QC174.8 .L87 2021 | DDC 530.13--dc23
LC record available at https://lccn.loc.gov/2020048146

ISBN: 978-0-367-68927-8 (hbk)
ISBN: 978-1-138-54297-6 (pbk)
ISBN: 978-1-003-13966-9 (ebk)

Typeset in Computer Modern font
by KnowledgeWorks Global Ltd.

Access the Support Material: www.routledge.com/9781138542976

Statistical Mechanics

To my mother, who likes orderliness, and to my father who didn't. They taught me without trying that the equilibrium state is one of lowest energy and maximum disorder. To my father, who taught me to love the wildness of nature, and to my mother, who taught me to seek order, which I imported into the world of ideas. To my wife, Lisa, who encouraged me, and who lends an order of her own to the written language. To my children, Jennifer and Jimmy, who represent the future and have brought me so much joy.

Contents

Preface

STATISTICAL mechanics, a branch of theoretical physics with applications in physics, chemistry, astronomy, materials science, and even biology, relates the observable properties of macroscopic systems to the dynamics of their microscopic constituents. A definition of *macroscopic* is elusive. Objects indiscernible to human senses—not ordinarily considered macroscopic—can contain enormous numbers of atoms,[1] and it's in that sense we consider a system macroscopic: when the number of its constituents is sufficiently large that the dynamics of individual particles cannot be tracked. At the quantum level that number is small,[2] $N \approx 40$. Whatever the size of systems that can be treated quantum mechanically, microscopic descriptions *are not possible* for $N \approx N_A$, where *Avogadro's number* $N_A = 6.022 \times 10^{23}$ is representative of the number of atoms in samples of matter.[3,4] Not only are microscopic descriptions impossible, they're *pointless*. Even if the wavefunction of N_A particles could be found, what would you do with it? Wouldn't you seek to reduce the complexity of the data through some type of averaging procedure? Statistical mechanics provides the means by which averages are calculated for systems in thermal equilibrium.

Historically, statistical mechanics arose from attempts to explain, on mechanistic grounds, the results of *thermodynamics*—a phenomenological theory of macroscopic systems in thermal equilibrium. A system is said to be in equilibrium when none of its properties appear to be changing in time. The *state* of equilibrium is specified by the values of *state variables*—quantities that can be measured on macroscopic systems.[5] *Equilibrium* (in thermodynamics) is the quiescent state where the properties of systems don't change, at least over the observation time of experiments; it's effectively a timeless state described by time-independent variables.[6] In statistical mechanics, however, the concept of equilibrium is altogether different, where measurable quantities *fluctuate* in time about mean values. Fluctuations are produced by random motions of the microscopic constituents of matter and are the reason statistical mechanics invokes probability as a fundamental tool.

What should the prerequisites be for a study of statistical mechanics? Thermodynamics, certainly. Thermodynamics, however, is a difficult subject, one that takes time to learn its scope and methods. Textbooks on statistical mechanics tend to assume a familiarity with thermodynamics that may not be warranted. In years past, thermodynamics and statistical mechanics were taught separately, a practice that's largely gone by the wayside. I have written a companion volume on thermodynamics, [3], a *précis* that emphasizes the structure of the theory, aimed at preparing stu-

---

[1]A cube of silicon 1 μm on a side contains on the order of 10^{10} atoms. It isn't just the number of particles, however. Critical phenomena (Chapter 7) are dominated by long-wavelength fluctuations where in principle the wavelength $\lambda \to \infty$; systems must be of sufficiently "macroscopic" dimensions to manifest critical fluctuations.

[2]At the present limits of computing technology, the quantum physics of ≈ 40 particles can be simulated.[1] Walter Kohn (1998 Nobel Prize in Chemistry for work on density functional theory) argued that the many-particle wavefunction $\psi(r_1, \cdots, r_N)$ is *not a legitimate scientific concept for* $N > N_0$, where $N_0 \approx 10^3$.[2]

[3]We assume you're familiar with the concept of mole, but we define it anyway. A *mole* of a chemical element consists of N_A atoms, such that its mass equals the element's atomic weight in grams. Such a number is possible because 1) the mass of an atom is almost entirely that of its nucleus and 2) nuclei contain integer numbers of essentially identical units, *nucleons* (protons or neutrons). $N_A = (1.6605 \times 10^{-24})^{-1} = 6.022 \times 10^{23}$, where 1.6605×10^{-24} g is the atomic mass unit.

[4]Even if one could conceive a computer large enough to solve for the dynamical evolution of N_A particles, where would one get the initial data with sufficient precision to permit such calculations?

[5]State variables include the volume V occupied by a system, pressure P, temperature T, internal energy U, and entropy S. The state of equilibrium could be described using Dirac notation, e.g., $|PVT\rangle$.

[6]We say *effectively* timeless because systems remain in equilibrium *until the environment changes*.

dents for statistical mechanics; I will refer to [3] when detailed results from thermodynamics are required. I begin the present book with a review of a full course in thermodynamics. Prepared students could start with Chapter 2, "From Mechanics to Statistical Mechanics", which lays out the problem considered by statistical mechanics and how to "get there" using classical mechanics.

This book was written for advanced undergraduate/beginning graduate students in the physical sciences. I don't hesitate to draw on materials from the standard courses in undergraduate physics curricula: analytical mechanics, quantum mechanics, electrodynamics, and mathematical methods (it's highly useful to know what an analytic function is). An exposure to solid-state physics (rudiments of crystallography, actually—unit cells, lattice constants, sublattices, etc.) would be helpful, but is not required. We restrict ourselves to systems in thermal equilibrium. The treatment of systems out of equilibrium—nonequilibrium statistical mechanics—would fill another book. SI units are used throughout.

Statistical mechanics is a theory that entails just what its name implies. It relies on general results from mechanics, classical and quantum, combined with averaging techniques derived from the theory of probability to calculate the equilibrium properties of macroscopic systems given information about their microscopic constituents. We emphasize *general* results from mechanics because statistical mechanics seeks to provide a theory as universal as thermodynamics—one that applies to any macroscopic system, independent of the nature of its particles and their interactions—and thus we must rely on conclusions that apply to all mechanical systems. Returning to prerequisites, a knowledge of analytical mechanics would be useful; a review of selected topics is given in Appendix C. The theory, however, is not just about systems described by classical dynamics. The theory must accomplish quite a bit—account for the macroscopic behavior of many particle systems—but in such in a way that it can incorporate the quantum nature of matter. One strategy for a text on statistical mechanics would be to work entirely within the framework of quantum mechanics. We develop the theory starting with classical mechanics and bring in quantum mechanics separately. In many cases, the resulting expressions from quantum statistical mechanics agree with those of the classical theory. More generally, quantum statistical mechanics agrees with the classical theory in the high-temperature regime, and thus logically the classical theory should stand in its own right.

If I were to design the ideal curriculum for physical scientists, it would include a good course in probability. Because that's typically not part of the preparation of students, Chapter 3, "Probability Theory", provides a review of standard topics required for statistical mechanics; prepared students could skip this chapter. We develop in Chapter 3 the *non-standard* topic of cumulants, which find use in the treatment of interacting systems (Chapters 6 and 8). While cumulants are an advanced topic, students who go on to courses in quantum field theory and many body physics will find an exposure to cumulants in the guise of statistical mechanics beneficial.

Statistical mechanics is a different kind of subject than other branches of physics, such as mechanics and electromagnetism with their precisely stated laws. While it's a hugely successful theory, as measured by its applications in disparate fields of scientific study, at the same time it's a theory the foundations of which are not as well understood as is our ability to apply it. The subject requires a certain maturity on the part of students to understand that, while our knowledge of its foundations represents the efforts of many physicists and mathematicians, there's still more to be learned. "The proof is in the pudding" encapsulates the attitude one should adopt toward statistical mechanics. The theory works, and works well, but its foundations are not as well understood as we might like. Chapter 2 presents a more careful look at foundations than might be found in other books, knowing that in the limited space of a textbook one cannot cover all topics to the depth one might want. The goal here is not to write a reference work, but instead something that stimulates students to think about the subject.

The book is divided into two parts. Part I (Chapters 1–4) provides the theoretical structure of statistical mechanics in terms of the ensemble formalism. Part II (Chapters 5–8) covers applications, divided into systems of noninteracting particles (Chapter 5), systems of interacting particles (Chapter 6), and phase transitions and critical phenomena (Chapters 7 and 8). We end the book with the renormalization group, rather fitting in my opinion, as it can treat systems *dominated* by fluctuations. The existence of fluctuations is the entrée into statistical mechanics—fluctuations motivate the concepts of ensembles and probability on which the subject is based (see Chapter 2). For most systems, the effects of fluctuations are small. The renormalization group is, in a sense, the completion of equilibrium statistical mechanics in its ability to handle large fluctuations. Chapters 1–5 would form the backbone of a fairly standard one-quarter, upper-division course on statistical physics. Chapters 6–8 could form the basis of a graduate-level course appropriate for some instructional programs. Ambitious programs could cover the entire book in one semester. From thermodynamics to the renormalization group, this book guides you on an arduous journey—one well worth the effort to undertake.

It's an understatement to say that statistical mechanics is a big subject, and the choice of topics for this book has been daunting. Statistical mechanics sits at the intersection of physics, chemistry, and mathematics, and for that reason can appear as a multi-headed hydra. I have sought to develop general principles of statistical mechanics, followed by representative applications. Students of condensed matter physics will apply statistical mechanics to superfluids and superconductors, the Kosterlitz-Thouless transition, thermionic emission, diamagnetism, Thomas-Fermi theory, phonons—the list is seemingly endless. Students of chemical physics will study polyatomic molecules, polymers, and reaction-rate theory. I have sought to include a handful of topics that illustrate the essential ideas. In keeping the emphasis to a relatively few topics, the topics that have been included are treated in some detail, especially the points that experience shows students find mystifying. An omission from this book is the use of computer simulations, which can add insight through visualization of the results of statistical mechanics that have otherwise been derived mathematically. Gould and Tobochnik[4] is recommended for its focus on simulations.

I thank the editorial staff at CRC Press, in particular Rebecca Davies and Dr. Kirsten Barr. I thank my family, for they have seen me too often buried in a computer. To their queries, "How's it coming?" came the usual reply: slow, glacially slow. My wife, Lisa, I thank for encouragement and countless conversations on how not to mangle the English language. I'd like to record my gratitude to Professor Gene F. Mazenko, who sparked my interest in statistical mechanics. To the students of NPS, I have learned from you, more than you know. Try to remember that every note counts in a song and that science is a "work in progress"; more is unknown than known.

James H. Luscombe

Monterey, California

I

Structure of Statistical Mechanics

CHAPTER 1

Thermodynamics

Equilibrium, energy, and entropy

W E present a review of thermodynamics, which of necessity must be rather condensed.

1.1 SYSTEMS, BOUNDARIES, AND VARIABLES

System is a generic term used to indicate that little (yet macroscopic[1]) part of the universe we consider to be of interest, such as a solid or a container of gas. Systems are separated from their *environment* or *surroundings* by *boundaries*, indicated schematically in Fig. 1.1. There must be a demarcation between a system and its environment—one can't model the entire universe. The boundary isn't a mental construct; it's an actual physical boundary, *not part of the system*.

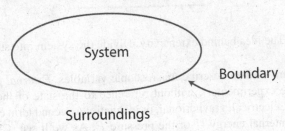

Figure 1.1: Boundary separating a thermodynamic system from its surroundings.

While boundaries (or *walls*) aren't part of systems, they determine the behavior of systems in terms of the interactions they permit with the environment;[2] see Table 1.1. *Diathermic* and *adia-*

Table 1.1: Systems, boundaries, and interactions

Type of system	Boundary	Interaction with environment
Isolated	Rigid adiabatic	None
Adiabatically isolated	Moveable adiabatic	Mechanical (adiabatic work only)
Closed	Diathermic	Thermal (flow of heat, mass fixed)
Open	Permeable	Permits flow of energy and matter

---

[1]The pitfalls in defining *macroscopic* are discussed in the Preface.
[2]As in other areas of mathematical physics, boundary conditions determine, or select the behavior of physical systems.

batic boundaries are those that do and do not conduct heat. Systems enclosed by rigid, immovable adiabatic walls are *isolated*—no interactions with the environment can happen. Adiabatic boundaries are an idealization, but they can be so well approximated in practice, as in a vacuum flask, that it's not much of a stretch to posit the existence of perfectly heat-insulating walls. *Adiabatically isolated* systems are enclosed by *moveable* adiabatic walls; such systems interact with the environment through mechanical means only. *Closed systems* have a fixed amount of matter. Systems separated by diathermic walls are said to be in *thermal contact*. *Open systems* are surrounded by *permeable* walls that permit the flow of matter and energy between system and surroundings.

A useful classification of the quantities appearing in macroscopic descriptions is the distinction between *intensive* and *extensive*.[3] Intensive quantities have the same values at different spatial positions in the system, such as pressure P, temperature T, or chemical potential,[4] μ, and are independent of the size of the system. Extensive quantities are proportional to the number of particles in the system N, and are characteristic of the system *as a whole*, such as entropy[5] S, volume V, or magnetization[6] M. To visualize extensivity, consider a system symbolically represented by the rectangle in Fig. 1.2. Divide the system into nine subsystems. Extensive quantities are *additive* over subsystems.[7] In this example, $V = 9V_{subsystem}$. Intensive quantities occur in the theory as partial derivatives of one extensive variable with respect to another (see Eq. (1.22) for example) and therefore extensive variables might seem more fundamental than intensive. Through the use of Legendre transformations (see Appendix C), intensive variables can equally well be considered fundamental.[3, p55]

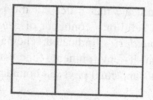

Figure 1.2: The idea behind extensivity: Divide the system into subsystems.

Three other terms are used to describe macroscopic variables. *External parameters* are set by the environment (or the experimenter), without reference to the state of the system, such as the values of external fields, electrical, gravitational, or magnetic. A second term consists of *mechanical quantities* such as the internal energy U or the pressure P. As we'll see (Chapter 4), mechanical quantities are calculated in statistical mechanics from averages over the microscopic dynamical (i.e., mechanical) variables of the system. A third, and distinctly thermodynamic category is that of *thermal quantities*, temperature and entropy. Thermal quantities have no microscopic interpretation and can only be associated with a macroscopic level of description.

---

[3]When we speak of macroscopic quantities, it's assumed we're referring to their values in the state of equilibrium.

[4]While P and T are familiar, the chemical potential μ is typically not. It's introduced in Section 1.6.

[5]If you're unfamiliar with entropy, you're not alone. An attempt is made here in Chapter 1 to shed light on the subject. Thermodynamics mainly *is* about entropy.

[6]In thermodynamics M is the total dipole moment of the system. In electromagnetic theory, M is the magnetization *density*, magnetic moment per volume. In thermodynamics M is an extensive quantity; in electromagnetism M is intensive.

[7]The dichotomy between intensive and extensive is fundamental to thermodynamics and statistical mechanics. While the distinction is introduced as physically reasonable, thermodynamics has no tools to prove the *existence* of extensive quantities. Using the methods of statistical mechanics (see Section 4.2), the existence of the *thermodynamic limit* can be established of $N \to \infty$, $V \to \infty$ with N/V fixed (the number density).

1.2 INTERNAL ENERGY: WORK AND HEAT

In the performance of *adiabatic work* W_{ad}—work on adiabatically isolated systems—it's found that transitions $i \to f$ produced between reproducible equilibrium states[8] (i, f) depend only on the amount of work and not on *how* it's performed. Regardless of how the proportions of the *types* of work are varied, the same *total amount* of adiabatic work results in the same transition $i \to f$. *This discovery is of fundamental importance.* If the transition $i \to f$ produced by adiabatic work is independent of the means by which it's brought about, it can depend only on the initial and final states (i, f). That implies the existence of a physical quantity associated with equilibrium states that "couples" to adiabatic work, the *internal energy* U, such that[9]

$$\Delta U = U_f - U_i = W_{ad} . \tag{1.1}$$

Internal energy is a state variable—one that depends only on the *state* of equilibrium of the system and *not on how the system was prepared in that state*. Adiabatic work done *on* a system increases its internal energy and is taken as a positive quantity. Adiabatic work done *by* a system (somewhere in the environment a weight is higher, a spring is compressed) is accompanied by a decrease in internal energy, and is taken as a negative quantity. *Changes in internal energy come at the expense of adiabatic work done on or by a system.* Equation (1.1) expresses *conservation of energy*: If we don't let heat escape, work performed on the system is *stored* in its internal energy, energy that can be recovered by letting the system do adiabatic work on the environment. We'll say that internal energy is the *storehouse* of adiabatic work.

Now, let work W be performed under nonadiabatic conditions. It's found, for the same transition $i \to f$ produced by W_{ad}, that $W \neq \Delta U$. *The energy of mechanical work is not conserved in systems with diathermic boundaries.* Energy conservation is one of the sacred principles of physics, and we don't want to let go of it. The principle can be restored by recognizing different *forms of energy*.[10] The *heat transferred to or from the system*, Q, is the difference in work

$$Q \equiv \Delta U - W \tag{1.2}$$

that effects the same change in state of systems with the two types of boundaries, $Q = W_{ad} - W$. If it takes more work W to produce the same change of state as that under adiabatic conditions, $Q < 0$: heat *leaves the system* by flowing through the boundary; $Q > 0$ corresponds to heat entering the system.[11] Equation (1.2) is the *first law of thermodynamics*. One might think it applies to closed systems only, based on how we've formulated it. It applies to open systems when we introduce another kind of work—*chemical work*—the energy required to change the amount of matter in the system (see Section 1.6). The point here is that *the nature of the boundaries allows us to classify different types of energy*.

The first law in differential form is written

$$dU = đQ + đW , \tag{1.3}$$

where đ signifies an *inexact differential*.[12] The notation on the left of Eq. (1.3) indicates that dU represents a small change in a physical quantity, the internal energy of the system. The notation on

---

[8]The term *equilibrium*, like macroscopic, is difficult to define precisely. See the discussion in the Preface and in [3, p4].

[9]Equation (1.1), $\Delta U = W_{ad}$, indicates that the change in a physical quantity, the internal energy U, is the same as the amount of adiabatic work done on the system. If that's all there was to internal energy, Eq. (1.1) would simply be a change of variables. But there *is* more to internal energy than W_{ad}. There are different *forms* of energy, with ΔU the sum of the work done on the system and the heat transferred to it, Eq. (1.2).

[10]Neutrinos were hypothesized for the purpose of preserving energy conservation in nuclear beta decay.

[11]We have adopted a consistent sign convention: Q and W are positive if they represent energy transfers *to* the system.

[12]An expression $P(x, y)dx + Q(x, y)dy$ is a *differential form*. If $\partial P/\partial y = \partial Q/\partial x$ (the *integrability condition*), differential forms represent *exact differentials* (differentials of integrable functions).[3, p6] Inexact differentials đg can be made exact through the use of *integrating factors*, $\lambda(x, y)$, such that $df = \lambda(x, y)đg$.

the right indicates that there do *not* exist quantities Q and W of which đQ and đW represent the differentials of; rather đQ and đW denote *modes* of small energy transfers to the system. Work is not a substance, neither is heat. Both are forms of energy. Inexact differentials cannot be integrated in the usual sense; the value ascribed to the integral of an inexact differential $h(x, y) = \int_C^{x,y} dh$ depends on the path of integration C—the "function" defined this way depends not just on (x, y), but on "how you get there," or in the language of thermodynamics, on the *process*. See Exercise 1.1. We have in Eq. (1.3) that the sum of inexact differentials is exact, a mathematical expression of the physical content of the first law that energy can be converted from one form to another. For finite changes, Eq. (1.3) is written as in Eq. (1.2), $\Delta U = Q + W$: There is a quantity, the internal energy U, such that changes ΔU are brought about by transferring heat Q and/or by the performance of work W.

Small amounts of work are expressed in terms of *generalized displacements* dX of extensive quantities X, the changes of which can be observed macroscopically, and their conjugate *generalized forces* Y, intensive quantities such that the product YX has the dimension of energy. For an infinitesimal expansion against pressure P, đ$W = -PdV$: If $dV > 0$ (work is done on the environment), $dU < 0$. The work done (on the system) in stretching a wire under tension J of length L by dL is JdL; the work done in expanding the surface area of a film by dA is σdA, where σ is the surface tension. These examples involve extensions of the boundaries of the system: volume, length, and area. More generally, *work involves any change in macroscopically observable properties* not necessarily involving the boundaries of the system. The latter entail interactions with externally imposed fields. The work done (on the system) in changing the magnetization of a sample by dM in a field H is (in SI units) $\mu_0 H \cdot dM$[3, p17]. The work done in changing the electric polarization P by dP in electric field E is $E \cdot dP$. The symbol đW represents all possible modes of infinitesimal energy transfers to the system (through work),

$$đW = \sum_i Y_i dX_i \,, \tag{1.4}$$

where Y_i denotes the generalized forces $(-P, J, \sigma, \mu_0 H, E)$ associated with generalized displacements dX_i, the differentials of extensible quantities,[13] (dV, dL, dA, dM, dP). A refinement of Eq. (1.3) is therefore[14]

$$dU = đQ + \sum_i Y_i dX_i \,. \tag{1.5}$$

Based on Eq. (1.5), can we say what the variables are that U depends on? Certainly U is a function of the extensive variables in the problem, $U = U(X_i)$. What about the heat term? Is there a state variable that couples to heat transfers?[15] Heat is extensive in the sense that *heat capacities* (see Section 1.8.3) $C_X \equiv (đQ/dT)_X$ are mass dependent, where X denotes what is held fixed in the process. The heat capacity is the heat required to change the temperature of an object from $T \to T + \Delta T$, and that quantity scales with the size of the system. Heat is extensive; it couples to the system as a whole. It does not, however, directly couple to the observable quantities associated with work,[16] which begs the question: *Is* there an extensive property of macroscopic systems (call it S) that heat transfers couple to, such that đ$Q = $ (conjugate intensive variable) $\times dS$? There is, and that property is *entropy*.

---

[13] An extensible quantity is (as the name suggests) one that can be extended, the change in an extensive variable.[3, p17]

[14] Equation (1.5) applies to closed systems having a fixed amount of matter; it generalizes to open systems that allow for the exchange of particles with the environment by the inclusion of chemical work; see Section 1.6.

[15] The heat required to bring a system from state A to state B, $\int_A^B đQ$, is not well defined: It depends on the path between A and B. The question of how much heat is contained in a system is therefore meaningless; heat is not a state variable.

[16] As heat is added to a system its *temperature* rises. Temperature, however, is an intensive quantity.

1.3 CLAUSIUS ENTROPY: IRREVERSIBILITY, DISORGANIZATION

Entropy—discovered through an analysis of the second law of thermodynamics—is an unexpected yet significant development.[17] The *Clausius inequality* is a consequence of the second law:[3, p33]

$$\oint \frac{dQ}{T} \leq 0 , \tag{1.6}$$

where the integral is over all steps of a *cyclic process* (one that returns a system to its initial state), T is the *absolute temperature*[18] at which heat transfer dQ occurs, and where equality in (1.6) holds for *reversible* heat transfers,[19]

$$\oint \frac{(dQ)_{\text{rev}}}{T} = 0 . \tag{1.7}$$

Differentials of state variables are exact.[20] Exact differentials dg have the property that $\oint_C dg = 0$ for any integration path[21] C. We infer from Eq. (1.7) the existence of a state variable, entropy, the differential of which is[22]

$$dS \equiv (dQ)_{\text{rev}}/T . \tag{1.8}$$

The quantity T^{-1} is the integrating factor[23] for $(dQ)_{\text{rev}}$. As a state variable, *entropy is defined only in equilibrium*. Changes in entropy between equilibrium states (A, B) are found by integrating its differential, Eq. (1.8),

$$S(B) - S(A) = \int_A^B (dQ)_{\text{rev}}/T$$

---

[17]The second law concerns the direction of heat flow. There are several equivalent ways of stating the second law[3, p139], but basically heat does not spontaneously flow from cold to hot. The first law places restrictions on the possible changes of state a system may undergo, those that conserve energy. There are processes, however, that while *conceivable* as energy conserving, are never observed. What distinguishes naturally occurring processes is the direction of heat flow.

[18]Absolute temperature is another consequence of the second law [3, p29]. *Temperature* is a property that systems in equilibrium have in common (zeroth law of thermodynamics). *Empirical temperatures* are based on the thermometric properties of substances, e.g., the height of a column of mercury [3, p12]. *Absolute* temperature is based on reversible heat transfers and is independent of the properties of materials; it's measured on the Kelvin scale.[3, p31]. We'll use T exclusively to denote absolute temperature. Absolute temperature gives meaning to statements such as one temperature is twice as large as another (which is a colder day, 0 °C or 0 °F?). It also indicates that $T \to 0$ K is possible. Whether $T = 0$ is experimentally achievable is another matter (the province of the third law of thermodynamics); the larger point is that absolute temperature provides a framework in which zero temperature can be discussed.

[19]Reversibility is a central concept of thermodynamics, on par with equilibrium. Equilibrium is determined by the environment (Section 1.12). If external conditions are altered slightly and then restored, the system will return to its original state. Reversible processes can be *exactly reversed through infinitesimal changes in the environment* [3, p14]. In reversible processes, the system *and* the environment are restored to their original conditions. An *irreversible* process is one in which the system cannot be restored to its original state without leaving changes in the environment. A change of state can occur reversibly if the forces between the system and the surroundings are essentially balanced at all steps of the process, and heat exchanges occur with essentially no temperature differences between system and surroundings. The concept of reversible process is an idealization, yet it can be operationalized: In any experiment there will be some small yet finite rate at which processes occur, such that *disequilibrium* has no observable consequences within experimental uncertainties.

[20]State variables characterize the state of equilibrium, wherein all memory is lost of *how* they have come to have their values. (In saying that a sample of water has a given temperature, one has no knowledge of how many times in the past week it has frozen and thawed out.) There are an unlimited number of possible "paths" $i \to f$ between equilibrium states (i, f). Because the values of state variables are independent of the history of the system, *changes* in state variables must be described independently of the manner by which change is brought about. That type of path independence is ensured mathematically by requiring differentials of state variables to be exact.

[21]A necessary and sufficient condition for dg to be exact is that the integral $\int_A^B dg = g(B) - g(A)$ be independent of the path of integration [3, p6]. If $B \equiv A$, $\oint_C dg = 0$ for any integration path C.

[22]Entropy makes its entrance onto the stage of physics in the form of a differential. We know what dS is, i.e., how to *change* entropy, but do we know what S is? Stay tuned.

[23]The conditions under which inexact differentials (such as dQ) possess integrating factors is fundamental to the theory of thermodynamics. The Carathéodory formulation centers on the existence of integrating factors.[3, p149]

for any path connecting A and B, such that heat transfers occur reversibly at all stages.[24]

Figure 1.3 shows a cycle consisting of an irreversible process from point A in state space[25] to

Figure 1.3: Irreversible process from $A \to B$, reversible process from $B \to A$.

point B, followed by a reversible process from B to A. The irreversible process is indicated as a dashed line because it cannot be represented in state space. Applying inequality (1.6) to the process in Fig. 1.3, $\int_A^B đQ/T + \int_B^A (đQ)_{\mathrm{rev}}/T \le 0$. Because the process from B to A is reversible,

$$\int_A^B (đQ)_{\mathrm{rev}}/T \equiv \int_A^B dS \ge \int_A^B đQ/T \,,$$

from which we infer the *Clausius inequality in differential form*:

$$dS \ge đQ/T \,. \tag{1.9}$$

The quantity $đQ$ in (1.9) is not restricted to reversible heat transfers (which apply in the case of equality). The closed-loop integral of inequality (1.9) reproduces inequality (1.6).

Inequality (1.9) informs us that $đQ/T$ *does not account for all contributions to* dS. There must be another *type* of entropy to close the gap between $đQ/T$ and dS. Define the difference

$$dS_i \equiv dS - đQ/T \,. \tag{1.10}$$

An equivalent form of the Clausius inequality is thus $dS_i \ge 0$. Heat transfers occur through system boundaries and can be *positive, negative, or zero*. Only for *reversible* heat transfers is $\Delta S = \int (đQ)_{\mathrm{rev}}/T$. Otherwise, in irreversible processes, *entropy is created,* $\Delta S_i > 0$. Entropy can vary for two reasons, and two reasons only: either from entropy transport to or from the surroundings by means of heat transfers, or by the *creation of entropy* from irreversible changes in state. *Irreversible processes create entropy.*[26] Equation (1.10) exhibits the same logic as the first law, Eq. (1.2), in that whereas the heat transferred to the system is the difference between the change in internal energy and the work done, $Q = \Delta U - W$, the entropy created through irreversibility is the difference between the change in entropy and that due to heat transfers, $\Delta S_i = \Delta S - \int đQ/T$.

An implication of inequality (1.9) is that processes exist in which entropy changes occur between equilibrium states *not due to heat exchanges with the environment*. The classic example is the *free expansion* (left side of Fig. 1.4), the flow of gas through a ruptured membrane that separates a gas from a vacuum chamber in an otherwise isolated system. Even without heat transfers to or from

---

[24]While the heat $\int_A^B đQ$ required to bring a system from state A to state B is not well defined, with entropy we have a quantity closely related to heat transfers, which *is* determined by the state of the system. For a system brought from A to B, the quantity $\int_A^B (đQ)_{\mathrm{rev}}/T$ has a value independent of path, so long as the path between A and B is reversible.

[25]*Thermodynamic state space* is the mathematical space of the values of state variables (which by definition are their equilibrium values). A state of equilibrium is represented as a point in this space [3, p5]. Reversible processes can be approximated as proceeding through a sequence of equilibrium states, and thus can be plotted as curves in state space [3, p15]. Irreversible processes cannot be so represented.

[26]We asked at the end of Section 1.2 whether an extensive system property exists that couples to heat transfers. With entropy we have found that, and then some. Entropy is created in irreversible processes between equilibrium states.

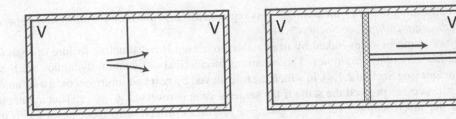

Figure 1.4: Irreversible free expansion (left); reversible adiabatic expansion (right).

the environment (adiabatically isolated system), there's an increase in entropy because the process is irreversible. With đ$Q = 0$ in (1.9),

$$dS \geq 0 \, . \qquad \text{(adiabatically isolated systems)} \qquad (1.11)$$

This is a remarkable state of affairs. For systems not interacting with the environment, *one may still affect changes between equilibrium states*! The "interaction" in a free expansion consists of ruptur-ing a membrane or opening a valve, acts that can be accomplished without transferring energy to the system. That transitions between equilibrium states can be induced without transferring energy indicates that *another state variable must exist*.[27] Entropy is that new state variable.

Inequality (1.11) is one of the most far-reaching results in physics: *The entropy of an isolated system can never decrease*, only increase. In reaching equilibrium, therefore, *the entropy of an iso-lated system has the maximum value it can have* (subject to constraints on the system). Time sneaks into the theory of thermodynamics through entropy because of the direction of time exhibited by irreversible processes—entropy is a proxy for time.[28] *The values of S can be put into correspon-dence with the time order established by irreversibility*. The laws of physics are usually *equalities* ($F = ma$, $E = mc^2$, $\Delta U = Q + W$), whereas (1.11) is an *inequality*. Therein lies its universality! The second law is *not* an equality, $A = B$. It specifies the time progression of spontaneous processes *regardless of the system*.

We've been led mathematically to the *existence* of entropy (starting from the Clausius inequality, (1.6)), yet we have no idea what it *is*.[29] Insight into its nature can be had by noting that the rupturing of the membrane in a free expansion is the *removal of a constraint*. With the system initially confined to volume V, by rupturing the membrane a constraint is removed and the system evolves to occupy a new volume, $2V$. Entropy captures the propensity of macroscopic systems to "explore" all available possibilities.[30] The removal of constraints is irreversible. Once removed, *constraints cannot be put back*, at least not without the performance of work, implying irreversibility. Entropy can thus be seen as a measure of the *degree of constraint*: The second law captures the tendency of systems to spontaneously evolve from states of more to less constraint. Because the entropy of isolated systems can only increase, states can be classified according to the *order* in which they occur in a process of progressively removing constraints. Constraints *restrict* the possibilities available to systems, and in that sense they *organize* them. A system undergoing a sequence of changes brought about by the

[27]Thus, there's more to the description of equilibrium states than internal energy and temperature, which one would infer from the first and zeroth laws of thermodynamics.

[28]In statistical mechanics, time rears its head through the phenomena of *fluctuations*. As we'll see in Chapter 2, entropy allows us to establish the *probability* of the occurrence of fluctuations of various magnitudes.

[29]Try not to obsess over the meaning of entropy, but instead get to know it. Asking "what is entropy?" presumes that an answer can be given in terms of familiar concepts. What if entropy is something *new*, not reducible to concepts gained through prior experience? Get to know its properties, then you'll know what *it* is. How does it change with temperature at constant volume? How does it change with temperature at constant pressure? If we know how it behaves, we know a great deal about what it *is*.

[30]Maxwell wrote that "The second law of thermodynamics has the same degree of truth as the statement that if you throw a tumblerful of water into the sea, you cannot get the same tumblerful of water out again." Quoted in [5].

removal of constraints becomes increasingly less organized. *There must be a connection between entropy and disorganization* (see Section 1.11).

The time ordering of states linked by irreversible processes is a distinctive feature of thermo-dynamics, *unlike any other in physics.* The equations of classical or quantum dynamics are *time-reversal invariant* (see Section 2.1.4), in which the role played by past and future occur in a symmet-ric manner;[31] processes proceed the same if the sense of time is reversed, $t \to -t$. That symmetry is broken at the macroscopic level where there's a *privileged direction of time*, an order to the sequence of spontaneous events, the *arrow of time.*[32] Why microscopic theories are time-reversal invariant while the macroscopic world is not, we cannot say.[33] The second law cannot be derived from microscopic laws of motion; *it's a separate principle.* Irreversibility is an undeniable feature of the macroscopic world, *our experience of which is codified in the second law.*[34]

The first and second laws establish internal energy and entropy as state variables.[35] The first law places restrictions on the possible states that can be connected by thermodynamic processes—those that conserve energy. The second law determines the *order* of states as they occur in spontaneous processes; whether the transition $A \to B$ is possible in an isolated system is determined by whether $S(A) < S(B)$, which is referred to as the *entropy principle.*

How does our newly found state variable fit in with the structure of thermodynamics? With $(đQ)_{\text{rev}} = T \mathrm{d}S$ and $đW = -P \mathrm{d}V$, we have from Eq. (1.3)

$$\mathrm{d}U = T\mathrm{d}S - P\mathrm{d}V \tag{1.12}$$

when it's understood that all infinitesimal changes are reversible. With Eq. (1.12) we have the first law expressed in terms of exact differentials. Equation (1.12) can be integrated to find ΔU for *any reversible process connecting initial and final states.* A little reflection, however, shows that Eq. (1.12) holds *for any change of state, however accomplished,* as long as there is *some* conceivable reversible path connecting the initial and final equilibrium states.[36]

As an example, consider the free expansion of an adiabatically isolated *ideal gas* that's allowed to expand into an evacuated chamber, doubling its volume in the process (left side of Fig. 1.4). The entropy change ΔS—which must be positive because the process is irreversible—can be calculated if we can conceive a reversible path connecting the initial and final states, say if a piston is slowly moved over, allowing the gas to double its volume (right side of Fig. 1.4). The internal energy of ideal gases is independent of volume and depends only on temperature—*Joule's law* (see Eq. (1.44)). In this process $\mathrm{d}U = 0$, implying that the temperature does not change.[37] From Eq. (1.12) with $\mathrm{d}U = 0$ and using the ideal gas equation of state[38] $PV = NkT$, where k is *Boltzmann's*

---

[31] Maxwell's equations of electrodynamics are also time-reversal symmetric.

[32] The term arrow of time (actually *time's arrow*) was introduced by Eddington.[6, p69]

[33] At the *really* microscopic level, time-reversal symmetry *is* broken. By the CPT theorem of quantum field theory, any Lorentz-invariant theory (commutation relations obey the spin-statistics connection) is invariant under the combined operations of C (charge conjugation, interchange particles for their antiparticles), P (parity, inversion of space axes through the origin), and T (invert time, $t \to -t$). The discovery of CP violations in weak decays implies (if CPT invariance is sacred) that time reversal symmetry is broken at the sub-nuclear level for processes involving weak interactions. At the *super* macroscopic level, we have an expanding universe, which provides a sense for the direction of time.

[34] Said differently, the second law is the explicit recognition in the pantheon of physical laws of the existence of irre-versibility. We could draw a line right here and say that the existence of irreversible processes along with the notion of entropy *is* the essential core of thermodynamics.

[35] The zeroth law of thermodynamics establishes temperature as a state variable [3, p11]. The third law of thermodynamics doesn't establish a state variable; it's concerned with entropy as T approaches absolute zero [3, p121].

[36] If entropy exists for one equilibrium state, it exists for all states connected to the first by a reversible path. Can every point in state space be reached by a reversible path? In the Carathéodory formulation of the second law [3, p149], the question is posed differently: Starting from a given point in state space, are there points *not* reachable by *adiabatic* processes, those with $đQ = 0$? Entropy exists if there are states *inaccessible* by adiabatic processes.

[37] It follows from the first law that $\mathrm{d}U = 0$ because $đQ = 0$ (adiabatic isolation), and no work is performed in expanding into a vacuum—because $P = 0$, $đW = 0$.

[38] An *equation of state* is a functional relation among the state variables of a system. The existence of equations of state is a consequence of the zeroth law of thermodynamics.[3, p12]

constant,[39] we have $T\Delta S = NkT \int dV/V$. Thus, $\Delta S = Nk \ln 2$. What has changed as a result of a free expansion? Not the energy, nor the number of particles. What's changed (in an increase in entropy) is the increased number of *ways* particles can be located in the increased system volume— see Section 1.11.

1.4 THERMODYNAMIC POTENTIALS

Equilibrium is specified by the values of state variables. Whether measured or postulated by the laws of thermodynamics, state variables are not independent of each other (peek ahead to Eq. (1.53)). Depending on the system, there may be variables that are not readily subject to experimental control and yet others that are. Legendre transformations (defined in Appendix C) provide a way of obtaining equivalent descriptions of the energy of a system known as *thermodynamic potentials*, which involve variables that may be easier to control.

Three Legendre transformations of $U(S, V)$ can be formed from the products of variables with the dimension of energy (when the number of particles is fixed): TS and PV (our old friends heat and work).[40] They are:

$$
\begin{aligned}
F &\equiv U - TS && \text{(Helmholtz free energy)} \\
H &\equiv U + PV && \text{(enthalpy)} \\
G &\equiv U - TS + PV = F + PV = H - TS. && \text{(Gibbs free energy)}
\end{aligned}
\tag{1.13}
$$

The relationships amongst these functions are shown in Fig. 1.5. Their physical interpretation is

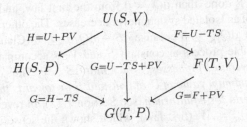

Figure 1.5: The four ways to say energy: Legendre transformations of the internal energy function (for fixed particles numbers).

discussed below. The *natural variables* for these functions are found by forming their differentials:

$$
\begin{aligned}
dU &= TdS - PdV & dH &= TdS + VdP \\
dF &= -SdT - PdV & dG &= -SdT + VdP.
\end{aligned}
\tag{1.14}
$$

Thus, $U = U(S, V)$, $H = H(S, P)$, $F = F(T, V)$, and $G = G(T, P)$. The differentials of U, H, F, and G each involve a different "mixture" of differentials of extensive and intensive quantities, multiplied respectively by intensive and extensive quantities. The transformation $U(S, V) \to F(T, V)$ (for example) shifts the emphasis from S as an independent variable to its associated slope $T = (\partial U/\partial S)_V$ as the independent variable. *The duality between points and slopes*

---

[39] All gases become ideal in the low-pressure limit [7]. (Low pressure implies low density, a limiting case in which particles don't interact; they can't interact if they only rarely encounter each other.) For n moles of any gas it's found that $\lim_{P \to 0}(PV/nT) = R = 8.314$ J (K mole)$^{-1}$, where R is the *gas constant*, a universal property of matter. The ideal gas is a fictional substance *defined* as having the equation of state $PV = nRT$ for all pressures. Boltzmann's constant is the gas constant divided by Avogadro's number, $k \equiv R/N_A = 1.381 \times 10^{-23}$ J K^{-1}. Note that $nR = Nk$.

[40] The three Legendre transformations of $U(S, V)$ bring to mind the three Legendre transformations in classical mechanics of the generating function of canonical transformations (see Section C.8).

*achieved by the Legendre transformation of convex functions is reflected in the formalism of ther-
modynamics as a duality between extensive and intensive variables.*[41]

The physical meaning of the thermodynamic potentials can be inferred from the Clausius inequality. It's important to recognize that U, H, F, and G are not "just" Legendre transformations
(possessing certain mathematical properties); they represent *energies* stored in a system under prescribed conditions. They are, as the name suggests, *potential energies.* Just as a knowledge of the
types of potential energy helps one solve problems in mechanics, knowing the interpretations of the
potentials can prove advantageous in efficiently working problems in thermodynamics. Combine
the Clausius inequality with the first law, Eq. (1.3):

$$T dS \geq dQ = dU - dW .$$

(1.15)

In what follows, we distinguish between the work due to changes in volume ($dW = -PdV$) and
all other forms of work, denoted dW': $dW \equiv -PdV + dW'$.

Internal energy

Consider a process in which no work is performed on the system, $dW = 0$. From (1.15),

$$T dS \geq dU .$$

(1.16)

We can interpret inequality (1.16) in two ways:[42]

$$[dS]_{U,V} \geq 0 \quad \text{or} \quad [dU]_{S,V} \leq 0 .$$

(1.17)

If U is fixed and no work is done, then $dQ = 0$ from the first law, and thus by inequality (1.9),
$[dS]_{U,V} \geq 0$; the entropy of an isolated system never decreases. The other part of (1.17), $[dU]_{S,V} \leq
0$, should be interpreted as $[dQ]_S \leq 0$ because $dS - dQ/T \geq 0$ (Clausius inequality), and thus
$[dU]_{S,V} \leq 0$. An irreversible process at constant S and V is accompanied by a *decrease* in U.
Heat leaves systems in irreversible processes,[43] such that $dS = 0$. Thus, *the system spontaneously
evolves either towards a minimum value of U at constant S or towards the maximum value of S at
constant U* (the *Gibbs criteria for equilibrium*, see Section 1.12). In reversible processes (the case
of equality in (1.15)), $[\Delta U]_S = W$ (first law), which shows the sense in which U is a potential
energy, the energy of work performed adiabatically.

Enthalpy

From inequality (1.15) and $dQ = dH - VdP - dW'$ (first law), we have

$$T dS \geq dQ = dH - dW' - VdP .$$

(1.18)

For a process at constant pressure and no other forms of work ($dW' = 0$),

$$T [dS]_P \geq [dH]_P .$$

(1.19)

Like inequality (1.16), we can interpret (1.19) in two ways:

$$[dS]_{H,P} \geq 0 \quad \text{or} \quad [dH]_{S,P} \leq 0 .$$

(1.20)

For constant H and P (and $dW' = 0$), $dH = 0 = dU + PdV$ implies $[dQ]_{H,P} = 0$ (first law);
thus, $[dS]_{H,P} \geq 0$ (Clausius inequality) is again the entropy of an isolated system can never decrease.[44] From (1.18), for constant P and $dW' = 0$, $dH = [dQ]_P$; *enthalpy is the heat added*

---

[41]Convex functions are defined in Appendix C; see also Section 4.3.
[42]The notation $[dX]_Y$ indicates a change in X holding Y fixed.
[43]In an irreversible process $dS_i > 0$; to keep $dS = 0$, heat must leave the system.
[44]We can treat a system for which H and P are held constant as isolated.

at constant pressure without other work performed. The word enthalpy is based on the Greek *enthalpein* ($\epsilon\nu\theta\alpha\lambda\pi\epsilon\iota\nu$)—"to put heat into." Enthalpy is referred to as the *heat function*. As for the other part of (1.20), $[dH]_{S,P} \leq 0$ is equivalent to $[dQ]_{S,P} \leq 0$ (inequality (1.18)). Again, heat leaves systems in irreversible processes at constant S. From (1.20), systems evolve spontaneously until equilibrium is achieved with H having a minimum value. For reversible processes, $[\Delta H]_{S,P} = W'_{ad}$; *enthalpy is the energy of "other" adiabatic work performed on the system.*

Helmholtz and Gibbs energies

By combining $TdS \geq dQ$ with the definitions of F and G, we infer the inequalities

$$dF \leq -SdT - PdV + dW' \qquad dG \leq -SdT + VdP + dW'.$$

The quantities F and G spontaneously decrease in value until a minimum is achieved for the equilibrium compatible with the constraints on the system ($dW' = 0$):

$$[dF]_{T,V} \leq 0 \qquad [dG]_{T,P} \leq 0.$$

For reversible processes $[\Delta F]_{T,V} = W'$ and $[\Delta G]_{T,P} = W'$; F and G are storehouses of the energy of other work performed under the conditions indicated.

The properties of the potentials are summarized[45] in Table 1.2. Additional interpretations of the Gibbs and Helmholtz energies are discussed in Sections 1.5 and 1.10.

Table 1.2: Thermodynamic potentials for closed systems (dW' is work other than PdV work)

Potential	Stored Energy	Spontaneous Change	In Equilibrium
Internal energy, $U = U(S, V)$ $dU = TdS - PdV + dW'$	$[\Delta U]_S = (W)_{adiabatic}$	$[\Delta U]_{S,V} \leq 0$	Minimum
Enthalpy, $H = H(S, P)$ $H = U + PV$ $dH = TdS + VdP + dW'$	$[\Delta H]_{S,P} = (W')_{adiabatic}$ $[\Delta H]_P = Q$	$[\Delta H]_{S,P} \leq 0$	Minimum
Helmholtz energy, $F(T, V)$ $F = U - TS$ $dF = -SdT - PdV + dW'$	$[\Delta F]_{T,V} = W'$ $[\Delta F]_T = W$	$[\Delta F]_{T,V} \leq 0$	Minimum
Gibbs energy, $G = G(T, P)$ $G = U - TS + PV$ $dG = -SdT + VdP + dW'$	$[\Delta G]_{T,P} = W'$	$[\Delta G]_{T,P} \leq 0$	Minimum

1.5 FREE ENERGY AND DISSIPATED ENERGY

Free energy and maximum work

By rewriting (1.15), we have the inequality $-dW \leq TdS - dU$. For there to be work done *by* the system (counted as a negative quantity), we must have $dU < TdS$, a generalization of $\Delta U = W_{ad}$: $\Delta U < 0$ if $W_{ad} < 0$. The *maximum* value of dW (counted as a negative quantity) is therefore $|dW|_{max} = TdS - dU$. Undoing the minus sign, $dW_{max} = dU - TdS = [dF]_T$. The Helmholtz energy is *the maximum obtainable work at constant T*: $W_{max} = [\Delta F]_T$.

---

[45] Other books adopt different sign conventions for heat and work; the inequalities in Table 1.2 may be reversed.

Thus, not *all* of the energy change ΔU is available for work if $\Delta S > 0$, which is the origin of the term *free energy*: *the amount of energy available for work*. For this reason F is called the *work function*. Enthalpy is the heat function, $[\Delta H]_P = Q$, and the Helmholtz energy is the work function, $[\Delta F]_T = W$. It's straightforward to show that $[\Delta H]_P + [\Delta F]_T = \Delta U$.

The Gibbs energy also specifies a free energy. With $đW = -PdV + đW'$, (1.15) implies $-đW' \leq TdS - (dU + PdV)$. To obtain other work *from* the system, we must have that $TdS > dU + PdV$. The maximum value of $đW'$ (counted as a negative quantity) is therefore $đW'_{\text{max}} = dU + PdV - TdS = [dG]_{T,P}$. The Gibbs energy is the maximum work obtainable from the system in a form other than PdV work: $W'_{\text{max}} = [\Delta G]_{T,P}$.

Energy dissipation

That not all energy is available for work is called *dissipation* of energy. Energy is not *lost* (which would be a violation of the first law), it's diverted into degrees of freedom not associated with work. Consider two experiments. In the first, heat Q is extracted from a heat reservoir[46] at temperature T_1 and is used in a reversible cyclic process (so that $\Delta S = 0$) in which work output W_1 is obtained, with the remaining energy, heat $Q - W_1$ transferred to a colder reservoir at temperature T_0 ($T_1 > T_0$). In the second experiment, Q is first allowed to flow spontaneously (irreversibly) between reservoirs at temperatures T_1 and T_2, where $T_1 > T_2 > T_0$. Entropy is created in this process,[47]

$$\Delta S = Q\left(\frac{1}{T_2} - \frac{1}{T_1}\right) > 0 \, .$$

Now let the same amount of heat Q be extracted from the reservoir at temperature T_2 and be used in a reversible cyclic process to produce work W_2 on the environment, with the remaining energy transferred to the reservoir at temperature T_0. It can be shown that $W_2 < W_1$ [3, p59]. *Energy has been dissipated*: Less work is obtained in a process with $\Delta S > 0$ than one for which $\Delta S = 0$. It can be shown that the dissipated energy (the energy not available for work) is related to the entropy increase, with $(\Delta U)_{\text{diss}} \equiv W_1 - W_2 = T_0 \Delta S$.

1.6 CHEMICAL POTENTIAL AND OPEN SYSTEMS

Work is the energy to change the macroscopically observable, extensive properties of systems. What work is required to change the number of particles, N? Open systems allow the exchange of matter as well as energy with the environment. The first law for open systems is[48]

$$dU = TdS - PdV + \mu dN \, , \tag{1.21}$$

where μ is the *chemical potential*—roughly the energy to add another particle of given chemical species to the system.[49,50] We can now answer the question posed at the end of Section 1.2. We see from Eq. (1.21) that U is a function of S, V, N: $U = U(S, V, N)$, and more generally $U = U(S, X_i, N)$. We can therefore identify the chemical potential as the derivative

$$\mu \equiv \left(\frac{\partial U}{\partial N}\right)_{S,V} , \tag{1.22}$$

[46] A *heat reservoir* is an object with a heat capacity sufficiently large that heat transfers to or from it occur with negligible temperature change. Throw an ice cube into the ocean—does its temperature change?

[47] *Where* is the created entropy? More entropy enters reservoir 2 than leaves reservoir 1.

[48] The extra term in Eq. (1.21) was introduced by J.W. Gibbs. Equation (1.21) is sometimes called *Gibbs's equation*.

[49] Equation (1.21) generalizes for *multicomponent* systems having several chemical species through the inclusion of a chemical potential for each species μ_j by adding $\sum_j \mu_j dN_j$ to the first law.

[50] Chemical *potential* is a confusing name because μ is an energy. It's an energy per particle, however, and in that sense it's a potential, just as electrostatic potential is the work per charge; chemical potential is work per particle, and particle number is a dimensionless quantity. It serves as a potential (and thus the name is apt): Matter flows from regions of higher chemical potential to regions of lower chemical potential.[3, p49]

the change in internal energy upon adding a particle, holding S, V fixed.[51] The chemical potential is often, but not always, a negative quantity [3, p39]. We *require* $\mu \leq 0$ for bosons, whereas there is no restriction on the sign of μ for fermions (Section 5.5). Only if inter-particle interactions are sufficiently repulsive does μ become positive. The Fermi energy is an example; the repulsive interaction in that case is the requirement of the Pauli exclusion principle. We'll examine the effects on μ of a *hard-core*, short-range repulsive inter-particle potential in Sections 6.4 and 7.3.

It's often better to write the first law in terms of entropy,[52]

$$dS = \frac{1}{T}dU - \frac{1}{T}\sum_j Y_j dX_j - \frac{\mu}{T}dN \,. \tag{1.23}$$

Entropy is therefore a function of the extensive variables in the problem: $S = S(U, X_i, N)$. From Eq. (1.23), intensive variables are related to derivatives of S with respect to extensive variables,

$$\frac{1}{T} = \left(\frac{\partial S}{\partial U}\right)_{\{X_i\},N} \qquad -\frac{Y_i}{T} = \left(\frac{\partial S}{\partial X_i}\right)_{U,\{X_{j\neq i}\},N} \qquad -\frac{\mu}{T} = \left(\frac{\partial S}{\partial N}\right)_{U,\{X_i\}} \,. \tag{1.24}$$

The potentials introduced in Section 1.4 pertain to closed systems (fixed particle number N). A thermodynamic potential suitable for open systems (as we'll see in Chapter 4) is another Legendre transformation, the *grand potential*

$$\Phi \equiv F - N\mu \,. \tag{1.25}$$

Which variables does Φ depend on? Form its differential:

$$d\Phi = -SdT - PdV - Nd\mu \,. \tag{1.26}$$

Thus, $\Phi = \Phi(T, V, \mu)$. It's shown in Exercise 1.11 that Φ represents the work done on the system at constant T, μ, V, $[\Delta\Phi]_{T,\mu,V} = W''$, where W'' excludes PdV *and* μdN work.

1.7 MAXWELL RELATIONS

Maxwell relations are equalities between partial derivatives of state variables that ensue from the exactness of differentials of thermodynamic potentials. For $\Phi = \Phi(x, y)$, $d\Phi = Adx + Bdy$ is exact when $\partial A/\partial y = \partial B/\partial x$. Maxwell relations are none other than the integrability conditions. Table 1.3 shows the three Maxwell relations[53] generated by the differential forms in Eq. (1.14) when μdN is added to each. Additional Maxwell relations are associated with other thermodynamic potentials (grand potential, magnetic Gibbs energy).

1.8 RESPONSE FUNCTIONS

Much of what we know about macroscopic systems is obtained from experiments where in essence we "tickle" them,[54] where the response to small variations in external parameters is measured. Underlying the seemingly quiescent state of macroscopic equilibrium are incessant and rapid transitions among the microscopic configurations of a system. In the time over which measurements are made, macroscopic systems "explore" a large number of microstates (those consistent with system constraints), and thus measurements represent time averages. Such averages are calculated in statistical mechanics using the microscopic energies of the system. In thermodynamics, relatively few

---

[51]The quantity μ is also the energy required to add another particle to the system at constant T and P; see Section 1.10.
[52]$U = U(S, V, N)$ is the *energy representation*; $S = S(U, V, N)$ is the *entropy representation* [3, p53]. Thermodynamics can be developed in either representation, whichever is convenient for a particular problem.
[53]A thermodynamic potential with n independent variables implies $\frac{1}{2}n(n-1)$ Maxwell relations.
[54]I thank Professor Steve Baker for the tickling metaphor.

Table 1.3: Maxwell relations associated with $U, H, F,$ and G

$$U = U(S, V, N)$$

$$\left(\frac{\partial T}{\partial V}\right)_{S,N} = -\left(\frac{\partial P}{\partial S}\right)_{V,N} \qquad -\left(\frac{\partial P}{\partial N}\right)_{V,S} = \left(\frac{\partial \mu}{\partial V}\right)_{N,S} \qquad \left(\frac{\partial \mu}{\partial S}\right)_{N,V} = \left(\frac{\partial T}{\partial N}\right)_{S,V}$$

$$H = H(S, P, N)$$

$$\left(\frac{\partial T}{\partial P}\right)_{S,N} = \left(\frac{\partial V}{\partial S}\right)_{P,N} \qquad \left(\frac{\partial V}{\partial N}\right)_{P,S} = \left(\frac{\partial \mu}{\partial P}\right)_{N,S} \qquad \left(\frac{\partial \mu}{\partial S}\right)_{N,P} = \left(\frac{\partial T}{\partial N}\right)_{S,P}$$

$$F = F(T, V, N)$$

$$\left(\frac{\partial S}{\partial V}\right)_{T,N} = \left(\frac{\partial P}{\partial T}\right)_{V,N} \qquad -\left(\frac{\partial P}{\partial N}\right)_{V,T} = \left(\frac{\partial \mu}{\partial V}\right)_{N,T} \qquad \left(\frac{\partial \mu}{\partial T}\right)_{N,V} = -\left(\frac{\partial S}{\partial N}\right)_{T,V}$$

$$G = G(T, P, N)$$

$$-\left(\frac{\partial S}{\partial P}\right)_{T,N} = \left(\frac{\partial V}{\partial T}\right)_{P,N} \qquad \left(\frac{\partial V}{\partial N}\right)_{P,T} = \left(\frac{\partial \mu}{\partial P}\right)_{N,T} \qquad \left(\frac{\partial \mu}{\partial T}\right)_{N,P} = -\left(\frac{\partial S}{\partial N}\right)_{T,P}$$

quantities are calculated. What thermodynamics does (spectacularly) is to establish *interrelations* such as Maxwell relations. Thermodynamics is a "consistency machine" that relates known information about systems to quantities that might be difficult to measure. To make the machine useful it must be fed information. *Response functions* quantify under controlled conditions how measurable quantities vary with respect to each other. We pause to develop some mathematics.

1.8.1 Mathematical interlude

Consider three variables connected through a functional relation, $f(x, y, z) = 0$. Any two can be taken as independent, and each can be considered a function of the other two: $x = x(y, z)$, $z = z(x, y)$, or $y = y(x, z)$. Form the differential of x in terms of the differentials of y and z:

$$\mathrm{d}x = \left(\frac{\partial x}{\partial y}\right)_z \mathrm{d}y + \left(\frac{\partial x}{\partial z}\right)_y \mathrm{d}z. \tag{1.27}$$

Now form the differential of z in terms of x and y,

$$\mathrm{d}z = \left(\frac{\partial z}{\partial x}\right)_y \mathrm{d}x + \left(\frac{\partial z}{\partial y}\right)_x \mathrm{d}y. \tag{1.28}$$

Substitute $\mathrm{d}z$ in Eq. (1.28) for that in Eq. (1.27). We find:

$$0 = \left[\left(\frac{\partial x}{\partial z}\right)_y \left(\frac{\partial z}{\partial x}\right)_y - 1\right] \mathrm{d}x + \left[\left(\frac{\partial x}{\partial y}\right)_z + \left(\frac{\partial x}{\partial z}\right)_y \left(\frac{\partial z}{\partial y}\right)_x\right] \mathrm{d}y. \tag{1.29}$$

For Eq. (1.29) to hold for arbitrary $\mathrm{d}x, \mathrm{d}y$, we have

$$1 = \left(\frac{\partial x}{\partial z}\right)_y \left(\frac{\partial z}{\partial x}\right)_y \tag{1.30}$$

$$-1 = \left(\frac{\partial x}{\partial y}\right)_z \left(\frac{\partial y}{\partial z}\right)_x \left(\frac{\partial z}{\partial x}\right)_y . \tag{1.31}$$

Equation (1.30) is the *reciprocity relation* and Eq. (1.31) the *cyclic relation*.[55]

Now consider that we have a quantity x that's a function of y and z, $x = x(y, z)$. As happens in thermodynamics, assume that x is *also* a function of y and w, $x = x(y, w)$. (Thus, $w = w(y, z)$.) Form the differential first of $x = x(y, z)$ and then of $x = x(y, w)$,

$$\mathrm{d}x = \left(\frac{\partial x}{\partial y}\right)_z \mathrm{d}y + \left(\frac{\partial x}{\partial z}\right)_y \mathrm{d}z \qquad \mathrm{d}x = \left(\frac{\partial x}{\partial y}\right)_w \mathrm{d}y + \left(\frac{\partial x}{\partial w}\right)_y \mathrm{d}w .$$

Divide both equations by $\mathrm{d}y$ and equate:

$$\left(\frac{\partial x}{\partial y}\right)_z + \left(\frac{\partial x}{\partial z}\right)_y \frac{\mathrm{d}z}{\mathrm{d}y} = \left(\frac{\partial x}{\partial y}\right)_w + \left(\frac{\partial x}{\partial w}\right)_y \frac{\mathrm{d}w}{\mathrm{d}y} .$$

Now let z be constant, which implies a relation among four variables that's occasionally useful:

$$\left(\frac{\partial x}{\partial y}\right)_z = \left(\frac{\partial x}{\partial y}\right)_w + \left(\frac{\partial x}{\partial w}\right)_y \left(\frac{\partial w}{\partial y}\right)_z . \tag{1.32}$$

Equation (1.32) is an application of the chain rule for $x = x(y, w)$ where $w = w(y, z)$.

1.8.2 Mechanical response functions: Expansivity and compressibility

The *thermal expansivity* and *isothermal compressibility* are by definition

$$\alpha \equiv \frac{1}{V}\left(\frac{\partial V}{\partial T}\right)_P \qquad \beta_T \equiv -\frac{1}{V}\left(\frac{\partial V}{\partial P}\right)_T . \tag{1.33}$$

These measure fractional changes in volume that occur with changes in T and P. It follows from the second law of thermodynamics that β_T is *always positive*,[56] whereas α has no definite sign. What about the third derivative from this trio of variables, $(\partial P/\partial T)_V$? Is it defined as a response function? Nope, no need. Using the cyclic relation, Eq. (1.31),

$$\left(\frac{\partial P}{\partial T}\right)_V = -\frac{(\partial V/\partial T)_P}{(\partial V/\partial P)_T} = \frac{\alpha}{\beta_T} . \tag{1.34}$$

Knowing α and β_T, we know the derivative on the left of Eq. (1.34).

We're free to assume that V, T, and P are connected by an equation of state $V = V(T, P)$. Thus,

$$\mathrm{d}V = \left(\frac{\partial V}{\partial T}\right)_P \mathrm{d}T + \left(\frac{\partial V}{\partial P}\right)_T \mathrm{d}P = \alpha V \mathrm{d}T - \beta_T V \mathrm{d}P . \tag{1.35}$$

Equation (1.35) exemplifies a "differentiate-then-integrate" strategy[3, p19]: The laws of thermodynamics typically involve the differentials of various quantities, which, with enough experimental data on α, β_T, can be integrated to find the equation of state.

---

[55] In my experience, even graduate students in mathematics have not seen the cyclic relation. And why should they have seen it? Ordinarily a functional relation $f = f(x, y)$ presumes that x, y are independent variables. We have in thermodynamics redundancy in the description of the equilibrium state; everything depends on everything else.

[56] The compressibility β_T is a fundamental response function which shows up in many equations in statistical mechanics; be on the lookout for it. The stability of the equilibrium state requires that β_T, C_V are positive quantities.[3, Section 3.10]

1.8.3 Thermal response functions: Heat capacities

The *heat capacity* $C_X \equiv (\mathrm{d}Q/\mathrm{d}T)_X$ measures the heat absorbed per change in temperature, where X signifies what's held fixed in the process. Typically measured are C_P and C_V. Heat capacities are tabulated as *specific heats*, the heat capacity per mole or per gram of material.[57] What does a large heat capacity signify? For a small temperature change ΔT, $Q = C_X \Delta T$: The heat capacity is the heat required to change the temperature of an object from $T \to T + \Delta T$. Systems having a large heat capacity require more heat to produce the same temperature change ΔT than systems with a small heat capacity.[58]

There's an important connection between C_P and C_V that we now derive (see Eq. (1.42)). Assume that[59] $U = U(T, V)$. Then, from the first law of thermodynamics,[60]

$$dU = \left(\frac{\partial U}{\partial T}\right)_{V,N} dT + \left(\frac{\partial U}{\partial V}\right)_{T,N} dV = \mathrm{d}Q - PdV . \tag{1.36}$$

Thus

$$\frac{\mathrm{d}Q}{\mathrm{d}T} = \left(\frac{\partial U}{\partial T}\right)_{V,N} + \left[P + \left(\frac{\partial U}{\partial V}\right)_{T,N}\right] \frac{dV}{dT} . \tag{1.37}$$

By holding V fixed,

$$C_V = \left(\frac{\mathrm{d}Q}{\mathrm{d}T}\right)_{V,N} = \left(\frac{\partial U}{\partial T}\right)_{V,N} . \tag{1.38}$$

Equation (1.38) is an example of a thermodynamic identity, one of many we'll encounter. A measurement of C_V is thus a measurement of $(\partial U/\partial T)_{V,N}$. Holding P fixed in Eq. (1.37),

$$C_P = \left(\frac{\mathrm{d}Q}{\mathrm{d}T}\right)_{P,N} = C_V + \left[P + \left(\frac{\partial U}{\partial V}\right)_{T,N}\right]\left(\frac{\partial V}{\partial T}\right)_{P,N} , \tag{1.39}$$

where we've used Eq. (1.38). Turning Eq. (1.39) around and using Eq. (1.33),

$$\left(\frac{\partial U}{\partial V}\right)_{T,N} = \frac{C_P - C_V}{\alpha V} - P . \tag{1.40}$$

Equation (1.40) relates the derivative $(\partial U/\partial V)_{T,N}$ (not easily measured) to quantities that are more readily measured. Using a Maxwell relation, it can be shown that (treat S as a function of T, V)

$$\left(\frac{\partial U}{\partial V}\right)_{T,N} = T\left(\frac{\partial P}{\partial T}\right)_{V,N} - P = T\frac{\alpha}{\beta_T} - P , \tag{1.41}$$

and we have another connection between a derivative and measurable quantities. With measurements of C_V, α, and β_T, we have the derivatives $(\partial U/\partial T)_{V,N}$ and $(\partial U/\partial V)_{T,N}$ (Eqs. (1.38) and (1.41)), from which $U(T, V)$ can be obtained by integrating Eq. (1.36). Equating (1.41) and (1.40),

$$C_P - C_V = \frac{\alpha^2}{\beta_T} TV . \tag{1.42}$$

Because $\beta_T > 0$, Eq. (1.42) implies[61] that $C_P \geq C_V$.

---

[57] A useful mnemonic is "heat capacity of a penny, specific heat of copper."

[58] The definition of heat capacity is analogous to the definition of electrical capacitance, $Q = CV$. A large capacitor requires a large amount of charge to raise the voltage by one volt, $Q = C\Delta V$.

[59] We're free to take U to be a function of whatever variables we choose.

[60] Equation (1.36) assumes that N is fixed. Variations in particle number are treated in Chapter 4; see Eq. (4.87).

[61] Heat added at constant V causes the temperature to rise ($C_V > 0$). Heat added at constant P (for the same amount of a substance) could entail changes in volume, allowing the system to do work. More heat at constant P would have to be supplied for the same temperature increase as for constant V, $C_P > C_V$.

For the ideal gas we have the special results, using Eqs. (1.41) and (1.42),

$$C_P - C_V = nR \,, \qquad \text{(ideal gas)} \tag{1.43}$$

and Joule's law

$$\left(\frac{\partial U}{\partial V}\right)_T = 0 \,. \qquad \text{(ideal gas)} \tag{1.44}$$

Joule's law can be taken as an alternate definition of the ideal gas: A gas whose internal energy is a function of temperature only, $U = U(T)$. The ideal gas ignores inter-particle interactions; its internal energy is independent of the volume. We return to the heat capacity of gases in Chapter 4.

1.9 HEAT CAPACITY OF MAGNETIC SYSTEMS

The work required to reversibly change the magnetization M in external field H is $dW = \mu_0 H \cdot dM$ [3, p17], which, because M is aligned with H, we can take to be $dW = \mu_0 H dM$, and, to simplify matters,[62] we set $B = \mu_0 H$. Let's combine $dW = BdM$, an "other work" term, with the first law:

$$dU = TdS + BdM \,, \tag{1.45}$$

where we drop the PdV work term.[63] Entropy and magnetization are difficult to control experimentally; use a Legendre transformation to shift emphasis to variables that are easier to control. Define the *magnetic Gibbs energy*, $G_m \equiv U - TS - BM$, for which

$$dG_m = -SdT - MdB \,. \tag{1.46}$$

Equation (1.46) implies the Maxwell relation

$$\left(\frac{\partial S}{\partial B}\right)_T = \left(\frac{\partial M}{\partial T}\right)_B \,. \tag{1.47}$$

The statistical mechanics of magnetic systems is treated in Sections 5.2 and 6.5 and in Chapters 7 and 8. Here we focus on the two types of heat capacity for magnets, C_M and C_B. We prove the inequality $C_B \geq C_M$, analogous to that for PVT systems, $C_P \geq C_V$, Eq. (1.42). Assume $S = S(T, M)$ and $M = M(T, B)$. Applying Eq. (1.32),

$$\left(\frac{\partial S}{\partial T}\right)_B = \left(\frac{\partial S}{\partial T}\right)_M + \left(\frac{\partial S}{\partial M}\right)_T \left(\frac{\partial M}{\partial T}\right)_B = \left(\frac{\partial S}{\partial T}\right)_M - \left(\frac{\partial B}{\partial T}\right)_M \left(\frac{\partial M}{\partial T}\right)_B \,. \tag{1.48}$$

The first equality in Eq. (1.48), derived from mathematics, can be arrived at through a more circuitous yet physical path, as outlined in Exercise 1.19, and the second equality follows from a Maxwell relation derived in Exercise 1.16. Using the cyclic relation, Eq. (1.31),

$$\left(\frac{\partial B}{\partial T}\right)_M = -\frac{(\partial M/\partial T)_B}{(\partial M/\partial B)_T} = -\frac{1}{\chi}\left(\frac{\partial M}{\partial T}\right)_B \,, \tag{1.49}$$

where

$$\chi \equiv (\partial M/\partial B)_T \tag{1.50}$$

---

[62] B or H, would the real magnetic field please stand up? As shown in Appendix E, the magnitude of magnetic moments μ is of the order of the Bohr magneton, μ_B, which has the units of Joule/Tesla. The question is settled—use B. Later, when we wish to compare with experimental results, we should use H, which is directly coupled to the currents producing laboratory magnetic fields; in that case set $B = \mu_0 H$.

[63] We do this for convenience, but it could be retained; see Exercise 1.15. At low temperatures, PdV work does not play a significant role (in fact, it must vanish as $T \to 0$ by the third law of thermodynamics), while the magnetic work term BdM serves as the basis for the *adiabatic demagnetization* technique of cyrogenic cooling.[3, Section 8.1]

defines the *magnetic susceptibility*,[64] another response function (akin to the compressibility, β_T). Combining Eq. (1.49) with Eq. (1.48), and using the results of Exercise 1.19, we have

$$C_B = C_M + \frac{T}{\chi}\left(\frac{\partial M}{\partial T}\right)_B^2 . \tag{1.51}$$

One should compare Eq. (1.51) with Eq. (1.42).[65] A stability analysis of the equilibrium state associated with magnetic systems shows that[66] $C_M > 0, \chi > 0$. Thus, $C_B \geq C_M$.

1.10 EXTENSIVITY OF ENTROPY, SACKUR-TETRODE FORMULA

Internal energy is extensive—it scales with the amount of matter in the system. Entropy is extensive because it's related to an extensive quantity—the heat capacity (see Exercise 1.7). The extensivity of S, however, is not *as* obvious as that for U. Internal energy is related to microscopic properties of matter, and thus scales with the mass of the system, but *entropy is not a microscopic property of matter*.[67] It's a property of the state of equilibrium specified by the values of macrovariables, the extensive quantities in the problem, X_j (those associated with work). *Non-extensive* entropies lead to a puzzle known as *Gibbs's paradox*, the resolution of which is that entropy must be extensive [3, p113]. In axiomatic formulations of thermodynamics, the extensivity of S is taken as a *postulate*.[8]

Thus we require that S be extensive. Referring to Fig. 1.2, extensivity requires that if we scale all extensive variables by a positive parameter λ, entropy must scale by the same factor:

$$S(\lambda U, \lambda V, \lambda N) = \lambda S(U, V, N) . \tag{1.52}$$

Functions satisfying scaling relations such as Eq. (1.52) are said to be *homogeneous functions*. A function F of k variables having the property $F(\lambda x_1, \cdots, \lambda x_k) = \lambda^p F(x_1, \cdots, x_k)$ is a *homogeneous function of order p*. *Euler's theorem*[3, p39] states that for homogeneous functions F of order p, $pF(x_1, \cdots, x_k) = \sum_{j=1}^{k}(\partial F/\partial x_j)x_j$. Example: $F(x) = x^p$; it's true that $pF = (\partial F/\partial x)x$.

Applying Euler's theorem to the first-order homogeneous function $S(U, V, N)$, and using Eq. (1.24),

$$S = \left(\frac{\partial S}{\partial U}\right)U + \left(\frac{\partial S}{\partial V}\right)V + \left(\frac{\partial S}{\partial N}\right)N = \frac{1}{T}U + \frac{P}{T}V - \frac{\mu}{T}N .$$

The extensivity of S therefore implies a connection between *seven* thermodynamic variables[68]

$$TS = U + PV - \mu N , \tag{1.53}$$

which we'll refer to as the *Euler relation*.

Combining Eq. (1.53) with the definition of Gibbs energy, Eq. (1.13),

$$G = U - TS + PV = N\mu . \tag{1.54}$$

Thus, *chemical potential is the Gibbs energy per particle*, $\mu = G/N$, and therefore, $[dG]_{T,P} = \mu[dN]_{T,P}$ ($[d\mu]_{T,P} = 0$, from the Gibbs-Duhem equation, (P1.1)). *The chemical potential is the energy to add another particle at constant* T, P, as well as at constant S, V (see Eq. (1.22)).

---

[64]In applications, it's often the case that χ is defined as the derivative in Eq. (1.50) evaluated at $B = 0$.

[65]Thermodynamic formulas pertaining to magnetic systems can sometimes (but not always) be found from their counterparts describing fluids through the substitutions (from the first law) $V \to M$ and $P \to -B$; see [3, p131].

[66]χ is positive for paramagnetic and ferromagnetic systems, but not for diamagnetic systems (not treated in this book).

[67]Likewise, temperature is not a property of the microscopic constituents of the system, and is a property possessed by macroscopic systems in thermal equilibrium. One can't speak of the temperature (or entropy) of a single particle.

[68]For a system with k "work variables" (see Eq. (1.5)), there would be an Euler relation (as in Eq. (1.53)) involving $5 + 2k$ variables, $TS = U - \sum_{i=1}^{k} Y_i X_i - \mu N$.

Equation (1.53) underscores that TS is the energy of heat, because *what is not work is heat:*[69]

$$\underbrace{TS}_{\substack{\text{heat added to}\\\text{the system}}} = \underbrace{U}_{\substack{\text{total}\\\text{internal energy}}} - \underbrace{(-PV + \mu N)}_{\substack{\text{work required to establish system}\\\text{in volume } V \text{ with } N \text{ particles}}} . \qquad (1.55)$$

The product TS has the dimension of energy. Now, absolute temperature "should" be a quantity having the dimension of energy (see [3, p31]), implying that entropy "should" be dimensionless. But, because absolute temperature is conventionally measured in a dimensionless unit (Kelvin), we require a conversion factor between Joules (SI unit of energy) and Kelvin, which is supplied by Boltzmann's constant, k:

$$TS = kT \times \left(\frac{S}{k}\right) .$$

Formulas for entropy always involve k (or the gas constant R if specified on a per-mole basis).

Can the entropy of the ideal gas, the simplest macroscopic system, be calculated?[70] Yes and no. The *Sackur-Tetrode formula*[3, p110] for the entropy of an ideal gas at temperature T having atoms of mass m is

$$S(U, V, N) = Nk\left[\frac{5}{2} + \ln\left(\frac{V}{N}\left(\frac{4\pi mU}{3Nh^2}\right)^{3/2}\right)\right] \qquad (1.56)$$

(note the occurrence of Planck's constant). Equation (1.56) obeys the scaling relation Eq. (1.52) (show this). All parts of Eq. (1.56) have been tested against experiment. From Eqs. (1.24) and (1.56),

$$\frac{1}{T} = \left(\frac{\partial S}{\partial U}\right)_{V,N} = \frac{3}{2}\frac{Nk}{U} \implies U = \frac{3}{2}NkT , \qquad (1.57)$$

an expression for the internal energy of the ideal gas. Equation (1.57) is consistent with the heat capacity of the noble gases (He, Ne, Ar, Kr, and Xe), which best approximate ideal gases. Combining Eq. (1.57) with Eq. (1.38),

$$C_V = \left(\frac{\partial U}{\partial T}\right)_V = \frac{3}{2}Nk . \qquad (1.58)$$

The molar specific heat of the noble gases is $1.50R$, consistent with Eq. (1.58). Using Eqs. (1.24) and (1.56),

$$\frac{P}{T} = \left(\frac{\partial S}{\partial V}\right)_{U,N} = \frac{Nk}{V} \implies PV = NkT , \qquad (1.59)$$

the ideal gas equation of state. Can the third relation in Eq. (1.24) be tested? Not directly, because we lack an independent way of specifying μ. Equations (1.57) and (1.59) would follow from any expression of the form $S = Nk\ln\left(VU^{3/2}\right)$. Everything else in Eq. (1.56) is required to reproduce the entropy of the vapor phase of a substance in equilibrium with its liquid.[71] The Sackur-Tetrode formula can be motivated from thermodynamics,[3, p108] but cannot be derived within the confines of thermodynamics because it involves Planck's constant.[72] A derivation of the Sackur-Tetrode formula is presented in Section 2.3.

An expression for the chemical potential of the ideal gas is given in Exercise 1.4, Eq. (P1.2). What should be noted is that $\mu \neq 0$, *even though there are no interactions between the atoms of an*

---

[69] See Eq. (1.2). Our interpretation in Eq. (1.55) follows from integrating Eq. (1.21) at constant T and P.

[70] To even ask such a question is a departure from thermodynamics, and the beginning of statistical mechanics.

[71] The entropy of a gas can be measured relative to that of its liquid phase through the *latent heat*, the heat released or absorbed upon a change of phase [3, Section 7.4]. *Phase*, as the term is used in thermodynamics, is a spatially homogeneous material body in a state of equilibrium; see Chapter 7 of this book.

[72] The Sackur-Tetrode formula, which dates from 1912, would appear to be the first time in the logical development of physics of associating Planck's constant with material particles. Before this time, Planck's constant was associated only with photons. The Sackur-Tetrode formula presages the de Broglie wavelength.

ideal gas. Chemical potential is the energy to add another particle at constant entropy (Eq. (1.22)), and entropy is not a microscopic property of matter; it's a property of the equilibrium state. Even the ideal gas is in a state of thermal equilibrium, a property shared by all particles of the gas. That $\mu \neq 0$ for the ideal gas underscores that it involves entropy, $\mu = (U + PV - TS)/N$.

1.11 BOLTZMANN ENTROPY: CONNECTION WITH MICROSCOPICS

Several considerations contribute to the view that entropy is related to *orderliness*. Work is an *organized* form of energy involving systematic extensions of macroscopic quantities; $\Delta S > 0$ reduces the effectiveness of energy to perform work (energy dissipation). Entropy increases upon the removal of constraints (irreversibility), and less constrained systems are less organized. The entropy of isolated systems is a maximum in equilibrium. The state of equilibrium should be the *most likely* state, by far: Equilibrium persists unchanged in time; amongst the myriad dynamical microstates of a system,[73] those consistent with equilibrium should occur with far greater likelihood. Disorder, maximum entropy, likelihood: we expect there is a connection between entropy and *probability*.

How to find such a connection? To get started, let W denote the probability of a specified macrostate, where obviously we must define what W represents.[74] We seek $S = f(W)$, where f is a universal function (the same for all systems). To determine f we are guided by two requirements. First, the entropy of a system of two independent subsystems is the sum of the subsystem entropies, $S = S_1 + S_2$, i.e., S is extensive. Second, by the rules of probability,[75] the probability W for the occurrence of two independent states 1 and 2 is the *product* of the separate probabilities W_1, W_2, $W = W_1 W_2$. With $S_1 = f(W_1)$ and $S_2 = f(W_2)$, the function f must be such that

$$f(W_1 W_2) = f(W_1) + f(W_2) . \tag{1.60}$$

Equation (1.60) is a *functional equation*. It can be turned into a differential equation by (noting that W_1, W_2 occur in Eq. (1.60) symmetrically), first differentiating Eq. (1.60) with respect to W_1 and then differentiating the result with respect to W_2. We find (show this)

$$W_1 W_2 f''(W_1 W_2) + f'(W_1 W_2) = 0 . \tag{1.61}$$

Solutions of $z f''(z) + f'(z) = 0$ are of the form $f(z) = k \ln z + f_0$, where k and f_0 are constants. We must set $f_0 = 0$ so that it satisfies the functional equation. The solution of Eq. (1.60) is therefore

$$S = k \ln W . \tag{1.62}$$

Equation (1.62), which defines the *Boltzmann entropy*, is one of the great achievements of theoretical physics. It provides, first, a physical interpretation of entropy beyond qualitative notions of disorder—W represents the "missing" degrees of freedom not accounted for in thermodynamics descriptions, it's the *bridge* between microscopic and macroscopic—and, second, a *formula* by which S can be calculated if we can learn to calculate W. We'll distinguish the *statistical entropy*, $S = k \ln W$, from the *thermodynamic entropy*, $\Delta S = \int (\mathrm{d}Q)_{\text{rev}} /T$. Of course, we'd like the two to agree in their predictions.

---

[73] Microstates are defined in Section 2.1.

[74] We follow tradition and use W for *Wahrscheinlichkeit*, German for probability (W here does not mean work!). Of course W must be defined as a physical concept; the language we use is immaterial. The quantity W is not a traditional probability in the sense that it's a number between zero and one. Max Planck distinguished *thermodynamic probability* (W) from *mathematical probability* (P), where the two are proportional but not equal [9, p120], see Section 2.4.2. W is *the number of microstates per macrostate*, an integer. To quote Planck:[10, p226] "Every macroscopic state of a physical system comprises a perfectly definite number of microscopic states of the system, and the number represents the thermodynamic probability or the statistical weight W of the macroscopic state." Serendipitously, W also stands for the English word, *ways*—W is the number of ways a given macrostate can be realized by the microstates of the system.

[75] Probability is covered in Chapter 3.

The scale factor k in Eq. (1.62) can be established by a simple argument.[76] The *change* in entropy between equilibrium states is, from Eq. (1.62):

$$\Delta S \equiv S_2 - S_1 = k \ln (W_2/W_1) \ . \tag{1.63}$$

Consider the free expansion of an ideal gas with N particles into an evacuated chamber where the volume doubles in the process (Section 1.3). While we don't yet know how to calculate W_1 and W_2 separately, the *ratio* W_2/W_1 is readily found in this case. After the system comes to equilibrium, it's twice as likely to find a given particle in either volume than it is to find the particle in the original volume. The particles of an ideal gas are noninteracting, and thus the same is true *for each particle*, implying $W_2/W_1 = 2^N$ (see Appendix F). By Eq. (1.63) therefore, $\Delta S = Nk \ln 2$. In Section 1.3, we calculated ΔS for the free expansion using thermodynamics, where we obtained $\Delta S = Nk \ln 2$. Agreement between the two formulae is obtained by choosing k to be Boltzmann's constant.

Equation (1.62) is encouraging, yet at this point it's on shaky ground: It relates a symbol S to another symbol W. The only properties used in its derivation are that S is additive over subsystems (like entropy), $S = S_1 + S_2$, and W behaves like a probability with $W = W_1 W_2$. Despite the best of intentions, using the symbol S does not make it entropy. Of course, that begs the question of what *is* entropy. We established the scale factor k by making ΔS from Eq. (1.63) agree with ΔS from thermodynamics for the same process. It's standard practice in theoretical physics to "nail down" unknown parameters by seeking agreement with previously established theories in their domains of validity. Such a practice does not *guarantee* the validity of Eq. (1.62); it merely asserts it's not obviously wrong in this one application. If Eq. (1.62) proves to be of general validity, then having determined k in this manner will suffice. Of greater importance: What allows us to call anything (such as S in Eq. (1.62)) entropy? What we know about entropy comes from thermodynamics:

$$\left(\frac{\partial S}{\partial U}\right)_{V,N} = \frac{1}{T} \qquad \left(\frac{\partial S}{\partial V}\right)_{U,N} = \frac{P}{T} \qquad \left(\frac{\partial S}{\partial N}\right)_{U,V} = -\frac{\mu}{T} \ . \tag{1.24}$$

If we agree to call "entropy" anything satisfying these identities, then Eq. (1.24) must be imposed as a requirement on Eq. (1.62) before we're entitled to call S in Eq. (1.62), entropy. That implies W is *a function of macroscopic variables*, $W = W(U,V,N)$, because, from Eq. (1.62), $W = \exp\left(S(U,V,N)/k\right)$. Calculating $W(U,V,N)$ is the province of statistical mechanics.

The Sackur-Tetrode formula is in the form of Eq. (1.62).[77] Combining Eqs. (1.57) and (1.56),

$$S = Nk \left[\frac{5}{2} + \ln\left(\frac{V}{N\lambda_T^3}\right)\right], \tag{1.64}$$

where λ_T is the *thermal wavelength*,

$$\lambda_T \equiv \frac{h}{\sqrt{2\pi mkT}} \ . \tag{1.65}$$

The thermal wavelength is (approximately) the mean of the de Broglie wavelengths of the atoms of a gas of mass m in equilibrium at temperature T (see Exercise 5.4). We show in Section 2.3, how to calculate W for the ideal gas, such that Eq. (1.62) reproduces the Sackur-Tetrode formula.

Writing the Sackur-Tetrode formula as in Eq. (1.64) gives insight into its meaning. A gas particle can be considered, on average, to be confined to the volume per particle, V/N. Each particle, however, has associated with it an *intrinsic* volume[78] λ_T^3. Thus, $V/(N\lambda_T^3)$ is the number of ways a particle can be situated in the average volume per particle. We're assuming $\lambda_T^3 \ll V/N$. There are two independent lengths in the problem (see Fig. 1.6), $(V/N)^{1/3}$, the average distance between

---

[76]The scale factor is dependent upon the *base* of the logarithm used in Eq. (1.62). We've chosen to work with the natural logarithm (standard practice in physics), $\ln x = \log_e x$, but we could work with $\log_b x = \ln x / \ln b$.

[77]The factor of 5/2 in Eq. (1.64) can be absorbed into the logarithm: $\frac{5}{2} = \ln e^{5/2}$.

[78]The intrinsic volume λ_T^3 is quantum in nature—it contains Planck's constant. λ_T^{-3} is called the *quantum concentration*.

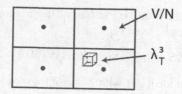

Figure 1.6: Two lengths: Average distance between particles and the thermal wavelength.

particles, and the thermal wavelength, λ_T. When $(V/N)^{1/3} \gg \lambda_T$, particles behave independently and a classical description suffices;[79] when $(V/N)^{1/3} \lesssim \lambda_T$, the quantum nature of particles becomes apparent (see Section 5.5). These two regimes are shown schematically in Fig. 1.7 where the higher the density $n \equiv N/V$, the higher is the temperature T up to which a quantum description is required.

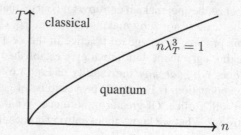

Figure 1.7: Particles behave classically (quantum-mechanically) for $n\lambda_T^3 \ll 1$ ($n\lambda_T^3 \gtrsim 1$).

1.12   FLUCTUATIONS AND STABILITY

Gibbs formulated two equivalent principles for characterizing equilibrium in isolated systems, the *entropy maximum principle* and the *energy minimum principle*:[11, p56]

1. Entropy is a maximum for fixed energy, *there are no fluctuations by which* $[\Delta S]_U > 0$;

2. Energy is a minimum for fixed entropy, *there are no fluctuations by which* $[\Delta U]_S < 0$.

A proof of these statements is given in [3, pp53–54]. We've used the word *fluctuation*. Fluctuations arise from the random motions of the microscopic constituents of macroscopic systems, a topic therefore that's seemingly outside the purview of thermodynamics.[80] Despite statements to the contrary, fluctuations *can* be introduced in thermodynamics through the device of *virtual variations*, where the isolation of a system is *conceptually* relaxed.[81]

---

[79] A refinement of this criterion is $V/N \gg \lambda_T^3/g$, where $g = 2S + 1$ is the spin degeneracy; see Eq. (5.83).

[80] Thermodynamics is a science of matter that—remarkably—makes no assumptions about the constitution of matter. It's immaterial to thermodynamics that matter is composed of atoms. That's both the strength and the weakness of thermodynamics: Its predictions in the form of interrelations between measurable quantities are independent of the details of physical systems, and hence are of great generality, yet it's incapable of saying *why* such quantities have the values they do. Statistical mechanics assumes from the outset that matter consists of a great many microscopic components, and that the macroscopic properties of matter are a manifestation of the laws governing the motions of its constituent parts.

[81] Treating fluctuations as virtual processes is similar to the virtual displacements used in analytical mechanics, "mathematical experiments" consistent with existing constraints, but occurring at a fixed time. In thermodynamics, fluctuations are conceivable variations of a system that would be induced by contact with its environment. We allow, conceptually, variations of state variables induced by random interactions with the environment, knowing that in a state of isolation, the entropy of an equilibrium system is a maximum, or the energy is a minimum. See [3, p45].

Equilibrium (which persists in time) must be *stable* against fluctuations. To analyze stability, we introduce an oft-used theoretical device in thermodynamics, the *composite system* consisting of subsystems A and B in contact with each other, and surrounded by rigid adiabatic walls (see Fig. 1.8). A and B are separated by a partition that's moveable, diathermic, and permeable. The subsys-

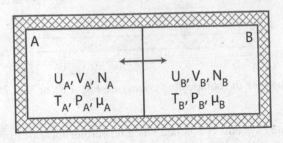

Figure 1.8: Composite system comprised of subsystems A and B.

tems can therefore exchange volume, energy, and particles, which we assume are the same chemical species. Define $V = V_A + V_B$, $N = N_A + N_B$, and $U = U_A + U_B$, which are *fixed* quantities. Let there be small variations in the energy, volume, and particle number of each subsystem, δU_A, δU_B, δV_A, δV_B, δN_A, δN_B. The variations are constrained because energy, volume, and particle number are conserved: $\delta U_A = -\delta U_B$, $\delta V_A = -\delta V_B$, and $\delta N_A = -\delta N_B$.

Consider changes in the *total* entropy of the composite system, $S_+ \equiv S_A + S_B$, under variations in U, V, N to first order in small quantities,

$$
\begin{aligned}
\delta S_+ &= \sum_{\alpha=A,B} \left[\left(\frac{\partial S_\alpha}{\partial U_\alpha} \right)_{V_\alpha,N_\alpha} \delta U_\alpha + \left(\frac{\partial S_\alpha}{\partial V_\alpha} \right)_{U_\alpha,N_\alpha} \delta V_\alpha + \left(\frac{\partial S_\alpha}{\partial N_\alpha} \right)_{U_\alpha,V_\alpha} \delta N_\alpha \right] \\
&= \sum_{\alpha=A,B} \left[\frac{1}{T_\alpha} \delta U_\alpha + \frac{P_\alpha}{T_\alpha} \delta V_\alpha - \frac{\mu_\alpha}{T_\alpha} \delta N_\alpha \right] \\
&= \left(\frac{1}{T_A} - \frac{1}{T_B} \right) \delta U_A + \left(\frac{P_A}{T_A} - \frac{P_B}{T_B} \right) \delta V_A - \left(\frac{\mu_A}{T_A} - \frac{\mu_B}{T_B} \right) \delta N_A ,
\end{aligned}
\tag{1.66}
$$

where we've used Eq. (1.23) in the second line of Eq. (1.66). The total system is isolated and hence S_+ is a maximum. Because δU_A, δV_A, and δN_A can be varied independently, $\delta S_+ = 0$ in Eq. (1.66) requires *the equality of subsystem temperatures, pressures, and chemical potentials* (the conditions of thermal, mechanical, and compositional equilibrium):

$$
T_A = T_B \qquad P_A = P_B \qquad \mu_A = \mu_B .
\tag{1.67}
$$

A similar analysis could be repeated for a system containing any number of chemical species. As long as the partition can pass a species, $\mu_{j,A} = \mu_{j,B}$ in equilibrium. If, however, the partition cannot pass species i, $\delta N_{i,A} = \delta N_{i,B} = 0$, and such a species cannot come to equilibrium between the subsystems, implying there's no reason for the chemical potentials to be equal, $\mu_{i,A} \neq \mu_{i,B}$.

Figure 1.8 can be redrawn, so that B surrounds A; see Fig. 1.9. In that case, B becomes the environment for A. In writing $TdS = dU + PdV - \mu dN$, U, V, and N are *conserved* quantities. The conditions for equilibrium, Eq. (1.67), require equality between system and environmental intensive variables, T, P, and μ, *those that are conjugate to conserved extensive quantities.* The equilibrium values of intensive quantities are "set" by the environment. Thus, what we termed generalized *forces* in Section 1.2 are connected with conserved quantities (temperature with energy, pressure with volume, chemical potential with particle number). The *uniformity* of forces (as intensive quantities) is a property of equilibrium; when such forces vary, there are changes in conserved quantities.

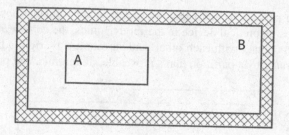

Figure 1.9: Redrawing the composite system (see Fig. 1.8), so that B surrounds A.

Equilibrium in isolated systems is attained when no further increases in entropy can occur by processes consistent with system constraints. As we've shown, $\delta S_+ = 0$ implies the conditions for equilibrium, Eq. (1.67), but it's insufficient to determine whether S_+ is a maximum. A generic function $S(U, V, N)$ has a maximum at (U_0, V_0, N_0), if $S(U_0, V_0, N_0) > S(U, V, N)$ for all (U, V, N) in a neighborhood of U_0, V_0, N_0. This familiar mathematical idea poses a conceptual problem, however, when applied to the state variables of thermodynamics. We can't speak of an entropy *surface* associated with a *fixed* equilibrium state, because state variables are constrained to have their equilibrium values! Enter the idea of fluctuations as virtual variations that produce $\Delta S_+ < 0$, even though *there are no macroscopic physical processes* that drive $\Delta S < 0$ for isolated systems. We require entropy to *decrease* during fluctuations—the *stability condition*—because otherwise entropy increases would imply an evolution to a new state of equilibrium. Fluctuations are temporary departures to *more ordered* (less disordered) system configurations (equilibrium is the state of maximum disorder); thus entropy decreases in fluctuations about the equilibrium state. Equilibrium is stable when perturbed by fluctuations: Entropy decreases engendered by the creation of fluctuations produce a "restoring force" that brings the system back to the state of maximum entropy. To second order in small quantities we require $\mathrm{d}^2 S_+ < 0$. The stability condition places physically sensible restrictions on the *sign* of response functions such as the heat capacity, $C_V > 0$, the condition of *thermal stability* [3, pp45–48]. A system having a negative heat capacity would *decrease* in temperature upon the addition of heat! The other implications of $\mathrm{d}^2 S < 0$ are that 1) the isothermal compressibility $\beta_T > 0$, the condition of *mechanical stability*, which if violated a substance would *expand* ($\mathrm{d}V > 0$) upon an increase in applied pressure, $\mathrm{d}P > 0$, and 2) $(\partial \mu / \partial N)_{T,P} > 0$, the condition for *compositional stability*, that matter flows from high to low chemical potential [3, p49]. We revisit the connection between fluctuations and stability in Section 2.4.

1.13 LIMITATIONS OF THERMODYNAMICS

Thermodynamics is a science of matter and energy of great utility and generality. Einstein wrote:

> A theory is the more impressive the greater the simplicity of its premises is, the more different kinds of things it relates, and the more extended is its area of applicability. Therefore the deep impression which classical thermodynamics made upon me. It is the only physical theory of universal content concerning which I am convinced that, within the framework of the applicability of its basic concepts, will never be overthrown.[12, p33]

Yet, as with all physical theories, thermodynamics has its domain of validity. We note three limitations of thermodynamics that are removed in statistical mechanics.

- Fluctuations do not easily fit into the framework of thermodynamics.[82] In the thermodynamic description, the equilibrium state is one where state variables have fixed values. Statistical mechanics is naturally able to treat fluctuations through its strong reliance on the theory of probability. The existence of fluctuations is the entrée into statistical mechanics (see Section 2.4).

- The state of thermodynamic equilibrium is time invariant. There is an extension of thermodynamics to systems slightly out of equilibrium, *non-equilibrium thermodynamics*, where the traditional state variables of thermodynamics are taken to be field quantities, well defined locally, yet varying in space and time [3, Chapter 14]. *Kinetic theory* is a branch of statistical mechanics that treats systems strongly out of equilibrium. Neither topic is treated in this book.

- Within the confines of the equilibrium state, perhaps the most glaring deficiency of thermodynamics is that it cannot systematically incorporate interactions between atoms, except phenomenologically through equations of state such as the virial expansion (see Chapter 6). Thermodynamics is best suited for systems having *independent* constituents (ideal gas, ideal paramagnet, and photon gas; see Chapter 5). Statistical mechanics naturally incorporates interactions between atoms, either in the guise of classical or quantum mechanics.

SUMMARY

We reviewed thermodynamics with the intent of preparing us for statistical mechanics, wherein a working knowledge is presumed of the internal energy U, absolute temperature T, entropy S, and chemical potential μ. They are related through the first law of thermodynamics,

$$\underbrace{dU}_{\text{internal energy}} = \underbrace{TdS}_{\text{heat}} + \underbrace{(-PdV + \mu dN)}_{\text{work}},$$

where N is the number of particles. Work involves changes in macroscopically observable, extensive properties of a system. Heat is the transference of energy to microscopic degrees of freedom not associated with work. Entropy is a measure of the number W of microscopic configurations consistent with the equilibrium state through the Boltzmann relation $S = k \ln W$.

EXERCISES

1.1 Consider the differential form $\omega \equiv (x^2 + y)dx + xdy$. Is it exact? Integrate ω from (0,0) to (1,1) along two paths: from (0,0) to (1,0) to (1,1), and then from (0,0) to (0,1) to (1,1). Is the value of the integral the same? Repeat using the differential form $(x^2 + 2y)dx + xdy$.

1.2 Does $x^2y^4dx + x^3y^3dy$ represent an exact differential? Can you find an integrating factor? (Integrating factors are not unique.) The goal is to find a function $\lambda(x, y)$ (the integrating factor) such that $\lambda \left(x^2y^4dx + x^3y^3dy \right)$ is an exact differential. Show that λ must satisfy the partial differential equation $\lambda = x(\partial\lambda/\partial x) - y(\partial\lambda/\partial y)$. Use the separation of variables method to show that $\lambda = x^\alpha y^{\alpha-1}$, where α is an arbitrary constant.

1.3 Differentiate all variables in the Euler relation Eq. (1.53) and combine with the first law, Eq. (1.21), to derive the *Gibbs-Duhem equation* in its most basic form

$$Nd\mu = -SdT + VdP. \tag{P1.1}$$

---

[82]It's often said that thermodynamics can't handle fluctuations, but, as we've seen, fluctuations can be introduced in thermodynamics as virtual variations. What's fair to say is that fluctuations are more naturally handled in statistical mechanics than in thermodynamics. One is reminded of analogous statements in the theory of relativity, that special relativity can't handle accelerated motion, which is incorrect [13, Chapter 12]. What's fair to say is that accelerated motion isn't handled in special relativity as naturally as inertial motion.

Thus, $\mu = \mu(T, P)$. The Gibbs-Duhem equation expresses the first law in terms of the differentials of intensive quantities, versus Eq. (1.21), the first law expressed in terms of differentials of extensive quantities.

1.4 a. Derive the formula for the chemical potential of the ideal gas using Eqs. (1.24) and (1.56):

$$\mu = -kT \ln \left(\frac{V}{N} \left(\frac{4\pi mU}{3Nh^2} \right)^{3/2} \right). \tag{P1.2}$$

Note that μ as given by Eq. (P1.2) is intensive, a *zeroth*-order homogeneous function of extensive variables.

b. Show that Eq. (P1.2) follows from Eq. (1.53) and Eqs. (1.56), (1.57), and (1.59).

c. Show that Eq. (P1.2) can be written

$$\mu = -kT \ln \left(\frac{V}{N\lambda_T^3} \right), \tag{P1.3}$$

where λ_T is the thermal wavelength, Eq. (1.65). The chemical potential of the ideal gas is negative so long as $(V/N) > \lambda_T^3$.

d. Show that

$$\left(\frac{\partial \mu}{\partial N} \right)_T = \frac{kT}{N}. \qquad \text{(ideal gas)} \tag{P1.4}$$

The three stability criteria of thermodynamics are $C_V > 0$, $\beta_T > 0$ (isothermal compressibility), and the derivative $(\partial \mu / \partial N)_T > 0$.

1.5 Equation (1.56) can be written schematically as $S = Nkf(U/N, V/N)$. One can *estimate* the entropy as $S \approx Nk$. Using this rule of thumb, estimate the entropy in a glass of wine. Make reasonable assumptions for whatever you need to know. Taking $S \propto N$, useful for purposes of guesstimates, cannot be taken literally. If S varies precisely linearly with N, the chemical potential is identically zero (see Eq. (1.22); what is $(\partial U / \partial N)_N$?).

1.6 Show that $[\Delta H]_P + [\Delta F]_T = \Delta U$.

1.7 We've stated that entropy is extensive because heat capacities are extensive. Let's show the connection. Assume[83] S and U are functions of (T, V), $S = S(T, V)$ and $U = U(T, V)$. From calculus, $\mathrm{d}S = (\partial S/\partial T)_V \, \mathrm{d}T + (\partial S/\partial V)_T \, \mathrm{d}V$. Show the identities

$$C_V = \left(\frac{\mathrm{d}Q}{\mathrm{d}T} \right)_V = T \left(\frac{\partial S}{\partial T} \right)_V = -T \left(\frac{\partial^2 F}{\partial T^2} \right)_V. \tag{P1.5}$$

By the same method of analysis, let $U = U(T, P)$, $S = S(T, P)$, and show that

$$C_P = \left(\frac{\mathrm{d}Q}{\mathrm{d}T} \right)_P = T \left(\frac{\partial S}{\partial T} \right)_P = -T \left(\frac{\partial^2 G}{\partial T^2} \right)_P. \tag{P1.6}$$

---

[83] We often use the term "natural variables of thermodynamics," such as $U = U(S, V)$, $F = F(T, V)$, $G = G(T, P)$ that we infer from Eq. (1.14), but we're free to use any variables we find convenient for a given problem. There is considerable redundancy in the description of the equilibrium state; through the equation of state, any variable depends on the rest.

1.8 Show that

$$\left(\frac{\partial C_V}{\partial V}\right)_T = T\left(\frac{\partial^2 P}{\partial T^2}\right)_V. \qquad \text{(P1.7)}$$

Hint: Use a Maxwell relation. Thus, for the ideal gas (no inter-particle interactions) and the van der Waals gas (Eq. (6.32), a mean-field theory), C_V is independent of V.

1.9 The Boltzmann entropy is defined so that it satisfies the functional equation, (1.60), which, as shown in Section 1.11, implies the differential equation, $zf''(z) + f'(z) = 0$, (1.61). Show that for the solution of the differential equation $f(z) = k\ln z + f_0$ to satisfy the functional equation we must set $f_0 = 0$.

1.10 Verify the cyclic relation, Eq. (1.31), using (P, V, T) for (x, y, z) and the ideal gas equation of state.

1.11 The grand potential $\Phi \equiv F - N\mu$ is a thermodynamic potential suitable for open systems. Using the method of analysis of Section 1.4, show that Φ satisfies the inequality

$$d\Phi \leq -SdT - PdV - Nd\mu + dW'',$$

where dW'' denotes forms of work other than PdV work *and* other than μdN work, i.e., $dW = -PdV + \mu dN + dW''$. For $dW'' = 0$, $[d\Phi]_{T,\mu,V} \leq 0$; systems spontaneously evolve so as to minimize Φ. For reversible processes, $[\Delta\Phi]_{T,\mu,V} = W''$. The grand potential is the energy of other work W'' stored in the system under conditions of constant (T, μ, V).

1.12 Work out the Maxwell relations associated with the grand potential, Φ. A:

$$\left(\frac{\partial S}{\partial V}\right)_{T,\mu} = \left(\frac{\partial P}{\partial T}\right)_{V,\mu} \qquad \left(\frac{\partial P}{\partial \mu}\right)_{V,T} = \left(\frac{\partial N}{\partial V}\right)_{\mu,T} \qquad \left(\frac{\partial N}{\partial T}\right)_{\mu,V} = \left(\frac{\partial S}{\partial \mu}\right)_{T,V}. \qquad \text{(P1.8)}$$

These relations apply to open systems where the particle number N is a variable, so that $N \equiv \langle N \rangle$, the average particle number, as we'll show in statistical mechanics.

1.13 Take $U = U(T, P)$ and $V = V(T, P)$.

a. Show from the first law of thermodynamics that

$$\frac{dQ}{dT} = \left(\frac{\partial U}{\partial T}\right)_P + P\left(\frac{\partial V}{\partial T}\right)_P + \left[\left(\frac{\partial U}{\partial P}\right)_T + P\left(\frac{\partial V}{\partial P}\right)_T\right]\frac{dP}{dT}.$$

b. Show that this result implies another set of thermodynamic identities

$$C_P = \left(\frac{\partial U}{\partial T}\right)_P + P\left(\frac{\partial V}{\partial T}\right)_P \qquad \left(\frac{\partial U}{\partial P}\right)_T = -\frac{\beta_T}{\alpha}(C_P - C_V) + P\beta_T V. \qquad \text{(P1.9)}$$

1.14 From the results of Exercise 1.13, and Eqs. (1.33), (1.38), (1.41), and (1.42), show that the first law can be written in either of the two ways

$$dQ = C_V dT + \frac{T\alpha}{\beta_T}dV \qquad\qquad dQ = C_P dT - \alpha V T dP.$$

1.15 Take the first law for a magnetic system to be $dU = TdS - PdV + BdM$. (We ignored the PdV term in Section 1.9, but let's keep it now.) Define the Helmholtz energy as usual, $F \equiv U - TS$, so that $F = F(T, V, M)$ (show this). Derive the three Maxwell relations associated with $F(T, V, M)$:

$$\left(\frac{\partial S}{\partial V}\right)_{T,M} = \left(\frac{\partial P}{\partial T}\right)_{V,M} \qquad -\left(\frac{\partial P}{\partial M}\right)_{V,T} = \left(\frac{\partial B}{\partial V}\right)_{M,T} \qquad \left(\frac{\partial B}{\partial T}\right)_{M,V} = -\left(\frac{\partial S}{\partial M}\right)_{T,V}.$$

1.16 Equation (1.47) is the Maxwell relation associated with the magnetic Gibbs energy, $G_m = F - MB$. Derive the Maxwell relations associated with U (using Eq. (1.45)), the magnetic enthalpy[84] $H = U - BM$, and the Helmholtz energy, $F = U - TS$. Show that:

$$\left(\frac{\partial T}{\partial M}\right)_S = \left(\frac{\partial B}{\partial S}\right)_M \quad -\left(\frac{\partial S}{\partial M}\right)_T = \left(\frac{\partial B}{\partial T}\right)_M \quad \left(\frac{\partial T}{\partial B}\right)_S = -\left(\frac{\partial M}{\partial S}\right)_B. \quad \text{(P1.10)}$$

1.17 Start with the first law in the form $dU = TdS - PdV + BdM$. Take U and S to be functions of T, V, and M, and derive the relation

$$\left(\frac{\partial U}{\partial M}\right)_{T,V} = B - T\left(\frac{\partial B}{\partial T}\right)_{M,V}.$$

Hint: Use a Maxwell relation derived in Exercise 1.15. This formula is the magnetic analog of Eq. (1.41).

1.18 a. Using $dU = đQ + BM$, show that $đQ$ can be written in two ways:

$$đQ = \left(\frac{\partial U}{\partial T}\right)_M dT + \left[\left(\frac{\partial U}{\partial M}\right)_T - B\right]dM$$

$$đQ = \left[\left(\frac{\partial U}{\partial T}\right)_B - B\left(\frac{\partial M}{\partial T}\right)_B\right]dT + \left[\left(\frac{\partial U}{\partial B}\right)_T - B\left(\frac{\partial M}{\partial B}\right)_T\right]dB.$$

Assume in the first case that $U = U(T, M)$, and in the second, $U = U(T, B)$ and $M = M(T, B)$. These equations are the magnetic analogs of the results in Exercise 1.14.

b. Derive the following expressions for the heat capacity:

$$C_M = \left(\frac{\partial U}{\partial T}\right)_M \qquad C_B = \left(\frac{\partial U}{\partial T}\right)_B - B\left(\frac{\partial M}{\partial T}\right)_B. \quad \text{(P1.11)}$$

The expression for C_M is the magnetic analog of Eq. (1.38) for C_V, and that for C_B is the analog of Eq. (P1.9) for C_P.

c. Derive the magnetic analog of Eq. (1.39),

$$C_B = C_M + \left[\left(\frac{\partial U}{\partial M}\right)_T - B\right]\left(\frac{\partial M}{\partial T}\right)_B. \quad \text{(P1.12)}$$

1.19 Now use the first law for magnetic systems in the form $dU = TdS + BdM$.

a. Assume $U = U(T, M)$ and $S = S(T, M)$. Show that

$$C_M = T\left(\frac{\partial S}{\partial T}\right)_M \qquad \left(\frac{\partial U}{\partial M}\right)_T = T\left(\frac{\partial S}{\partial M}\right)_T + B. \quad \text{(P1.13)}$$

b. With U, S, M functions of T, B, show that

$$C_B = T\left(\frac{\partial S}{\partial T}\right)_B \qquad \left(\frac{\partial U}{\partial B}\right)_T = T\left(\frac{\partial S}{\partial B}\right)_T + B\left(\frac{\partial M}{\partial B}\right)_T. \quad \text{(P1.14)}$$

c. Use the results of this exercise together with Eq. (P1.12) to show that

$$\left(\frac{\partial S}{\partial T}\right)_B = \left(\frac{\partial S}{\partial T}\right)_M + \left(\frac{\partial S}{\partial M}\right)_T\left(\frac{\partial M}{\partial T}\right)_B, \quad \text{(P1.15)}$$

the same as what would be found using Eq. (1.32) with $S = S(T, M)$ and $M = M(T, B)$.

---

[84]What we've called the magnetic Gibbs energy is also called the *free enthalpy*, $G_m = H - TS$.

CHAPTER 2

From mechanics to statistical mechanics

S TATISTICAL mechanics generalizes the usual problem of classical dynamics, of finding particle trajectories associated with given initial conditions, to the broader problem of a great many particles differing in their initial conditions.[1] Its goal is not to seek the motions of individual particles, but rather to determine how a collection of particles is *distributed* over conceivable positions and velocities, given that such a distribution has been specified at one time. Equations of motion for individual particles are known. We seek an equation that tells us the rate of change of the number of particles that fall within prescribed limits on positions and velocities (peek ahead to *Liouville's equation*, (2.55)). For systems in thermal equilibrium, the problem can be characterized as that of *the many and the few*: Systems with enormous numbers of microscopic degrees of freedom are, in equilibrium, described by only a handful of variables, the state variables of thermodynamics. We will be guided in our development of statistical mechanics by the requirement that it be consistent with the laws of thermodynamics. In this chapter, we present the conceptual steps involved in setting up statistical mechanics.

2.1 MICROSTATES—THE MANY

2.1.1 Classical microstates

Consider N identical particles where each particle has f *degrees of freedom*, implying the system has fN degrees of freedom.[2] Denote the fN independent *generalized coordinates*, $q_1, \ldots, q_{fN}$. The Newtonian equations of motion for particles of mass m are (in an inertial reference frame)

$$m \frac{d^2}{dt^2} q_i = -\frac{\partial}{\partial q_i} V(q_1, \ldots, q_{fN}), \qquad i = 1, \ldots, fN \qquad (2.1)$$

---

[1] We base our initial development of statistical mechanics on classical mechanics because: it's often a good approximation; it's mathematically simpler than quantum mechanics; and the form of the micro-dynamics is immaterial. Statistical mechanics is not simply an application of classical or quantum mechanics to large systems. Quantum statistical mechanics is not substantively different from classical statistical mechanics (see Section 4.4). As noted by Wolfgang Pauli (1945 Nobel Prize in Physics for discovery of the exclusion principle): "The otherwise so important difference between classical and quantum mechanics is not relevant in principle for thermodynamic considerations." Quoted in [14, pxvi].

[2] The number of degrees of freedom of a particle or molecule is the minimum number of independent parameters required to specify its geometrical configuration. Point particles in three dimensions have three degrees of freedom: Coordinates x, y, and z can each be independently varied. For a rigid dumbbell, two particles maintained at a fixed separation d, with $d^2 = (x_1 - x_2)^2 + (y_1 - y_2)^2 + (z_1 - z_2)^2$, the coordinates cannot be independently varied, there being an equation of constraint amongst them. A rigid dumbbell has five degrees of freedom—three translational degrees of freedom of its center of mass, and two rotational degrees of freedom to describe its orientation about the center of mass.

where $V(q_1, \ldots, q_{fN})$ is the potential energy function representing inter-particle forces, forces from outside the system (electric field, for example), and forces of constraint, e.g., those constraining the system to occupy the volume it has.[3] A *microstate* is the set of fN coordinates, $\{q_i\}_{i=1}^{fN}$, and time derivatives, $\{\dot{q}_i\}_{i=1}^{fN}$, specified at an instant of time. The *generalized velocities* $\{\dot{q}_k\}$ are independent of coordinates, in the sense that their values can be freely chosen at an instant of time for any set of coordinates $\{q_k\}$ (see Section C.3). The microstate at a future time could be calculated if we could solve the equations in (2.1), and if we knew the state at one time with sufficient accuracy. Microstates can be represented as points in a $2fN$-dimensional space, which we can call *dynamical space*,[4] the mathematical space of all possible values of coordinates and speeds. The time evolution of microstates can be visualized as trajectories in dynamical space.

2.1.2 Phase space, the arena of classical statistical mechanics

Newtonian mechanics has the drawback that it obscures the connection with quantum mechanics, a disadvantage absent from Hamiltonian mechanics (reviewed in Appendix C), an equivalent formulation of classical mechanics that provides a "platform" for the transition to quantum mechanics, and a few other theoretical advantages for statistical mechanics. In the Hamiltonian framework, there's a change in emphasis from velocities $\{\dot{q}_i\}$ to their associated *canonical momenta*

$$p_k \equiv \frac{\partial L}{\partial \dot{q}_k}, \qquad k = 1, \cdots, fN \tag{C.30}$$

where $L = L(q_j, \dot{q}_j)$ is the Lagrangian function (the difference between the kinetic and potential energy functions, $L(q_i, \dot{q}_i) = T(\dot{q}_i) - V(q_i)$, Eq. (C.21)). Hamilton's equations of motion are

$$\dot{p}_k = -\frac{\partial H}{\partial q_k} \qquad \dot{q}_k = \frac{\partial H}{\partial p_k}. \qquad k = 1, \cdots, fN \tag{C.34}$$

where $H = H(p_j, q_j)$, the Hamiltonian function (the sum of the kinetic and potential energy functions, $H(p_i, q_i) = T(p_i) + V(q_i)$, Eq. (C.32)), is the Legendre transformation of the Lagrangian with respect to the velocities, Eq. (C.31). Any set of variables $\{p_k, q_k\}$ satisfying Hamilton's equations are termed *canonical variables*. Hamilton's equations provide, instead of fN second-order differential equations as in Eq. (2.1), $2fN$ first-order differential equations.

Microstates are thus represented as points in another $2fN$-dimensional space, *phase space* [15]. denoted Γ-space, the mathematical space of all possible values of canonical variables, the values of $\{p_k, q_k\}_{k=1}^{fN}$ at an instant of time.[5] We'll refer to the point in Γ-space that represents an N-particle system as the *system point* or the *phase point*. In the course of time, the phase point moves in Γ-space, creating a *phase trajectory* (see Excercise 2.1). Make sure you understand that the time development of an N-particle system is represented by the motion of a *single vector* $(q_1(t), \cdots, q_{fN}(t), p_1(t), \cdots, p_{fN}(t))$ locating the system point in a high-dimensional space. Phase trajectories are *unique*: Trajectories associated with different initial conditions *never cross*.[6,7]

---

[3]Such forces are assumed to propagate instantaneously, and hence our treatment is explicitly nonrelativistic.

[4]Not quite phase space, which we'll introduce presently.

[5]Spaces of the same dimension are *isomorphic*—there's a one-to-one correspondence between the elements of isomorphic sets. The connection between p_i and $\dot{q}_i$ in Eq. (C.30) is invertible (because the kinetic energy function is positive-definite), and thus there's an isomorphism between $\{p_i\}$ and $\{\dot{q}_i\}$.

[6]Trajectories in Γ-space cannot cross because of basic uniqueness theorems on the solutions of differential equations. For that reason, trajectories don't cross in dynamical space either. The advance represented by Hamiltonian over Lagrangian mechanics is the simpler structure of the equations of motion—derivatives with respect to time appear only on the left sides of the equations because H does not contain any derivatives of q_i or p_i. Moreover, the Hamiltonian (and not the Lagrangian) is the generator of time translations (see Section C.10).

[7]As we proceed, observe that statistical mechanics can be formulated without knowledge of the *specific* form of the potential energy function V, which is perhaps surprising. The Hamiltonian $H(p, q)$ is determined only when we know the

Every point in Γ-space evolves under the action of Hamilton's equations to another, uniquely determined point in the same space—the *natural motion* of phase space—a one-to-one mapping of Γ-space onto itself at different instants of time.[8] Because the displacement of a point in Γ-space over a time interval Δt depends only on the initial point and the magnitude of Δt, but does not depend on the choice of the initial time, the momenta of particles depend uniquely on their positions in Γ-space. The natural motion is *stationary*; it does not change in the course of time.

The theoretical advantage of Γ-space as the "arena" of statistical mechanics (over dynamical space) is that the Hamiltonian generates the time evolution of the system point (see Section C.10). Consider a *function* of phase-space coordinates, a *phase-space function*, $A(p, q, t) \equiv A(p_1(t), \cdots, p_n(t), q_1(t), \cdots, q_n(t), t)$, where we allow for the possibility of an *explicit* time dependence in A (the time dependence of (p, q) is implicit). Form its total time derivative:

$$\frac{dA}{dt} = \frac{\partial A}{\partial t} + \sum_k \left(\frac{\partial A}{\partial q_k} \dot{q}_k + \frac{\partial A}{\partial p_k} \dot{p}_k \right) = \frac{\partial A}{\partial t} + \sum_k \left(\frac{\partial A}{\partial q_k} \frac{\partial H}{\partial p_k} - \frac{\partial A}{\partial p_k} \frac{\partial H}{\partial q_k} \right)$$

$$\equiv \frac{\partial A}{\partial t} + [A, H] , \tag{2.2}$$

where we've used Hamilton's equations and we've defined the *Poisson bracket* in the second line. The Poisson bracket between two phase-space functions $u(p, q)$ and $v(p, q)$ is defined as

$$[u, v] \equiv \sum_k \left(\frac{\partial u}{\partial q_k} \frac{\partial v}{\partial p_k} - \frac{\partial u}{\partial p_k} \frac{\partial v}{\partial q_k} \right) . \tag{2.3}$$

Poisson brackets have the algebraic property[9] that $[A, B] = -[B, A]$, and hence $[A, A] = 0$. An immediate consequence is, if H has no explicit time dependence, then, from Eq. (2.2),

$$\frac{d}{dt} H(p, q) = [H, H] = 0 .$$

Thus, *the Hamiltonian is a constant of the motion* if $\partial H / \partial t = 0$, an assumption that applies in equilibrium statistical mechanics.

Assuming no explicit time dependence of $A(p, q)$ (i.e., $\partial A / \partial t = 0$), the equation of motion (2.2) can be written

$$\frac{dA}{dt} = -[H, A] \equiv iLA , \tag{2.4}$$

where the *Liouville operator* L is an operator waiting to act on a function,

$$L \cdot \equiv i[H, \cdot] = i \sum_k \left(\frac{\partial H}{\partial q_k} \frac{\partial \cdot}{\partial p_k} - \frac{\partial H}{\partial p_k} \frac{\partial \cdot}{\partial q_k} \right) , \tag{2.5}$$

where the "dot" ($\cdot$) indicates a placeholder for the function that L acts on. The factor of i (unit imaginary number) is included in the definition of L to make it self-adjoint, $L = L^\dagger$ (see Section C.11). The formal solution[10] of Eq. (2.4) (which is a first-order differential equation in time) is

$$A(p(t), q(t)) = e^{iLt} A(p(0), q(0)) . \tag{2.6}$$

---

potential energy function, yet non-intersecting phase trajectories exist no matter what the form of the interactions that give rise to V. Statistical mechanics seeks to provide a theory as universal as thermodynamics, in that it reaches conclusions given only the microscopic (atomic) picture of matter, irrespective of the nature of atoms and their interactions. All we assume in setting up statistical mechanics is that the systems it applies to are mechanical systems. Statistical mechanics must rely on general conclusions from analytical mechanics that apply to all mechanical systems.

[8]The mapping is one-to-one because Hamilton's equations are time-reversal symmetric (Section 2.1.4): The phase point at time t_1 uniquely determines its position at time t_2 regardless of whether $t_2 > t_1$ or $t_2 < t_1$.

[9]Poisson brackets have the same *algebraic* properties as the commutators of quantum mechanics: Anticommutativity ($[f, g] = -[g, f]$); bilinearity ($[af + bg, h] = a[f, h] + b[g, h]$, $[h, af + bg] = a[h, f] + b[h, g]$ for constants a, b); Leibniz's rule ($[fg, h] = [f, h]g + f[g, h]$); and the Jacobi identity ($[f, [g, h]] + [g, [h, f]] + [h, [f, g]] = 0$).

[10]The exponential of an operator B, e^B, is an operator defined by the Taylor series, $e^B = \sum_{n=0}^{\infty} B^n / n!$.

2.1.3 Hilbert space, the arena of quantum statistical mechanics

At the quantum level, microstates are specified by a vector $|\psi\rangle$ in *Hilbert space*,[11] represented by a wavefunction $\psi(q_1, \cdots, q_{fN}; t)$. It's time evolution is provided by *Schrödinger's equation*,

$$i\hbar \frac{\partial}{\partial t}|\psi\rangle = H|\psi\rangle, \tag{2.7}$$

where now H signifies the Hamiltonian *operator* obtained from the Hamiltonian function by the replacement $p_k \to -i\hbar\partial/\partial q_k$. Schrödinger's equation is a first-order partial differential equation in time, and thus its solution requires specifying ψ at an instant of time. The wavefunction evolves in time according to the time-displacement operator (Eq. (2.8) holds if H is independent of time),

$$\psi(q, t) = e^{-iH(t-t_0)/\hbar}\psi(q, t_0). \tag{2.8}$$

Note the formal similarity between Eqs. (2.6) and (2.8), which allows us to consider simultaneously the classical and quantum pictures. Table 2.1 summarizes the descriptions of microstates.

Table 2.1: Description of microstates, classical, and quantum

	Classical	Quantum
Space of microstates	Γ-space $2fN$-dimensional phase space	Hilbert space Infinite-dimensional vector space
Microstate represented by	Point in Γ-space	Wave function $\psi(q_1, \cdots, q_{fN})$
Dynamics generated by	Hamilton's equations	Schrödinger equation

2.1.4 Time-reversal symmetry

Consider the time-inversion mapping $t \to t' = -t, p_i \to p_i' = -p_i, q_i \to q_i' = q_i$ ($i = 1, \cdots, fN$). Under time inversion, the canonical momentum of every particle becomes its negative (see Section C.5). The Liouville operator transforms as $L(p', q') = L(-p, q) = -L(p, q)$ (show this). Thus, the classical time evolution operator $\exp(iLt)$ has the property under time reversal:

$$\exp(iL(-p, q)(-t)) = \exp(iL(p, q)t).$$

The time-reversed solution of Eq. (2.6) is therefore

$$A(p(-t), q(-t)) = \exp(iLt)A(-p(0), q(0)). \tag{2.9}$$

The system has time-reversal symmetry.

For the quantum state, take the complex conjugate of Eq. (2.7),

$$-i\hbar\frac{\partial\psi^*}{\partial t} = i\hbar\frac{\partial\psi^*}{\partial(-t)} = H^*\psi^* = H\psi^*, \tag{2.10}$$

where we've used that the Hamiltonian operator is real ($H^* = H$). Comparing Eqs. (2.10) and (2.7), for every solution $\psi(t)$, there is a solution $\psi^*(-t)$. The system is therefore dynamically reversible.

---

[11] We assume familiarity with Hilbert space, a complete, infinite-dimensional inner-product space.[16, Chapter 1]

2.1.5 Single-particle density of states function, $g(E)$

Counting is a big part of statistical physics—the number of ways that various physical quantities can be arranged or processes occur. A fundamental quantity in statistical mechanics (see Section 2.6) is the *density of energy levels*, the number of energy levels per unit energy range.[12] We derive in Chapter 4, the *general* properties of density-of-state functions without regard to their specific form. We reserve $\Omega(E)$ to denote the density of states function for interacting particles. In many cases, however, it turns out that the *single-particle* density of states function is all we need to know; we'll use $g(E)$ to denote the density of states for noninteracting, i.e., *free particles*.

A free particle of mass m is described by wavefunctions satisfying the time-independent Schrödinger equation $-(\hbar^2/(2m))\nabla^2\psi(r) = E\psi(r)$ in the form of plane waves, $\psi(r) = \exp(i\boldsymbol{k} \cdot r)$, where $k \equiv \sqrt{2mE/\hbar^2}$, with the direction of the wavevector $\boldsymbol{k}$ the direction of propagation. How many single-particle energy levels are available between E and $E + dE$?

To answer that question, boundary conditions must be specified on $\psi(r)$. For electrons in solids (an important case), there are confining potentials at the edges of the system keeping them within the system. Confining potentials present a complication to analyzing the density of states. Unless we're specifically interested in the physics of surface phenomena, we'd rather not have to take into account the modification of allowed energies engendered by confining potentials. If L is a characteristic length of a macroscopic sample, and if there are N particles in volume $\approx L^3$, and if the ratio of surface area to volume scales as L^{-1}, then the ratio of the number of atoms near a surface to the number in bulk scales as $N^{-1/3}$ (10^{-8} for Avogadro's number). Is there a way to ignore surface effects while retaining the finite volume V of a sample?

Periodic boundary conditions are a way to do that which is easy to implement and give accurate results. Assume a particle is confined to a cube of volume V of side $L = V^{1/3}$. Imagine that each face of the cube "wraps around" to adjoin the face opposite to it, so that a particle approaching one of the faces appears at the corresponding point on the opposite face. In one dimension, the point at $x = L$ coincides with the point at $x = 0$; the system is a circle of circumference L. A square of side L embodying these boundary conditions is a torus (show this). A three-dimensional system satisfying periodic boundary conditions can't be visualized by denizens of three dimensions such as us, but we can easily write down the mathematical requirements:

$$\left.\begin{array}{r}\psi(x, y, z + L) \\ \psi(x, y + L, z) \\ \psi(x + L, y, z)\end{array}\right\} = \psi(x, y, z) \,. \tag{2.11}$$

By treating $\boldsymbol{k}$ as a vector, with $\boldsymbol{k} = k_x\hat{\boldsymbol{x}} + k_y\hat{\boldsymbol{y}} + k_z\hat{\boldsymbol{z}}$, Eq. (2.11) implies

$$e^{ik_x L} = e^{ik_y L} = e^{ik_z L} = 1 \,. \tag{2.12}$$

To satisfy Eq. (2.12), k_x, k_y, k_z must be restricted to the values

$$k_x = \frac{2\pi}{L}n_x \qquad k_y = \frac{2\pi}{L}n_y \qquad k_z = \frac{2\pi}{L}n_z \,, \tag{2.13}$$

where n_x, n_y, n_z are integers (positive, negative, or zero). The energy levels permitted by periodic boundary conditions are therefore of the form (using Eq. (2.13))

$$E_{n_x,n_y,n_z} = \frac{\hbar^2}{2m}\left(k_x^2 + k_y^2 + k_z^2\right) = \frac{2\pi^2\hbar^2}{mL^2}\left(n_x^2 + n_y^2 + n_z^2\right) \,. \tag{2.14}$$

Periodic boundary conditions allow us to mathematically "fake out" a particle so that it never experiences the effects of surfaces, yet which builds in the finite size of a system utilizing wavefunctions appropriate to a uniform potential energy environment.[13]

---

[12]Known more loosely in physics jargon as the *density of states*.

[13]Studies have shown that the bulk properties of matter are independent of the boundary conditions. Pathria[17] presents numerical results on the number of states associated with various boundary conditions.

To *count* energy levels it suffices to count the allowed values of k_x, k_y, k_z, knowing that for every allowed k-vector there is an associated energy, Eq. (2.14). The three-dimensional space with Cartesian axes k_x, k_y, and k_z is known as *k-space* (see Fig. 2.1). Many calculations in statistical physics rely on a certain dexterity with the use of k-space.[14] Allowed wavevectors have components given by integer multiples of $2\pi/L$. Figure 2.1 indicates a small part of the infinite *lattice* of allowed points in k-space, each separated by $2\pi/L$ from neighboring points in each of the three directions.[15] Each allowed k-vector locates a point in k-space that's uniquely associated with a small cubical *volume* of k-space, the *unit cell* of volume $(2\pi/L)^3 = 8\pi^3/V$.

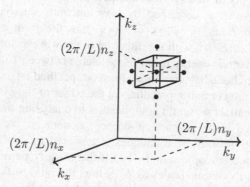

Figure 2.1: Lattice of allowed points in k-space as imposed by periodic boundary conditions, shown as dots (●). The unit cell of volume $8\pi^3/V$ of k-space is uniquely associated with a single point.

How many allowed k-points are inside a sphere[16] of radius k? The answer is simple: the volume of a sphere divided by the volume of the unit cell, $8\pi^3/V$,

$$N(k) \equiv N(|\mathbf{k}| \leq k) = \frac{4\pi k^3/3}{8\pi^3/V} = \frac{V}{6\pi^2}k^3 . \tag{2.15}$$

The number of k-points in a spherical shell between k and $k + \mathrm{d}k$ is therefore, from Eq. (2.15),

$$N(k + \mathrm{d}k) - N(k) = \frac{\mathrm{d}N}{\mathrm{d}k}\mathrm{d}k = \frac{V}{2\pi^2}k^2\mathrm{d}k . \tag{2.16}$$

Particle states with energies specified by Eq. (2.14) can be *degenerated*—there can be quantum numbers not accounted for in the free-particle Schrödinger equation, such as spin. Denote by g_d the number of degenerate states having energies specified by Eq. (2.14).[17] We're now in a position to derive what we want—the number of energy states per energy range. Define

$$N(E + \mathrm{d}E) - N(E) = \frac{\mathrm{d}N}{\mathrm{d}E}\mathrm{d}E \equiv g(E)\mathrm{d}E = g_d\frac{\mathrm{d}N}{\mathrm{d}k}\frac{\mathrm{d}k}{\mathrm{d}E}\mathrm{d}E .$$

The density of single-particle states, $g(E)$, is therefore

$$g(E) = g_d\frac{\mathrm{d}N}{\mathrm{d}k}\left(\frac{\mathrm{d}E}{\mathrm{d}k}\right)^{-1} . \tag{2.17}$$

Equation (2.17) is quite general; what distinguishes the density of states for different systems lies in the connection between energy and wavevectors, $E = E(k)$.

---

[14]Thus, get to know k-space. As I'm fond of telling students, k-space is your friend—it wants to help you do calculations.

[15]A knowledge of crystal lattices and associated terminology (lattice constants, unit cells) is presumed; Ashcroft and Mermin is a good source [18]. The lattice in k-space we're referring to is the *reciprocal lattice*.

[16]A sphere in k-space corresponds to the lowest-energy configuration of the system.

[17]For particles of spin S, $g_d = 2S + 1$.

In three spatial dimensions, we find by combining Eqs. (2.14), (2.16), and (2.17),

$$g(E) = g_d \frac{V}{4\pi^2} \left(\frac{2m}{\hbar^2} \right)^{3/2} \sqrt{E} . \qquad \text{(three spatial dimensions)} \qquad (2.18)$$

A free particle in an n-dimensional space would have an n-dimensional k-vector. The generalization of Eq. (2.18) is, using the volume for an n-dimensional sphere, Eq. (B.14), and the volume of the unit cell $(2\pi/L)^n$,

$$g(E) = g_d \frac{V\pi^{n/2}}{(2\pi)^n \Gamma(n/2)} \left(\frac{2m}{\hbar^2} \right)^{n/2} E^{(n/2)-1} , \qquad (n \text{ spatial dimensions}) \qquad (2.19)$$

where $V \equiv L^n$. The properties of the gamma function $\Gamma(x)$ are reviewed in Appendix B.

Equation (2.17) applies to photon states ("waves is waves"). Using the Planck relation $E = \hbar c k$ in Eq. (2.17), we have the density of states for photons:

$$g(E) = g_d \frac{V}{2\pi^2(\hbar c)^3} E^2 . \qquad \text{(photons)} \qquad (2.20)$$

One would take $g_d = 2$ for polarization degeneracy. Equation (2.20) also applies for extreme relativistic particles; see Exercise 4.3.

2.2 STATE VARIABLES—THE FEW

Whereas microstates incessantly evolve in time,[18] equilibrium is described by the time-independent state variables of thermodynamics.[19] The first law of thermodynamics, Eq. (1.2), $\Delta U = Q + W$, indicates the two ways by which internal energy is changed: heat transfers and/or the performance of work. In differential form, Eq. (1.3), $dU = đQ + đW$, where đ signifies an inexact differential. The notation indicates we're *not* implying small changes in previously existing quantities Q and W (which would be written dQ and dW), but rather that $đQ$ and $đW$ denote process-dependent modes of energy transfer.[20]

Work involves observable changes in the extensive properties of a system,[21] X_i. For small amounts of work, $đW = \sum_i Y_i dX_i$, the sum of products of generalized displacements dX_i with their conjugate generalized forces Y_i (see Section 1.2). The energy to change the number of particles by dN is expressed in terms of the chemical potential μ, $đW = \mu dN$ (see Section 1.6). Thus,

$$dU = đQ + \sum_i Y_i dX_i + \mu dN . \qquad (2.21)$$

Relatively, few variables appear in thermodynamic descriptions, yet there are myriad microscopic degrees of freedom that contribute to the energy of a system (its Hamiltonian). *Where are microscopic degrees of freedom represented in thermodynamics?* The division of internal energy into work and heat lines up with the distinction between macroscopic and non-macroscopic, i.e., microscopic. What is not work is heat, energy transfers to microscopic degrees of freedom. Heat is extensive (see Section 1.2), yet it's not a state variable (see Section 1.3). Entropy, an extensive state variable related to heat transfers, is defined through a differential relation $dS \equiv (đQ)_{\text{rev}}/T$, Eq. (1.8). Equation (2.21) can therefore be written (including only PdV work), $dU = TdS - PdV + \mu dN$. To answer our question, *microscopic degrees of freedom are represented as entropy.* Entropy is a proxy for the remaining degrees of freedom after the observable state variables are accounted for.

---

[18] The time evolution of microstates is generated by the Hamiltonian; if a system has a Hamiltonian, it evolves in time.

[19] Systems in equilibrium stay in equilibrium, a timeless state characterized by the values of state variables.

[20] Notation such as $đQ$ should be interpreted as a mnemonic device to indicate a small heat transfer in which the process must be specified. It does *not* indicate the action of a differential operator on some function Q.

[21] Thus, X_1 could denote volume V, X_2, magnetization M, etc. Extensive variables are defined in Section 1.1.

2.3 ENTROPY, THE BRIDGE BETWEEN MICRO AND MACRO

We therefore have two ways of describing macroscopic systems: mechanics, which provides the time evolution of microstates, and thermodynamics, a theory of the equilibrium state in which time does not explicitly appear; see Table 2.2. The mechanical approach is too general for our purposes; we're not seeking the dynamics of arbitrary collections of particles, but those in thermal equilibrium—that aspect of the problem must be taken into account. What the two approaches have in common is not time, but *energy*. Entropy, through the Boltzmann formula[22] $S = k \ln W$ (Eq. (1.62)), provides the connection with microscopic degrees of freedom of systems *in thermal equilibrium*,[23] where W is the number of *ways* macrostates specified by state variables can be realized from microstates. We must learn to calculate $W = W(U, X_i, N)$, especially its dependence on internal energy U. We now show how that calculation works for the ideal gas.

Table 2.2: Microscopic and macroscopic descriptions

Description	Internal energy represented by	Time development
Microscopic (classical mechanics)	Hamiltonian $H(p_1, \cdots p_n, q_1, \cdots, q_n)$	Generated by Hamiltonian
Macroscopic (thermal equilibrium)	$U(S, X_i, N)$ (first law of thermodynamics)	None (time independent)

The energy of N identical, noninteracting particles of mass m is[24]

$$U = \frac{1}{2m} \sum_{i=1}^{3N} p_i^2 \equiv \frac{1}{2m} R^2 , \tag{2.22}$$

where $R = \sqrt{2mU}$ is the radius of a $3N$-dimensional sphere in *momentum space*. All states of energy U lie at the $(3N - 1)$-dimensional surface of this sphere.[25] How *many* states are there? To calculate that number, microstates must have a property that's *countable*, something *discrete* in how we describe states. Bound states in quantum mechanics have a naturally countable feature in the discreteness of energy eigenvalues. For *classical* particles, it turns out we require the existence of Planck's constant to count states.[26] It's found,[3, Section 7.4] for the Sackur-Tetrode formula, Eq. (1.56), to agree with experiment, there must be a *granularity* in Γ-space, depicted in Fig. 2.2. To count states in Γ-space, we rely on a result consistent with the Heisenberg uncertainty principle, that *microstates can't be specified with greater precision than to lie within a region of phase space of size*[27] $\Delta x \Delta p_x = h$. Agreement with the experimentally verified Sackur-Tetrode formula requires this phase-space cell size to be h—not $\hbar$, nor $\frac{1}{2}\hbar$, but h [3, p110]. Taking into account the other

---

[22]There are two equivalent accounts of entropy: Clausius, Eq. (1.8), and Boltzmann, Eq. (1.62).

[23]There is a generalization of entropy, *information*, of which entropy is a special case for equilibrium systems [3, Chapter 12]. Entropy is information, but information is not necessarily entropy. We won't develop information theory in this book.

[24]Equation (2.22) holds for an N-particle gas in three spatial dimensions—we live in a three-dimensional world. There are systems, however, where particles are confined spatially to two or even one dimension. Electrons in MOSFETS, for example, are confined to a thin layer—they can move freely in two dimensions, but not in the third. One must acquire, in applications of statistical mechanics, a certain dexterity in handling systems of various dimensionalities.

[25]A *surface* is a two-dimensional object (a generalization of a plane that need not be flat) as viewed from three-dimensional space—the surface of an apple, for example. Higher-dimensional "surfaces" are termed *hypersurfaces*—for $n > 3$, $(n - 1)$-dimensional objects (manifolds) embedded in n-dimensional space. Higher-dimensional spheres are termed hyperspheres. For simplicity, we will refer to hypersurfaces as surfaces and hyperspheres as spheres.

[26]There are many instances where thermodynamics anticipates quantum mechanics (summarized in Ref. [3]). One can't scratch the surface of thermodynamics very deeply without finding quantum mechanics lurking beneath.

[27]The uncertainty principle specifies $\Delta x \Delta p_x \geq \frac{1}{2}\hbar$, consistent with $\Delta x \Delta p_x = h$.

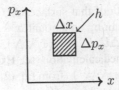

Figure 2.2: Granularity in phase space: Cells the size of Planck's constant, $\Delta x \Delta p_x = h$.

spatial dimensions, phase-space cells for a single particle are of size h^3 ($\Delta x \Delta p_x \Delta y \Delta p_y \Delta z \Delta p_z = h^3$).

The number of configurations of N identical particles is calculated as follows:

$$W = \frac{1}{h^{3N} N!} \times (\text{volume of accessible } N\text{-particle phase space}) . \tag{2.23}$$

The factor of h^{3N} "quantizes" Γ-space, providing the discreteness we require to be able to count.[28] The factor of $N!$ in Eq. (2.23) recognizes that *permutations of identical particles are not distinct states*.[29] Evidently Gibbs (in 1902) was the first to recognize that identical particles should be treated as *indistinguishable*:

> If two phases differ only in that certain entirely similar particles have changed places with one another, are they to be regarded as identical or different phases? If the particles are regarded as indistinguishable, it seems in accordance with the spirit of the statistical method to regard the phases as identical.[19, p187]

By *phase*, Gibbs means a phase point. He therefore made a break with classical physics in treating identical particles as indistinguishable. Entropy is a property of the *entire* system, with the interchange of identical particles having no consequence. The *accessible* volume of phase space in Eq. (2.23) means the volume of Γ-space (the *phase volume*) subject to constraints on the system—the spatial volume V the system occupies and its total energy U.

The Sackur-Tetrode formula can quickly be derived starting from Eq. (2.23). Let

$$W = \frac{1}{h^{3N} N!} \int d^3 x_1 \int d^3 p_1 \cdots \int d^3 x_N \int d^3 p_N = \frac{V^N}{h^{3N} N!} \int d^3 p_1 \cdots d^3 p_N , \tag{2.24}$$

where the integration over momentum space must be consistent with the total energy, $2mU = \sum_{i=1}^{3N} p_i^2$. The energy constraint specifies a $3N$-dimensional sphere in momentum space of radius $\sqrt{2mU}$, the volume of which is given by Eq. (B.14). Combining Eqs. (2.24) and (B.14),

$$W(U, V, N) = \frac{1}{N!} \frac{V^N}{\Gamma(\frac{3N}{2} + 1)} \left(\frac{2m\pi U}{h^2} \right)^{3N/2} . \tag{2.25}$$

Equation (2.25) leads to the Sackur-Tetrode formula when combined with Eq. (1.62) and the Stirling approximation is invoked; see Exercise 2.5. Is Eq. (2.25) the correct formula to use, however? Equation (2.25) accounts for *all* states of the system with energies up to and including U. Shouldn't we use the *surface area* of a $3N$-dimensional sphere (Eq. (B.17))? Phase trajectories are constrained to lie on a constant-energy surface, a $(6N - 1)$-dimensional space. Indeed we should, but to do so

---

[28]Why isn't the normalization factor in Eq. (2.23) Nh^3 rather than h^{3N}? There's a difference between N single-particle states, and the (single) state of an N-particle system in thermal equilibrium[3, p104]. Even the ideal gas, which ignores interactions between particles, is maintained in a state of equilibrium (through inter-particle collisions), a condition shared by all particles of the gas. Note that Nh^3 does not have the correct dimension to quantize N-particle Γ-space.

[29]Permutations are reviewed in Section 3.3. There are $N!$ permutations of N objects.

requires that we know the energy *width* of that surface, ΔU. Because the volume of Γ-space occupied by free-particle states is uncertain by an amount h^{3N}, the location of the energy surface is uncertain by an amount ΔU, where[30] $\Delta U/U \approx O(1/\sqrt{N})$. It turns out to be immaterial whether we use Eq. (2.25) for W or a more refined calculation (see Exercise 2.6); the density of states at the surface of the hypersphere is so large for $N \gg 1$ (from Eq. (2.19), $g(E) \sim E^{3N/2-1}$) that it doesn't matter what estimate for ΔU we use, even $\Delta U = U$.

2.4 FLUCTUATIONS: GATEWAY TO ENSEMBLES AND PROBABILITY

The transition to statistical mechanics requires that we generalize the central concept of thermodynamics: equilibrium. In thermodynamics, state variables have fixed values, corresponding to observable properties that persist unchanged in time. At the microscopic level, no system is quiescent. Due to the random motions of atoms, the local density (and other quantities) fluctuates about a mean value. Figure 2.3 shows, as evidence of random motions, the time record of the position of a small

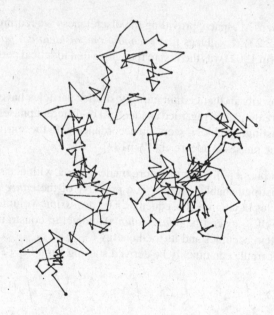

Figure 2.3: Brownian motion, the positions of a particle suspended in water (small dots) recorded at regular time intervals. From [20, p116]. The length scale is approximately 3 μm.

particle (size $\approx 0.5\ \mu$m) suspended in water, *Brownian motion*. The small dots indicate the position of the particle recorded every 30 seconds. These data were published by Jean Perrin in 1916.[31][20]. Perrin simply connected the dots with straight lines, but one mustn't interpret these paths literally. To quote Perrin, "If, in fact, one were to mark second by second, each of the straight line segments would be replaced by a polygon path of 30 sides." The jagged path in Fig. 2.3 is evidence of random impacts from fast-moving particles of the fluid.[32]

---

[30]We're using "big O" notation to indicate that $\Delta U/U$ involves terms of the order of magnitude specified by $1/\sqrt{N}$.

[31]Perrin received the 1926 Nobel Prize in Physics for work on the atomic nature of matter.

[32]How fast? The speeds of atoms in gases at room temperature are on the order of 500 m/s (see Section 5.1), but liquids are not gases. Motions of atoms in liquids are *diffusive* in character, a topic not treated in this book. The *self-diffusion coefficient* of water $D \approx 10^{-9}$ m^2/s, and with a near-neighbor distance $a \approx 10^{-10}$ m between water molecules (obtained from the density of water), one infers an average speed $D/a \approx 10$ m/s.

2.4.1 Time averages, ensemble averages, and the ergodic hypothesis

Let $\hat{A}$ denote a measurable quantity. In thermodynamics, the measured value of $\hat{A}$, denoted A, is the result of a single measurement on a system. In statistical mechanics, we must *interpret the value of $\hat{A}$ to be the mean value $\overline{A}$ of a large number of measurements of $\hat{A}$.* We can envision a series of repeated measurements of $\hat{A}$ *in time*, in which case we speak of *time averages*. The quantity $\hat{A}$, measured at times t_i, $i = 1, \cdots, N$, having values $A(t_i)$, has the average value

$$\overline{A} \equiv \frac{1}{N} \sum_{i=1}^{N} A(t_i) \stackrel{N \to \infty}{\longrightarrow} \frac{1}{T} \int_0^T A(t)\mathrm{d}t , \qquad (2.26)$$

where we allow for a continuous reading of $A(t)$ as $N \to \infty$. Birkhoff's theorem[21] (see Section 2.6) proves the *existence* of time averages[33] in the limit $T \to \infty$, when it's understood that the time dependence of the integrand is generated by the natural motion of phase space, $\int_0^T A(t)\mathrm{d}t \equiv \int_0^T A(p(t), q(t))\mathrm{d}t$. To *calculate* time averages requires that we know the phase trajectories[34]—just what we don't know!

For that reason, statistical mechanics employs another kind of average: *ensemble averages*. Imagine a collection of identical copies of a system—an *ensemble*—with each copy prepared in the same way, *subject to known information about the system*. If, for example, the system occupies volume V at temperature T, the same is true of every copy. The ensemble is prepared consistent with known macroscopic information, but of course there are myriad microvariables not subject to our control. The ensemble average is the mean $\overline{A}$ obtained from simultaneous measurements of $\hat{A}$ on all members of the ensemble; *it's an average over initial conditions*. A tenet of statistical mechanics, the *ergodic hypothesis*, asserts the equality of time and ensemble averages.[35] We'll give a mathematical expression shortly (Eq. (2.53)), but essentially the rest of this book is devoted to calculating ensemble averages.

2.4.2 Probability of fluctuations: Einstein fluctuation theory

A measurement of $\hat{A}$ on a randomly selected member of an ensemble will, in general, return a value different from $\overline{A}$. Can we deduce the *probability* $P(A)\mathrm{d}A$ that the result lies in the range $(A, A + \mathrm{d}A)$? Thermodynamics posits a connection between probability and entropy, and we're going to reach for that. When $A \neq \overline{A}$ is obtained, we say that at the time of measurement the system was *exploring* a configuration in which $\hat{A}$ has the value A. The entropy associated with the state of a system momentarily exploring a given fluctuation is *less* than the entropy of the equilibrium state, the maximum entropy a system can have subject to macroscopic constraints. Equilibrium fluctuations temporarily produce *decreases* in entropy—the stability condition (see Section 1.12).

Referring to the composite system in Fig. 1.9, the entropy deviation $\Delta S_+ \equiv S_+ - S_+^0$ (S_+ (S_+^0) denotes the entropy of the composite system after (before) the fluctuation). Equation (1.62) provides the connection between entropy and probability, except that we'll write not W (for ways) but P for the probability of a fluctuation,

$$P(\Delta S_+) = K e^{\Delta S_+/k} , \qquad (2.27)$$

---

[33]One might think that *of course* time averages exist (from an experimental point of view), and thus Birkhoff's theorem is a mathematical "nicety"—we certainly wouldn't want to formulate concepts that can be shown *not* to exist. To the contrary, Birkhoff's theorem is one of the two main theorems on which the construction of statistical mechanics relies (the other is Liouville's theorem). A user-friendly proof of Birkhoff's theorem is given in Farquhar[22, Appendix 2].

[34]Equation (2.26) doesn't indicate *which* phase trajectory we should follow, i.e., which trajectory associated with which initial conditions, and therein lies the explanatory power of Birkhoff's theorem. It shows that time averages exist independently of the trajectory. See Section 2.6.

[35]*Ergodic theory*, which had its origins in statistical mechanics, has become a branch of mathematics that studies the long-time properties of dynamical systems. See Lebowitz and Penrose[23]. The word *ergodic* was coined by Boltzmann as a combination of the Greek words $\epsilon\rho\gamma o\nu$ (*ergon*—"work") and $o\delta\acute{o}\varsigma$ (*odos*—"path").

where K is a normalization factor (produces a probability as a number between zero and one). Note the change in emphasis: Probability is treated as fundamental in Eq. (1.62), from which entropy can be calculated; entropy is treated as fundamental in Eq. (2.27), from which probability can be found.

Energy fluctuations

Let's apply the Einstein formula, Eq. (2.27), to energy fluctuations. Let energy ΔU be transferred from the environment to the system, where V and N remains fixed. Suppressing the V and N dependence of $S(U, V, N)$, and referring to Fig. 1.9,

$$\Delta S_+ = \overbrace{S_A(U_A + \Delta U) + S_B(U_B - \Delta U)}^{S_+} - \overbrace{(S_A(U_A) + S_B(U_B))}^{S_+^0} .$$

Note that the energy of the composite system remains fixed. For small energy transfers, we have to second order in a Taylor expansion of ΔS_+:

$$\Delta S_+ \approx \Delta U \left(\frac{\partial S_A}{\partial U_A}\right)_0 - \Delta U \left(\frac{\partial S_B}{\partial U_B}\right)_0 + \frac{1}{2}(\Delta U)^2 \left[\left(\frac{\partial^2 S_A}{\partial U_A^2}\right)_0 + \left(\frac{\partial^2 S_B}{\partial U_B^2}\right)_0\right]$$

$$= \Delta U \left(\frac{1}{T_A} - \frac{1}{T_B}\right) - \frac{1}{2}(\Delta U)^2 \left(\frac{1}{C_V^A T_A^2} + \frac{1}{C_V^B T_B^2}\right) = -\frac{(\Delta U)^2}{2T^2}\left(\frac{1}{C_V^A} + \frac{1}{C_V^B}\right)$$

$$\approx -\frac{(\Delta U)^2}{2T^2 C_V} , \tag{2.28}$$

where in the first line of Eq. (2.28), the subscript zero refers to derivatives evaluated in equilibrium, and in the second line, we've used $(\partial S/\partial U)_{V,N} = 1/T$, $(\partial^2 S/\partial U^2)_{V,N} = -1/(C_V T^2)$ (see Exercise 2.7) and that, in equilibrium, $T_A = T_B \equiv T$, Eq. (1.67), and, finally, in the third line that the heat capacity of system B (nominally the environment), C_V^B, is much larger than that of system A, C_V^A, implying $C_V^A/C_V^B \ll 1$. The final result in Eq. (2.28) therefore involves quantities referring solely to system A. Note that ΔS_+ in Eq. (2.28) is negative, as required by the basic theory.

Utilizing Eq. (2.28) in Eq. (2.27), we have an expression for the probability of an energy fluctuation of size u ($\equiv \Delta U$),

$$P(u) = Ke^{-\alpha u^2} , \tag{2.29}$$

where $\alpha \equiv (2kT^2 C_V)^{-1}$. The quantity K can be found through normalization (see Appendix B),

$$\int_{-\infty}^{\infty} P(u)\mathrm{d}u = 1 = K \int_{-\infty}^{\infty} e^{-\alpha u^2}\mathrm{d}u = K\sqrt{\frac{\pi}{\alpha}} . \tag{2.30}$$

Thus, $K = \sqrt{\alpha/\pi} = 1/\sqrt{2\pi kT^2 C_V}$. Armed with $P(u)$, we can calculate its *moments*.[36] The first moment, the average size of u, $\bar{u} = \int_{-\infty}^{\infty} uP(u)\mathrm{d}u = 0$ ($P(u)$ is an even function). Thus, *positive energy transfers are as equally likely to occur as negative*—such as the nature of equilibrium, but it underscores that equilibrium is anything but a "quiet place"; energy transfers of all magnitudes between system and environment are continually occurring. To capture the magnitude of the average energy fluctuation, consider the second moment:

$$\overline{(\Delta U)^2} = \int_{-\infty}^{\infty} u^2 P(u)\mathrm{d}u = K \int_{-\infty}^{\infty} u^2 e^{-\alpha u^2}\mathrm{d}u = K\sqrt{\frac{\pi}{\alpha}}\frac{1}{2\alpha} = kT^2 C_V ; \tag{2.31}$$

we've used Eq. (B.9). Thus,

$$\sqrt{\overline{(\Delta U)^2}} = \sqrt{kC_V T} . \tag{2.32}$$

---

[36] The nth moment of a probability distribution $P(x)$ is the integral $\int x^n P(x)\mathrm{d}x$. We're invoking probability ideas here without formally introducing the subject; see Chapter 3.

Equation (2.32) is a milestone achievement in transitioning from thermodynamics to a microscopic theory. Thermodynamics, a theory that nominally can't "handle" fluctuations (see Section 1.12), has guided us, through the use of the entropy principle, to a nontrivial result concerning energy fluctuations. If we apply Eq. (2.32) to the ideal gas (with $C_V = \frac{3}{2}Nk$ and $U = \frac{3}{2}NkT$, Eqs. (1.57) and (1.58)), we find the *relative* magnitude of the fluctuation,

$$\frac{\sqrt{\overline{(\Delta U)^2}}}{U} = \frac{1}{\sqrt{3N/2}} \sim \frac{1}{\sqrt{N}} . \qquad \text{(ideal gas)} \qquad (2.33)$$

Equation (2.33) has been derived for the ideal gas, yet it's a typical result (see Chapter 4).

General analysis

The entropy of the composite system $S_+ = S_{\text{sys}} + S_{\text{env}}$ is the sum of the system entropy S_{sys} and that of the environment, S_{env}. Clearly, $\Delta S_+ = \Delta S_{\text{sys}} + \Delta S_{\text{env}}$. Because the environment is far larger than the system, the values of *environmental* intensive variables will be unaffected by system fluctuations. By integrating $dS = (dU + PdV - \mu dN)/T$,

$$\Delta S_{\text{env}} = \int \left[\frac{1}{T}(dU + PdV - \mu dN) \right]_{\text{env}} = \left[\frac{1}{T}(\Delta U + P\Delta V - \mu \Delta N) \right]_{\text{env}}$$

$$= -\left[\frac{1}{T}(\Delta U + P\Delta V - \mu \Delta N) \right]_{\text{sys}} ,$$

where we've used $\Delta U_{\text{sys}} = -\Delta U_{\text{env}}$, and so on ($U$, V, and N are conserved). Thus,

$$\Delta S_+ = \Delta S - \frac{1}{T}(\Delta U + P\Delta V - \mu \Delta N) , \qquad (2.34)$$

where all quantities on the right side of Eq. (2.34) refer to system variables. Equation (2.34) is quite useful because the terms *linear* in fluctuating quantities $(\Delta U + P\Delta V - \mu \Delta N)$ cancel the same terms in an expansion of ΔS to first order. Expand ΔS to second order:

$$\Delta S = \frac{1}{T}\Delta U + \frac{P}{T}\Delta V - \frac{\mu}{T}\Delta N + \frac{1}{2}\Delta^2 S, \qquad (2.35)$$

where $\Delta^2 S$ involves all second-order terms in a Taylor expansion of $S(U, V, N)$. Combining Eqs. (2.34) and (2.35),

$$\Delta S_+ = \frac{1}{2}\Delta^2 S . \qquad (2.36)$$

To develop an expression for $\Delta^2 S$, it's helpful to adopt a notation for second derivatives. Let $S_{UV} \equiv \partial^2 S / \partial U \partial V$, and so on. Then, all second order terms in a Taylor expansion of S are

$$\Delta^2 S = S_{UU}(\Delta U)^2 + 2S_{UV}\Delta U\Delta V + 2S_{UN}\Delta U\Delta N + S_{VV}(\Delta V)^2 + 2S_{VN}\Delta V\Delta N + S_{NN}(\Delta N)^2 . \qquad (2.37)$$

Equation (2.37) can be rewritten,

$$\Delta^2 S = \Delta U \left[S_{UU}\Delta U + S_{UV}\Delta V + S_{UN}\Delta N \right] + \Delta V \left[S_{VU}\Delta U + S_{VV}\Delta V + S_{VN}\Delta N \right]$$
$$+ \Delta N \left[S_{NU}\Delta U + S_{NV}\Delta V + S_{NN}\Delta N \right] . \qquad (2.38)$$

Equation (2.38) simplifies when it's recognized that each term in square brackets is the first-order variation of a *derivative* about it's equilibrium value. For example,

$$\Delta \left(\frac{\partial S}{\partial U} \right) = S_{UU}\Delta U + S_{UV}\Delta V + S_{UN}\Delta N .$$

Thus, Eq. (2.38) is equivalent to

$$\Delta^2 S = \Delta U \Delta \left(\frac{\partial S}{\partial U} \right) + \Delta V \Delta \left(\frac{\partial S}{\partial V} \right) + \Delta N \Delta \left(\frac{\partial S}{\partial N} \right) . \tag{2.39}$$

Equation (2.39) involves fluctuations of extensive quantities multiplied by fluctuations of intensive quantities. Using Eq. (1.24),

$$\Delta^2 S = \Delta U \Delta \left(\frac{1}{T} \right) + \Delta V \Delta \left(\frac{P}{T} \right) - \Delta N \Delta \left(\frac{\mu}{T} \right)$$

$$= -\frac{1}{T} \left(-\Delta V \Delta P + \Delta N \Delta \mu \right) - \frac{\Delta T}{T^2} \overbrace{\left(\Delta U + P \Delta V - \mu \Delta N \right)}^{T \Delta S}$$

$$= -\frac{1}{T} \left[\Delta T \Delta S - \Delta P \Delta V + \Delta \mu \Delta N \right] . \tag{2.40}$$

We require $\Delta^2 S < 0$ (stability condition), and thus in any fluctuation the terms in square brackets in Eq. (2.40) must be positive, $\Delta T \Delta S - \Delta P \Delta V + \Delta \mu \Delta N > 0$. Equation (2.40) follows from an analysis of how S responds to small variations in U, V, and N. Yet there are *six* types of variations in Eq. (2.40), *including* ΔS. The variations ΔU, ΔV, and ΔN account for the variation ΔS *as well as* the variations ΔT, ΔP, and $\Delta \mu$. The six fluctuations indicated in Eq. (2.40) are therefore not independent. We can work with *any set of three independent variations* as a starting point for a stability analysis.

Equation (2.40) can be written in the general form

$$\Delta^2 S = - \sum_{ij} A_{ij} \Delta X_i \Delta X_j ,$$

where the terms $A_{ij} \equiv -\partial^2 S / \partial X_i \partial X_j$ are the elements of a symmetric matrix (the *Hessian matrix*). The probability distribution can then be expressed as

$$P(\Delta X_k) = K \exp \left(-\frac{1}{2k} \sum_{ij} A_{ij} \Delta X_i \Delta X_j \right) . \tag{2.41}$$

It's shown in Appendix B how to evaluate the normalization integral, Eq. (B.12).

Fluctuations in T and V

To illustrate Eq. (2.40), choose T and V as the independent variables and assume that N is fixed, $\Delta N = 0$. We need to express the variations of S and P in terms of T and V:[37]

$$dS = \left(\frac{\partial S}{\partial T} \right)_V dT + \left(\frac{\partial S}{\partial V} \right)_T dV = \frac{C_V}{T} dT + \left(\frac{\partial P}{\partial T} \right)_V dV , \tag{2.42}$$

and

$$dP = \left(\frac{\partial P}{\partial T} \right)_V dT + \left(\frac{\partial P}{\partial V} \right)_T dV = \left(\frac{\partial P}{\partial T} \right)_V dT - \frac{1}{\beta_T V} dV , \tag{2.43}$$

where in Eq. (2.42) we've used the result of Exercise 1.7 and a Maxwell relation, and in Eq. (2.43), β_T is the isothermal compressibility, Eq. (1.33). Using Eqs. (2.42) and (2.43),

$$\Delta T \Delta S - \Delta P \Delta V = \frac{C_V}{T} (\Delta T)^2 + \frac{1}{\beta_T V} (\Delta V)^2 , \tag{2.44}$$

---

[37]Entropy is most naturally a function of U, V, N (see Eq. (1.23)), but we can take entropy (or other state variables) to be a function whatever variables we find useful; there is considerable redundancy in how equilibrium is described.

which is positive ($C_V > 0$, $\beta_T > 0$). Combining Eqs. (2.44), (2.40), and (2.36) with Eq. (2.27), we have the joint probability for fluctuations in T and V,

$$P(\Delta T, \Delta V) = K \exp\left(-\frac{C_V}{2kT^2}(\Delta T)^2 - \frac{1}{2kT\beta_T V}(\Delta V)^2\right). \qquad (2.45)$$

Note that Eq. (2.45) has the form of Eq. (2.41). To normalize, it's useful to introduce abbreviations: Let $t \equiv \Delta T$, $v \equiv \Delta V$, $a \equiv C_V/(2kT^2)$, and $b \equiv 1/(2kT\beta_T V)$. We seek the normalization factor,

$$\int P(t,v)dtdv = K \int_{-\infty}^{\infty} e^{-at^2}e^{-bv^2}dtdv = K\sqrt{\frac{\pi}{a}}\sqrt{\frac{\pi}{b}} = K\frac{\pi}{\sqrt{ab}}, \qquad (2.46)$$

and hence $K = \sqrt{ab}/\pi$. Note how we've summed over all possible fluctuations in Eq. (2.46).

We can now calculate the magnitude of the average temperature and volume fluctuations.

$$\overline{(\Delta T)^2} = \int P(t,v)t^2 dtdv = K \int_{-\infty}^{\infty} e^{-at^2}t^2 dt \int_{-\infty}^{\infty} e^{-bv^2} dv = \frac{1}{2a} = \frac{kT^2}{C_V}. \qquad (2.47)$$

In the same way,

$$\overline{(\Delta V)^2} = \int P(t,v)v^2 dtdv = \frac{1}{2b} = kT\beta_T V. \qquad (2.48)$$

We can also calculate the *correlation* between temperature and volume fluctuations,

$$\overline{\Delta T \Delta V} = \int P(t,v)tv\,dtdv = 0, \qquad (2.49)$$

that is, fluctuations in T and V are *uncorrelated*. In understanding correlation functions, it's useful to ask, under what conditions would $\overline{\Delta T \Delta V} \neq 0$? If a positive temperature fluctuation ΔT is associated with a positive volume fluctuation ΔV, then the average of the product of ΔT and ΔV would be nonzero. As it is, $\overline{\Delta T \Delta V} = 0$ in Eq. (2.49) indicates a lack of correlation between the occurrence of fluctuations in temperature and volume—they are independent fluctuations.

Discussion

Fluctuations in extensive quantities are proportional to the square root of the size of the system (Eqs. (2.32) and (2.48)), whereas those of intensive quantities are *inversely* proportional to the square root of the system size, Eq. (2.47). In either case, the *relative* root mean square fluctuation of any thermodynamic quantity X is inversely proportional to the square root of the size of the system,

$$\frac{\sqrt{\overline{(\Delta X)^2}}}{X} \propto \frac{1}{\sqrt{\text{size of the system}}}.$$

For macroscopic systems fluctuations are typically negligible. There are systems, however, dominated by fluctuations (those near critical points; see Chapter 7).

The *product* of the fluctuation in an extensive quantity with that of an intensive quantity is therefore *independent of the size of the system*. Combining Eqs. (2.32) and (2.47),

$$\sqrt{\overline{(\Delta U)^2}}\sqrt{\overline{(\Delta T)^2}} = kT^2. \qquad (2.50)$$

Equation (2.50) has the form of an uncertainty principle: *One cannot simultaneously reduce fluctuations in both by increasing the size of the system*—fluctuations in T and U are fundamentally linked. Equation (2.50) indicates a *complementarity* between the two *modes of description* that statistical mechanics is based on: microscopic, associated with dynamical reversibility, and macroscopic, associated with thermodynamic irreversibility.[38]

---

[38]This argument is due to Neils Bohr [24], who wrote "··· thermodynamical irreversibility, as exhibited by the leveling of temperatures, does not mean that a reversal in the course of events is impossible, but that the prediction of such a reversal cannot be part of any description involving a knowledge of the temperature of various bodies."

2.5 ENSEMBLE FLOWS IN PHASE SPACE, LIOUVILLE'S THEOREM

The ergodic hypothesis (equality of time and ensemble averages) is unproven except in special cases. We can either wait for its status to be clarified or we can move on. We will be guided methodologically by the requirement that *statistical mechanics be the simplest mechanical model consistent with the laws of thermodynamics*. We are free to adopt whatever conditions we require as long as they're consistent with the overriding requirement that the laws of thermodynamics emerge from the statistical theory. Thus, we're going to *sidestep* the difficulties associated with the ergodic hypothesis. We assert that *thermodynamic quantities correspond to appropriate ensemble averages*.

2.5.1 Phase-space probability density, ensemble average

Members of ensembles are prepared in the same way subject to macroscopic constraints. Microscopically, however, systems (elements of an ensemble), even though prepared as identically as possible, have system points in Γ-space that are *not* identical. While the same macroscopic constraints are met by all members of an ensemble, they differ in their phase points.[39] Ensembles are represented as *collections* of system points in Γ-space (see Fig. 2.4), where, we emphasize, each point is the phase

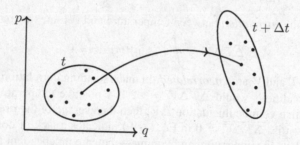

Figure 2.4: A collection of system points in Γ-space at time t (an ensemble) flows under Hamiltonian dynamics to the ensemble at time $t + \Delta t$. Axes p, q represent collectively the $3N$ axes associated with p_i, q_i for $i = 1, \cdots 3N$.

point of an N-particle system. The ensemble is thus a *swarm* of points in Γ-space. As time progresses, the swarm moves under the natural motion of phase space (indicated in Fig. 2.4). The *flow* of the ensemble turns out to be that of an incompressible fluid (Exercise 2.12). We can now pose the analogous question we asked about fluctuations: What is the probability that the phase point of a randomly selected member of an ensemble lies in the range (p, q) to $(p + dp, q + dq)$ (where p and q denote $(p_1, \cdots, p_{3N})$ and $(q_1, \cdots, q_{3N})$, with $dp \equiv dp_1 \cdots dp_{3N}$ and $dq \equiv dq_1 \cdots dq_{3N}$)? Such a probability will be proportional to dp and dq, which we can write as

$$\rho(p, q) \frac{dp\, dq}{h^{3N}} \equiv \rho(p, q) d\Gamma ,\tag{2.51}$$

where the factor of h^{3N} ensures that the *phase-space probability density*, $\rho(p, q)$, is dimensionless. Whatever the form of $\rho(p, q)$, it must be normalized,

$$\int_\Gamma \rho(p, q) d\Gamma = 1 ,\tag{2.52}$$

which simply indicates that any element of the ensemble has *some* phase point. If $A(p, q)$ is a generic phase-space function, its ensemble average is, by definition (see Chapter 3),

$$\overline{A} \equiv \int_\Gamma A(p, q) \rho(p, q) d\Gamma .\tag{2.53}$$

---

[39]Many microscopic configurations are consistent with the same macroscopic state—the message of Boltzmann entropy.

2.5.2 Liouville's theorem and constants of the motion

Our goal is to find the phase-space probability density $\rho(p, q)$, a task greatly assisted by *Liouville's theorem*. If at time t we denote a given volume in Γ-space as Ω_t, such as the ellipse indicated in Fig. 2.4, and $\mathcal{N}_t$ as the number of ensemble points within Ω_t, then the *density* of ensemble points in Ω_t is $\mathcal{N}_t/\Omega_t$. An important property of ensembles is that the phase trajectories of system points never cross (Section 2.1). Assume that the $\mathcal{N}_t$ ensemble points are within the interior of Ω_t, i.e., not on the boundary on Ω_t. As time progresses, the number of ensemble points interior to Ω_{t+dt} is the same as in Ω_t (see Fig. 2.4), $\mathcal{N}_{t+dt} = \mathcal{N}_t$. Points interior to Ω_t (not necessarily ensemble system points) evolve to points interior to Ω_{t+dt}, because points on the boundary of Ω_t evolve to points on the boundary of Ω_{t+dt} and trajectories never cross. The *volumes* are also equal, $\Omega_{t+dt} = \Omega_t$; as shown in Appendix C, the volume element $d\Gamma$ is invariant under the time evolution imposed by Hamilton's equations,[40] $d\Gamma_{t+dt} = d\Gamma_t$. The density $\mathcal{N}_t/\Omega_t$ is thus a dynamical invariant.

The density of ensemble points, however, is none other than the phase-space probability density for ensemble points, $\rho(p, q)$. We therefore arrive at Liouville's theorem: $\rho(p, q)$ is constant along phase trajectories,

$$\frac{d}{dt}\rho(p, q) = 0 .$$

(2.54)

The function $\rho(p, q)$ is certainly a phase-space function, and thus, combining Eqs. (2.2) and (2.54),

$$\frac{\partial \rho}{\partial t} + [\rho, H] = 0 ,$$

(2.55)

where we allow for a possible explicit time dependence to $\rho(p, q, t)$. Equation (2.55), *Liouville's equation*, is the fundamental equation of classical statistical mechanics, on par with the status of the Schrödinger equation in quantum mechanics.

A *constant of the motion* is a phase-space function $g(p, q, t)$ for which $dg/dt = 0$, i.e., the value of g stays constant in time when its arguments p and q are replaced by solutions of Hamilton's equations of motion. Constants of the motion are therefore solutions to Liouville's equation. Conversely, solutions of Liouville's equation are constants of the motion. A function is a solution of Liouville's equation *if and only if* it's a constant of the motion. *The most general solution of Liouville's equation is an arbitrary function of the constants of the motion.*

In thermal equilibrium we require that $\rho(p, q, t)$ be *stationary*, $\partial \rho_{eq}/\partial t = 0$. Thus, the distribution functions we seek are such that ($[\cdot, \cdot]$ denotes Poisson bracket)

$$[\rho_{eq}, H] = 0 .$$

(2.56)

The equilibrium distribution $\rho_{eq}(p, q)$ is therefore a function of *time-independent* constants of the motion. How many are there? Quite a few! A closed N-particle system[41] has $6N - 1$ constants of the motion;[42][25, p13] most, however, are not relevant for our purposes.[43] Many constants of motion might be found for a dynamical system, call them $g_1, g_2, \cdots$, each such that $[g_i, H] = 0$. To be useful, constants of the motion should possess certain properties. They should be *independent*—it should not be possible to express g_3 in terms of g_1 and g_2, for example. For independent constants of the motion, we expect the phase trajectory to lie in the *intersection* of all surfaces for which

---

[40]Time evolution in phase space, as governed by Hamilton's equations, is equivalent to a *canonical transformation* that maps the variables (p, q) at time t as satisfying Hamilton's equations, into (p, q) at time $t + dt$ that also obey Hamilton's equations (see Section C.10). The Jacobian of any canonical transformation is unity (Section C.9).

[41]A *closed* mechanical system is one in which the particles of the system interact through internal forces only, and not from forces produced by bodies external to the system.

[42]The "minus one" arises because for a closed system the equations of motion do not explicitly involve time, and thus the origin of time may be chosen arbitrarily. Time translational invariance implies energy conservation.

[43]On the other hand, if one knew *all* the constants of motion available to a system, then, because the solution of Eq. (2.56) is an arbitrary function of the constants of the motion, knowledge of the most general solution of Liouville's equation is equivalent (in principle) to a complete solution of the dynamical system.

$g_i(p, q) =$ constant, which are termed *isolating* constants of the motion.[26, p96] Suppose one has found k independent, isolating constants of the motion, including the Hamiltonian. The phase trajectory is confined to a $(6N - k)$-dimensional subspace of Γ-space. Consider two constants of the motion, g_1 and g_2. One could follow a curve C_1 in this subspace along which g_1 is constant. One would like, at any point of C_1, that it be intersected by another curve C_2 along which g_2 is constant, in such a way that a surface is formed by a mesh of curves of type C_1 and C_2. It turns out that can happen only if $[[g_1, g_2], H] = 0$; the Poisson bracket of constants of the motion must itself be a constant of the motion—*Poisson's theorem*.[27, p216] The most important constants of the motion are associated with conservation laws, which derive from fundamental *symmetries* of physical systems, homogeneity and isotropy of space and time, from which we infer conservation of energy, momentum, and angular momentum.[44] Constants of the motion associated with conservation laws have the important property of being *additive* over the particles of the system. As a further distinction, constants of the motion are as a rule *non-algebraic* functions of the canonical variables.[45] The *Bruns-Painlevé theorem* states that every algebraic constant of the motion of an N-particle system is a combination of *ten* constants of the motion: energy, three components of angular momentum, and six components of the position and momentum vectors of the center of mass.[46] These ten constitute a "basis" for generating other algebraic constants of the motion.

If k isolating constants of the motion are known to exist, then phase trajectories are restricted to a manifold of dimension $(6N - k)$. It's not known in general how many isolating constants of the motion a given Hamiltonian has (other than the Hamiltonian itself). Statistical mechanics proceeds on the assumption that *the Hamiltonian is the only isolating constant of the motion*, a procedure that's justified *a posteriori* through comparison of the subsequent inferences of the theory with experimental results. We arrive at the working assumption that ρ_{eq} is a function of the Hamiltonian:

$$\rho_{eq}(p, q) = F(H(p, q)) \,. \tag{2.57}$$

It remains to determine the function F, which we do in Chapter 4. We note that in seeking a theory of equilibrium that goes beyond thermodynamics, *all microscopic information about the components of a system enters through the Hamiltonian*.

2.6 BIRKHOFF'S THEOREM

In this section we take a deeper look at the two theorems on which statistical mechanics rests, those of Birkhoff and Liouville. The working idea of statistical mechanics is that phase trajectories of isolated systems are confined to *ergodic surfaces*, $(6N - 1)$-dimensional hypersurfaces in Γ-space.[47] We said in Section 2.5 that the ergodic hypothesis can be bypassed: We asserted that thermodynamic variables correspond to ensemble averages of microscopic quantities, and the proof is in the pudding—statistical mechanics is a successful theory in terms of the predictions it makes. *It is not necessary to link the construction of statistical mechanics with the ergodic hypothesis*. Nevertheless, it is one of two broad problems associated with the logical foundations of the subject, the other being the methods we use to calculate ensemble averages (Chapter 4).

Just for the purposes of Section 2.6, we introduce the idea of *measure*, a mathematical generalization of *extent* (volume, length, or area). Measure theory is an area of mathematics to which

---

[44]Conservation of linear momentum, angular momentum, and energy follow directly from Newton's second law. A more general way to establish conservation laws, one not limited to nonrelativistic mechanics, involves space-time coordinate transformations that leave the action integral invariant. See for example [13, p140].

[45]*Algebraic functions* can be formed using the algebraic operations of addition, multiplication, division; they're often defined as functions obtained as the root of a polynomial equation. Transcendental functions "transcend" algebra, and do not satisfy a polynomial equation.

[46]The theorem is examined in detail in Whittaker[28, Chapter 14].

[47]Ergodic surfaces have an *odd* number of dimensions—it's not that we've eliminated one coordinate axis and one momentum axis; the ergodic surface passes through $3N$ momentum axes and $3N$ coordinate axes. The energy surface of the one-dimensional harmonic oscillator, for example—an ellipse, intersects the p and q axes.

students of the physical sciences are not typically exposed; it is used in the foundations of the theory of probability.[48] Consider the length of an interval $I = [b, a]$ of the real line, the set[49] $\{x \in I | a \le x \le b\}$. The measure of I, denoted $\mu(I)$, may be defined as its length $\mu(I) = b - a$. What is the "length" of a set with one element, $\{x | x = a\}$? Generalizing the length of an interval, $\mu(a) = a - a = 0$, a set that has *measure zero*. Consider the function

$$f(x) \equiv \begin{cases} 1 & \text{for } x \neq 0 \\ 0 & \text{for } x = 0 . \end{cases}$$

Is $f(x)$ continuous? Clearly not everywhere. Yet it *is* continuous if the point $x = 0$ is excluded. Such a function is said to be continuous *almost everywhere*. Theorems utilizing measure theory (such as Birkhoff's theorem) are rife with conclusions that are true almost everywhere, that the result of the theorem may not hold on a set of measure zero.[50]

Let A be a measurable set of points in Γ-space. Let the measure of A be defined by the integral

$$\mu(A) \equiv \int_A d\Gamma . \tag{2.58}$$

If A_t denotes the set of phase points associated with A at time t, then by Liouville's theorem $\mu(A_t) = \mu(A_{t=0})$—the measure of a set in Γ-space is *constant* as its points evolve under the natural motion of Γ-space. Such a set is said to have *invariant measure*. Invariant measures play an essential role in the mathematical foundations of statistical mechanics.

We can now state Birkhoff's theorem more precisely than in Section 2.4.1. For V an invariant part[51] of Γ-space, and for $f(P)$ a phase-space function ($P \equiv (p, q)$ denotes a phase point) determined at all points $P \in V$, then the theorem states that the limit (the time average)

$$\overline{f} \equiv \lim_{T \to \infty} \frac{1}{T} \int_0^T f(P(t))dt \tag{2.59}$$

exists for all points $P \in V$, *except on a set of measure zero*—the theorem holds almost everywhere on V. The limit also exists (almost everywhere) for $T \to -\infty$; it's also independent of the starting time: $\lim_{T \to \infty} (1/T) \int_{t_0}^{T+t_0} f(p(t), q(t))dt$ exists independently of t_0. A general problem in dynamics is the *sensitivity* of initial conditions—to what extent can the initial data be varied without qualitatively altering dynamical trajectories. For our purposes, we'd like to know to what extent the values of phase functions $f(P \in V)$ "almost" do not depend on initial conditions. Birkhoff's theorem asserts that for a great majority of the trajectories (except on a set of measure zero), the values of a phase function $f(P(t))$ are very nearly equal to a constant.[52]

Birkhoff's theorem has a corollary which has a bearing on the ergodic hypothesis, one that applies if an invariant part V of Γ-space is *metrically indecomposable* (equivalently *metrically transitive*)—V cannot be divided into two invariant subsets V_1 and V_2 of finite measure. If V is metrically indecomposable, either one of the components (V_1 or V_2) has measure zero (in which case the other has the measure $\mu(V)$), or both are not measurable. The theorem states that, for almost everywhere on V,

$$\overline{f} = \frac{1}{\mu(V)} \int_V f(P)dV , \tag{2.60}$$

---

[48] A user-friendly introduction to measure theory is given in Cramer[29, Chapters 4–7].

[49] Measure theory is concerned with assigning non-negative numbers to *sets* of points (the measure of a set) in such a way that it captures the *size* of a set. A set that has measure is said to be a *measurable set*.

[50] And that's the reason we venture to mention the concept of measure—to get accustomed to the mathematical notion that assertions can apply "almost everywhere."

[51] An *invariant part* of Γ-space is a region V such that a point in V remains in V through the natural motion of Γ-space.

[52] See the comment about the law of large numbers in Section 3.6.

where $\bar{f}$ is the time average defined in Eq. (2.59), i.e., the theorem asserts the equality of time averages with the *phase averages* on the right side of Eq. (2.60) for almost all points $P \in V$ (for V metrically transitive). Metric indecomposability is a necessary and sufficient condition for the equality (almost everywhere) of time and phase averages. The ergodic problem reduces to that of determining whether V is metrically indecomposable, a task of equal, if not greater difficulty.[53]

We have one more task. Liouville's theorem guarantees an invariant measure of Γ-space under the natural motion of $6N$-dimensional phase space. Birkhoff's theorem requires an invariant part V of Γ-space, which we want to apply to a surface of dimension $6N - 1$. What is an invariant measure of an ergodic hypersurface? We can't use Eq. (2.58) applied to the lower-dimensional surface—*it's not invariant under the natural motion of Γ-space.* Liouville's theorem applies to the *volume* of Γ-space, not the area of one of its hypersurfaces. To apply Birkhoff's theorem to trajectories on energy surfaces, we must be in possession of a measure of the lower-dimensional surface that's invariant under the natural motion of Γ-space, in which it's embedded.

Figure 2.5 depicts a surface S_E of constant energy E (the set of points for which $H(p, q) = E$),

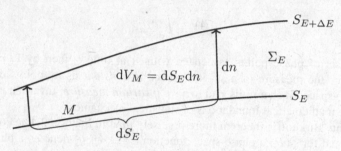

Figure 2.5: Energy shell Σ_E, region of Γ-space between energy surfaces $S_{E+\Delta E}$ and S_E. For $M \subset S_E$, $dV_M = dS_E dn$ is the volume of a small part of Σ_E associated with M.

and a neighboring energy surface $S_{E+\Delta E}$, where $\Delta E > 0$ so that S_E lies entirely within $S_{E+\Delta E}$. The region of Γ-space between S_E and $S_{E+\Delta E}$ is an *energy shell*, Σ_E; Σ_E is invariant under the natural motion of Γ-space (Liouville's theorem). For $M \subset S_E$, let $V_M \subset \Sigma_E$ be the volume of Σ_E associated with M that's bounded by S_E, $S_{E+\Delta E}$, and the vectors dn normal to S_E that extend to $S_{E+\Delta E}$. For $dS_E \subset M$ the volume element of S_E, $dV_M = dS_E dn$. The invariant measure $\mu(V_M) = \int_{V_M} dV_M = \int_{V_M} dS_E dn$. The magnitude of dn is obtained from the familiar idea that for a change df in a scalar field $f(r)$ over a displacement dr, $df = \nabla f \cdot dr$, where the gradient ∇f is orthogonal to surfaces of constant values of f. For the energy difference dE separating S_{E+dE} and S_E, we have for dn orthogonal to S_E, $dE = \|\nabla_\Gamma H\| dn$, and thus $dn = dE/\|\nabla_\Gamma H\|$, where ∇_Γ denotes the gradient operator in Γ-space $\nabla_\Gamma \equiv (\partial/\partial p_1, \cdots, \partial/\partial p_{3N}, \partial/\partial q_1, \cdots, \partial/\partial q_{3N})$, with the norm

$$\|\nabla_\Gamma H\| = \sqrt{\sum_{i=1}^{3N} \left[\left(\frac{\partial H}{\partial p_i} \right)^2 + \left(\frac{\partial H}{\partial q_i} \right)^2 \right]}.$$

The measure of V_M is therefore

$$\mu(V_M) = \int_{V_M} dS_E dn = \int_{V_M} dS_E \frac{dE}{\|\nabla_\Gamma H\|} = \int_E^{E+\Delta E} dE \int_{S_E} \frac{dS_E}{\|\nabla_\Gamma H\|}.$$

By Liouville's theorem, $\mu(V_M)$ is invariant, and thus so is $\mu(V_M)/\Delta E$. The quantity

$$\mu(S_E) \equiv \lim_{\Delta E \to 0} \left(\frac{\mu(V_M)}{\Delta E} \right) = \int_{S_E} \frac{dS_E}{\|\nabla_\Gamma H\|} \tag{2.61}$$

---

[53]Mark Kac wrote: "As is well known, metric transitivity plays an important part in ergodic theory; however it is almost impossible to decide what Hamiltonians give rise to metrically transitive transformations."[30, p67]

is an invariant measure of S_E.

We can apply this idea to phase averages where the invariant part is now S_E. First,

$$\int_{V(E)} f(P)\mathrm{d}\Gamma = \int_0^E \mathrm{d}E \left(\int_{S_E} f(P)\frac{\mathrm{d}S_E}{\|\nabla_\Gamma H\|} \right), \tag{2.62}$$

where $V(E)$ denotes the volume of Γ-space associated with energies up to and including[54] E. By differentiating Eq. (2.62),

$$\frac{\mathrm{d}}{\mathrm{d}E} \int_{V(E)} f(P)\mathrm{d}\Gamma = \int_{S_E} f(P)\frac{\mathrm{d}S_E}{\|\nabla_\Gamma H\|}. \tag{2.63}$$

Setting $f(P) = 1$ in Eq. (2.63), we have that the total measure of an energy surface

$$\frac{\mathrm{d}}{\mathrm{d}E}V(E) = \int_{S_E} \frac{\mathrm{d}S_E}{\|\nabla_\Gamma H\|} = \mu(S_E), \tag{2.64}$$

where we've used Eq. (2.61). Thus, the volume of the ergodic surface S_E is the derivative of the phase volume, the *density of states*,[55] $\mu(S_E) = V'(E)$. For phase averages on energy surface S_E,

$$\overline{f} = \frac{1}{\mu(S_E)} \int_{S_E} f(P)\frac{\mathrm{d}S_E}{\|\nabla_\Gamma H\|} = \frac{1}{\mu(S_E)}\frac{\mathrm{d}}{\mathrm{d}E}\left(\int_{V(E)} f(P)\mathrm{d}\Gamma \right), \tag{2.65}$$

where we've used Eqs. (2.60) and (2.63). Equation (2.65) is the starting point of Chapter 4.

2.7   THE ROLE OF PROBABILITY

Statistical mechanics arises from the interplay between two approaches to describing macroscopic systems: as *mechanical* systems, the dynamics of which is provided by classical or quantum mechanics, and as *thermodynamic* systems, the *observational* state of which is specified by macrovariables. Classically, microstates evolve according to Hamilton's equations of motion. Two systems having the same Hamiltonian that are in the same dynamical state at an instant of time have the same dynamical state for all times; such is the *determinism* of classical physics. In classical (but not quantum) mechanics, the dynamical state implies the observational state; two systems in the same dynamical state give rise to the same observational state. The converse is not true, however. Two systems in the same observational state are almost certainly not in the same dynamical state; *dynamical states cannot be inferred through observation of macroscopic systems.*

Observational states can change in time (like microstates), but, unlike microstates, they do not change in deterministic ways: The observational state at one time does not imply that at another. As an illustration, Fig. 2.6 depicts the observational state of the number of "dots" (people, cars, etc.) $N_i(t)$ on city blocks at two times. While each dot may move according to its own internal dynamics, our knowledge is limited to the number of dots $N_i(t)$ per block at a given time. Based only on what we see in the two parts of Fig. 2.6, can we say which dot moved to which street? The lack of determinism makes it impossible to predict future observational states. Experience shows, however, that we can make *statistical* predictions about the behavior of large numbers of systems. *Probability plays the analogous role in formulating the laws of change of observational states as does the dynamics of particles in describing microstates.*

The distinction between deterministic motion of microstates and non-deterministic change of observational states carries over to microstates described by quantum mechanics. At the quantum

---

[54]We're assuming the lower bound of the energy is $E = 0$.
[55]Thus we see the *fundamental* significance of the density of states in statistical mechanics.

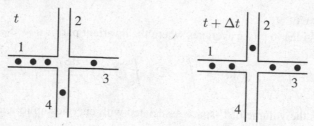

Figure 2.6: Occupation numbers $N_i(t)$ of city blocks. At time t, $N_1 = 3$, $N_2 = 0$, $N_3 = 1$, and $N_4 = 1$. At time $t + \Delta t$, $N_1 = 2$, $N_2 = 1$, $N_3 = 2$, and $N_4 = 0$.

level microstates are specified by the wave function $\psi(q, t)$, the dynamics of which is generated by Schrödinger's equation. Despite its reputation, solutions of the Schrödinger equation are perfectly deterministic. Two systems in the same microstate at one time will have the same dynamical state for all times, *if the system is undisturbed*[56] (the same as in classical mechanics). In this sense, quantum mechanics is as deterministic as classical mechanics. When, however, we make measurements on quantum systems, microstates *do not* determine observational states (see Section 4.4). Measurements on quantum systems reveal only a range of possibilities, the eigenvalues of the Hermitian operator representing the measurable quantity. The observational state of a quantum system at one time cannot be predicted from that at a previous time, yet we can make statistical predictions using the standard probability interpretation of quantum mechanics. The basic principles of quantum statistical mechanics do not differ *conceptually* from those of classical statistical mechanics, despite the conceptual differences between classical and quantum mechanics. While quantum mechanics introduces an extra "layer" of *indeterminism* in the transition from dynamical to observational states, statistical mechanics relies on the same statistical techniques to formulate the laws of change of observational states whether based on classical or quantum mechanics.

SUMMARY

Statistical mechanics is a theory by which one can calculate observable properties of equilibrium, macroscopic systems given information about their microscopic constituents.[57] It utilizes general results from mechanics combined with a suitable averaging technique to calculate properties of systems in such a way that the laws of thermodynamics are obeyed (this program is instituted in Chapter 4). In this chapter, we outlined the conceptual ingredients involved in setting up the subject.

• **Phase space**: The mathematical "arena" of classical statistical mechanics is phase space, the space of all possible canonical variables. For N point particles in three spatial dimensions, we require a $6N$-dimensional phase space known as Γ-space. Every point in Γ-space represents a possible microstate of an N-particle system. Phase space is the natural arena because each point of Γ-space evolves under the action of the Hamiltonian; the Hamiltonian generates the time evolution of arbitrary functions of canonical variables, in addition to the points of Γ-space itself. The Hamiltonian is a constant of the motion, arguably the most important one.

• **Ensembles**: Ensembles are collections of identical copies of physical systems, each prepared in the same way subject to macroscopic constraints. This abstract concept is introduced so that we can calculate the properties of systems from an average over microscopic initial conditions. Such a step is necessary because calculating the motion of N interacting particles is not possible.

---

[56] A nominally undisturbed quantum system may undergo spontaneous transitions if we allow for vacuum fluctuations, a piece of physics not included in the Schrödinger equation.

[57] Just to reiterate, thermodynamics is a theory that excels at providing interrelations between measurable properties of systems in thermal equilibrium, but which provides few calculations of the magnitudes of such quantities.

• **Phase-space probability**: Even though the elements of ensembles are identical copies of a system in terms of their macroscopic properties, microscopically the phase point of each member of the ensemble is different. That's the basic message of thermodynamics: There are numerous microscopic ways of achieving the same macroscopically specified state. The ensemble is a swarm of points in Γ-space, which evolve in time as an incompressible fluid. It's at this point that probability can be introduced: What is the probability $\rho(p, q)dpdq$ that the phase point of a randomly selected member of an ensemble lies in the range (p, q) to $(p+dp, q+dq)$? If one was in possession of the probability density $\rho(p, q)$, one could calculate ensemble averages, as in Eq. (2.53): $\overline{A} = \int_\Gamma A(p, q)\rho(p, q)d\Gamma$ when $\int_\Gamma \rho(p, q)d\Gamma = 1$.

• **Liouville's theorem**: Liouville's theorem (Eq. (2.54)) is that $\rho(p, q)$ is a constant of the motion along the phase trajectories generated by the natural motion of the points of Γ-space. The Liouville equation, (2.55), is the fundamental equation of classical statistical mechanics, on par with the status of the Schrödinger equation. For systems in thermal equilibrium, we require that ensemble probability densities be independent of time, Eq. (2.56), implying that the Hamiltonian is independent of time. The most general solution of Liouville's equation is an arbitrary function of the constants of the motion. Standard statistical mechanics proceeds on the assumption that $\rho(p, q)$ is a function solely of the Hamiltonian, $\rho(p, q) = F(H(p, q))$, where F will be determined in Chapter 4. All microscopic information about a system enters the theory through its Hamiltonian function.

• **The need for probability**: Microstates evolve in time deterministically, but observational states do not. Observed statistical regularities of the behavior of systems in the same observational state can be modeled using the methods of probability theory. Probability plays the analogous role in formulating the laws of change of observational states that particle dynamics does for microstates. Our next order of business (Chapter 3) is to introduce probability theory. Prepared students could skip to Chapter 4.

EXERCISES

2.1 Draw the phase trajectory for the one-dimensional motion of a particle of mass m in a uniform gravitational field g oriented in the negative z-direction. Consider a particle initially projected upward from position $z = z_0$ with momentum $p = p_0$.

a. What is the Lagrangian? What is the canonical momentum?

b. What are the equations of motion of $p(t)$ and $z(t)$?

c. Eliminate time between these equations to find the phase curve, $p(z)$.
   A: $p^2 = p_0^2 - 2m^2g(z - z_0)$ for $z \le z_0 + p_0^2/(2m^2g)$.

2.2 Show, using the definition of the Poisson bracket and Hamilton's equations, that

$$[q_i, H] = \dot{q}_i \qquad [p_i, H] = \dot{p}_i \qquad [q_i, p_j] = \delta_{ij} .$$

2.3 Calculate the phase volume for:

a. A one-dimensional harmonic oscillator of angular frequency ω having energy E. The phase volume is the area enclosed by the phase trajectory of energy E. Hint: The area of an ellipse of semi-minor and semi-major axes a and b is πab. A: $2\pi E/\omega$.

b. A relativistic particle moving in a three-dimensional space of volume V having energy E. A: $V \frac{4\pi}{3}p^3(E)$, where $p(E)c = \sqrt{E^2 - (mc^2)^2}$.

2.4 Discuss time-reversal symmetry in the case of Eq. (2.9) applied to the one-dimensional harmonic oscillator. Which way does the phase point move on the elliptical phase trajectory under time reversal?

2.5 a. Show using Eq. (2.25), that when Stirling's approximation is invoked (Section 3.3.2),

$$\ln W = N \left[\frac{5}{2} + \ln \left(\frac{V}{N} \left(\frac{4\pi m U}{3Nh^2} \right)^{3/2} \right) \right] \tag{P2.1}$$

and thus the Sackur-Tetrode formula, Eq. (1.56), is recovered (use Eq. (1.62)).

b. Repeat the calculation by *omitting* the factor of $N!$ from Eq. (2.25). Show that 1) you don't get the factor of $\frac{5}{2}$ in the Sackur-Tetrode formula, rather a factor of $\frac{3}{2}$, and 2) the expression for S is not extensive. The Sackur-Tetrode formula correctly accounts for the entropy of a gas in contact with its liquid phase; the factor of $\frac{5}{2}$ is *required* [3, p110]. The difference between $\frac{5}{2}$ and $\frac{3}{2}$ is seemingly small, but actually it's *huge*: It implies an increase in W in Eq. (1.62) by a factor of e^N.

2.6 Denote as the density of states, the derivative of $W(U, V, N)$ in Eq. (2.25) with respect to U,

$$W'(U) \equiv \frac{\partial}{\partial U} W(U, V, N) . \tag{P2.2}$$

The number of *accessible* states to a gas of fixed particle number N in a fixed volume V, of energy U is then

$$\Omega(U, \Delta U) \equiv W'(U)\Delta U , \tag{P2.3}$$

where ΔU is the uncertainty in the energy. In systems in thermal contact with the environment, ΔU would arise from energy exchanges between system and environment. We don't allow such interactions here, but there is still an uncertainty in how we count states due to the "binning" of microstates into phase-space cells of volume h^{3N}.

a. Show, using Eq. (2.25), that

$$\Omega(U, \Delta U) = \frac{3N}{2} \left(\frac{\Delta U}{U} \right) W(U) . \tag{P2.4}$$

b. Show that (using the result of Exercise 2.5 and invoking Stirling's approximation)

$$\ln \Omega = N \left\{ \left[\frac{5}{2} + \ln \left(\frac{V}{N} \left(\frac{4\pi m U}{3Nh^2} \right)^{3/2} \right) \right] + \frac{1}{N} \ln \left(\frac{3N\Delta U}{2U} \right) \right\} . \tag{P2.5}$$

Noting that $\lim_{N\to\infty}(\ln N)/N = 0$, the energy width ΔU is immaterial—we could set $\Delta U = U$ and not affect the expression for entropy for large N. This fact justifies using the *volume* of a $3N$-dimensional hypersphere in deriving Eq. (2.25) instead of the more correct picture provided by Ω in Eq. (P2.3). The density of states for a system of N free particles scales like $U^{(3N/2)-1}$, ostensibly identical to $U^{3N/2}$ for $N \gg 1$.

2.7 Prove the identities used in Eq. (2.28), $(\partial S/\partial U)_{V,N} = T^{-1}$ and $(\partial^2 S/\partial U^2)_{V,N} = -(C_V T^2)^{-1}$. Hints: Eqs. (1.24) and (1.58) and the reciprocity relation for partial derivatives, $(\partial x/\partial z)_y = (\partial z/\partial x)_y^{-1}$, Eq. (1.30).

2.8 Choose T and N as independent variables in a system of fixed volume (and a single chemical species). Show that

$$\overline{(\Delta N)^2} = \frac{kT}{(\partial \mu / \partial N)_T} = kT \frac{N^2}{V} \beta_T \,,$$

where the last equality is established in Exercise 4.32. Follow the steps that begin after Eq. (2.40), and use a Maxwell relation, $(\partial S / \partial N)_{T,V} = -(\partial \mu / \partial T)_{N,V}$.

2.9 Assume an ideal gas in the preceding problem. Show that $\dfrac{\sqrt{\overline{(\Delta N)^2}}}{N} = \dfrac{1}{\sqrt{N}}$. Hint: Use Eq. (P1.4).

2.10 Choose S and P as independent variables in a system where N is fixed. Show that

a.

$$\overline{(\Delta S)^2} = kC_P \qquad \overline{(\Delta P)^2} = \frac{kT}{\beta_S V} \qquad \overline{\Delta S \Delta P} = 0 \,,$$

where β_S denotes the *isentropic compressibility*, $\beta_S \equiv -(1/V)(\partial V / \partial P)_S$ ($\beta_S > 0$; it can be shown that $\beta_S = C_V \beta_T / C_P$ [3, p51]). Hint: Use $(\partial S / \partial T)_P = C_P / T$ [3, p39] and a Maxwell relation, $(\partial T / \partial P)_S = (\partial V / \partial S)_P$.

b. Argue that the product of fluctuations $\sqrt{\overline{(\Delta S)^2}} \sqrt{\overline{(\Delta P)^2}}$ is independent of the size of the system.

2.11 It's not *mandatory* to use Hamiltonian over Lagrangian mechanics, it's just simpler. Suppose we use $\{\dot{q}_i, q_i\}$ as variables. Show that the volume element $d\Gamma$ of dynamical space (see Section 2.1) cannot be written $d\Gamma = \text{constant} \times d\dot{q}_1 \cdots d\dot{q}_{3N} dq_1 \cdots dq_{3N}$. Start with canonical variables $\{p_i^*, q_i^*\}$, for which it's known that $d\Gamma = h^{-3N} dp_1^* \cdots dp_{3N}^* dq_1^* \cdots dq_{3N}^*$ (Eq. (2.51)) is a dynamical invariant (Liouville's theorem), the choice of variables $\{p_i^*, q_i^*\}$ being at our disposal as long as they're canonical. Make a change of variables (not a canonical transformation) $q_k^* = q_k$, $p_k^* = f_k(\dot{q}, q) \equiv \partial L / \partial \dot{q}_k$. In that case, $\partial q_k^* / \partial q_i = \delta_{ki}$, $\partial q_k^* / \partial \dot{q}_i = 0$, $\partial p_k^* / \partial q_i = \partial^2 L / \partial q_i \partial \dot{q}_k$, and $\partial p_k^* / \partial \dot{q}_i = \partial^2 L / \partial \dot{q}_i \partial \dot{q}_k$. Show that

$$d\Gamma = h^{-3N} D d\dot{q}_1 \cdots d\dot{q}_{3N} dq_1 \cdots dq_{3N} \,,$$

where $D = \det \partial^2 L / \partial \dot{q}_i \partial \dot{q}_k$. There is no reason to expect that $D = \text{constant}$. Hint: What a great opportunity to use the properties of Jacobians (see for example Eq. (C.57)):

$$\frac{\partial(p^*, q^*)}{\partial(\dot{q}, q)} = \frac{\partial(p^*, q^*)}{\partial(\dot{q}, q^*)} \frac{\partial(\dot{q}, q^*)}{\partial(\dot{q}, q)} = \frac{\partial(p^*)}{\partial(\dot{q})} \frac{\partial(q^*)}{\partial(q)} \,.$$

2.12 An equivalent way of stating Liouville's theorem is that *the flow of the ensemble in Γ-space is that of an incompressible fluid*. In this exercise we're going to walk through that statement.

a. The equations of fluid mechanics simplify if we introduce a special time derivative, the *convective derivative*, $D/dt \equiv \partial / \partial t + \boldsymbol{v} \cdot \nabla$, where $\boldsymbol{v}$ is the velocity field of the fluid as seen from the "lab" frame (*Eulerian observer*). To an observer moving *with* the fluid (*Lagrangian observer*), $D/dt = \partial / \partial t$. Consider a vector field $\boldsymbol{A}(\boldsymbol{r}, t)$ that changes in space and time. Show that the Taylor expansion of $\boldsymbol{A}(\boldsymbol{r}, t)$ to first order in dt and $d\boldsymbol{r}$ is

$$\boldsymbol{A}(\boldsymbol{r} + d\boldsymbol{r}, t + dt) \approx \boldsymbol{A}(\boldsymbol{r}, t) + dt(\partial \boldsymbol{A} / \partial t) + (d\boldsymbol{r} \cdot \nabla)\boldsymbol{A} \,.$$

Show that in the limit $dt \to 0$ we have as the total time derivative:

$$\frac{d\boldsymbol{A}}{dt} \equiv \frac{D\boldsymbol{A}}{dt} = \frac{\partial \boldsymbol{A}}{\partial t} + (\boldsymbol{v} \cdot \boldsymbol{\nabla})\boldsymbol{A}.$$

Show that the convective derivative applies equally well to scalar fields, $\rho(\boldsymbol{r}, t)$.

b. The *continuity equation* describes *conserved* quantities. Let $\rho(\boldsymbol{r}, t)$ be the density of a quantity (mass, energy, or probability in quantum mechanics). If ρ is conserved, then

$$\frac{\partial \rho}{\partial t} = -\boldsymbol{\nabla} \cdot \boldsymbol{J}, \tag{P2.6}$$

where $\boldsymbol{J}$ denotes the *current density*. For convective transport, $\boldsymbol{J} = \rho\boldsymbol{v}$. Equation (P2.6) indicates that the local time rate of change of ρ is balanced by the negative of the divergence of the current density vector field. If $\boldsymbol{\nabla} \cdot \boldsymbol{J}$ is positive (negative), it implies that ρ is decreasing (increasing) in time. That defines a conserved quantity: The only way for ρ to change locally is if the quantity it represents flows away—it can't just "disappear." Show that the two statements, 1) ρ is constant along the streamlines of the fluid, $D\rho/dt = 0$, and 2) ρ is locally conserved, imply

$$\boldsymbol{\nabla} \cdot \boldsymbol{v} = 0. \tag{P2.7}$$

Equation (P2.7) defines an *incompressible fluid*.

c. Let's invent a special notation for the components of the position vector in Γ-space: $\boldsymbol{\Gamma} \equiv (p_1, \cdots, p_{3N}, q_1, \cdots, q_{3N})$. Moreover, denote the components of the velocity in Γ-space as $\boldsymbol{v}_\Gamma \equiv \dot{\boldsymbol{\Gamma}} = (\dot{p}_1, \cdots, \dot{p}_{3N}, \dot{q}_1, \cdots, \dot{q}_{3N})$. Finally, define the components of a gradient operator in Γ-space, $\boldsymbol{\nabla}_\Gamma \equiv (\partial/\partial p_1, \cdots, \partial/\partial p_{3N}, \partial/\partial q_1, \cdots, \partial/\partial q_{3N})$. Show, using Hamilton's equations of motion, that

$$\boldsymbol{\nabla}_\Gamma \cdot \boldsymbol{v}_\Gamma = 0.$$

Thus, an equivalent statement of Liouville's theorem is that the flow of ensemble points in Γ-space is incompressible—the density is constant along streamlines and the number of ensemble points is conserved.

2.13 Show that a linear combination of two constants of the motion g_1 and g_2, say $g_3 \equiv \alpha g_1 + \beta g_2$, is a constant of the motion. Hint: Poisson brackets are linear in their arguments.

2.14 The equilibrium phase-space probability distribution ρ_{eq} has a vanishing Poisson bracket with the Hamiltonian, $[\rho_{eq}, H] = 0$, Eq. (2.56). Show explicitly, from the definition of the Poisson bracket, that Eq. (2.56) is identically satisfied by any differentiable function of the Hamiltonian, $\rho_{eq}(p, q) = F(H(p, q))$. This statement is the equivalent to what you've seen about the solutions of the wave equation, that, for example, $\partial^2 f(x, t)/\partial x^2 - (1/c^2)\partial^2 f(x, t)/\partial t^2 = 0$ is solved by any function of the form $f(x, t) = f(x - ct)$.

CHAPTER 3

Probability theory

W HAT are the odds a tossed coin shows "heads," H? No doubt you're thinking "1 in 2." If we knew the forces on a coin in its trajectory, and if we knew the initial conditions precisely, we could predict which side of the coin will show. That information might be difficult to obtain, but there's no reason *not* to suppose that sufficiently precise measurements could be made that would allow us to make an informed choice. Because there are only two possibilities, and if there's no reason to suspect that both don't occur with equal likelihood, then we have the odds, 1 in 2. What if 1000 coins were tossed? What are the odds that 500 show heads? Still 1 in 2? What about 501 coins showing heads? How do we calculate those odds? In this chapter we introduce concepts from probability and combinatorics needed for statistical mechanics.

3.1 EVENTS, SAMPLE SPACE, AND PROBABILITY

Probabilities involve a ratio of two numbers: 1) the number of times a given *event* occurs (outcome of an experiment) and 2) the number of possible outcomes the experiment has, where we envision experiments that can be repeated many times. The word *trial* is used for a single experiment.

Definition. *The set of all possible outcomes of an experiment is called its* sample space, *where every outcome of the experiment is completely described by one and only one* sample point *of the set.*

Example. The sample space for the toss of a six-sided die is the set $\Omega = \{1, 2, 3, 4, 5, 6\}$. The sample space for the roll of two dice has 36 elements, the set $\Omega = \{(1, 1), \cdots, (6, 6)\}$.

The sample space of two coins tossed is $\Omega = \{HH, HT, TH, TT\}$. One way to *represent* these outcomes would be to assign H the number 1 and T the number 0, so that they're given by the points $(1, 1), (1, 0), (0, 1), (0, 0)$ in the xy plane; see Fig. 3.1. One doesn't *have* to depict the sample space as in Fig. 3.1—one could mark off any four points on the x-axis for example. For three coins tossed there are eight outcomes; one could depict the sample space using a three-dimensional Cartesian space, or, again, mark off any eight points on the x-axis. Sample space is a useful mathematical concept for discussing probability. How we display these sets is a matter of convenience. It's often simpler to display the sample space in an abstract manner. The right part of Fig. 3.1 shows the 36 elements of Ω for the roll of two dice simply as points in a box.

Experiments that produce a finite number of outcomes, such as the roll of a die, have *discrete* sample spaces where the events can be represented as isolated points, as in Fig. 3.1. Probabilities defined on discrete sample spaces are referred to as *discrete probabilities*. Not every sample space is discrete. *Continuous* sample spaces are associated with experiments that produce a continuous range of possibilities, such as the heights of individuals in a certain population. Probabilities defined on continuous sample spaces are referred to as *probability densities*.

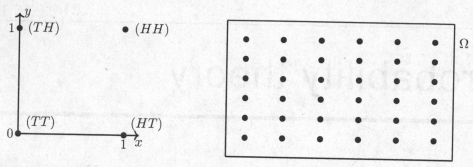

Figure 3.1: Left: Sample space for two tossed coins represented in the xy plane. Right: Abstract sample space (set of 36 points) associated with all outcomes of tossing two dice.

The individual elements of Ω are *elementary events*.[1] The word *event* (not elementary event) is reserved for *subsets* of Ω, *aggregates* of sample points. A subset A of Ω is a set such that every element of A is an element of Ω, a relationship indicated $A \subset \Omega$. In tossing two coins, the event A might be the occurrence of TT or HH; $A \subset \Omega$ is then the set of elementary events $A = \{TT, HH\}$, where $\Omega = \{TT, HH, TH, HT\}$. The terms "sample point" and "event" have an intuitive appeal, that, once specified for a given experiment, can be treated using the mathematics of point sets.

Example. From the sample space $\Omega = \{1, 2, 3, 4, 5, 6\}$, the event having any even number is the subset $E = \{2, 4, 6\}$.

The left part of Fig. 3.2 indicates the event A consisting of one dot showing on the first die in

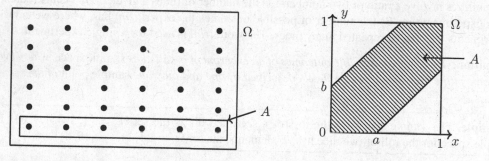

Figure 3.2: Left: Abstract sample space. Event A consists of all rolls of two dice having 1 on the first die. Right: Continuous sample space and event A such that $-b \leq x - y \leq a$.

the throw of two dice. The right part shows a continuous sample space, $\Omega = \{x, y | 0 \leq x \leq 1, 0 \leq y \leq 1\}$. Event A is the subset $A = \{x, y | -b \leq x - y \leq a\}$, where $0 \leq a, b \leq 1$.

Events (defined as subsets) can have elementary events in common. The left part of Fig. 3.3 shows, for the sample space Ω, the *intersection* of $A \subset \Omega$ and $B \subset \Omega$—elementary events contained in A *and* B, denoted $A \cap B$. The right part shows the *union* of A and B, the set of sample points in A *or* B, denoted $A \cup B$. Consider the sample space $\Omega = \{1, 2, 3, 4, 5, 6\}$. As subsets, we have the events consisting of all even and all odd integers: $E = \{2, 4, 6\}$ and $O = \{1, 3, 5\}$. The union of these sets is the sample space, $E \cup O = \Omega$. (Two sets are equal if and only if they have precisely the same elements.) Sets E and O have no elements in common, however. Their intersection, which contains no elements, is termed the *empty set*, a unique set denoted $\{\}$; also $\emptyset$. Thus, $E \cap O = \emptyset$.

---

[1]An elementary event is one that can't be a union of other events.

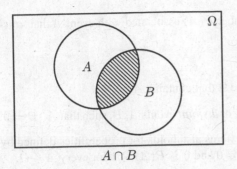 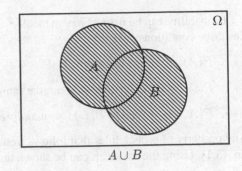

$$A \cap B \qquad\qquad\qquad\qquad A \cup B$$

Figure 3.3: Left: The intersection of events A and B (sample points in A and B, in sample space Ω), $A \cap B \subset \Omega$. Right: The union of A and B (points in A or B), $A \cup B \subset \Omega$.

We can now re-state the definition of probability, this time more precisely.

Definition. *The probability $P(A)$ that event A occurs is the ratio of the number of sample points N_A that correspond to the occurrence of A to the total number of points in the sample space, N_Ω,*

$$P(A) \equiv \frac{N_A}{N_\Omega}. \tag{3.1}$$

Equation (3.1) embodies the *frequency interpretation* of probability.[2] It presumes that in experiments with otherwise well-controlled conditions, the property under observation does not always occur with the same value—so no deterministic regularity—but in such a way that over the course of many trials there is an empirically observed *statistical regularity* that A occurs a fraction of the time as given by Eq. (3.1). Our *experience* of statistical regularity suggests a law of nature could be at work—witness quantum mechanics. Max Born wrote, on chance versus causality:[3]

> The conception of chance enters into the very first steps of scientific activity, in virtue of the fact that no observation is absolutely correct. I think chance is a more fundamental conception than causality; for whether in a concrete case a cause-effect relation holds or not can only be judged by applying the laws of chance to the observations.[32, p47]

We needn't be concerned unduly whether the substrata of the physical world are inherently probabilistic; it's enough to note that the theory of probability provides definite *laws of chance* that we can apply to the observed statistical regularities of nature. Once we invoke *randomness* as a proxy for our imprecise knowledge of initial conditions, the methods of probability can be brought to bear on statistical mechanics. Indeed, we make the stronger statement that the problem of statistical mechanics *reduces* to one in the theory of probability (see Chapter 4).

Our definition of probability, Eq. (3.1), is predicated on the *equi-likelihood* of the elementary events underlying event A, the principle of *equal a priori probabilities*. Microstates require the specification at one time of all canonical coordinates of the system particles. A specification of just a few variables is therefore an *incomplete* description. In the face of our ignorance of microscopic coordinates, or rather our inability to control them, we have no basis on which *not* to assume that microscopic coordinates occur with equal likelihood (subject to macroscopic constraints being satisfied). This hypothesis is often stated as a *postulate* of statistical mechanics.[4][33, p59].

---

[2]There are other interpretations of probability in the marketplace of ideas. R. Carnap suggested the two main classes of probability interpretation should be referred to as probability$_1$ and probability$_2$ [31]. Probability$_1$ is concerned with methods of inductive inference in the nature of scientific proofs. Probability$_2$ is concerned with the study of repetitive events possessing the property that their relative frequency of occurrence in a large number of trials has a definite value.

[3]Born received the 1954 Nobel Prize in Physics for the statistical interpretation of the wave function.

[4]The principle of equal a priori probabilities is not a mechanical, but rather is a statistical assumption. The role of mechanics is exhausted once we choose the phase-space probability density to be a function solely of the Hamiltonian.

Probability can be defined axiomatically[5] as a number $P(A)$ associated with event A that satisfies three conditions:[34, p6]

1. $P(A) \geq 0$ for every event A,

2. $P(\Omega) = 1$ for the *certain event*, the sample space Ω that contains A,

3. $P(A \cup B) = P(A) + P(B)$ for *mutually exclusive* or *disjoint* events A, B, such that $A \cap B = \emptyset$.

Any property of probabilities that follows from these axioms also holds for probabilities defined by Eq. (3.1). Using the axioms it can be shown that $P(\emptyset) = 0$ and $0 \leq P(A) \leq 1$ for every $A \subseteq \Omega$.

3.2 COMBINING PROBABILITIES, CONDITIONAL PROBABILITY

3.2.1 Adding probabilities—"or" statements, mutually exclusive events

Consider events A and B (such as in Fig. 3.3), which have N_A and N_B sample points (elementary events). In $A \cup B$ there are $N_{A \cup B} = N_A + N_B - N_{A \cap B}$ elements, where $N_{A \cap B}$ is the number of elements of the intersection $A \cap B$, which must be subtracted to prevent overcounting.[6] We then have using Eq. (3.1) the analogous formula for probabilities,

$$P(A \cup B) = P(A) + P(B) - P(A \cap B).$$ (3.2)

If A and B have no sample points in common (mutually exclusive), $A \cap B = \emptyset$. In that case,

$$P(A \cup B) = P(A) + P(B). \qquad (A, B \text{ mutually exclusive})$$ (3.3)

Equation (3.3) is used frequently in applications—it tells us that the probability of A *or* B is the sum of the probabilities when A, B are mutually exclusive. It pays to get in the habit of noticing how many calculations stem from questions of the form "what is the probability of the occurrence of this or that or that?" There's often an implicit "or" statement underlying calculations in physics. Equation (3.3) easily generalizes to more than two mutually exclusive events.

3.2.2 Multiplying probabilities—"and" statements, independent events

How is $P(A \cap B)$ in Eq. (3.2) calculated? To answer that, it's necessary to introduce another kind of probability, the *conditional probability*, denoted $P(A|B)$, the probability of A occurring, *given* that B has occurred. Referring to Fig. 3.3, we're interested in the probability that A occurs given that B has definitely occurred, a type of problem where the sample space has changed—in this case the *certain* event is B. The probability we want is the ratio of the number of sample points in the intersection, $N_{A \cap B}$, to that in B:

$$P(A|B) = \frac{N_{A \cap B}}{N_B} = \frac{N_{A \cap B}}{N_\Omega} \frac{N_\Omega}{N_B} = \frac{P(A \cap B)}{P(B)},$$

or

$$P(A \cap B) = P(A|B)P(B).$$ (3.4)

In words, Eq. (3.4) indicates that the probability of A and B is the probability of A given that B has occurred, multiplied by the probability that B occurs. This relation is symmetrical between A and B: $P(A \cap B) = P(B|A)P(A)$, implying $P(B|A) = P(A|B)P(B)/P(A)$.

---

[5]Many concepts in mathematics are given axiomatic definitions. Vectors, for example, are elements of vector spaces in which the rules of combining vectors are defined axiomatically.[16, p3] In probability theory, a *probability space* is defined as a construct having three parts: A sample space (the set of all possible outcomes), a set of events containing zero or more outcomes, and a rule that assigns probabilities to the events.

[6]We must be ever vigilant to avoid overcounting.

Suppose A and B are such that $P(A|B) = P(A)$. In that case A is said to be *independent* of B—the probability of A occurring is independent of the condition that B has occurred. For independent events, Eq. (3.4) reduces to

$$P(A \cap B) = P(A)P(B) . \quad \text{(independent events)} \tag{3.5}$$

For independent events, the probability of A *and* B is the product of the probabilities. Many problems in physics implicitly assume independent events; many problems implicitly ask for the probability of "this and that and that." Be on the lookout for how statements are worded; there may be implied "ands." Thus, for mutually exclusive events, probabilities are added, Eq. (3.3), whereas for independent events, probabilities are multiplied, Eq. (3.5). In Section 3.4, we give examples of how to calculate probabilities using these rules. First we must learn to count.

3.3   COMBINATORICS

Calculating probabilities is an exercise in *counting* (the number of elementary events in various sets). That sounds simple enough—counting is something we've done all our lives—yet there are specialized techniques for counting large sets. *Combinatorics* is a branch of mathematics devoted to just that. Its basic ideas—permutations, combinations, and binomial coefficients—involve factorials. Factorials, however, become difficult to handle for large integers—there's no easy way to multiply factorials, for example. *Stirling's approximation* allows one to treat factorials using ordinary means of algebra. In this section we consider these topics.

3.3.1   Permutations, combinations, and binomial coefficients

Suppose there are N numbered balls in a bag, and you pull them all out, one at a time. How many different ways are there of performing that experiment? There are N ways to choose the first ball, $N - 1$ ways to choose the second, and so on: There are $N!$ *permutations* of N items.

Example. How many permutations are there of the letters A, B, C? Suppose you have three boxes, and you must put a different letter in each box. The first box can be filled in three ways. Having put a letter in the first box, you have two choices for the second box, and in the third you have but one choice. There are $3! = 6$ permutations: $ABC, ACB, BAC, BCA, CAB, CBA$. Alternatively, one could start with a reference sequence, ABC, and perform pairwise swaps of the letters.

How many ways are there to bring out *two* balls, when we don't care about the order in which they're displayed? That is, pulling out ball 3 and then ball 7 is the same as pulling out 7 and then 3. There are N ways to pull out the first ball and $N - 1$ ways to pull out the second. Once the two balls are out, there are $2!$ ways of arranging them, which we've deemed to be equivalent. Thus there are

$$\frac{N(N-1)}{2!} = \frac{N!}{2!(N-2)!}$$

distinct ways of choosing two balls out of N when order is not important.

Example. Four choose two. There are $N = 4$ items, numbered $(1, 2, 3, 4)$. There are $N(N-1) = 12$ ways of pulling two items out of the collection of four: $(1, 2), (1, 3), (1, 4), (2, 1), (2, 3), (2, 4),$ $(3, 1), (3, 2), (3, 4), (4, 1), (4, 2),$ and $(4, 3)$. If we don't care about the order in which the pairs are displayed, so that, for example, $(4, 2)$ is considered the same as $(2, 4)$, there are six distinct ways of choosing two items out of the four: $(1, 2), (1, 3), (1, 4), (2, 3), (2, 4),$ and $(3, 4)$.

Generalizing to the number of ways of choosing k balls out of N when order is not important, we define the *binomial coefficient* ("N choose k")

$$\binom{N}{k} \equiv \frac{N!}{k!(N-k)!} . \tag{3.6}$$

Equation (3.6) indicates the number of *combinations* of N items taken k at a time; $\binom{N}{k}$ is called the binomial coefficient because the same term shows up in the *binomial theorem*, which for N a positive integer, is

$$(x+y)^N = \sum_{k=0}^{N} \binom{N}{k} x^{N-k} y^k . \tag{3.7}$$

We note the special result that follows from Eq. (3.7) with $x = 1$ and $y = 1$:

$$2^N = \sum_{k=0}^{N} \binom{N}{k} . \tag{3.8}$$

We will use binomial coefficients ("N choose k") frequently in this book.

Example. Lotto 6/49 is a Canadian lottery game in which, to win, one must pick the six numbers that are drawn from a set of 49, in any order. How many ways can one choose six numbers from 49 when order is immaterial? A: $\binom{49}{6} = 13{,}983{,}816$.

Equation (3.7) can be written in an equivalent, alternative form where the upper limit of the sum extends to infinity:

$$(x+y)^N = \sum_{k=0}^{\infty} \binom{N}{k} x^{N-k} y^k \tag{3.9}$$

because

$$\binom{N}{k} = \frac{N!}{k!(N-k)!} = \frac{\Gamma(N+1)}{\Gamma(k+1)\Gamma(N+1-k)} ,$$

which vanishes for $k = N+1, N+2, \cdots$ (see Appendix B). The gamma function $\Gamma(x)$ diverges for $x = 0, -1, -2, \cdots$, and thus terms in Eq. (3.9) for $k \geq N+1$ vanish.[7] Equation (3.9) allows for a generalization to negative values of N:

$$\frac{1}{(1-x)^N} = \sum_{k=0}^{\infty} \binom{-N}{k} (-x)^k = \sum_{k=0}^{\infty} \binom{N+k-1}{k} x^k , \qquad (|x| < 1) \tag{3.10}$$

where we've used

$$\binom{-N}{k} = (-1)^k \binom{N+k-1}{k} ,$$

which follows from an application of Eq. (B.4). Set $N = 1$ in Eq. (3.10); it generates the geometric series for $(1-x)^{-1}$. Equation (3.10) is used in Chapter 5 in the derivation of the Bose-Einstein distribution.

We now ask a different question. How many ways are there to choose N_1 objects from N (where order is immaterial), followed by choosing N_2 objects from the remainder, where these objects are kept separate from the first collection of N_1 objects? Clearly that number is:

$$\frac{N!}{N_1!(N-N_1)!} \times \frac{(N-N_1)!}{N_2!(N-N_1-N_2)!} = \frac{N!}{N_1!N_2!(N-N_1-N_2)!} . \tag{3.11}$$

---

[7]There's no harm in adding an infinite number of zeros to Eq. (3.7).

Generalize to the number of ways of choosing from N objects, N_1 kept separate, N_2 kept separate, $\ldots$, N_r kept separate, where $\sum_{k=1}^{r} N_k = N$ exhausts the N objects. The number of ways of distributing N objects among r containers, each containing N_k objects with $\sum_{k=1}^{r} N_k = N$, is given by the *multinomial coefficient*

$$\binom{N}{N_1, N_2, \ldots, N_r} \equiv \frac{N!}{N_1! N_2! \ldots N_r!}. \tag{3.12}$$

Equation (3.12) appears in the *multinomial theorem*,

$$(x_1 + x_2 + \cdots + x_m)^N = \sum_{k_1 + k_2 + \cdots + k_m = N} \binom{N}{k_1, k_2, \cdots, k_m} \prod_{i=1}^{m} x_i^{k_i}, \tag{3.13}$$

where the sum is over all combinations of the indices $k_1, \ldots, k_m$, nonnegative integers such that the sum of all k_i is N.

3.3.2 Stirling's approximation

In its simplest form, Stirling's approximation is, for $n \gg 1$,

$$\ln n! = n (\ln n - 1) + O(\ln n), \tag{3.14}$$

where $O(\ln n)$ indicates that terms of order $\ln n$ have been neglected (which are negligible compared to n for large n). Equation (3.14) is one of those results that *should* work only for $n \to \infty$, but which is accurate for relatively small values of n ($n \approx 10$); see Exercise 3.8. Equation (3.14) is surprisingly easy to derive: $\ln n! = \sum_{k=1}^{n} \ln k \approx \int_1^n \ln x\,dx = (x \ln x - x)\big|_1^n \approx n \ln -n$. The $O(\ln n)$ remainder is evaluated below.

A more accurate version of Stirling's approximation is

$$n! \overset{n \to \infty}{\sim} \sqrt{2\pi n} \left(\frac{n}{e}\right)^n, \tag{3.15}$$

where the notation $\sim$ indicates asymptotic equivalence.[8] Equation (3.15) can be derived from $\Gamma(x)$ (see Eq. (B.1)):

$$\Gamma(n+1) = n! = \int_0^\infty x^n e^{-x} dx \overset{x=ny}{=} nn^n \int_0^\infty e^{n(\ln y - y)} dy. \tag{3.16}$$

The integral on the right of Eq. (3.16) can be approximated using the method of steepest descent[16, p233] for large n:

$$\int_0^\infty e^{n(\ln y - y)} dy \sim \sqrt{\frac{2\pi}{n}} e^{-n}. \tag{3.17}$$

Combining Eqs. (3.17) and (3.16) yields Eq. (3.15). By taking the logarithm of Eq. (3.15), we see that the remainder term in Eq. (3.14) is $\frac{1}{2} \ln(2\pi n)$.

Sometimes we require the logarithm of the gamma function (the log-gamma function), $\ln \Gamma(x)$. From the recursion relation, Eq. (B.3), $\ln \Gamma(x+1) = \ln x + \ln \Gamma(x)$, and thus

$$\ln \Gamma(x) = \ln \Gamma(x+1) - \ln x. \tag{3.18}$$

Use Stirling's approximation,

$$\Gamma(x+1) \sim \sqrt{2\pi x} \left(\frac{x}{e}\right)^x. \tag{3.19}$$

---

[8]*Asymptotic equivalence $f \sim g$ indicates that for functions $f(x)$ and $g(x)$, the ratio $f(x)/g(x) \to 1$ as x approaches some value, usually zero or infinity. Example: $\sinh x \sim \frac{1}{2} e^x$ as $x \to \infty$.*

Take the logarithm of Eq. (3.19) and combine with Eq. (3.18):

$$\ln \Gamma(x) \sim x\,(\ln x - 1) - \frac{1}{2}\ln(x/2\pi)\ . \tag{3.20}$$

For large x, $\ln(x/2\pi)$ may be neglected in comparison with x.

3.4 EXAMPLES INVOLVING DISCRETE PROBABILITIES

• Two cards are drawn from a 52-card deck, with the first being replaced before the second is drawn. What is the probability that both cards are spades? Let A be the event of drawing a spade, with B the event of drawing another spade after the first has been replaced in the deck. This is an "and" kind of problem: What is the probability of a spade being drawn *and* another spade being drawn. $P(A) = P(B) = 13/52 = 1/4$. The two events are independent, and thus from Eq. (3.5), $P(A \cap B) = P(A)P(B) = 1/16$.

• What is the probability of *at least* one spade in drawing two cards, when the first is replaced? The slick way to work this problem is to calculate the probability of *not* drawing a spade—the probability of at least one spade is the *complement* of the probability of no spades in two draws. The probability of no spades (not drawing a spade *and* not drawing another one) is $(39/52)^2 = 9/16$ (independent events). The probability of at least one spade is then $1 - P(\text{no spades}) = 7/16$. The direct approach is to treat this as an "or" problem: What is the probability of drawing one *or* two spades? Let A be the event of drawing a spade *and* not drawing a spade on the other draw, with B the event of drawing two spades. The probability of at least one spade is $P(A \text{ or } B) = P(A) + P(B)$ (mutually exclusive). $P(A) = P(\text{spade on one draw } and \text{ not a spade on the other}) = (1/4)(3/4) = 3/16$ (independent). There are two ways to realize the first experiment, however, draw a spade and then not, *or* not draw a spade and then a spade, so we add the probabilities: The probability of one spade is $2 \times (3/16)$. The probability of two spades, $P(B) = (1/4)^2 = 1/16$. The probability of at least one spade is $2 \times (3/16) + (1/16) = 7/16$, in agreement with the first answer.

• Two cards are drawn from a deck, but now suppose the first is not put back. What is the probability that both are spades? This is an "and" problem, the probability of drawing a spade and drawing another one. The events are independent. Thus, $P = (13/52) \times (12/51) = 1/17$.

• What is the probability that the second card is a spade, when it's not known what the first card was? Let B be the event of drawing a spade on the second draw. All we know about the first event is that a card was drawn and not replaced. There are two mutually exclusive possibilities: The first card was a spade or not, call these events A and $\overline{A}$. Then, $P(A \cap B) + P(\overline{A} \cap B) = P(A)P(B) + P(\overline{A})P(B) = (P(A) + P(\overline{A}))\,P(B) = P(B)$. Thus, $P(B) = 1/4$. The probability of a spade on the second draw, when the result of the first draw is unknown, is the probability of a spade on the first draw.

• A bridge hand of 13 cards is chosen from a 52-card deck. What is the probability that the dealt hand will have 7 spades? We don't care about the order in which the cards are dealt. What is the sample space? There are $\binom{52}{13}$ ways to choose 13 cards from 52. That's the size of the set of all 13-card hands when order is not important. There are 13 spades in a deck, and thus there are $\binom{13}{7}$ ways to have 7 spades in these hands (which happen to have 13 cards). The other 6 cards are not spades, and thus there are $\binom{39}{6}$ ways of *not* having 6 spades. The total number of 13-card hands with 7 spades is therefore $\binom{13}{7} \times \binom{39}{6}$. The probability we seek is

$$P = \binom{13}{7}\binom{39}{6} \div \binom{52}{13} \approx 0.009\ .$$

• What is the probability that when a coin is tossed 5 times, H occur 3 times and T twice? This is an "and" problem; the coin tosses are independent events—a sequence $HHHTT$ is asking for the probability of H and H and H and T and T, which is $P(H)^3 P(T)^2$. Because the probabilities are equal, $P(H) = P(T) = 1/2$, the probability of *any* string of 5 coin flips is the same, $(1/2)^5$. There are $\binom{5}{2} = 10$ ways to arrange the two tails in the string of 5. Thus, $P = 10/(2^5) = 5/16$.

• Consider N people in a room. What is the probability at least two have the same birthday, meaning the same day and month? This problem is best solved by first seeking the probability that none have the same birthdays, i.e., they all have distinct birthdays. Pick one person. The probability that another person has a different birthday is $364/365$. The probability that a third person has a distinct birthday from the first two is then $363/365$. These are independent events (your birthday has no bearing on someone else's birthday). The probability that 3 people have distinct birthdays is therefore

$$\frac{364}{365} \times \frac{363}{365} \equiv \frac{365 \times 364 \times 363}{(365)^3},$$

which we've written in a way that allows generalization. The probability of the complementary event, that at least two persons out of N have a common birthday, is therefore

$$P(N) = 1 - \frac{365 \times 364 \times 363 \times \cdots \times (365 - N + 1)}{(365)^N}. \tag{3.21}$$

For $N = 22$, $P = 0.476$; for $N = 23$, $P = 0.507$. There is better than even odds that in any gathering of 23 people, at least two will have the same birthday.

• Consider a family with two children. What is the *conditional* probability that both children are boys, given that: 1) the older child is a boy, and 2) at least one of the children is a boy? Let A be the event that the older child is a boy, and B that the younger child is a boy. It may be helpful to refer to Fig. 3.1, which can be recast for this problem. $A \cup B$ is the event that at least one of the children is a boy, while $A \cap B$ is the event that both children are boys. To answer the first question, we can use Eq. (3.5) because the events are independent. Then,

$$P(A \cap B | A) = \frac{P((A \cap B) \cap A)}{P(A)} = \frac{P(A \cap B)}{P(A)} = \frac{1/4}{1/2} = \frac{1}{2},$$

where we've used that for any sets $(A \cap B) \cap A = A \cap B$. Applying the same reasoning to the second question,

$$P(A \cap B | A \cup B) = \frac{P((A \cap B) \cap (A \cup B))}{P(A \cup B)} = \frac{P(A \cap B)}{P(A \cup B)} = \frac{1/4}{3/4} = \frac{1}{3},$$

where we've used for any sets $(A \cap B) \cap (A \cup B)) = A \cap B$.

3.5 RANDOM VARIABLES AND PROBABILITY DISTRIBUTIONS

As noted in Section 3.1, the sample space $\{HH, HT, TH, TT\}$ can be represented by points in the x-y plane, $\{(1,1), (1,0), (0,1), (0,0)\}$. How we display the sample space is immaterial, yet some ways are better than others. We'd like to represent the sample space in a way that's aligned with the business of calculating averages. One could, for example, define a variable x associated with the *number* of heads observed in the toss of two coins. Let x have the value 0 for no heads, TT, 1 for one head, TH or HT, and 2 for two heads, HH. Under such an assignment, the probabilities associated with the points of Ω are represented as a line graph in Fig. 3.4. As another example, let x denote the *sum* of the dots showing in the roll of two dice, with $f(x)$ the probability associated with the value of x. Enumeration of the possibilities shows that $f(2) = f(12) = 1/36$, $f(3) = f(11) = 2/36$, $f(4) = f(10) = 3/36$, $f(5) = f(9) = 4/36$, $f(6) = f(8) = 5/36$, and $f(7) = 6/36$. These are displayed as a line graph in Fig. 3.5.

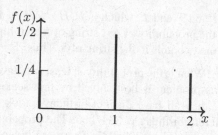

Figure 3.4: Probability distribution for the number of heads in a two-coin toss.

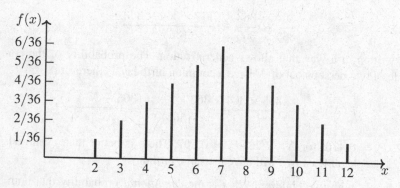

Figure 3.5: Probability distribution for the number of dots in a roll of two dice.

Definition. *A random variable x is an assignment of a real number to each point of sample space.*[9]

3.5.1 Probability distributions on discrete sample spaces

The collection of probabilities associated with the range of values of a random variable is known as a *probability distribution*.[10] For each value x_j of a random variable x, the aggregate of sample points associated with x_j form the event for which $x = x_j$; its probability is denoted $P(x = x_j)$. From Fig. 3.4, for example, $f(1) = 1/2$ is associated with the event TH or HT.

Definition. *A function $f(x)$ such that $f(x_j) = P(x = x_j)$ is the probability distribution of x.*

For the range of values $\{x_j\}$ of x, $f(x_j) \geq 0$ and $\sum_j f(x_j) = 1$; see Figs. 3.4 and 3.5.

There can be more than one random variable defined on the same sample space. Consider random variables x and y that take on the values $x_1, x_2, \ldots$ and $y_1, y_2, \ldots$, and let the corresponding probability distributions be $f(x_j)$ and $g(y_k)$. The aggregate of events for which the two conditions $x = x_j$ and $y = y_k$ are satisfied forms the event having probability denoted $P(x = x_j, y = y_k)$.

Definition. *A function $p(x, y)$ for which $p(x_j, y_k) = P(x = x_j, y = y_k)$ is called the joint proba-bility distribution of x and y.*

---

[9]What is referred to as a random variable x is actually a real-valued *function*, the domain of which is the sample space, $x : \Omega \to \mathbb{R}$. The term is confusing, therefore a better term would be *random function*. The word *random* is used because the elements of the sample space are associated with physical experiments in which the outcome of any one experiment is uncertain and evidently associated with chance. Once the sample space is established, random variables can be treated as ordinary variables of mathematics. Note that a random variable inherits the *ordering* of the points of the real line $\mathbb{R}$.

[10]The term *probability distribution* is ambiguous; beware. We take it to be a function $f(x)$ defined on the range of the random variable x such that $f(x_j) = P(x = x_j)$. It can also mean, however, in the statistics literature the *cumulative probability up to and including the value x, $F(x) \equiv \sum_{x_i \leq x} f(x_i)$, a quantity known as the (probability) *distribution function*. We'll use probability distribution in the first sense.

Clearly, $p(x_j, y_k) \geq 0$ and $\sum_{jk} p(x_j, y_k) = 1$. Moreover, for fixed x_j,

$$\sum_k p(x_j, y_k) = f(x_j), \tag{3.22}$$

while for fixed y_k

$$\sum_j p(x_j, y_k) = g(y_k). \tag{3.23}$$

That is, adding the probabilities for all events y_k for fixed x_j produces the probability distribution for x_j, and adding the probabilities for all events x_j produces the probability distribution for y_k.

Example. Let x be the number of spades drawn from a deck of 52, with y the number of spades from a second draw, without the first card being replaced. The quantity $p(0,0)$ is the probability of the occurrence of two events: no spade in the draw of a card from a deck of 52 and no spade from the draw of a card from a deck of 51. Thus, $p(0,0) = (39/52)(38/51)$ (independent events). Because no spade was obtained on the first draw, there are 13 spades in the deck of 51. Other joint probabilities are $p(1,0) = (13/52)(39/51)$, $p(0,1) = (39/52)(13/51)$, and $p(1,1) = (13/52)(12/51)$. Equation (3.23) is satisfied: $p(0,0) + p(1,0) = 3/4 = g(0)$—the probability $g(0)$ on the second draw equals the probability of no spade on the first draw, when the result of the first draw is unknown. Similarly, $p(0,1) + p(1,1) = 1/4 = g(1)$. Equation (3.22) is satisfied: $p(1,0) + p(1,1) = 1/4 = f(1)$—the probability of a spade on the first draw is independent of what happens on the second. Likewise, $p(0,0) + p(0,1) = 3/4 = f(0)$, the probability of no spade on the first draw.

Example. Let x_1, x_2, and x_3 be random variables associated with the numbers of ones, twos, and threes obtained in N throws of a die. With $p_1 = p_2 = p_3 = 1/6$ and for $1/2$ the probability of no ones, twos, or threes, we have, using Eq. (3.11), the joint probability distribution

$$p(k_1, k_2, k_3) = \frac{N!}{k_1! k_2! k_3! (N - k_1 - k_2 - k_3)!} \left(\frac{1}{6}\right)^{k_1} \left(\frac{1}{6}\right)^{k_2} \left(\frac{1}{6}\right)^{k_3} \left(\frac{1}{2}\right)^{N - k_1 - k_2 - k_3}. \tag{3.24}$$

By summing Eq. (3.24) over the possible values $k_3 = 0, 1, \ldots, N - k_1 - k_2$, we find using the binomial theorem Eq. (3.7),

$$\sum_{k_3=0}^{N-k_1-k_2} p(k_1, k_2, k_3) = \frac{N!}{k_1! k_2! (N - k_1 - k_2)!} \left(\frac{1}{6}\right)^{k_1+k_2} \left(\frac{2}{3}\right)^{N-k_1-k_2} \equiv p(k_1, k_2). \tag{3.25}$$

Definition. *If the joint probability distribution* $f(x_1, x_2, \cdots, x_n)$ *can be factored in the form* $f(x_1, x_2, \cdots, x_n) = f_1(x_1) f_2(x_2) \cdots f_n(x_n)$, *where* $f_i(x_i)$ *is the probability distribution of* x_i, *then the random variables* $x_1, x_2, \cdots, x_n$ *are said to be* independently distributed.

3.5.2 Probability densities on continuous sample spaces

A broad class of experiments involve continuous sample spaces on which *continuous* random variables are defined.[11] The probability distribution of a continuous random variable x is called a *probability density function* $f(x)$, of which $f(x)\,dx$ represents the probability that its value lies between

---

[11]Continuous random variables have cumulative probability distributions $\int_{-\infty}^{x} f(y)\,dy$ that are everywhere continuous.

x and $x + dx$. No measurement of a continuous quantity is ever perfectly precise; one can only specify a probability that a continuous random variable lies within a *window* $[x, x + dx]$. How tall are you, *exactly*? Whatever answer you give, it can only lie within experimental uncertainties. You can only specify a continuous quantity within so many decimal places of accuracy; round-off errors are incurred at the stated level of precision.

Definition. *A probability density function for a continuous random variable x is a function $f(x)$ having the properties:*

$$f(x) \geq 0 \qquad \int_{-\infty}^{\infty} f(x)dx = 1 \qquad \int_{a}^{b} f(x)dx = P(a < x < b) . \qquad (3.26)$$

Example. The Maxwell speed distribution $f(v)$ (see Chapter 5) is the probability of finding a particle of mass m in a gas in equilibrium at absolute temperature T having speed between v and $v + dv$:

$$f(v)dv = \left(\frac{m}{2\pi kT}\right)^{3/2} 4\pi v^2 \exp(-mv^2/(2kT))dv . \qquad (3.27)$$

It satisfies the criteria in Eq. (3.26). Because $v \geq 0$, it's normalized such that $\int_0^\infty f(v)dv = 1$.

Probability distributions for several continuous random variables can be defined.

Definition. *A probability distribution for n continuous random variables $x_1, x_2, \cdots, x_n$ is a function $f(x_1, x_2, \cdots, x_n)$ having the properties:*

$$f(x_1, x_2, \cdots, x_n) \geq 0$$
$$\int_{-\infty}^{\infty} \cdots \int_{-\infty}^{\infty} f(x_1, x_2, \cdots, x_n)dx_1 dx_2 \cdots dx_n = 1 \qquad (3.28)$$
$$\int_{a_n}^{b_n} \cdots \int_{a_1}^{b_1} f(x_1, x_2, \cdots, x_n)dx_1 dx_2 \cdots dx_n = P(a_1 < x_1 < b_1, \cdots, a_n < x_n < b_n) .$$

3.5.3 Moments of distributions

Definition. *The kth-moment (about the origin) of a probability distribution $f(x)$ is found from*

$$\mu_k' = \langle x^k \rangle \equiv \sum_j (x_j)^k f(x) , \qquad (k = 0, 1, 2, \cdots) \qquad (3.29)$$

where the prime on μ_k indicates the moment is defined about the "origin" of the range of values of the random variable x (assumed to be of the "counting" type, so that $x \geq 0$).

Moments characterize the *shape* of probability distributions. If we know the moments of a distribution, we know the distribution itself. It can happen that the sum in Eq. (3.29) fails to exist for some value $k = r$ (see Section 3.5.4). When the rth-moment exists, all moments for $k \leq r$ exist. The moment $\mu_0' = 1$ is the normalization of the distribution.

Definition. *The moment associated with $k = 1$ is known as the* average, *or the* mean, *or the* expectation value *of a random variable:*

$$\mu_1' = \bar{x} = \langle x \rangle \equiv \sum_j x_j f(x_j) , \qquad (3.30)$$

where both notations are used in physics, $\bar{x}$ and $\langle x \rangle$.

Equation (3.30) embodies what you'd do, for example, to calculate the average number of dots showing on the roll of a die: Add up the number of dots and divide by six: ($P(n) = 1/6$ for all n)

$$\overline{n} = \frac{1}{6}\sum_{n=1}^{6} n = \sum_{n=1}^{6} nP(n) = \frac{21}{6}.$$

Equation (3.30) generalizes to any *function* of a random variable, because a function $\phi(x)$ of a random variable x is a new random variable! Thus,

$$\overline{\phi} = \langle \phi \rangle = \sum_j \phi(x_j) f(x_j). \tag{3.31}$$

Equations (3.29) and (3.31) generalize to continuous variables where sums are replaced by integrals,

$$\mu'_k = \int x^k f(x)\mathrm{d}x \qquad \overline{\phi} = \langle \phi \rangle = \int \phi(x) f(x)\mathrm{d}x. \tag{3.32}$$

Definition. *The kth moment about the mean of a probability distribution is*

$$\mu_k \equiv \sum_j (x_j - \overline{x})^k f(x_j). \tag{3.33}$$

In describing probability distributions, the first moment about the origin, μ'_1, and the square root of the second moment about the mean, $\sqrt{\mu_2}$, occur so often they're given special symbols: μ and σ. Thus, $\mu \equiv \mu'_1$ and $\sigma \equiv \sqrt{\mu_2}$. In evaluating μ_2, it's usually more convenient to evaluate the first two moments about the origin, rather than from Eq. (3.33) directly,

$$\mu_2 = \sum_j (x_j - \overline{x})^2 f(x_j) = \sum_j (x_j)^2 f(x_j) - 2\overline{x}\sum_j x_j f(x_j) + \overline{x}^2 \sum_j f(x_j)$$
$$= \mu'_2 - 2\overline{x}^2 + \overline{x}^2 = \mu'_2 - \overline{x}^2 \equiv \langle x^2 \rangle - \langle x \rangle^2 \equiv \sigma^2. \tag{3.34}$$

The quantity μ_2 is known as the *variance* of the distribution, with σ (its positive square root) the *standard deviation*.

Example. The second moment of the *flat* probability distribution for the throw of a single die is

$$\mu'_2 \equiv \overline{n^2} = \frac{1}{6}\sum_{n=1}^{6} n^2 = \frac{91}{6}.$$

Because $\langle n \rangle = 21/6$,

$$\mu_2 = \langle n^2 \rangle - \langle n \rangle^2 = \frac{35}{12},$$

with the standard deviation $\sigma = \sqrt{\mu_2} \approx 1.71$.

To calculate the moments, one can always use Eq. (3.29). There's a systematic method, however, which can be easier than direct calculations. This method is similar to a technique we'll use in statistical mechanics (the partition function).

Definition. *The moment generating function of a random variable x with probability distribution $f(x)$ is defined as*

$$M(\theta) = \sum_j e^{\theta x_j} f(x_j) = \langle e^{\theta x} \rangle, \tag{3.35}$$

where θ is a real parameter. For continuous random variables, $M(\theta) = \int e^{\theta x} f(x)\mathrm{d}x$.

The quantity θ has no physical meaning; it's a mathematical device to facilitate the calculation of moments. Using the Maclaurin series for the exponential, it's readily shown from Eq. (3.35),

$$M(\theta) = \sum_{n=0}^{\infty} \mu_n' \frac{\theta^n}{n!} \,, \tag{3.36}$$

when the moments exist. By differentiation, therefore,

$$\mu_k' = \frac{\mathrm{d}^k M}{\mathrm{d}\theta^k}\bigg|_{\theta=0} \,. \tag{3.37}$$

Example. For the probability distribution in Fig. 3.4, we have using Eq. (3.35):

$$M(\theta) = f(0) + e^\theta f(1) + e^{2\theta} f(2) = \frac{1}{4}\left(1 + e^\theta\right)^2 \,.$$

Clearly, $M(\theta = 0) = 1$, indicating the probability distribution is normalized. Using Eq. (3.37),

$$\mu_1' = \frac{\mathrm{d}M}{\mathrm{d}\theta}\bigg|_{\theta=0} = \frac{1}{2}(1 + e^\theta)e^\theta\bigg|_{\theta=0} = 1 \qquad \mu_2' = \frac{\mathrm{d}^2 M}{\mathrm{d}\theta^2}\bigg|_{\theta=0} = \frac{1}{2}e^\theta(1 + 2e^\theta)\bigg|_{\theta=0} = \frac{3}{2} \,.$$

These moments can be calculated directly from Eq. (3.29).

We've assumed that θ is real. If we allow it to be pure imaginary, $\theta = it$, the moment generating function is the Fourier transform of the probability distribution. In this guise $M(it)$ is known as the *characteristic function* of the distribution (see Section 3.7). If the characteristic function is known, then the distribution can be found through inverse Fourier transformation.

3.5.4 Examples of probability distributions

3.5.4.1 Binomial distribution

Probability thrives on the *repeatability* of experiments. Much can be learned about random processes realized through repeated measurements of a quantity that produces only a few, perhaps just two, outcomes. Consider a *pair* of coins that's tossed 200 times. What is the probability that x of the 200 tosses shows two heads (x is an integer)? Let S denote the probability of "success" in obtaining two heads in a given trial, with F the probability of "failure." Referring to the sample space of Fig. 3.1, $S = 1/4$ and $F = 3/4$. The tosses are independent and thus the probability of *any* realization of x successes and $(200 - x)$ failures is *the same*: $S^x F^{200-x}$. There are $\binom{200}{x}$ ways that x successes can occur among the 200 outcomes. Thus, we have the probability distribution (x is a random variable)

$$f(x) = \frac{200!}{x!(200 - x)!}\left(\frac{1}{4}\right)^x \left(\frac{3}{4}\right)^{200-x} \,. \tag{3.38}$$

Equation (3.38) readily lends itself to generalization. Let the probability of success in an individual trial be p, with the probability of failure $q = 1 - p$, and let there be N trials.[12] The probability distribution $f(x)$ of x successes (whatever "success" refers to) in N trials is

$$f(x) = \binom{N}{x} p^x q^{N-x} = \frac{N!}{x!(N - x)!} p^x q^{N-x} \,. \tag{3.39}$$

---

[12]Repeated independent trials are called *Bernoulli trials* if there are only two possible outcomes for each trial, and if their probabilities remain constant throughout the trials.

Equation (3.39) is the *binomial distribution*; it applies to many problems involving a discrete variable x where the probability p is known. Is it normalized—is $\sum_{x=0}^{N} f(x) = 1$? That is indeed the case, as can be seen by applying the binomial theorem, Eq. (3.7):

$$1 = (p+q)^N = \sum_{x=0}^{N} \binom{N}{x} p^x q^{N-x} . \tag{3.40}$$

Example. If a die is rolled 5 times, what is the probability that two of the rolls show one dot? Probability of success, $p = 1/6$, with $q = 5/6$. Thus, from Eq. (3.39),

$$f(2) = \binom{5}{2} \left(\frac{1}{6}\right)^2 \left(\frac{5}{6}\right)^3 \approx 0.161 .$$

What is the probability of obtaining *at most* two ones in five rolls of a die? The answer, $P(x \leq 2) = f(0) + f(1) + f(2)$, requires we calculate $f(0)$ and $f(1)$. In this case,

$$f(0) = \binom{5}{0}\left(\frac{1}{6}\right)^0 \left(\frac{5}{6}\right)^5 = f(1) = \binom{5}{1}\left(\frac{1}{6}\right)^1 \left(\frac{5}{6}\right)^4 \approx 0.402 .$$

Thus, $P(x \leq 2) = 0.965$.

Example. Find the moment generating function of the binomial distribution. Combine Eqs. (3.35) and (3.39),

$$M(\theta) = \sum_{x=0}^{N} e^{\theta x} \binom{N}{x} p^x q^{N-x} = \sum_{x=0}^{N} \binom{N}{x} \left(pe^\theta\right)^x q^{N-x} = \left(q + pe^\theta\right)^N , \tag{3.41}$$

where we've used the binomial theorem. Thus, using Eqs. (3.37) and (3.41),

$$M'(\theta) = Np\left(q + pe^\theta\right)^{N-1} e^\theta \qquad\qquad \implies \mu = Np$$

$$M''(\theta) = Npe^\theta \left(q + pe^\theta\right)^{N-2} \left(q + pe^\theta + p(N-1)\right) \implies \sigma = \sqrt{Npq} . \tag{3.42}$$

One should try to calculate the same moments using Eq. (3.29). See Exericse 3.21.

3.5.4.2 Poisson distribution

When N becomes large, direct calculations using Eq. (3.39) become unwieldy. In that case having *approximate* expressions is quite useful. We develop the *Poisson distribution*,

$$\lim_{\substack{N \to \infty \\ Np = \mu}} f(x = k) = \frac{\mu^k}{k!} e^{-\mu} , \tag{3.43}$$

which holds for $p \ll 1$, such that $Np \equiv \mu$ is fixed. The Poisson distribution is normalized; $\sum_{k=0}^{\infty} f(k) = 1$. We can let $k \to \infty$ because we've already let $N \to \infty$. A formula like Eq. (3.43) is known as a *limit theorem* or as an *asymptotic theorem*; see Section 3.6.

To derive Eq. (3.43), first note that for fixed x, (see Exercise 3.22)

$$\binom{N}{x} \underset{N \to \infty}{\sim} \frac{N^x}{x!} . \tag{3.44}$$

From Eq. (3.39),

$$f(x) \sim \frac{N^x}{x!} p^x q^{N-x} = \frac{\mu^x}{x!}(1-p)^{N-x} = \frac{\mu^x}{x!}\left(1 - \frac{\mu}{N}\right)^{N-x},$$

where we've used $\mu = Np$. Equation (3.43) follows in the limit $N \to \infty$ when we make use of the Euler form of the exponential, $e^y = \lim_{N \to \infty}(1 + y/N)^N$.

Example. In a certain population, there is a one percent chance of finding left-handed individuals. What is the probability there are at least four left-handed individuals in a sample of 200 (from among this population). We can apply the Poisson distribution with $p = 0.01$ and $\mu = pN = 2$. The probability of least four is, using Eq. (3.43), $e^{-\mu}\sum_{k=4}^{\infty}\mu^k/k! = e^{-\mu}\left(e^{\mu} - (1 + \mu + \mu^2/2 + \mu^3/6)\right)$. With $\mu = 2$, $P \approx 0.143$.

3.5.4.3 Gaussian distribution

For which value of x is $f(x)$ in Eq. (3.39) maximized? To answer that question, we develop an approximation assuming that N and x are sufficiently large that Stirling's approximation is valid, but where no restriction is placed on p. We find, applying Eq. (3.14) to Eq. (3.39):

$$\ln f(x) \approx N \ln\left(\frac{Nq}{N-x}\right) + x \ln\left(\frac{N-x}{x}\frac{p}{q}\right) \equiv g(x). \tag{3.45}$$

The first and second derivatives of $g(x)$ are:

$$g'(x) = \ln\left(\frac{(N-x)}{x}\frac{p}{q}\right) \qquad g''(x) = -\frac{N}{x(N-x)}. \tag{3.46}$$

Thus, $g(x)$ has an extremum (a maximum) at $x = Np = \mu$, i.e., $g'(x = Np) = 0$ (show this). Moreover, the function itself has the value zero at $x = Np$, $g(x = Np) = 0$. The Taylor series of $g(x)$ about $x = \mu$ is therefore at lowest order

$$g(x) \approx -\frac{1}{2Npq}(x-\mu)^2 = -\frac{1}{2\sigma^2}(x-\mu)^2, \tag{3.47}$$

where $\sigma^2 \equiv Npq$. Combining Eqs. (3.47) and (3.45),

$$f(x) = e^{g(x)} \approx \exp\left(-\frac{1}{2\sigma^2}(x-\mu)^2\right). \tag{3.48}$$

Distributions in the form of Eq. (3.48) are known as the *Gaussian* or *normal* distribution. Its shape is shown in Fig. 3.6.

We would expect that Eq. (3.48), an approximate expression derived from a second-order Taylor expansion, would provide accurate predictions only for sufficiently small values of $x - \mu$, say for $|x - \mu| \lesssim \sqrt{N}$. A theorem, the *DeMoivre-Laplace theorem* states that Eq. (3.48) holds for *all* x in the limit $N \to \infty$. For $A\sqrt{N} \leq x - \mu \leq B\sqrt{N}$, where A,B are *arbitrary numbers* with $A < B$, we have the normalized probability distribution:

Theorem. DeMoivre-Laplace.

$$\lim_{N \to \infty} \binom{N}{x} p^x q^{N-x} = \frac{1}{\sqrt{2\pi\sigma^2}}\exp\left(-(x-\mu)^2/2\sigma^2\right). \tag{3.49}$$

We omit the proof[13] of Eq. (3.49). It should be noted that the Poisson distribution holds for $p \ll 1$ such that $Np = \mu = $ constant for large N, whereas the Gaussian distribution holds for $Np \gg 1$.

---

[13] A proof is given in, for example, Sinai.[34, p30]

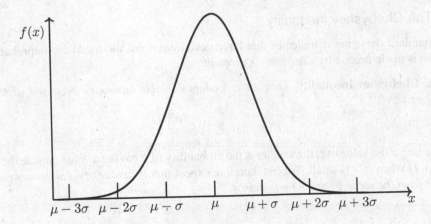

Figure 3.6: Shape of the Gaussian distribution.

3.5.4.4 Cauchy distribution

The Cauchy distribution[14] of a random variable x is, by definition,

$$f(x) = \frac{1}{\pi} \frac{1}{1+x^2} . \qquad (3.50)$$

While it's normalized, $\int_{-\infty}^{\infty} f(x)dx = 1$, it has no higher moments. From the indefinite integral $\int x dx/(1+x^2) = \frac{1}{2}\ln(1+x^2)$, the "one-sided" first moment doesn't exist: $\int_0^R x dx/(1+x^2) \to \infty$ as $R \to \infty$. The Cauchy distribution is an example of a normalized probability distribution that decays too slowly ("long tails") as $x \to \pm\infty$ for its moments to exist; see Fig. 3.7.

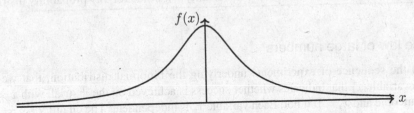

Figure 3.7: The Cauchy distribution, with its "long tails."

3.6 CENTRAL LIMIT THEOREM, LAW OF LARGE NUMBERS

Two important theorems in the theory of probability play to our hand in statistical mechanics: the *central limit theorem* and the *law of large numbers*. A probability distribution is determined by its moments, all of them (when they exist); knowledge of just the mean and the variance are in general insufficient. Yet the Poisson distribution Eq. (3.43) is characterized by the mean μ, and the Gaussian distribution Eq. (3.49) is characterized by two parameters,[15] μ and σ. The central limit theorem states that the value of *the sum of a large number of independent random variables is normally distributed*, i.e., according to the Gaussian distribution The law of large numbers states that *the average of a large number of random variables is certain to converge to its expected value.* We first prove an inequality obeyed by probability distributions.

---

[14]Also called the Lorentz distribution, the shape of the resonance curve of a damped harmonic oscillator.
[15]It should not be construed that the Poisson and Gaussian distributions do not possess higher moments, e.g., μ_4. Such moments exist, but they're not independent—they're functions of μ and σ.

3.6.1 The Chebyshev inequality

A *small* standard deviation σ indicates that large deviations from the mean are improbable. This expectation is made precise by *Chebyshev's inequality*:

Theorem. Chebyshev inequality. *Let x be a random variable having mean μ and variance σ^2. Then, for any $t > 0$,*

$$P(|x - \mu| \geq t) \leq \frac{\sigma^2}{t^2} . \tag{3.51}$$

That is, the larger the value of t, the smaller is the probability of x having a value outside the interval $(\mu - t, \mu + t)$ when σ/t is small. We can therefore expect that *the value of a random variable will not deviate from the mean by much more than σ.*

Proof.

$$P(|x - \mu| \geq t) = \int_{-\infty}^{\mu-t} f(x)\mathrm{d}x + \int_{\mu+t}^{\infty} f(x)\mathrm{d}x = \int_{|x-\mu|\geq t} f(x)\mathrm{d}x .$$

But,

$$\sigma^2 = \int_{-\infty}^{\infty} (x-\mu)^2 f(x)\mathrm{d}x \geq \int_{|x-\mu|\geq t} (x-\mu)^2 f(x)\mathrm{d}x \geq t^2 \int_{|x-\mu|\geq t} f(x)\mathrm{d}x = t^2 P(|x-\mu| \geq t) .$$

□

Example. For the throw of a die, with $P(n) = 1/6$ for $n = 1, \cdots, 6$, $\mu = 7/2$ and $\sigma^2 = 35/12$. In this case, $P(x > 6) = 0$, but Chebyshev's inequality, (3.51), indicates only that $P(|x - 3.5| \geq 2.5) \leq \sigma^2/(2.5)^2 \approx 0.47$. Chebyshev's inequality may not provide the best *numerical* bound for estimating probabilities; its value lies in its universality—it holds for *any* probability distribution.

3.6.2 The law of large numbers

Consider, in the sequence of experiments underlying the binomial distribution, that we define a new random variable x_i that indicates whether success is achieved on the ith-trial, with $x_i = 1$ for a successful outcome and $x_i = 0$ if not. Each variable x_i is independent. The quantity $S_N = \sum_{i=1}^{N} x_i$ records the number of successful outcomes achieved in N trials. We would expect that after many trials the ratio $S_N/N \to p$, the probability for success on a single trial. This expectation is made precise by the law of large numbers.

Theorem. Law of large numbers. *For $\{x_i\}$ a sequence of mutually independent random variables having a common distribution, then for every $\epsilon > 0$,*

$$P\left(\left|\frac{x_1 + \cdots + x_N}{N} - p\right| > \epsilon\right) \xrightarrow{N\to\infty} 0 . \tag{3.52}$$

Said differently, the probability that $(x_1 + \cdots + x_N)/N$ differs from p by *less* than an arbitrarily prescribed number ϵ tends to unity as $N \to \infty$.

Proof. Using Chebyshev's inequality, we have, for any $t > 0$,

$$P(|S_N - \mu| \geq t) \leq \frac{N\sigma^2}{t^2} ,$$

where the variance of $S_N = \sum_{i=1}^{N} x_i$ is $N\sigma^2$, with σ^2 the variance of x_i. For $t = N\epsilon$, the right side becomes $\sigma^2/(N\epsilon^2)$, which tends to zero as $N \to \infty$. This accomplishes the proof of Eq. (3.52). □

The law of large numbers is a powerful result. For sufficiently many trials, the relative frequency of an outcome, will, with probability one, converge to the probability p of a single trial. It can be considered as supporting (but not proving) the ergodic hypothesis: The value associated with a phase trajectory (on a metrically indecomposable region of Γ-space of constant energy) is, on average, a constant, independent of initial conditions on the energy surface (see Eq. (2.59)).

3.6.3   The central limit theorem

Given N mutually independent random variables $\{x_i\}_{i=1}^N$ with $\langle x_i \rangle = \mu_i$ and $\sigma_i^2 = \langle (x_i - \mu_i)^2 \rangle$, their sum $x \equiv \sum_{i=1}^N x_i$ is a random variable with mean $\mu \equiv \sum_{i=1}^N \langle x_i \rangle$ and variance $\sigma^2 = \sum_{i=1}^N \langle (x_i - \mu_i)^2 \rangle$ (see Exercise 3.26).

Theorem. Central limit theorem. *For $\{x_i\}_{i=1}^N$ mutually independent random variables having mean μ_i and finite variance σ_i^2, the distribution $f(x)$ of the sum $x \equiv \sum_{i=1}^N x_i$ approaches, as $N \to \infty$, the Gaussian distribution of x about the mean $\mu \equiv \sum_{i=1}^N \mu_i$ with variance $\sigma^2 = \sum_{i=1}^N \sigma_i^2$,*

$$f(x) \xrightarrow{N \to \infty} \frac{1}{\sqrt{2\pi\sigma^2}} e^{-(x-\mu)^2/2\sigma^2} . \tag{3.53}$$

We omit a proof, which is fairly involved.[16] A better name (than central limit theorem) might be the "normal convergence theorem." By this theorem, the probability that a sum $x = \sum_{i=1}^N x_i$ of random variables deviates from the mean μ normally distributed. The central limit theorem is stronger than the law of large numbers:[17] It gives a quantitative estimate for the probability that $|x - \mu|$ is larger than σ, whereas the law of large numbers indicates "merely" the inevitability that $\langle x \rangle \to \mu$ as $N \to \infty$. A proviso of the central limit theorem is that the variance σ^2 is finite; it does not apply if σ^2 is not finite.

Example. Assume that a random variable x (which can be taken to be a sum of other random variables) has the normalized probability density

$$f(x) = \frac{c^n}{(n-1)!} x^{n-1} e^{-cx} , \tag{3.54}$$

where c is a positive constant. Find an *approximate* expression for $f(x)$ using the central limit theorem. To do so, we must find the parameters μ and σ for the distribution of x. It's readily shown that, from Eq. (3.54), $\mu = n/c$ and $\sigma^2 = n/c^2$. We find, using Eq. (3.53) (see Exercise 3.28),

$$f(x) \approx \frac{c}{\sqrt{2\pi n}} e^{-(cx-n)^2/2n} . \tag{3.55}$$

3.7   CUMULANTS AND CHARACTERISTIC FUNCTIONS

A more general way of describing probability distributions $f(x)$ (than with their moments) is with their *characteristic functions* $\Phi(\omega)$, their Fourier transforms:

$$\Phi(\omega) \equiv \int_{-\infty}^{\infty} f(x) e^{i\omega x} \, \mathrm{d}x = \langle e^{i\omega x} \rangle , \tag{3.56}$$

---

[16] See for example Papoulis[35, p218].

[17] A pure mathematician might disagree—the law of large numbers applies even when the random variables x_i have no finite variance (required in the central limit theorem), so that (in this regard) it's more general than the central limit theorem.

where ω is a real variable. The quantity $\Phi(\omega)$ is thus the expectation value of the random variable[18] $e^{i\omega x}$, $\Phi(\omega) = \langle e^{i\omega x} \rangle$. Sometimes it turns out to be easier to find $\Phi(\omega)$ than $f(x)$, in which case $f(x)$ is obtained through inverse Fourier transformation; it's also easier to develop approximations for $\Phi(\omega)$ than it is for $f(x)$. Note that $\Phi(\omega = 0) = \int f(x)\mathrm{d}x = 1$, i.e., $\Phi(0)$ is the normalization integral on $f(x)$. One can show[19] that $|\Phi(\omega)| \le 1$ for $-\infty < \omega < \infty$.

Using Eq. (3.56),

$$\langle x^k \rangle \equiv \int x^k f(x)\mathrm{d}x = (-\mathrm{i})^k \frac{\mathrm{d}^k \Phi(\omega)}{\mathrm{d}\omega^k}\bigg|_{\omega=0}. \tag{3.57}$$

Thus, $\Phi(\omega)$ *must be differentiable at $\omega = 0$ for the moments to exist.* The Cauchy distribution, for example, has no moments, yet $\Phi(\omega) = e^{-|\omega|}$ (Exercise 3.30), which we see is not differentiable at $\omega = 0$. When all moments exist, we have from Eq. (3.56)

$$\Phi(\omega) = \sum_{n=0}^{\infty} \frac{(\mathrm{i}\omega)^n}{n!} \langle x^n \rangle. \tag{3.58}$$

Cumulants are certain functions of the moments, defined as follows. Write, using Eq. (3.58),

$$\Phi(\omega) = \langle e^{i\omega x} \rangle \equiv e^{C(\omega)} \implies C(\omega) = \ln \Phi(\omega) = \ln \left(\sum_{n=0}^{\infty} \frac{(\mathrm{i}\omega)^n}{n!} \langle x^n \rangle \right). \tag{3.59}$$

The function $C(\omega)$—the *cumulant generating function*—is assumed analytic so that it has a power-series representation,

$$\ln \langle e^{i\omega x} \rangle = C(\omega) = \sum_{n=1}^{\infty} \frac{C_n}{n!} (\mathrm{i}\omega)^n, \tag{3.60}$$

where the nth-order cumulant, C_n, is calculated from[20]

$$C_n = (-\mathrm{i})^n \frac{\partial^n}{\partial \omega^n} C(\omega)\bigg|_{\omega=0} = (-\mathrm{i})^n \left(\frac{\partial^n}{\partial \omega^n} \ln \Phi(\omega) \right)\bigg|_{\omega=0}. \tag{3.61}$$

Note that the sum in Eq. (3.60) starts at $n = 1$: $C(0) = \ln \Phi(0) = 0$. Using Eqs. (3.58) and (3.61), we find the first few cumulants:

$$\begin{aligned}
C_1 &= \langle x \rangle \\
C_2 &= \langle x^2 \rangle - \langle x \rangle^2 \equiv \sigma^2 \\
C_3 &= \langle x^3 \rangle - 3\langle x^2 \rangle\langle x \rangle + 2\langle x \rangle^3 \\
C_4 &= \langle x^4 \rangle - 3\langle x^2 \rangle^2 - 4\langle x \rangle\langle x^3 \rangle + 12\langle x \rangle^2\langle x^2 \rangle - 6\langle x \rangle^4.
\end{aligned} \tag{3.62}$$

Expressions for the first 10 cumulants in terms of moments can be found in [36]. Note that C_1 and C_2 are the mean and variance. If the cumulants C_n become progressively smaller sufficiently rapidly with increasing n, a good approximation to $\ln \Phi(\omega)$ can be had by retaining only the first few cumulants in Eq. (3.60). Cumulants beyond second order are related to higher-order fluctuations (see Exercise 3.33), and, because fluctuations typically tend to be small in macroscopic systems (but not always), an approximation scheme is to set cumulants beyond a given order to zero. If the

---

[18] A function of a random variable is itself a random variable.

[19] *Proof*: $|\Phi(\omega)| = \left| \int \mathrm{d}x e^{i\omega x} f(x) \right| \le \int \mathrm{d}x \left| e^{i\omega x} \right| f(x) = \int f(x)\mathrm{d}x = 1$, which uses that $f(x)$ is real and positive.

[20] One might wonder why the imaginary number has been introduced in defining $\Phi(\omega)$, rather than the moment generating function $M(\theta) = \int e^{\theta x} f(x)\mathrm{d}x = \langle e^{\theta x} \rangle$, Eq. (3.35). The reason is that $\langle e^{i\omega x} \rangle$ is well defined for all values of ω, whereas $\langle e^{\theta x} \rangle$ may not be defined for all values of θ when the probability distribution $f(x)$ doesn't decay sufficiently rapidly for large x. For the Cauchy distribution (which has no moments), $M(\theta)$ does not exist, whereas $\Phi(\omega) = e^{-|\omega|}$.

probability density is a delta function, $f(x) = \delta(x - x_0)$, so that the event $x = x_0$ occurs with certainty and with no variance, all cumulants beyond the first vanish, $C_{n>1} = 0$. Cumulants are defined for *general* probability densities—the average symbols here $\langle \rangle$ do not necessarily refer to the equilibrium, *thermal* probability densities that we'll derive in upcoming chapters; cumulants apply to any averaging process (see Section 6.3 or Section 8.3.3). They have the useful feature that if the moments *factorize* such that $\langle x^m \rangle = \langle x \rangle^m$, then all cumulants $C_{n>1}$ vanish, as can be seen from Eq. (3.62). A property that makes them useful in statistical mechanics (Chapter 6) is that a cumulant of nth-order is represented by moments of nth *and lower*-order, not higher-order moments.

The advantage of working with Eq. (3.60) over Eq. (3.58) is seen by considering *two* random variables, x and y. The cumulants for the sum, $x + y$, are obtained from Eq. (3.60),

$$\ln \left\langle e^{i\omega(x+y)} \right\rangle \equiv \sum_{n=1}^{\infty} \frac{C_n^{(x+y)}}{n!} (i\omega)^n , \tag{3.63}$$

where the notation $C_n^{(x+y)}$ indicates, these are the cumulants associated with the random variable $x + y$. We could have labeled the cumulants in Eq. (3.62) $C_n^{(x)}$ to indicate their association with the random variable x. If x and y are *independent* random variables, so that the joint probability distribution $f(x, y)$ factorizes, $f(x, y) = f(x)f(y)$ (see Section 3.5.1), the left side of Eq. (3.63) factorizes:

$$\ln \left\langle e^{i\omega(x+y)} \right\rangle = \ln \left(\langle e^{i\omega x} \rangle \langle e^{i\omega y} \rangle \right) = \ln \left\langle e^{i\omega x} \right\rangle + \ln \left\langle e^{i\omega y} \right\rangle . \tag{3.64}$$

Combining Eqs. (3.63) and (3.60) with Eq. (3.64), we see that cumulants for independent random variables simply add:

$$C_n^{(x+y)} = C_n^{(x)} + C_n^{(y)} . \tag{3.65}$$

The importance of Eq. (3.65) is that cross terms drop out of the cumulants associated with the sum of random variables. We can see this for the second cumulant:

$$C_2^{(x+y)} = \langle (x + y)^2 \rangle - \langle x + y \rangle^2 = \left(\langle x^2 \rangle + 2\langle x \rangle \langle y \rangle + \langle y^2 \rangle \right) - \left(\langle x \rangle^2 + 2\langle x \rangle \langle y \rangle + \langle y \rangle^2 \right)$$
$$= \langle x^2 \rangle - \langle x \rangle^2 + \langle y^2 \rangle - \langle y \rangle^2 = C_2^{(x)} + C_2^{(y)} ,$$

where we've used that x and y are independently distributed, $\langle xy \rangle = \langle x \rangle \langle y \rangle$. Equation (3.65) generalizes for the sum of any number of independent random variables $\{x_i\}$,

$$C_n^{(\sum_i x_i)} = \sum_i C_n^{(x_i)} . \tag{3.66}$$

SUMMARY

This chapter provided an introduction to probability theory and combinatorics. You should understand the concept of sample space and the notion of events as subsets of sample space. The probability of an event A is the ratio of the number of sample points N_A associated with A to the number of points in sample space, N_Ω. Calculating probabilities relies on the ability to count the number of elementary events in various sets. The basic concepts of combinatorics—permutations, combinations, and binomial coefficients—are specified in terms of factorials. Stirling's approximation provides an algebraic formula for the approximate value of factorials, an approximation that becomes more accurate the larger the number. We reviewed the basic rules of probability in Section 3.2, with representative examples given in Section 3.4. Applications typically invoke the concepts of mutually exclusive and mutually independent events. For mutually exclusive events (those that have no sample points in common), one adds the respective probabilities of the individual events. Such problems are often based on an implicit "or"—what is the probability of this or that or that. It pays to scrutinize the wording of problems to infer the often hidden, implicit *ors*. For mutually independent events (where $P(A|B) = P(A)$), one multiplies the probabilities of the individual events. Such problems are often based on an implicit use of "and"—what is the probability of this and that and that. Again, it pays to scrutinize the wording of problems to infer the implicit *ands*. Random variables $\{x_j\}$ assign real numbers to events in sample space, and probability distributions are functions $f(x)$ such that $f(x_j) = P(x = x_j)$ is the probability of the event characterized by x_j. The binomial distribution, Eq. (3.39), specifies the probability of x successes (whatever "success" refers to) in N trials. There are two limit theorems associated with the binomial distribution as $N \to \infty$: the Poisson distribution associated with events having sufficiently small elementary probability p such that $pN = \mu = $ constant, and the Gaussian distribution which places no restrictions on p. The Gaussian distribution is completely characterized by two numbers, the mean μ and the standard deviation σ. We introduced two important theorems, the law of large numbers and the central limit theorem. If an event A with $P(A) = p$ occurs k times in a large number of trials N, then $p \approx k/N$. The law of large numbers is that for sufficiently many trials (as $N \to \infty$), the probability $P(|(k/N) - p| \le \epsilon)$ converges to unity for any $\epsilon > 0$. The central limit theorem shows, for mutually independent random variables x_i, that, as $N \to \infty$, deviations of the sum $\sum_{i=1}^{N} x_i$ from its expectation value are normally distributed.

EXERCISES

3.1 For $T = \{2\}$ and $S = \{2, 3\}$, show that $T \cap S = T$ and $T \cup S = S$.

3.2 In a certain city, everyone has a name with initials given by three letters of the English alphabet. How many people must this city have to make it inevitable that at least two residents have the same three initials? A: $(26)^3 + 1 = 17,577$.

3.3 a. How many straight lines can be drawn through seven points such that each line passes through two points (no three points are co-linear)? A: $\binom{7}{2} = 21$.

  b. How many different poker hands of five cards can be dealt from a deck of 52 cards?

3.4 Explicitly verify Eq. (3.8) for $N = 2$.

3.5 Show that the binomial coefficients have the symmetry $\binom{N}{k} = \binom{N}{N-k}$.

3.6 Use the multinomial theorem, Eq. (3.13), to find an expression for the third power of the trinomial $(a + b + c)^3$.

3.7 Evaluate $\Gamma(7/2)$. A: $\frac{15}{8}\sqrt{\pi} \approx 3.32$. Note that $\Gamma(7/2)$ lies between $\Gamma(3) = 2!$ and $\Gamma(4) = 3!$.

3.8 a. Plot $(n \ln n - n)/(\ln n!)$ versus n. For what value of n are you comfortable replacing $\ln n!$ with Stirling's approximation in the form of Eq. (3.14)?

  b. Plot $(\sqrt{2\pi n}(n/e)^n)/(n!)$ versus n. To what extent does the extra factor in Eq. (3.15) improve the approximation?

3.9 In the game of bridge, each of four players is dealt a hand of 13 cards. Show that the number of four-player bridge hands can be expressed in terms of the multinomial coefficient Eq. (3.12),

$$\binom{52}{13}\binom{39}{13}\binom{26}{13}\binom{13}{13} = \binom{52}{13, 13, 13, 13}$$

Evaluate this number. A: $\approx 5 \times 10^{28}$. Hint: Use Stirling's approximation, Eq. (3.15).

3.10 Consider a lottery where one must pick five numbers (in any order) that are drawn from a set of 47 and an additional number (the ω-number) from another set of 27 numbers.

a. How many ways are there of choosing numbers in this fashion when order is immaterial? This is the size of the same space. A: $\Omega = \binom{47}{5} \times \binom{27}{1}$. The probability P of winning is $1/\Omega$. The *odds* of winning are 1 in $(1/P)$.

b. What are the odds of correctly picking the five numbers, but not the ω-number? Show that the probability of this outcome is

$$P = \frac{1}{\Omega}\binom{5}{5} \times \binom{26}{1}.$$

c. What are the odds of correctly picking the ω-number and any 4 out of the 5? Show that the probability is

$$P = \binom{5}{4}\binom{42}{1} \div \binom{47}{5}.$$

3.11 Supply the steps leading to Eq. (3.21). Write a computer program to evaluate $P(N)$ for arbitrary N. What is the probability that among 50 people, at least two have a common birthday? A: 0.970.

3.12 Three cards are drawn from a deck of 52 cards.

a. Find the probability there will be exactly one ace among them. A: ≈ 0.204.

b. Find the probability there will be at least one ace among them. A: ≈ 0.217. You are encouraged to work this problem in two ways to test your understanding.

3.13 Show that the probability of a bridge hand consisting of five spades, four hearts, three diamonds, and one club is

$$\binom{13}{5}\binom{13}{4}\binom{13}{3}\binom{13}{1} \div \binom{52}{13}.$$

Evaluate this number. A: ≈ 0.005.

3.14 What is the probability, in a family of four children, to have a boy followed by three girls, $BGGG$, precisely in that order? What is the probability of four girls, $GGGG$? Assume equal probabilities of having a boy or a girl. What is the probability, in a family of four children, of having a boy and three girls, in any order? A: $\frac{1}{16}, \frac{1}{16}, \frac{1}{4}$.

3.15 Show that the act of drawing an ace from a deck of cards is independent of drawing a spade. Let A denote drawing an ace, with B drawing a spade. Is $P(A|B) = P(A)$?

3.16 15 students go hiking. Five get lost, eight get sunburned, and six return home without incident. What is the probability that a sunburned hiker gets lost? (A: 1/2.) What is the probability a lost hiker gets sunburned? (A: 4/5.)

3.17 An integer n is drawn at random from the integers $1, \cdots, 15$. Are the statements n is odd and $n > 10$ independent?

3.18 The probability distribution for a certain sample space is given by $f(n) = C/n!$ for $n = 0, 1, 2, \cdots$, where C is a constant. What should the value of C be? Calculate $P(n = 2)$. Calculate $P(n < 2)$. A. $1/(2e)$, $2/e$.

3.19 In the example on page 71, what is the probability of obtaining one dot showing in five successive rolls of a die? A: 1.3×10^{-4}.

3.20 Derive Eq. (3.25). Give an argument why the result of the summation is the joint probability distribution for k_1 and k_2. Show that by summing Eq. (3.25) over all possible values $k_2 = 0, 1, \ldots, N - k_1$ yields the distribution function of x_1 for the probability $p = 1/6$, i.e.,

$$\sum_{k_2=0}^{N-k_1} p(k_1, k_2) = p(k_1).$$

3.21 The results in Eq. (3.42) for the moments of the binomial distribution were derived using the powerful method of the moment generating function. Show that they follow directly from Eq. (3.29). Hint: Liberally make changes to dummy indices and use the binomial theorem. For the second moment, make use of the identity $x^2 = x(x - 1) + x$.

3.22 Derive the result in Eq. (3.44) for large N and fixed x when $N \gg x$.

3.23 Derive the moment generating function for the Poisson distribution. Find the mean and the standard deviation. A: $M(\theta) = \exp\left(\mu(e^\theta - 1)\right)$, $\mu_1' = \mu$, $\sigma = \sqrt{\mu}$. For the Poisson distribution, the standard deviation is not an independent parameter but is given in terms of the mean.

3.24 The manufacturer of a certain part reports that the probability of a defective part is 10^{-3}. What is the probability $P(k > 5)$ that in an order of 3000 parts, there will be more than five defective parts? Hint: $P(k > 5) = 1 - P(k \leq 5)$. A: $P(k > 5) = 0.084$.

3.25 Derive the results in Eqs. (3.45) and (3.46).

3.26 Let $\{x_i\}_{i=1}^N$ be a set of mutually independent random variables having a common distribution, where, for each variable, $\langle x_i \rangle = \mu$ and $\sigma^2 = \langle (x_i - \mu)^2 \rangle$. If the variables x_i are independently distributed, then for $j \neq i$, $\langle x_i x_j \rangle = \langle x_i \rangle \langle x_j \rangle$. Let the sum of these random variables be denoted $\Sigma_N \equiv \sum_{i=1}^N x_i$. Show that $\langle \Sigma_N \rangle = N\mu$ and $\langle (\Sigma_N)^2 \rangle = N\sigma^2 + N^2\mu^2$.

3.27 a. Show that the moment generating function for the Gaussian distribution, the right side of Eq. (3.49), is

$$M(\theta) = \frac{1}{\sqrt{2\pi\sigma^2}} \int_{-\infty}^{\infty} e^{\theta x} e^{-(x-\mu)^2/(2\sigma^2)} \, dx = e^{\theta\mu + \sigma^2\theta^2/2}.$$

Hint: Complete the square and make use of the integrals in Appendix B.

b. Show for the Gaussian distribution that $\mu_1' = \mu$ and $\mu_2' = \sigma^2 + \mu^2$ and thus the standard deviation is σ. The Gaussian distribution has the same first two moments as does the binomial distribution.

c. Derive μ_3 for the binomial and Gaussian distributions. Are they the same? Hint: Use the moment generating function in each case. A: $\mu_3 = 0$ for the Gaussian distribution and $\mu_3 = \mu q(p - q)$ for the binomial distribution.

3.28 Fill in the steps from Eq. (3.54) to Eq. (3.55) (Hint: Make use of the gamma function). Make a plot of the two functions for various values of n (for simplicity, set $c = 1$). For what value of n would you say that $f(x)$ in Eq. (3.55) is a good approximation of $f(x)$ in Eq. (3.54)? For what value of x does $f(x)$ in Eq. (3.54) have a maximum? Where is the maximum of $f(x)$ in Eq. (3.55) located?

3.29 Show that the second moment of the Cauchy distribution does not exist. Hint:

$$\int \frac{x^2}{1 + x^2} \, dx = \int \left[1 - \frac{1}{1 + x^2}\right] dx.$$

3.30 Show that the characteristic function associated with the Cauchy distribution is

$$\Phi(\omega) = e^{-|\omega|} .$$

Thus, even though the Cauchy distribution has no moments, it has a Fourier transform. Hint: One way to show this result is with the methods of contour integration: $\int dx/(1+x^2) = \int dx/[(x-i)(x+i)]$—close the contours appropriately. Note that while $\Phi(\omega)$ is continuous, its derivative is not continuous at $\omega = 0$; it doesn't have a first moment—its derivative at $\omega = 0$ doesn't exist.

3.31 Show, from the definition in Eq. (3.56), that $\Phi^*(\omega) = \Phi(-\omega)$. Then show that

$$\Phi(\omega) - \Phi(-\omega) = \int_{-\infty}^{\infty} e^{i\omega x} \left(f(x) - f(-x) \right) dx .$$

Conclude that if $f(x)$ is even about the origin, then $\Phi(\omega)$ is real and even. This reasoning should be familiar from the Fourier cosine transform of an even function.

3.32. Show that

$$\int_{-\infty}^{\infty} \Phi(\omega) d\omega = 2\pi f(0) .$$

Hint: $\delta(x) = \frac{1}{2\pi} \int_{-\infty}^{\infty} dk e^{ikx}$.

3.33 Show that the third cumulant C_3 is the same as the third fluctuation moment, $\langle (x - \langle x \rangle)^3 \rangle$. Is the same true of C_4? A: No. $C_4 = \langle (x - \langle x \rangle)^4 \rangle - 3\langle (x - \langle x \rangle)^2 \rangle$.

3.34 a. Evaluate the moments μ'_3 and μ'_4 for the Gaussian distribution. Hint: Use the generating function evaluated in Exercise 3.27.

b. Evaluate the cumulants C_3 and C_4 for the Gaussian distribution using Eq. (3.62). A: $C_3 = C_4 = 0$.

c. Based on the results just derived, one might surmise that all cumulants $C_{n>2} = 0$ for the Gaussian distribution. Show that this is a general result. Hint: Evaluate the characteristic function $\Phi(\omega) = M(i\omega)$. Then, what is the expression for the cumulant expansion, $C(\omega) = \ln \Phi(\omega)$, Eq. (3.60)? You should see immediately that only the first two cumulants are nonzero for the Gaussian distribution. It can be shown (not here) that the Gaussian distribution is the only distribution whose cumulant generating function is a polynomial, i.e., the only distribution having a finite number of nonzero cumulants.

3.35 a. Derive an expression for the cumulant expansion $C(\omega)$ associated with the Poisson distribution. A: $C(\omega) = \mu \left(e^{i\omega} - 1 \right)$.

b. Show that, in general, $C_n = \mu$ for the Poisson distribution for all n.

3.36 Do any cumulants exist for the Cauchy distribution? Show your reasoning. A: No.

CHAPTER 4

Ensemble theory

ESUMING where we left off in Chapter 2, if we had the phase-space probability density $\rho(p,q)$, we could calculate averages of phase-space functions, $\langle A \rangle = \int_\Gamma A(p,q)\rho(p,q)\mathrm{d}\Gamma$, Eq. (2.53). For equilibrium systems, $\rho(p,q) = F(H(p,q))$, Eq. (2.57), where F is an unknown function. In this chapter, we show that F depends on the types of interactions the systems comprising the ensemble have with the environment (see Table 1.1). We distinguish three types of ensemble:

1. **Isolated systems**: systems having a well-defined energy and a fixed number of particles. An ensemble of isolated systems is known as the *microcanonical* ensemble;[1]

2. **Closed systems**: systems that exchange energy with the environment but which have a fixed number of particles. An ensemble of closed systems is known as the *canonical ensemble*;

3. **Open systems**: systems that allow the exchange of matter and energy with the environment. An ensemble of open systems is known as the *grand canonical ensemble*.

4.1 CLASSICAL ENSEMBLES: PROBABILITY DENSITY FUNCTIONS

4.1.1 The microcanonical ensemble: Systems of fixed U, V, N

Phase trajectories of N-particle, isolated systems are restricted to a $(6N - 1)$-dimensional hypersurface S_E embedded in Γ-space (the surface for which $H(p,q) = E$). Referring to Fig. 2.5, for M a set of points on S_E, the probability $P(M)$ that the phase point lies in M is, using Eq. (2.64),

$$P(M) = \frac{1}{\Omega(E)} \int_{M \cap S_E} \frac{\mathrm{d}S_E}{\|\nabla_\Gamma H\|} , \tag{4.1}$$

where $\Omega(E) = \int_{S_E} \mathrm{d}S_E/\|\nabla_\Gamma H\|$ is the volume[2] of S_E, Eq. (2.64). The expectation value of a phase-space function $\phi(p,q)$ is, using Eq. (2.65),

$$\langle \phi \rangle = \frac{1}{\Omega(E)} \int_{S_E} \phi(p,q)\frac{\mathrm{d}S_E}{\|\nabla_\Gamma H\|} = \frac{1}{\Omega(E)} \int_\Gamma \delta(E - H(p,q))\,\phi(p,q)\mathrm{d}\Gamma \equiv \int_\Gamma \rho(p,q)\phi(p,q)\mathrm{d}\Gamma , \tag{4.2}$$

where we've used the property of the Dirac delta function, that for a function $g : \mathbb{R}^N \to \mathbb{R}$, $\int_{\mathbb{R}^N} \delta(g(r))f(r)\mathrm{d}^N r = \int_{g^{-1}(0)} \mathrm{d}\sigma f(r)/\|\nabla g\|$, where $\mathrm{d}\sigma$ is the volume element of the $(N-1)$-dimensional surface specified by $g^{-1}(0)$. The phase-space probability density for the microcanonical ensemble, the *microcanonical distribution*, is, from Eq. (4.2),

$$\rho(p,q) = \frac{1}{\Omega(E)} \delta(E - H(p,q)) . \tag{4.3}$$

---

[1]The terms microcanonical and canonical were introduced by Gibbs.[19]
[2]We've changed notation, away from *measure* $\mu(S_E)$, to the volume of S_E, $\Omega(E)$. The two are synonymous.

Equation (4.3) indicates for an ensemble of isolated systems each having energy E, that a point (p, q) in Γ-space has probability zero of being on S_E unless $H(p, q) = E$, and, if $H(p, q) = E$ is satisfied, $\rho(p, q)$ is independent of position on S_E.

We're guided by the requirement that ensemble averages stand in correspondence with elements of the thermodynamic theory (Section 2.5). In that way, the macroscopic behavior of systems is an outcome of the microscopic dynamics of its components. The consistency of the statistical theory with thermodynamics is established in Section 4.1.2.8. Thermodynamic relations in each of the ensembles are most naturally governed by the thermodynamic potential having the same variables that define the ensemble. The Helmholtz energy $F = F(T, V, N)$ is the appropriate potential for the closed systems of the canonical ensemble (Section 4.1.2.9), and the grand potential $\Phi = \Phi(T, V, \mu)$ is appropriate for the open systems of the grand canonical ensemble (Section 4.1.3). We'll see that the normalization factor on the probability distribution in the canonical ensemble, the canonical partition function Z_{can}, Eq. (4.53), is related to F by $Z_{\text{can}} = \exp(-F/kT)$ (see Eq. (4.57)); likewise in the grand canonical ensemble, the grand partition function Z_G, Eq. (4.77), is related to Φ by $Z_G = \exp(-\Phi/kT)$ (see Eq. (4.76)). With that said, which quantity governs thermodynamics in the microcanonical ensemble? Answer: Consider the variables that specify the microcanonical ensemble. Entropy $S = S(U, V, N)$ is that quantity, with $\Omega = \exp(S/k)$ in Eq. (4.3).

4.1.2 The canonical ensemble: Systems of fixed T, V, N

We now find the Boltzmann-Gibbs distribution, the equilibrium probability density function for an ensemble of closed systems in thermal contact with their surroundings. The derivation is surprisingly involved—readers uninterested in the details could skip to Eq. (4.31).

4.1.2.1 The assumption of weak interactions

We take a composite system (system A interacting with its surroundings B; see Fig. 1.9) and consider it an isolated system of total energy E. Let Γ denote its phase space with canonical coordinates $\{p_i, q_i\}$, $i = 1, \cdots, 3N$, with A having coordinates $\{p_i, q_i\}$ for $i = 1, \cdots, 3n$ and B having coordinates $\{p_i, q_i\}$ for $i = 3n + 1, \cdots, 3N$, where $N \gg n$. We can write the Hamiltonian of the composite system in the form $H(A, B) = H_A(A) + H_B(B) + V(A, B)$, where $H_A(H_B)$ is a function of the canonical coordinates of system $A(B)$, and $V(A, B)$ describes the interactions between A and B involving both sets of coordinates. The energies $E_A \equiv H_A$, $E_B \equiv H_B$ far exceed the energy of interaction $V(A, B)$ because E_A and E_B are proportional to the *volumes* of A and B, whereas $V(A, B)$ is proportional to the surface area of contact between them (for short-range forces). For macroscopic systems, $|V(A, B)|$ is negligible in comparison with E_A, E_B. Thus, we take

$$E = E_A + E_B , \tag{4.4}$$

the assumption of *weak interaction* between A and B (even though "no interaction" might seem more apt). We can't take $V(A, B) \equiv 0$ because A and B would then be isolated systems. Equilibrium is established and maintained through a continual process of energy transfers between system and environment; taking $V(A, B) = 0$ would exclude that possibility. For systems featuring short-range interatomic forces, we can approximate $E \approx E_A + E_B$ when the surface area of contact between A and B does not increase too rapidly in relation to bulk volume (more than $V^{2/3}$ is too fast).[3] No matter how small in relative terms the energy of interaction $V(A, B)$ might be, it's *required* to establish equilibrium between A and B. We're not concerned (in equilibrium statistical mechanics) with *how* a system comes to be in equilibrium (in particular how much time is required). We assume that, in equilibrium, $E_A, E_B \gg |V(A, B)|$, with Eq. (4.4) as a consequence.

---

[3]Such an argument might not work if the system features long-range interactions. Lebowitz and Lieb[37] showed that the thermodynamic limit (see Section 4.2) exists for systems interacting through Coulomb forces.

4.1.2.2 Density of states function for composite systems

The phase volume of the composite system associated with energy E is found from the integral

$$V(E) = \int_{\{E_A < E\}} \mathrm{d}\Gamma_A \int_{\{E_B < E - E_A\}} \mathrm{d}\Gamma_B \equiv \int_{\{E_A < E\}} V_B(E - E_A)\mathrm{d}\Gamma_A , \qquad (4.5)$$

where $\{E_A < E\}$ indicates all points of Γ_A for which $E_A < E$, and $\{E_B < E - E_A\}$ indicates the points of Γ_B for which $E_B < E - E_A$. As an example of the integral in Eq. (4.5), suppose one wanted to find the area of the xy-plane for which $x + y \leq E$, where $x, y \geq 0$. One would have the integral $\int_{0 < x < E} \mathrm{d}x \int_{0 < y < E - x} \mathrm{d}y = \int_0^E \mathrm{d}x (E - x)$. From Eq. (2.64), $\mathrm{d}\Gamma_A = \Omega_A(E_A)\mathrm{d}E_A$, and thus Eq. (4.5) can be written $V(E) = \int_0^E \Omega_A(E_A)V_B(E - E_A)\mathrm{d}E_A$. Because $V_B(E - E_A) = 0$ for $E_A > E$, the limit of integration can be extended to infinity:

$$V(E) = \int_0^\infty \Omega_A(y)V_B(E - y)\mathrm{d}y . \qquad (4.6)$$

By differentiating Eq. (4.6) with respect to E, we have from Eq. (2.64) a composition rule for density-of-state functions when the total energy is conserved and shared between subsystems:

$$\Omega(x) = \int_0^\infty \Omega_A(y)\Omega_B(x - y)\mathrm{d}y \equiv \Omega_A * \Omega_B . \qquad (4.7)$$

The density of states function of a composite system Ω' is a *convolution* of the density-of-state functions of the subsystems, Ω_A and Ω_B.

Example. The density of states function for a free particle in n spatial dimensions, Eq. (2.19), which we denote $g^{(n)}(E)$, satisfies Eq. (4.7),

$$g^{(n+m)}(E) = \int_0^E g^{(n)}(y)g^{(m)}(E - y)\mathrm{d}y , \qquad (4.8)$$

where the limit of integration is finite because $g(E - y) = 0$ for $y > E$. See Exercise 4.4.

Equation (4.7) generalizes to a system of n subsystems, where x_i denotes the energy of the ith subsystem:

$$\Omega(x) = \int_0^\infty \left[\prod_{i=1}^{n-1} \Omega_i(x_i)\mathrm{d}x_i \right] \Omega_n \left(x - \sum_{i=1}^{n-1} x_i \right) , \qquad (4.9)$$

where the multiple integration is over the $(n - 1)$-dimensional space for which $x_i > 0$. The limits of integration can be extended to infinity because $\Omega_n(y < 0) = 0$. Equation (4.9) is a key result in the development of the theory; it follows by induction from n to $n + 1$ by decomposing Ω_n into two components and using Eq. (4.7).

4.1.2.3 Probability distribution for subsystems

For an isolated system of energy E consisting of system A together with its environment B, let M_A be a set of points in Γ_A having canonical coordinates $\{p_i, q_i\}$, $i = 1, \cdots, 3n$. Let M be a set of points in Γ-space *containing* M_A, such that the first $3n$ pairs of its canonical coordinates $\{p_k, q_k\}$, $k = 1, \cdots, 3N$, belong to M_A. (In math speak, M_A is *embedded* in M). Thus, phase points having coordinates $\{p_i, q_i\}$, $i = 1, \cdots, 3n$ belong to M_A if and only if the phase point of Γ-space belongs to M. The probability that a phase point of A lies in M_A coincides with the probability that the phase point of Γ lies in M. Using Eq. (4.1), which pertains to an isolated system,

$$P(M_A) = P(M) = \frac{1}{\Omega(E)} \int_{M \cap S_E} \frac{dS_E}{\|\nabla_\Gamma H\|} = \frac{1}{\Omega(E)} \int_{S_E} \mathbb{1}_M \frac{dS_E}{\|\nabla_\Gamma H\|} = \frac{1}{\Omega(E)} \frac{d}{dE} \int_{V(E)} \mathbb{1}_M d\Gamma ,$$
(4.10)

where we've introduced the *indicator function* $\mathbb{1}_M$,

$$\mathbb{1}_M(x) \equiv \begin{cases} 1 & x \in M \\ 0 & x \notin M , \end{cases}$$

and we've used Eq. (2.65). The final integral in Eq. (4.10) directs us to find the phase volume of M for energies up to and including E. Thus (compare with Eq. (4.5)),

$$\int_{V(E)} \mathbb{1}_M d\Gamma = \int_{\Gamma_A} \mathbb{1}_M d\Gamma_A \int_{\{E_B < E - E_A\}} d\Gamma_B = \int_{\Gamma_A} \mathbb{1}_M d\Gamma_A V_B(E - E_A) = \int_{M_A} V_B(E - E_A) d\Gamma_A .$$
(4.11)

Combining Eqs. (4.10) and (4.11),

$$P(M_A) = \frac{1}{\Omega(E)} \frac{d}{dE} \int_{M_A} V_B(E - E_A) d\Gamma_A = \frac{1}{\Omega(E)} \int_{M_A} \Omega_B(E - E_A) d\Gamma_A .$$
(4.12)

We therefore have, from Eq. (4.12), the phase-space probability density for system A:

$$\rho_A(p, q) = \frac{\Omega_B(E - E_A)}{\Omega(E)} ,$$
(4.13)

where $E_A = H_A(p, q)$. Equation (4.12) implies an *energy* distribution by setting $d\Gamma_A = \Omega_A(E_A) dE_A$. Thus,

$$\rho(E_A) = \frac{1}{\Omega(E)} \Omega_A(E_A) \Omega_B(E - E_A) .$$
(4.14)

Equations (4.13) and (4.14) are the *canonical distribution functions*, subsystem probability distributions determined by the density of states functions for the subsystem and the environment.

It might seem that we're done, but we're only part way there. These formulas are not in a form we can use for calculations because we lack expressions for the density-of-state functions.[4] The interpretation of these formulas is quite physical, however. The probability that system A has energy E_A is specified by a ratio of two numbers: the number of states system A has available at energy E_A multiplied by the number of states system B has at energy $E - E_A$, to the number of states the composite system has at energy[5] E. Our strategy in what follows is to develop an expression for $\Omega(x)$ appropriate to systems having large numbers of components; see Section 4.1.2.5.

4.1.2.4 *Partition function: Laplace transform of the density of states function*

The canonical distribution requires that we know the density-of-state functions of the system and the surroundings. Density-of-states functions satisfy a convolution relation, Eq. (4.9). By the convolution theorem,[6] the integral transform T of a convolution integral is equal to the product of

---

[4]While we might understand the density of states function for the system of interest, system A, we won't in general have any knowledge of $\Omega(x)$ for the environment.

[5]To reinforce that interpretation (a ratio of two numbers), Eq. (4.14) can be written $\rho(E_A) dE_A = [\Omega_A(E_A) dE_A] \times [\Omega_B(E - E_A) dE_B]/\Omega(E) dE$ because, for given E_A, $dE_B = dE$.

[6]See any text on mathematical methods of physics, such as [16].

the transforms of the functions appearing in the convolution, $T(f * g) = T(f)T(g)$. The Laplace transform of $\Omega(x)$ is known as the *partition function*:[7]

$$Z(\alpha) \equiv \int_0^\infty e^{-\alpha x}\Omega(x)\mathrm{d}x , \qquad (4.15)$$

where $\alpha > 0$. The Laplace transform is the natural choice of transform (as opposed to Fourier) because $\Omega(x \le 0) = 0$. We'll soon give α a physical interpretation, but for now we treat it as a mathematical parameter.[8]

Partition functions obey a simple composition law. By taking the Laplace transform of Eq. (4.9), we find for a system composed of n subsystems ($Z_i(\alpha)$ is the Laplace transform of $\Omega_i(x)$),

$$Z(\alpha) = \prod_{i=1}^{n} Z_i(\alpha) , \qquad (4.16)$$

where the parameter α applies to all subsystems.[9] Equation (4.16) implies a useful feature of partition functions. We can consider *each molecule* of a system as a subsystem! For a gas of N identical molecules, if $z(\alpha)$ is the partition function of a single molecule (Laplace transform of its density of states function), $Z(\alpha) = [z(\alpha)]^N$.

The partition function is always positive. A positive function $Z(\alpha)$ is *log-convex* if $\ln Z(\alpha)$ is convex (see Section C.1 for the definition of convex function). From Eq. (4.15), $Z(\alpha)$ is log-convex:

$$\frac{\mathrm{d}^2}{\mathrm{d}\alpha^2} \ln Z(\alpha) = \frac{Z''(\alpha)}{Z(\alpha)} - \left(\frac{Z'(\alpha)}{Z(\alpha)}\right)^2 = \frac{1}{Z(\alpha)}\int_0^\infty \left(x + \frac{Z'(\alpha)}{Z(\alpha)}\right)^2 e^{-\alpha x}\Omega(x)\mathrm{d}x > 0 . \quad (4.17)$$

In the next subsection, we use the fact that, based on the convexity of $\ln Z$, there is but one solution to the differential equation ($a > 0$):[10]

$$\frac{\mathrm{d}}{\mathrm{d}\alpha} \ln Z(\alpha) = \frac{Z'(\alpha)}{Z(\alpha)} = -a . \qquad (4.18)$$

4.1.2.5 Density of states function from the central limit theorem

One can devise a probability density having the partition function as its normalizing constant. Define, for $\alpha > 0$,

$$U(\alpha, x) \equiv \begin{cases} \dfrac{1}{Z(\alpha)}e^{-\alpha x}\Omega(x) & x > 0 \\ 0 & x \le 0 . \end{cases} \qquad (4.19)$$

The functions $\{U(\alpha, x)\}$ are a family of probability densities (each meets the requirements in Eq. (3.26)), one for each value of α. We'll single out a member of the family as physically relevant through the requirement that it generate the energy expectation value.[11]

---

[7] All moments of $\Omega(x)$ can be obtained by taking derivatives of $Z(\alpha)$; see Exercise 4.5. The symbol Z stands for the German word *Zustandssumme*. The partition function is the moment generating function of the density of states.

[8] For a system in equilibrium at absolute temperature T, we'll show that $\alpha = 1/(kT)$.

[9] Because α applies to each subsystem, we shouldn't be surprised to find that α is related to temperature, what equilibrium systems in thermal contact have in common (zeroth law of thermodynamics[3, p11]).

[10] *Proof*: Consider a family of functions labeled by parameter a: $Z_a(\alpha) \equiv e^{a\alpha}Z(\alpha)$. As $\alpha \to 0$, $Z_a(\alpha) \to \infty$ for any a (from Eq. (4.15), $Z(\alpha) \to \infty$ as $\alpha \to 0$). As $\alpha \to \infty$, $Z_a(\alpha) \to \infty$ for $a > 0$ ($Z(\alpha)$ is a decreasing function of α). Thus, $\ln Z_a(\alpha) \to \infty$ for $\alpha \to 0$ as well as for $\alpha \to \infty$ ($a > 0$). Because $\ln Z_a(\alpha)$ is convex (show this), it has a single minimum (a property of strictly convex functions), precisely where Eq. (4.18) is satisfied.

[11] Every member of the family $\{U(\alpha, x)\}$ is determined by $\Omega(x)$ and hence $U(\alpha, x)$ is *associated* with the system for any value of α. We'll use $U(\alpha, x)$ to develop an expression for $\Omega(x)$; Eq. (4.28). It turns out, as we show in Section 4.1.2.6, that Eq. (4.19) is equivalent to Eq. (4.13) when certain conditions are satisfied. The form of $U(\alpha, x)$ is in fact the Boltzmann distribution, the *goal* of this section.

Using $U(\alpha, x)$ as a probability density,

$$\langle E \rangle = \int_0^\infty x U(\alpha, x) \mathrm{d}x = \frac{1}{Z(\alpha)} \int_0^\infty x e^{-\alpha x} \Omega(x) \mathrm{d}x = -\frac{Z'(\alpha)}{Z(\alpha)} = -\frac{\mathrm{d}}{\mathrm{d}\alpha} \ln Z(\alpha), \quad (4.20)$$

where the middle equality follows from Eq. (4.15). If we knew $Z(\alpha)$, we would have the first moment of $U(\alpha, x)$, and there is but one solution of Eq. (4.20) for $\langle E \rangle > 0$. Among the functions $U(\alpha, x)$, *there is precisely one that generates a given expectation value*. Moreover, the variance of $U(\alpha, x)$ is (see Eq. (3.35))

$$\langle x^2 \rangle - \langle x \rangle^2 = \int (x - \langle x \rangle)^2 U(\alpha, x) \mathrm{d}x = \frac{Z''(\alpha)}{Z(\alpha)} - \left(\frac{Z'(\alpha)}{Z(\alpha)} \right)^2 = \frac{\mathrm{d}^2}{\mathrm{d}\alpha^2} \ln Z(\alpha), \quad (4.21)$$

where we've used Eq. (4.17). The first two moments of $U(\alpha, x)$ are known if $Z(\alpha)$ is known.

For a system of n subsystems, we find by combining Eq. (4.19) with Eq. (4.9),

$$U(\alpha, x) = \int \left[\prod_{i=1}^{n-1} U_i(\alpha, x_i) \mathrm{d}x_i \right] U_n \left(\alpha, x - \sum_{i=1}^{n-1} x_i \right), \quad (4.22)$$

where we've used Eq. (4.16). Thus, $U(\alpha, x)$ is composed of the functions $U_i(\alpha, x_i)$ for the subsystems, $i = 1, \cdots, n-1$, including that of the environment, U_n, the "keeper" of the total energy (here denoted by x). The convolution form of Eq. (4.22) is a consequence of the convolution property of $\Omega(x)$, Eq. (4.9), because of the special form of $U(\alpha, x)$, with its dependence on the exponential function and because it's normalized by the Laplace transform of $\Omega(x)$.

The upshot of Eq. (4.22) is that $U(\alpha, x)$ represents the probability of a sum of $n-1$ random variables, the probability that the energy of the combined system is the sum of the energies of its subsystems (when the latter are considered independent random variables). Consider, for n mutually independent random variables $\{x_i\}_{i=1}^n$, governed by probability densities $u_i(x_i)$ with moments $a_i \equiv \int x_i u_i(x_i) \mathrm{d}x_i$ and $b_i \equiv \int (x_i - a_i)^2 u_i(x_i) \mathrm{d}x_i$, that the central limit theorem provides, for large n, the form of the probability density for the sum $x \equiv \sum_{i=1}^n x_i$, which we may write as $U_n(x)$:

$$U_n(x) \xrightarrow{n \gg 1} \frac{1}{\sqrt{2\pi B_n}} \exp \left(-\frac{(x - A_n)^2}{2B_n} \right), \quad (4.23)$$

where $A_n \equiv \sum_{k=1}^n a_k$ and $B_n \equiv \sum_{k=1}^n b_k$. The form of Eq. (4.23) relies only on the *number* of components a system has and not on the physical laws governing them. That plays to our hand in establishing statistical mechanics, which we want to apply to *any* macroscopic physical system.

Equation (4.23) is an asymptotic expression, valid for $n \to \infty$, but which in practice provides satisfactory results for relatively small values of n (see for example Exercise 3.28). What if we use the asymptotic form, Eq. (4.23), in Eqs. (4.20) and (4.21)? Agreement with the results of Eqs. (4.20) and (4.21) (which are based on $U(\alpha, x)$) can be had if we stipulate the correspondences[12]

$$A_n \leftrightarrow -\frac{\mathrm{d}}{\mathrm{d}\alpha} \ln Z(\alpha) \equiv \mu \qquad B_n \leftrightarrow \frac{\mathrm{d}^2}{\mathrm{d}\alpha^2} \ln Z(\alpha) \equiv \sigma^2. \quad (4.24)$$

Using μ and σ^2 from Eq. (4.24) in Eq. (4.23), we have a normalized probability density having the same first two moments as $U(\alpha, x)$,

$$U(x) \equiv \frac{1}{\sqrt{2\pi\sigma^2}} \exp \left(-\frac{(x - \mu)^2}{2\sigma^2} \right). \quad (4.25)$$

---

[12]We have a minor notational issue—denoting the mean value of a quantity as μ is standard in the theory of probability. In thermodynamics, μ denotes the chemical potential. The intended meaning of the symbol μ should be clear from context.

The function $U(x)$ is seemingly independent of $\Omega(x)$, whereas $U(\alpha, x)$ explicitly depends on it (Eq. (4.19)). The two are connected, however, μ and σ^2 are derivatives of $\ln Z(\alpha)$, where $Z(\alpha)$ is the Laplace transform of $\Omega(x)$. The identity of the moments (first and second) obtained using the different functional forms of Eqs. (4.19) and (4.25) implies that the location of the maximum of $e^{-\alpha x}\Omega(x)$ and its shape near its maximum approximates the same for $U(x)$—the location of its mean and its shape near the mean. See the example on page 75.

The function $U(x)$ provides an approximation of $U(\alpha, x)$ that reproduces its first two moments. A loose end here is the value of α, which is uniquely determined through Eq. (4.20). Denote the value of α such that Eq. (4.20) is satisfied as β:

$$\mu = \left(-\frac{d}{d\alpha}\ln Z(\alpha)\right)\Big|_{\alpha=\beta} = \langle E \rangle . \tag{4.26}$$

Likewise,

$$\sigma^2 = \left(\frac{d^2}{d\alpha^2}\ln Z(\alpha)\right)\Big|_{\alpha=\beta} . \tag{4.27}$$

With Eq. (4.25), we can invert Eq. (4.19) to obtain an approximate expression for $e^{-\beta x}\Omega(x)$ in the vicinity of the mean energy, μ:

$$e^{-\beta x}\Omega(x) = \frac{Z(\beta)}{\sqrt{2\pi\sigma^2}}e^{-(x-\mu)^2/2\sigma^2} . \tag{4.28}$$

In particular, for $x = \mu = \langle E \rangle \equiv E$,

$$\Omega(E) = \frac{Z(\beta)}{\sqrt{2\pi\sigma^2}}e^{\beta E} . \tag{4.29}$$

4.1.2.6 Putting it together: The Boltzmann probability distribution

Referring to Fig. 1.9, let A have n particles and B have $N - n$ particles, with $N \gg n$. From Eq. (4.24), $\mu = \mu_A + \mu_B$ and $\sigma^2 = \sigma_A^2 + \sigma_B^2$, where μ_A, σ_A^2 are of order n (they each refer to a sum of n quantities), while μ_B, σ_B^2 are of order $N - n \approx N$. Combining Eqs. (4.13), (4.28), and (4.29),

$$\rho_A(p,q) = \frac{1}{\Omega(E)} \times \Omega_B(E - E_A) = \frac{\sqrt{2\pi\sigma^2}}{Z(\beta)e^{\beta E}} \times \frac{Z_B(\beta)}{\sqrt{2\pi\sigma_B^2}}e^{\beta(E-E_A)}e^{-(E-E_A-\mu_B)^2/2\sigma_B^2}$$

$$= \sqrt{\frac{\sigma^2}{\sigma_B^2}}\frac{Z_B(\beta)}{Z(\beta)}e^{-\beta E_A}e^{-(E_A-\mu_A)^2/2\sigma_B^2} \xrightarrow{N \gg n} \frac{e^{-\beta E_A}}{Z_A(\beta)}e^{-(E_A-\mu_A)^2/2\sigma_B^2} , \tag{4.30}$$

where we've used $E = \mu_A + \mu_B$, $\sigma^2/\sigma_B^2 = 1 + \sigma_A^2/\sigma_B^2 \approx 1$ for $N \gg n$, and $Z(\beta) = Z_A(\beta)Z_B(\beta)$ from Eq. (4.16). Note that the total system energy E drops out of Eq. (4.30).

The exponential in the final term of Eq. (4.30) can be approximated as unity: for $|E_A - \mu_A| \lesssim \sigma_A$, $\exp(-(E_A - \mu_A)^2/2\sigma_B^2) \approx 1$ because $\sigma_B \gg \sigma_A$. If we let $N \to \infty$, the final exponential in Eq. (4.30) has the value $e^0 = 1$. When A is much smaller than B we have the canonical phase-space distribution (where we erase the label A)

$$\rho(p,q) = \frac{1}{\Omega(E)}\Omega_B(E - E_A) \xrightarrow{N \to \infty} \frac{1}{Z(\beta)}e^{-\beta H(p,q)} , \tag{4.31}$$

where $E_A = H_A(p,q)$. The right side of Eq. (4.31) is the Boltzmann probability distribution; it depends on the size of system B (the surroundings, before the limit $N \to \infty$ is taken), but not on the size of system A. It's valid then even for a one-particle system! It remains to relate β, defined so that Eq. (4.26) is satisfied, to a thermodynamic parameter, a task we take up in Section 4.1.2.8. As an energy distribution, we have, combining Eqs. (4.31) and (4.14),

$$\rho(E) = \frac{1}{Z(\beta)} e^{-\beta E} \Omega(E),$$ (4.32)

which has the form of Eq. (4.19).

4.1.2.7 Getting to Boltzmann: A discussion

We've not taken the most direct path to arrive at the Boltzmann distribution. Our derivation has *almost* been a deductive process, but it's not the case you can start at "line one" and arrive at Eq. (4.31) purely through deduction. A genuinely new equation in physics can't be derived from something more fundamental and is justified only *a posteriori* by the success of the theory based on it.[13] One might say that the appearance of Eq. (4.19) in our derivation was fortuitous, but the form $e^{-\beta E} \Omega(E)$ presents itself naturally as a probability density having the partition function Z as its normalizing constant, given that Z presents itself naturally as the Laplace transform of the convolution relation for the density of states, Eq. (4.9). We've taken this path to support the claim (made on page 59) that the problem of statistical mechanics reduces to one in the theory of probability.[14] In this subsection, we review some other ways to arrive at the Boltzmann distribution.

• **The Boltzmann transport equation.** Perhaps the easiest way is to consider stationary solutions of the *Boltzmann transport equation*, which models the *nonequilibrium* phase-space probability density $\rho(p, q, t)$. This approach is outside the intended scope of this book, and the Boltzmann equation is not without its own issues that we can't explore here. Suffice to say that Eq. (4.31) occurs as the steady-state solution of the Boltzmann equation, appropriate to the state of thermal equilibrium.

• **As a postulate.** In his formulation of statistical mechanics, Gibbs simply started with the form of Eq. (4.31), known as the *Gibbs distribution*. He assumed[15] $P = e^{\eta}$ (because it's "...the most simple case conceivable"), where $\eta = (\psi - \epsilon)/\Theta$ is a combination of three functions: ψ, related to the normalization, $e^{\psi/\Theta} \equiv Z^{-1}$, ϵ the energy, and Θ which he termed the modulus of the distribution.[19, p33] The linear dependence of η on ϵ determines an ensemble which Gibbs called *canonically distributed*.[16] The form $P = e^{(\psi - \epsilon)/\Theta}$ was "line one" for Gibbs.

• **The most probable distribution.** In equilibrium, entropy has the maximum value it can have subject to macroscopic constraints. The state of equilibrium is the *most likely* state, by far. Divide a system into M subsystems, and let each subsystem have n_i particles such that $\sum_{i=1}^{M} n_i = N$, a fixed number. Assume the energy is the sum of the subsystem energies, $E = \sum_{i=1}^{M} E_i$. The number of ways the system can have a configuration specified by a given set of numbers $\{n_i\}$ is, from Eq. (3.12),

$$W[n_1, n_2, \cdots, n_k, \cdots] = \frac{N!}{n_1! n_2! \cdots n_k! \cdots}$$

One finds the numbers n_i that maximize W, from which emerges the canonical distribution, Eq. (4.31). See Appendix F.

• **The microcanonical ensemble.** The canonical probability density can be derived from the microcanonical ensemble. We treat a composite system (system A together with its surroundings B)

---

[13] As with any idea in physics, its veracity is ascertained by its consequences, with experiment being the arbiter of truth.

[14] We've followed the approach of Khinchin.[38] See also the article by Grad.[39]

[15] Such an assumption is not unwarranted. The Boltzmann entropy, Eq. (1.62), is a connection between probability and a thermodynamic quantity, $P = e^{S/k}$.

[16] To quote Gibbs[19, pxi]: "This distribution, on account of its unique importance in the theory of statistical equilibrium, I have ventured to call *canonical* ...."

as an isolated system, for which we know the probability density, Eq. (4.3), and then sum over the environmental phase-space coordinates to obtain the probability density for a system that interacts with its surroundings. The microcanonical probability density for the combined system is a *joint probability distribution* which satisfies Eq. (3.22):

$$\rho_{\text{can}}(A) = \int_{\Gamma_B} \rho_{\text{micro}}(A, B) d\Gamma_B = \frac{1}{\Omega(E)h^{3N}} \int_{\Gamma_B} \delta\left(E - H_A - H_B\right) dp_B dq_B ,\qquad (4.33)$$

where we've used Eqs. (4.4), (4.3), and (2.51) ($d\Gamma_B = dp_B dq_B / h^{3N}$), with N the number of particles in system B. The Hamiltonian H_B is a sum of two terms, $H_B(p_B, q_B) = K(p_B) + V(q_B)$, the kinetic and potential energy functions for system B. First integrate over the momentum space associated with Γ_B:

$$\int_{\Gamma_B} \delta\left(E - H_A - K(p_B) - V(q_B)\right) dp_B dq_B \equiv \int_{q_B} \Omega\left(E - H_A - V(q_B)\right) dq_B ,\qquad (4.34)$$

where Ω here is the volume of the hypersurface in momentum space associated with kinetic energy $K = y$:

$$\Omega(y) = \int_{p_B} \delta\left(y - K(p_B)\right) dp_B = \frac{d}{dy}\left(\int_{\{K(p_B) \leq y\}} dp_B\right),$$

where we've used Eq. (2.64). It's straightforward to show, for $K = (1/2m)\sum_{i=1}^{3N} p_i^2$, that

$$\Omega(y) = \frac{(2m\pi)^{3N/2}}{\Gamma(3N/2)} y^{(3N/2)-1} \equiv C_N y^{(3N/2)-1}.\qquad (4.35)$$

Combining Eqs. (4.33), (4.34), and (4.35), and for $M \equiv (3N/2) - 1$,

$$\rho_{\text{can}}(A) = \frac{C_N}{\Omega(E)h^{3N}} \int_{q_B} [E - H_A - V(q_B)]^M dq_B$$

$$= \frac{C_N}{\Omega(E)h^{3N}} E^M \left(1 - \frac{H_A}{E}\right)^M \int_{q_B} \left[1 - \frac{V(q_B)}{E - H_A}\right]^M dq_B .\qquad (4.36)$$

To make progress, we introduce two simplifying assumptions.[17]

1. The energy per particle is a constant (in the notation introduced by Gibbs):

$$\frac{E}{N} = \frac{3}{2}\Theta = \text{constant} .\qquad (4.37)$$

Equation (4.37) embodies *equipartition of energy*, that $\frac{3}{2}\Theta$ is the mean kinetic energy per particle. We'll show that $\Theta = kT$, implying that the specific heat ($N^{-1}(\partial E/\partial T)_V$) is *independent of temperature*—a supposition having experimental support. It's found (the *law of Dulong and Petit*) that the specific heats of chemical elements in the solid state have approximately the same value, independent of temperature, if the temperature is not too low. With Eq. (4.37), energy equipartition is *built in* to statistical mechanics from the outset.[18] The third law of thermodynamics requires that heat capacities vanish as $T \to 0$,[3, Chapter 8] an observed property of matter at low temperature, and one for which quantum mechanics accounts

---

[17] We've already made the assumption of weak interaction between system and reservoir, Eq. (4.4).
[18] That is, energy equipartition is not something we discover "later" as a feature of statistical mechanics—it's built in.

for in terms of the breakdown of energy equipartition. We anticipate that classical statistical mechanics will require revision.[19]

2. We assume $H_A \ll E$—the assumption that system A is much smaller than the environment.

Substituting $E = M\Theta$ in Eq. (4.36) (for $N \gg 1$), and ignoring H_A in relation to E, we have

$$\rho_{\text{can}}(A) \approx \frac{C_N}{\Omega(E)h^{3N}} E^M \left(1 - \frac{H_A}{M\Theta}\right)^{M'} \int_{q_B} \left[1 - \frac{V(q_B)}{M\Theta}\right]^M dq_B .$$

Using the Euler form of the exponential function, $e^x = \lim_{n\to\infty}(1 + x/n)^n$, we have for large M,

$$\rho_{\text{can}}(A) \approx \frac{C_N}{\Omega(E)h^{3N}} E^M e^{-H_A/\Theta} \int_{q_B} e^{-V(q_B)/\Theta} dq_B .$$

Now let $M \to \infty$. The terms multiplying $e^{-H_A/\Theta}$ approach a limiting value $D(\Theta)$ depending only on the parameter Θ:

$$\rho_{\text{can}}(A) \xrightarrow{M\to\infty} D(\Theta)e^{-H_A/\Theta} , \tag{4.38}$$

where D is the inverse normalization constant,

$$D^{-1} \equiv Z = \int_{\Gamma_A} e^{-H_A(p,q)/\Theta} d\Gamma_A . \tag{4.39}$$

Thus we arrive at the Boltzmann distribution, Eq. (4.31), where $\Theta \equiv \beta^{-1}$. In what follows, we'll use the canonical distribution as in Eq. (4.31), parameterized by β.

4.1.2.8 Consistency with thermodynamics

A requirement on statistical mechanics is that it reproduce the laws of thermodynamics, a demand ensured by equating macroscopically measurable quantities with appropriate ensemble averages (Section 2.5). As we now show, the framework we've established is consistent with thermodynamics if we identify $\beta = (kT)^{-1}$ and if we modify the partition function with a multiplicative factor.

Internal energy is the energy of adiabatic work and heat is the difference between work and adiabatic work (Section 1.2). Adiabatically isolated systems interact with their surroundings through mechanical means only. For such systems, the internal energy U is the conserved energy of mechanical work done on the system, which is the same as the value of the Hamiltonian H. Thus, we equate U with the ensemble average of H:

$$U = \langle H \rangle = \int \rho(p,q)H(p,q)d\Gamma = -\left(\frac{\partial}{\partial\beta}\ln Z\right)_V , \tag{4.40}$$

where the final equality follows from Eq. (4.39) with $\Theta = \beta^{-1}$, and which will be recognized as Eq. (4.20) with $\alpha = \beta$. *Equation (4.40) is perhaps the most useful formula in statistical mechanics.*

---

[19] As an historical aside, Gibbs, writing in 1901, knew something was amiss with physics [19, pix]. "In the present state of science, it seems hardly possible to frame a dynamical theory of molecular action, which shall embrace the phenomena of thermodynamics, of radiation, and of the electrical manifestations which accompany the union of atoms. Yet any theory is obviously inadequate which does not take account of all these phenomena. Even if we confine our attention to the phenomena distinctively thermodynamic, we do not escape difficulties in as simple a matter as the number of degrees of freedom of a diatomic gas. It is well known that while theory would assign to the gas six degrees of freedom per molecule, in our experiments on specific heat we cannot account for more than five. Certainly, one is building on an insecure foundation, who rests his work on hypotheses concerning the constitution of matter."

Work entails variations in a system's extensive external parameters $\{X_i\}$, $\delta W = \sum_i Y_i \delta X_i$; Eq. (1.4). Adiabatic work δW associated with variations δX_i is reflected in changes δH in the value of the Hamiltonian, provided the variations occur at an infinitesimal rate.[20] Thus,[21]

$$\delta W = \langle \delta H \rangle = \sum_i \langle Y_i \rangle \delta X_i = \sum_i \left(\int \frac{\partial H}{\partial X_i} \rho d\Gamma \right) \delta X_i = -\frac{1}{\beta} \sum_i \left(\frac{\partial}{\partial X_i} \ln Z \right) \delta X_i . \quad (4.41)$$

The Hamiltonian must therefore be a function of the external parameters[22] (as well as the canonical coordinates (p, q)), with

$$\langle Y_i \rangle = \int \frac{\partial H}{\partial X_i} \rho d\Gamma = \left\langle \frac{\partial H}{\partial X_i} \right\rangle = -\frac{1}{\beta} \frac{\partial}{\partial X_i} \ln Z . \quad (4.42)$$

Thus, if we have the partition function $Z(\beta, X_i)$, we can calculate the internal energy U and the intensive quantities Y_i from logarithmic derivatives of Z, Eqs. (4.40) and (4.42).

By the first law, the heat δQ transferred to a system is the difference between changes in energy δU, brought about by any means, and the adiabatic work done on the system δW. Thus, we *define*

$$\delta Q \equiv \delta U - \delta W = \delta \langle H \rangle - \langle \delta H \rangle . \quad (4.43)$$

In Chapter 1, we discussed the classification of variables that occur in thermodynamic descriptions. As another term, the quantities $\{X_i\}$ are referred to as *deformation coordinates*, because work involves "deformations" or extensions of the extensive variables associated with a system. For every equilibrium system, there is an additional, *non-deformation* or *thermal* coordinate. It's possible to impart energy to isolated systems by means not involving changes in deformation coordinates, e.g., stirring a fluid of fixed volume adiabatically isolated from the environment. In any thermodynamic description there must be a thermal coordinate and at least one deformation coordinate.[3, p12] We assume therefore that the parameter β is the thermal coordinate.[23] In what follows, we'll use a variation symbol δ, so that, for a function $f(\beta, X_i)$,

$$\delta f \equiv \frac{\partial f}{\partial \beta} \delta \beta + \sum_i \delta X_i \frac{\partial f}{\partial X_i} . \quad (4.44)$$

Using Eqs. (4.40), (4.41), (4.43), and (4.44),

$$\delta Q = \delta U - \delta W = \frac{\partial}{\partial \beta} \left(-\frac{\partial}{\partial \beta} \ln Z \right) \delta \beta + \sum_i \frac{\partial}{\partial X_i} \left(-\frac{\partial \ln Z}{\partial \beta} \right) \delta X_i + \frac{1}{\beta} \sum_i \left(\frac{\partial}{\partial X_i} \ln Z \right) \delta X_i$$

$$= -\frac{\partial^2 \ln Z}{\partial \beta^2} \delta \beta - \sum_i \frac{\partial}{\partial X_i} \left(\frac{\partial \ln Z}{\partial \beta} - \beta^{-1} \ln Z \right) \delta X_i = \frac{1}{\beta} \delta \left(-\beta^2 \frac{\partial}{\partial \beta} \left(\beta^{-1} \ln Z \right) \right) ,$$

where we've used the identities developed in Exercise 4.16. Thus,

$$\beta \delta Q = \delta \left(-\beta^2 \frac{\partial}{\partial \beta} \left(\beta^{-1} \ln Z \right) \right) . \quad (4.45)$$

---

[20] In the language of thermodynamics, we're describing a reversible, quasistatic process.[3, pp14–15]

[21] Recall the sign convention we have adopted for work. Note that δW is equal to an ensemble average.

[22] The external parameters X_i are the same for all members of the ensemble; they enter in the potential energy function of the system. See Exercise 4.13. An exception is the vector potential associated with an external magnetic field that would be part of the canonical momenta, i.e., the vector potential would enter in the kinetic energy part of the Hamiltonian.

[23] Gibbs gives heuristic arguments that his parameter Θ has the requisite properties of temperature.[19, Chapter 4] Because we've presupposed equilibrium, the zeroth law is automatically satisfied, which establishes the existence of temperature.[3, p11]

That is, $\beta\delta Q$ is equal to a total differential, implying that β *is an integrating factor for δQ.* Reaching for contact with thermodynamics,[24] we infer from Eq. (1.8) that β is proportional to the inverse absolute temperature, T^{-1}. It's quite remarkable that small variations in a heat-like quantity (defined in Eq. (4.43) which mimics the first law of thermodynamics) have, within the framework of statistical mechanics, an integrating factor β as required by the second law of thermodynamics.[25] It's not obvious this development could have been foreseen; it seems we're on to something.

Let's provisionally take $\beta = (kT)^{-1}$. In that case, we can equate the quantity in parentheses in Eq. (4.45) with entropy:

$$S = k\frac{\partial}{\partial T}(T\ln Z) + \text{constant} . \tag{4.46}$$

We find ourselves in an analogous situation to what we faced in Section 1.11: What's the constant k (setting the scale of entropy) and what's the constant in Eq. (4.46) (setting the zero of entropy)?

Let's first address the multiplicative constant, k. And, just as in Section 1.11, we reach for the ideal gas. Use Eq. (4.39) (with $\Theta = \beta^{-1} = kT$) to calculate Z for an ideal gas contained in V:

$$Z(N,V,T) = \frac{1}{h^{3N}}\int_V dq \int_{-\infty}^{\infty} dp\, e^{-\beta H} = V^N\left(\frac{2\pi mkT}{h^2}\right)^{3N/2} = (V/\lambda_T^3)^N , \tag{4.47}$$

where $H = (1/2m)\sum_{i=1}^{N} p_i^2$, λ_T is the thermal wavelength, Eq. (1.65), and we've used $3N$ copies of Eq. (B.7). Note that we did *not* make use of the volume of a hypersphere, as we've done in previous calculations. The energy is not fixed in the canonical ensemble; we must sum over all possible momenta. Using Eq. (4.46) to calculate the entropy, with Z given by Eq. (4.47),

$$S = Nk\left(\frac{3}{2} + \ln\left(V/\lambda_T^3\right)\right) + \text{constant} . \qquad \text{(wrong!)} \tag{4.48}$$

Equation (4.48) does not agree with the experimentally verified Sackur-Tetrode formula, Eq. (1.64). Let's sidestep that issue for now, because where Eq. (4.48) gets it wrong is in the dependence on particle number. The pressure can be calculated from S using Eq. (1.24) (at constant N),

$$\frac{P}{T} = \left(\frac{\partial S}{\partial V}\right)_{T,N} = \frac{Nk}{V} , \tag{4.49}$$

where we've used Eq. (4.48) for S. Equation (4.49) is of course the equation of state of the ideal gas, indicating that the constant k in Eq. (4.46) is indeed the Boltzmann constant.[26]

The additive "constant" in Eq. (4.48) is a function of particle number. The Clausius definition of entropy, which Eq. (4.45) is a model of, involves reversible heat transfers between *closed* systems and their environment. For open systems, there is an additional contribution to entropy from the diffusive flow of particles not accounted for in the Clausius definition[3, Section 14.3], *which is therefore an incomplete specification of entropy.* As noted by E.T. Jaynes, "As a matter of elementary logic, no theory can determine the dependence of entropy on the size N of a system unless it makes some statement about a process where N changes." [40]. Our "statement" is that *entropy is extensive,* which Eq. (4.48) does not exhibit (show this).[27] Let's add an unknown, N-dependent function to

---

[24]From the Clausius inequality, for reversible quasistatic heat transfers, $(dQ)_{\text{rev}}/T$ is an exact differential. The Clausius inequality is a consequence of Carnot's theorem, which in turn is a consequence of the second law of thermodynamics [3, p33]. With Eq. (4.45), we've established that statistical mechanics is consistent with the second law.

[25]Differential forms involving two variables always possess an integrating factor, implying the existence of entropy may be inferred independently of the second law *in that case.* Equation (4.45) involves an arbitrary number of external parameters X_i, in which case the second law of thermodynamics is required to establish the existence of entropy. See [3, Chapter 10].

[26]We reiterate (because of the quip that k is a "units thing"): The Boltzmann constant is an experimentally determined property of matter, the gas constant divided by Avogadro's number.

[27]Entropy is extensive because heat capacities scale with the mass of the system. Yet standard calculations using the Clausius definition lead to non-extensive expressions such as Eq. (4.48). Extensivity is not a consequence of the Clausius definition, and must be imposed as a separate requirement. In axiomatic formulations of thermodynamics, extensivity of entropy is taken as a postulate [8]. Extensivity is built into the definition of the Boltzmann entropy; see Eq. (1.60).

Eq. (4.48):

$$S = Nk \left(\frac{3}{2} + \ln \left(V/\lambda_T^3 \right) \right) + k\phi(N) . \tag{4.50}$$

For S in Eq. (4.50) to obey the scaling law $S(\lambda V, \lambda N) = \lambda S(V, N)$ (see Eq. (1.52)), ϕ must satisfy the scaling relation

$$\phi(\lambda N) = \lambda \phi(N) - \lambda N \ln \lambda . \tag{4.51}$$

Note the identity for $\lambda = 1$. Differentiate Eq. (4.51) with respect to λ and set $\lambda = 1$; one finds the differential equation $N\phi' = \phi - N$, the solution of which is $\phi(N) = \phi(1)N - N \ln N$. Combined with Eq. (4.50),

$$S = Nk \left(\frac{3}{2} + \phi(1) + \ln \left(\frac{V}{N\lambda_T^3} \right) \right) . \tag{4.52}$$

This form of S exhibits extensivity, but $\phi(1)$ cannot be established by this method of analysis. If we take $\phi(1) = 1$, Eq. (4.52) becomes identical to the Sackur-Tetrode formula. In that case, $\phi(N) = N - N \ln N$, which we note is the Stirling approximation for $\phi(N) = -\ln N!$.

We introduced a factor of $N!$ in our derivation of the Sackur-Tetrode formula (Section 2.3) to prevent overcounting configurations that are equivalent under permutations of N identical particles. The reason we obtained an incorrect expression of S in Eq. (4.48) is not because Eq. (4.46) is suspect, but because Eq. (4.47) is incorrect. The additive factor of $-\ln N!$ required to achieve agreement between Eq. (4.48) and the Sackur-Tetrode formula would occur automatically if the partition function were to include it. We define the *canonical partition function*

$$Z_{\text{can}}(N, V, T) \equiv \frac{1}{N!} \int d\Gamma e^{-\beta H} , \tag{4.53}$$

where $\beta = (kT)^{-1}$. Thus, $Z_{\text{can}} = Z/N!$, where Z is specified in Eq. (4.39) or in Eq. (4.15). In many cases, "plain old" Z suffices—where the quantity of interest is obtained from a logarithmic derivative of Z (such as Eqs. (4.40) and (4.42)) because the constant factor of $N!$ drops out. The expression for entropy, however, Eq. (4.46), does *not* involve a logarithmic derivative of the partition function and we must use Z_{can}. We can write, therefore,

$$\rho(p, q) = \frac{1}{Z} e^{-\beta H(p,q)} = \frac{1}{N! Z_{\text{can}}} e^{-\beta H(p,q)} . \tag{4.54}$$

The canonical distribution in the form of Eq. (4.54) is "more correct" than expression Eq. (4.31), which we obtained in a direct calculation, or the expressions Eqs. (4.38) and (4.39), which we derived from the microcanonical distribution. What did we overlook in those calculations? Simply put, we took every point in Γ-space to represent a distinct microstate; we did not take into account that the state of N identical particles is $N!$-fold degenerate under permutations. The expression for entropy, Eq. (4.46), is valid if we use Z_{can}:

$$S = k \frac{\partial}{\partial T} (T \ln Z_{\text{can}}) + S_0 . \qquad (S_0 = 0) $$

The constant S_0 is the province of the third law of thermodynamics, which states that entropy changes ΔS in physical processes vanish[28] as $T \to 0$. This experimentally observed property of matter can be interpreted to mean that entropy $S(T)$ approaches a constant $S(T) \to S_0$ as $T \to 0$. The value of S_0 is, by convention, zero for many substances. We'll take $S_0 = 0$. There are materials,

---

[28]This version of the third law is known as the *Nernst heat theorem*—there are several way of stating the third law.[3, Chapter 8] One does not have to achieve "heroically" cryogenic temperatures to see that $\Delta S \to 0$ at low temperature. The trends are noticeable even for "warm" temperatures of $T \approx 30$ K. Nernst received the 1920 Nobel Prize in Chemistry for work on the third law of thermodynamics.

however, that achieve a finite-entropy configuration[29] as $T \to 0$. As a consistency check, it must be ascertained that (if $S_0 = 0$),

$$\lim_{T \to 0} \frac{\partial}{\partial T} \left(T \ln Z_{\text{can}} \right) = 0 . \tag{4.55}$$

We anticipate that Eq. (4.55) won't hold for partition functions evaluated with classical statistical mechanics, which are based on energy equipartition.

4.1.2.9 Connection with the Helmholtz energy

The microcanonical ensemble is composed of systems having precise values of E, V, N; for that reason it's known as the NVE ensemble. The canonical ensemble is composed of systems having precise values of N and V, but a variable energy having the average value U. Temperature is a proxy for variable energy, and the canonical ensemble is known as the NVT ensemble. The Helmholtz energy $F = F(N, V, T)$ is a function of N, V, T (see Eq. (1.14)), and, as we now show, there is a simple connection between $Z_{\text{can}}(N, V, T)$ and the Helmholtz energy $F(N, V, T)$, Eq. (4.57).

From Eq. (4.45),

$$S = -k\beta^2 \frac{\partial}{\partial \beta} \left(\beta^{-1} \ln Z_{\text{can}} \right) = k \left[\beta \left(-\frac{\partial \ln Z}{\partial \beta} \right) + \ln Z_{\text{can}} \right] = \frac{U}{T} + k \ln Z_{\text{can}} , \tag{4.56}$$

where we've used Eq. (4.40). Equation (4.56) implies

$$F = U - TS = -kT \ln Z_{\text{can}} \implies Z_{\text{can}}(N, V, T) = e^{-\beta F(N,V,T)} . \tag{4.57}$$

Equation (4.57) provides the link between thermodynamics and statistical mechanics in the canonical ensemble. With it, we can write the canonical distribution (from Eq. (4.54)) $\rho = e^{\beta(F-H)}/N!$, the form assumed by Gibbs, except for the factor of $N!$.

With the connection between Z_{can} and F in Eq. (4.57), the standard formulas of thermodynamics can be used:

$$\mu = \left(\frac{\partial F}{\partial N} \right)_{T,V} = -kT \frac{\partial}{\partial N} \ln Z_{\text{can}} \Big|_{T,V} \qquad P = -\left(\frac{\partial F}{\partial V} \right)_{T,N} = kT \frac{\partial}{\partial V} \ln Z_{\text{can}} \Big|_{T,N}$$

$$S = -\left(\frac{\partial F}{\partial T} \right)_{V,N} = k \frac{\partial}{\partial T} \left(T \ln Z_{\text{can}} \right) \Big|_{V,N} . \tag{4.58}$$

Given information about N, V, T *and the Hamiltonian*, we can calculate μ, P, and S.

Consider the logarithm of ρ (which is dimensionless)

$$\ln \rho = \ln \left(e^{-\beta H}/Z \right) = -\beta H - \ln Z = -\beta H - \ln Z_{\text{can}} - \ln N! . \tag{4.59}$$

If we multiply Eq. (4.59) by ρ, integrate over Γ-space, and make use of Eq. (4.57), we find

$$S = -k \int \rho \ln \left(N! \rho \right) d\Gamma . \tag{4.60}$$

Entropy is therefore not the mean value of a mechanical quantity, it's a property of the ensemble.

---

[29] If the ground-state configuration is degenerate as $T \to 0$, it implies a nonzero entropy. Solid carbon monoxide, for example, achieves a random mixture of CO and OC molecules, depending on whether C or O is "up." See [3, Chapter 8].

4.1.2.10 Fluctuations in the canonical ensemble

We can calculate the internal energy from Eq. (4.40), $U = -\partial \ln Z/\partial\beta$, but we can also address fluctuations. From

$$\frac{\partial^2}{\partial\beta^2} \ln Z = \frac{1}{Z}\frac{\partial^2 Z}{\partial\beta^2} - \left(\frac{1}{Z}\frac{\partial Z}{\partial\beta}\right)^2 = \langle H^2 \rangle - \langle H \rangle^2 , \tag{4.61}$$

we infer

$$\langle H^2 \rangle - \langle H \rangle^2 = \frac{\partial}{\partial\beta}\left(\frac{\partial \ln Z}{\partial\beta}\right) = -\left(\frac{\partial U}{\partial\beta}\right)_V = kT^2\left(\frac{\partial U}{\partial T}\right)_V = kT^2 C_V , \tag{4.62}$$

the same as from the Einstein fluctuation theory, Eq. (2.31). *The heat capacity is a measure of energy fluctuations.* The relative root-mean-square fluctuation is

$$\frac{\sqrt{\langle (H - \langle H \rangle)^2 \rangle}}{\langle H \rangle} = \frac{\sqrt{kT^2 C_V}}{U} . \tag{4.63}$$

For macroscopic systems, $U \sim O(N)$ and $C_V \sim O(N)$, and thus the relative fluctuation scales as $N^{-1/2}$. A system in the canonical ensemble has, for all practical purposes, an energy equal to the mean for large N, which we expect from the law of large numbers.

It's not an accident that Eq. (4.63) agrees with Eq. (2.31). The form of Eq. (2.29) is a property of the canonical ensemble. Using the approximate density of states function Eq. (4.28), we have for the Boltzmann probability density, Eq. (4.32):

$$P(E) = e^{-\beta E}\Omega(E) \sim e^{-\beta F} \exp\left(-\frac{(E-U)^2}{2\sigma^2}\right) = e^{-\beta(U-TS)} \exp\left(-\frac{(E-U)^2}{2kT^2 C_V}\right) ,$$

where we've used Eq. (4.57), $Z = e^{-\beta F}$, and where $\sigma^2 = \partial^2 \ln Z/\partial\beta^2$ is the variance of the distribution given in Eq. (4.27), a quantity that we just derived in Eq. (4.62), $\sigma^2 = kT^2 C_V$. Thus, the distribution of energy among the systems of the ensemble is Gaussian, centered about the mean value U with variance $\sigma^2 = kT^2 C_V$, the same as Eq. (2.29).

4.1.2.11 The equipartition theorem

Energy equipartition was assumed in the derivation of the Boltzmann distribution, that the kinetic energy per particle is a constant, proportional to the temperature, Eq. (4.37). The *equipartition theorem* holds that energy equipartition applies not just to kinetic energy but to any *quadratic degree of freedom*. A quadratic degree of freedom is one which adds a quadratic term to the Hamiltonian associated with that degree of freedom: $p^2/(2m)$ or $\frac{1}{2}m\omega^2 x^2$.

To develop the equipartition theorem, we first show that

$$\left\langle x_i \frac{\partial H}{\partial x_j} \right\rangle = \delta_{ij} kT , \tag{4.64}$$

where x_i denotes any of the canonical coordinates (p_k, q_k) that enter into the Hamiltonian. Consider the integration over Γ-space,

$$\int x_i \frac{\partial H}{\partial x_j} e^{-\beta H} d\Gamma = -\frac{1}{\beta}\int x_i \left(\frac{\partial}{\partial x_j} e^{-\beta H}\right) d\Gamma = -\frac{1}{\beta}\int \left(x_i e^{-\beta H}\Big|_{x_j^{(1)}}^{x_j^{(2)}} - \int \frac{\partial x_i}{\partial x_j} e^{-\beta H} dx_j\right) d\Gamma^{[j]}$$

$$= \frac{1}{\beta}\int \frac{\partial x_i}{\partial x_j} e^{-\beta H} d\Gamma = \frac{1}{\beta}\delta_{ij}\int e^{-\beta H} d\Gamma = \frac{1}{\beta}\delta_{ij} Z(\beta) ,$$

where in the second equality we've integrated by parts, where $x_j^{(1)}$ and $x_j^{(2)}$ are the limits of integration of x_j, and where $d\Gamma^{[j]}$ indicates the volume element $d\Gamma$ with dx_j removed. The integrated part vanishes: For x_j a position coordinate, the range of integration extends over the volume of the system where the potential energy becomes infinite at the boundaries, and for x_j a momentum component it takes the values $\pm\infty$ at the "endpoints" of momentum space; in either case the integrated part vanishes. With this result established, Eq. (4.64) follows.[30]

Let's consider Hamiltonians featuring only quadratic degrees of freedom, which we can write

$$H = \sum_{i=1}^{Nf} b_i x_i^2 \,,$$

where x_i is any canonical coordinate and b_i is a constant. For example, the Hamiltonian of a harmonic oscillator with isotropic spring constant is in the form $H = \sum_{i=1}^{3}\left(p_i^2/(2m) + \frac{1}{2}m\omega^2 q_i^2\right)$. A Hamiltonian containing only quadratic degrees of freedom is a second-order homogeneous function, which by Euler's theorem (see Section 1.10), can be written

$$2H = \sum_{i=1}^{Nf} x_i \frac{\partial H}{\partial x_i} \,. \tag{4.65}$$

Applying Eq. (4.65) to Eq. (4.64),

$$2\langle H \rangle = \sum_{i=1}^{Nf} \left\langle x_i \frac{\partial H}{\partial x_i} \right\rangle = NfkT \,,$$

which implies

$$\langle H \rangle = \frac{1}{2}NfkT \,. \tag{4.66}$$

Equation (4.66) is the equipartition theorem: Each quadratic degree of freedom in the Hamiltonian makes a contribution of $\frac{1}{2}kT$ towards the internal energy of the system, and hence a contribution of $\frac{1}{2}k$ to the specific heat.

Example. Consider a *torsion pendulum*, a disk suspended from a torsion wire attached to its center. A torsion wire is free to twist about its axis. As it twists, it causes the disc to rotate in a horizontal plane. For θ the angle of twist, there is a restoring torque $\tau = -\kappa\theta$, and thus $I\ddot{\theta} = \tau$, where I is the moment of inertia of the disk. It's a standard exercise in classical mechanics to show that the Hamiltonian of this system is

$$H = \frac{1}{2I}p_\theta^2 + \frac{1}{2}\kappa\theta^2 \,,$$

where $p_\theta = I\dot{\theta}$. The Hamiltonian has two quadratic degrees of freedom. By the equipartition theorem, assuming the system is in equilibrium with a heat bath at temperature T,

$$\left\langle \frac{1}{2I}p_\theta^2 \right\rangle = \frac{1}{2}kT \implies \langle \dot{\theta}^2 \rangle = kT/I$$

$$\left\langle \frac{1}{2}\kappa\theta^2 \right\rangle = \frac{1}{2}kT \implies \langle \theta^2 \rangle = kT/\kappa \,.$$

---

[30]Note that $\partial x_i/\partial x_j = \delta_{ij}$ follows from the independence of canonical coordinates.

4.1.3 The grand canonical ensemble: Systems of fixed T, V, μ

We now consider the grand canonical ensemble,[31] an ensemble of open systems with fixed values of μ, V, T, in which N and E vary, but have the average values $\langle N \rangle$ and U. The first law of thermodynamics for open systems is (see Eq. (1.21))

$$dU = TdS + \sum_i Y_i dX_i + \sum_j \mu_j dN_j , \qquad (4.67)$$

where μ_j is the chemical potential of the jth type of particle, of which there are N_j particles. Thus, $U = U(N_j, X_i, S)$. Through the use of Legendre transformations, thermodynamic potentials characterize macroscopic systems under given external conditions (see Section 1.4). For the Helmholtz energy,[32]

$$dF = -SdT + \sum_i Y_i dX_i + \sum_j \mu_j dN_j . \qquad (4.68)$$

Thus, $F = F(N_j, X_i, T)$—it's easier to control T than S.

Is there a thermodynamic potential suited to open systems? The grand potential is a Legendre transformation of the Helmholtz energy[33]

$$\Phi \equiv F - \sum_j \mu_j N_j = \sum_i Y_i X_i , \qquad (4.69)$$

where we've used the Euler relation, Eq. (1.53), generalized to multicomponent systems,

$$TS = U - \sum_i Y_i X_i - \sum_j \mu_j N_j . \qquad (4.70)$$

By forming the differential of Φ, using Eqs. (4.69) and (4.68), we find

$$d\Phi = -SdT + \sum_i Y_i dX_i - \sum_j N_j d\mu_j . \qquad (4.71)$$

Hence, $\Phi = \Phi(\mu_j, X_i, T)$ is the thermodynamic potential for systems in which T and X_i are held fixed, but particle numbers are allowed to vary. From Eq. (4.71),

$$N_j = -\left(\frac{\partial \Phi}{\partial \mu_j}\right)_{T, X_i, \mu_{k \neq j}} \qquad Y_i = \left(\frac{\partial \Phi}{\partial X_i}\right)_{T, \mu_j, X_{k \neq i}} \qquad S = -\left(\frac{\partial \Phi}{\partial T}\right)_{X_i, \mu_j} . \qquad (4.72)$$

Note the familiar pattern: Extensive quantities (S and N_j) are obtained from partial derivatives with respect to intensive quantities (T and μ_j), while intensive quantities Y_i are found from partial derivatives with respect to the extensive quantities X_i. Compare Eq. (4.72) with Eq. (4.58).

The grand canonical ensemble is a collection of identically prepared systems that allow not only the exchange of energy with the surroundings, but particles as well. We ask for the probability $\rho_N d\Gamma_N$ that a randomly selected member of the ensemble has N particles and has its phase point lying within the volume element $d\Gamma_N$ for the phase space of an N-particle system. If ρ_N is known, the grand canonical ensemble average of a phase-space function $A(p, q, N)$ is defined as

$$\langle A \rangle \equiv \sum_{N=0}^{\infty} \int_{\Gamma_N} A(p, q, N) \rho_N(p, q) d\Gamma_N . \qquad (4.73)$$

---

[31] To recap, we considered the microcanonical distribution in Section 4.1.1, which applies to isolated systems with fixed N, V, E, and the canonical distribution in Section 4.1.2, which applies to closed systems with fixed N, V, T.

[32] Equation (4.68) is more general than Eq. (1.14).

[33] The grand potential is a Legendre transformation of F, $\Phi \equiv F - \sum_j \mu_j N_j = \sum_i Y_i X_i$. The Gibbs energy is a different Legendre transformation of F: $G = F - \sum_i Y_i X_i = \sum_j \mu_j N_j$.

For $A = 1$, we have the normalization requirement

$$1 = \sum_{N=0}^{\infty} \int_{\Gamma_N} \rho_N(p, q) d\Gamma_N . \tag{4.74}$$

We don't have to do a lengthy calculation for ρ_N because we already know what it is! Equation (4.54) specifies the phase-space probability distribution for a single-component system of N particles at fixed T and V,

$$\rho_N(p, q) = \frac{1}{N!} e^{\beta(F_N - H_N)} = \frac{1}{N!} e^{\beta(\Phi + N\mu - H_N)} \equiv \frac{1}{N! Z_G} e^{\beta(N\mu - H_N)} , \tag{4.75}$$

where F_N, H_N are the Helmholtz energy and Hamiltonian of an N-particle system, we've used $F_N = \Phi + N\mu$ from Eq. (4.69), and we've defined the *grand partition function*,

$$Z_G(T, V, \mu) = e^{-\beta\Phi(T,V,\mu)} \implies \Phi(T, V, \mu) = -kT \ln Z_G(T, V, \mu) . \tag{4.76}$$

Equation (4.76) is the link between thermodynamics and statistical mechanics in the grand canonical ensemble (compare with Eq. (4.57)). Equation (4.74), the normalization condition, is satisfied with

$$Z_G(\mu, V, T) = \sum_{N=0}^{\infty} Z_{\text{can}}(N, V, T) z^N , \tag{4.77}$$

where $z \equiv e^{\beta\mu}$ is the *fugacity*.[34,35] It should be noted that the value of z is *fixed* for a given grand canonical ensemble prescribed by values of (μ, V, T).

Using Eqs. (4.72) and (4.76),

$$N = kT \frac{\partial \ln Z_G}{\partial \mu}\bigg|_{T,V} = z \frac{\partial \ln Z_G}{\partial z}\bigg|_{T,V} \qquad P = kT \frac{\partial \ln Z_G}{\partial V}\bigg|_{T,\mu} \qquad S = k \frac{\partial}{\partial T}(T \ln Z_G)\bigg|_{V,\mu} . \tag{4.78}$$

The thermodynamic symbols N, P, S in Eq. (4.78) correspond to the ensemble averages $\langle N \rangle, \langle P \rangle, \langle S \rangle$. The formulas for P and S in Eq. (4.78) have the same form of the analogous quantities in Eq. (4.58), with Z_{can} replaced by Z_G. In the grand canonical ensemble, one solves for $\langle N \rangle$ given μ; in the canonical ensemble, one solves for μ given N.

Example. Evaluate the grand partition function for the ideal gas. Combining Eqs. (4.77) and (4.47) (with $Z_{\text{can}} = Z/N!$),

$$Z_G(\mu, V, T) = \sum_{m=0}^{\infty} z^m Z_{\text{can}}(m, V, T) = \sum_{m=0}^{\infty} z^m \frac{1}{m!} \left(\frac{V}{\lambda_T^3}\right)^m = \sum_{m=0}^{\infty} \frac{1}{m!} \left(\frac{zV}{\lambda_T^3}\right)^m = \exp\left(zV/\lambda_T^3\right) . \tag{4.79}$$

The average particle number is (from Eq. (4.78))

$$N = z \frac{\partial}{\partial z}\left(\frac{zV}{\lambda_T^3}\right)\bigg|_{T,V} = \frac{zV}{\lambda_T^3} = e^{\beta\mu} \frac{V}{\lambda_T^3} .$$

---

[34] Your inner mathematician may worry about convergence of the infinite series in Eq. (4.77). Often μ is negative and the series converges (see Section 1.6). A positive chemical potential is usually for systems in which the assumptions of classical statistical mechanics no longer apply. We'll see in Chapter 5 that for bosons $\mu \leq 0$, whereas there is no such restriction for fermions. The Pauli exclusion principle, however, restricts the sum in Eq. (4.77) to include only the terms for $N = 0, 1$.

[35] Fugacity isn't a word invented to denote $z \equiv e^{\beta\mu}$, it's used in thermodynamics to model pressure in gases.[41, p122] The chemical potential of the ideal gas is a linear function of $\ln(P)$, $\mu(T, P) = \mu(T, P_0) + RT \ln(P/P_0)$.[3, p59] Fugacity (as the term is used in thermodynamics) is defined so that the chemical potential of a real gas has the same form, $\mu = \mu^0 + RT \ln z$, such that $z \sim P$ as $P \to 0$. We show in Section 5.5.3 that z behaves this way in the classical limit.

Solving for μ, we find $\mu = -kT \ln \left(V/(N\lambda_T^3) \right)$, the same as in Eq. (P1.3), the only difference being that now N refers to the average number of particles in the gas. The other two formulas in Eq. (4.78) lead to the known expressions for the entropy and the equation of state of the ideal gas (see Exercise 4.30).

Consider a *multicomponent system*, a system with k distinct chemical species. The grand partition function for such a system is straightforward to define and it suffices to simply write down the result. Let $H_{N_1,\cdots,N_k}$ denote the Hamiltonian of a system with N_1 particles of type 1, N_2 particles of type 2, and so on. Define

$$Z_{N_1,\cdots,N_k} \equiv (N_1! N_2! \cdots N_k!)^{-1} \int e^{-\beta H_{N_1 \cdots N_k}} \, d\Gamma_1 \cdots d\Gamma_k \, ,$$

where $d\Gamma_i$ is the volume element for the phase space of the ith species. The grand partition function is then

$$Z_G = \sum_{N_1=0}^{\infty} \sum_{N_2=0}^{\infty} \cdots \sum_{N_k=0}^{\infty} \left(e^{\beta \mu_1} \right)^{N_1} \left(e^{\beta \mu_2} \right)^{N_2} \cdots \left(e^{\beta \mu_k} \right)^{N_k} Z_{N_1,\cdots,N_k} \, , \qquad (4.80)$$

from which various thermodynamic functions are obtained in the usual way by differentiation of $Z_G = e^{-\beta \Phi}$ and the use of Eq. (4.72)

4.1.4 Fluctuations in the grand canonical ensemble

We now consider fluctuations in particle number (which we can only do in the grand canonical ensemble). From the result of Exercise 4.31,

$$\langle N^2 \rangle - \langle N \rangle^2 = kT \frac{\partial \langle N \rangle}{\partial \mu} \bigg|_{T,V} = \frac{kT}{V} \langle N \rangle^2 \beta_T \, , \qquad (4.81)$$

where the final equality is established in Exercise 4.32. *The isothermal compressibility is a measure of fluctuations in particle number.* With $V \sim N$, the relative root-mean-square fluctuation in particle number is of order $N^{-1/2}$.

The calculation of the internal energy in the grand canonical ensemble can be confusing, so let's be clear. Suppose we try the same formula that we've used now several times, Eq. (4.40):

$$U \stackrel{?}{=} -\frac{\partial \ln Z_G}{\partial \beta} \bigg|_{\mu,V} = \langle H \rangle - \mu \langle N \rangle \, .$$

The formula from the canonical ensemble does *not* produce an expression for the internal energy in the grand canonical ensemble, $\langle H \rangle$. We can devise a method for generating $\langle H \rangle$ by defining a new kind of derivative,

$$U = -\left(\frac{\partial \ln Z_G}{\partial \beta} \right)_{z,V} = \langle H \rangle \, . \qquad (4.82)$$

That is, we differentiate with respect to β, holding the fugacity $z = e^{\beta \mu}$ fixed.

We therefore have the fluctuation in energy (borrowing from Eq. (4.61)):

$$\left(\frac{\partial^2}{\partial \beta^2} \ln Z_G \right)_{z,V} = \frac{1}{Z_G} \left(\frac{\partial^2 Z_G}{\partial \beta^2} \right)_{z,V} - \left(\frac{1}{Z_G} \frac{\partial Z_G}{\partial \beta} \right)_{z,V}^2 = \langle H^2 \rangle - \langle H \rangle^2 \, ,$$

and therefore

$$\langle H^2 \rangle - \langle H \rangle^2 = -\left(\frac{\partial U}{\partial \beta} \right)_{z,V} \, . \qquad (4.83)$$

Compare with Eq. (4.62) for the analogous formula in the canonical ensemble. To make sense of the right side of Eq. (4.83), we have to relate the derivative holding z fixed to other, more standard thermodynamic derivatives. This is a little tricky. The internal energy is a function of N, V, T, but N is a function of z, V, T. Thus, we can write $U = U(T, V, N(z, V, T))$. Using Eq. (1.32),

$$\left(\frac{\partial U}{\partial T}\right)_{z,V} = \left(\frac{\partial U}{\partial T}\right)_{N,V} + \left(\frac{\partial U}{\partial N}\right)_{T,V} \left(\frac{\partial N}{\partial T}\right)_{z,V}. \tag{4.84}$$

The first derivative is the heat capacity, $C_V = (\partial U/\partial T)_V$. The derivative $(\partial U/\partial N)_{T,V}$ can be found from the Helmholtz energy (see Eq. (4.58)),

$$\mu = \left(\frac{\partial F}{\partial N}\right)_{T,V} = \left(\frac{\partial}{\partial N}(U - TS)\right)_{T,V} = \left(\frac{\partial U}{\partial N}\right)_{T,V} - T\left(\frac{\partial S}{\partial N}\right)_{T,V} = \left(\frac{\partial U}{\partial N}\right)_{T,V} + T\left(\frac{\partial \mu}{\partial T}\right)_{V,N},$$

where we've used a Maxwell relation. Therefore,

$$\left(\frac{\partial U}{\partial N}\right)_{T,V} = \mu - T\left(\frac{\partial \mu}{\partial T}\right)_{V,N}.$$

The last derivative in Eq. (4.84) must be found in the same way that we found $(\partial U/\partial T)_{z,V}$. Noting that $N = N(\mu(T, z), T, V)$, we have using Eq. (1.32),

$$\left(\frac{\partial N}{\partial T}\right)_{z,V} = \left(\frac{\partial N}{\partial T}\right)_{\mu,V} + \left(\frac{\partial N}{\partial \mu}\right)_{T,V} \left(\frac{\partial \mu}{\partial T}\right)_{z,V}. \tag{4.85}$$

The last derivative in Eq. (4.85) simplifies using $\mu = kT \ln z$, implying $(\partial \mu/\partial T)_z = \mu/T$. The first derivative in Eq. (4.85) can be found using the cyclic relation, Eq. (1.31):

$$\left(\frac{\partial N'}{\partial T}\right)_\mu = -\left(\frac{\partial \mu}{\partial T}\right)_N \left(\frac{\partial N}{\partial \mu}\right)_T.$$

Thus,

$$\left(\frac{\partial N}{\partial T}\right)_{z,V} = \frac{1}{T}\left(\frac{\partial N}{\partial \mu}\right)_{T,V}\left[\mu - T\left(\frac{\partial \mu}{\partial T}\right)_{N,V}\right]. \tag{4.86}$$

Combining Eqs. (4.86), (4.84), and (4.83), we have the final result:

$$\langle H^2\rangle - \langle H\rangle^2 = kT^2 C_V + kT \left(\frac{\partial N}{\partial \mu}\right)_{T,V} \left(\frac{\partial U}{\partial N}\right)_{T,V}^2$$

$$= kT^2 C_V + \langle(N - \langle N\rangle)^2\rangle \left(\frac{\partial U}{\partial N}\right)_{T,V}^2, \tag{4.87}$$

where we've used Eq. (4.81). The mean-square fluctuation in energy in the grand canonical ensemble is therefore equal to the value it would have in the canonical ensemble, Eq. (4.62), plus a contribution from fluctuations in particle number. For most systems, the relative root-mean-square fluctuation in energy is negligible, of order $N^{-1/2}$ ($\sim 10^{-12}$ for Avogadro's number).

4.2 THE THERMODYNAMIC LIMIT: EXISTENCE OF EXTENSIVITY

Having worked through the ensemble formalism of statistical mechanics and its connection with thermodynamics (a big task!), we can now prove the existence of *extensivity*. We used the extensivity of $\langle H\rangle$, C_V, and V (quantities proportional to N) in Eqs. (4.63) and (4.81) to conclude that the relative deviations of energy and particle number from their mean values are of order $N^{-1/2}$. It was used to argue for the factor of $N!$ in the partition function, Eq. (4.53). Extensive expressions for entropy occur in the form $S(N, V, T) = Ns(n, T)$ where $n \equiv N/V$ (see Eq. (4.52)), because (more precisely) the limit

$$\lim_{\substack{N, V \to \infty \\ n \text{ fixed}}} N^{-1} S(N, V, T) = s(n, T) \tag{4.88}$$

exists. The *thermodynamic limit* is the limiting process in Eq. (4.88) of $N \to \infty$, $V \to \infty$, with $n = N/V$ finite. Thermodynamic relations derived in different ensembles will not in general agree for finite systems, but they become identical in the thermodynamic limit. The basic dichotomy in thermodynamics of intensive and extensive variables (Section 1.1) relies on the *existence* of this limit, yet thermodynamics lacks the tools to prove it. Much of what follows in this book utilizes the thermodynamic limit. We sketch a proof of its existence using the canonical ensemble.[36]

The proof assumes certain generic properties of the potential energy function,[37] $\phi(r)$, of two particles separated by a distance r. For suitable interparticle potentials, we'll show that the limit

$$\lim_{\substack{N, V \to \infty \\ n = N/V \text{ fixed}}} N^{-1} F(N, V, T) = f(n, T) \tag{4.89}$$

exists and is a function only of n and T. Equation (4.88) follows from the usual thermodynamic formulas (such as in Eq. (4.58)) if Eq. (4.89) holds. We proceed by analyzing a macroscopic system as a sequence of nested domains Ω_k, $k = 1, 2, \cdots$, with volumes $V_k < V_{k+1}$ containing N_k particles such that N_k/V_k is fixed[38] (see Fig. 4.1). The partition function for the kth domain is

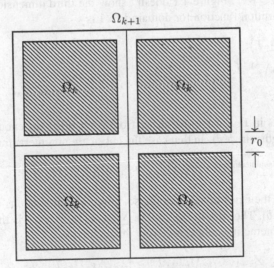

Figure 4.1: Domain Ω_{k+1} is composed of eight domains Ω_k (third dimension not shown). Shading indicates the volume accessible to integration in evaluating the partition function.

$$Z_k(N_k, \Omega_k, T) = \left(N_k! \lambda_T^{3N_k}\right)^{-1} \int_{\Omega_k} \cdots \int d\mathbf{r}_1 \cdots d\mathbf{r}_{N_k} \exp\left[-\beta \sum_{1 \le i < j \le N_k} \phi(|\mathbf{r}_i - \mathbf{r}_j|)\right]. \tag{4.90}$$

---

[36]Existence theorems for the thermodynamic limit in the classical and quantum-mechanical microcanonical, canonical, and grand canonical ensembles are given in Ruelle[42].

[37]We made reference to the potential energy function in the derivation of the Boltzmann distribution from the microcanonical ensemble, Section 4.1.2.7, but we did not need to specify its form. Here restrictions are placed on the allowed form of $\phi(r)$ such that the thermodynamic limit exists.

[38]More precisely, as $N_k \to \infty$ and $V_k \to \infty$, $N_k/V_k \to n$ as $k \to \infty$.

The notation $\sum_{i<j}$ indicates to sum over i and j such that $i < j$, which ensures that interactions between particles i and j are counted only once. Equation (4.90) should be compared with Eq. (4.47) where the thermal wavelength λ_T occurs as a result of integrating the momentum variables. Integration over spatial coordinates is trivial in Eq. (4.47) because it pertains to free particles. *The difficult step in evaluating partition functions is always taking into account inter-particle interactions.* In what follows, we'll use a quantity ψ_k related to the free energy per particle,

$$\psi_k \equiv \frac{1}{N_k} \ln Z_k = -\beta \left(F_{N_k} / N_k \right) . \tag{4.91}$$

For which potentials $\phi(r)$ and which domains Ω_k does $\lim_{k\to\infty} \psi_k = \psi(n, T)$ exist? We won't dwell on the issue of domains—almost any shape will work (we'll use cubes), as long as the surface area doesn't increase too rapidly (more than $V^{2/3}$ is too fast). The function $\phi(r)$ must have a repulsive component at short distances together with a long-range attractive component.[39] For simplicity, we assume that ϕ has a *hard core*, $\phi(r) = \infty$ for $r \leq r_0$, and $\phi(r) = 0$ for $r \geq b$:

$$\phi(r) = \begin{cases} \infty & r \leq r_0 \\ < 0 & r_0 < r < b \\ 0 & r \geq b . \end{cases} \tag{4.92}$$

Potentials having the form of Eq. (4.92) are known as *Van Hove potentials*.[40]

The domain Ω_{k+1} is composed of eight identical cubes Ω_k, each containing N_k particles; thus $N_{k+1} = 8N_k$ and $V_{k+1} = 8V_k$. Figure 4.1 doesn't show the third dimension; there are four more cubes not shown. The partition function for domain $k + 1$ is

$$Z_{k+1}(N_{k+1}, \Omega_{k+1}, T)$$
$$= \left(N_{k+1}! \lambda_T^{3N_{k+1}} \right)^{-1} \int \cdots \int_{\Omega_{k+1}} d\mathbf{r}_1 \cdots d\mathbf{r}_{N_{k+1}} \exp\left[-\beta \sum_{1 \leq i < j \leq N_{k+1}} \phi(|\mathbf{r}_i - \mathbf{r}_j|) \right] . \tag{4.93}$$

Imagine letting the cubes in Fig. 4.1 move away from each other far enough that the inter-cube distance exceeds the length b in $\phi(r)$. In that case the cubes are noninteracting, and

$$Z_{k+1}^{\text{noninteracting}}(N_{k+1}, \Omega_{k+1}, T) = [Z_k(N_k, \Omega_k, T)]^8 .$$

Now restore the cubes to their positions. In doing so, attractive interactions develop between them ($\phi(r) < 0$ for $r_0 < r < b$). The sum in Eq. (4.93), $\sum_{i<j} \phi(|\mathbf{r}_i - \mathbf{r}_j|)$, is therefore *more negative* than it would be for noninteracting cubes, implying the inequality

$$Z_{k+1}(8N_k, \Omega_{k+1}, T) > [Z_k(N_k, \Omega_k, T)]^8 . \tag{4.94}$$

Inequality (4.94), combined with Eq. (4.91), implies $\psi_{k+1} > \psi_k$, a key part of the proof. The sequence $\{\psi_k\}_{k=1}^\infty$ is an *increasing* sequence which has a limit if it's bounded above.[41] The sequence will be naturally bounded if, for all configurations of the coordinates $\mathbf{r}_1, \dots, \mathbf{r}_N$,

$$\sum_{1 \leq i < j \leq N} \phi(|\mathbf{r}_i - \mathbf{r}_j|) > -NB , \tag{4.95}$$

---

[39] We have in mind functions with the qualitative form of the Lennard-Jones potential between neutral atoms (see Fig. 6.1) or the effective potential energy function of the classical two-body problem under an attractive force.

[40] After Leon Van Hove, who showed that the thermodynamic limit exists for this type of potential.[43]

[41] We're relying on a piece of mathematics, the *monotone convergence theorem* (see for example, Rudin[44, p47]) that if a sequence is monotone and bounded, it has a limit. We get to infer the existence of a limit without knowing what it is!

where $B > 0$ is independent of N, i.e., if *there is a minimum potential-energy configuration of the system* (a reasonable request). Making use of inequality (4.95), we have an inequality on Z:

$$(N!\lambda_T^{3N}) \times Z(N,V,T) = \int \cdots \int d\mathbf{r}_1 \cdots d\mathbf{r}_N \exp\left[-\beta \sum_{1 \le i < j \le N} \phi(|\mathbf{r}_i - \mathbf{r}_j|)\right] < V^N e^{\beta N B}.$$

(4.96)

Using inequality (4.96) and Stirling's approximation,

$$\psi \equiv \lim_{k \to \infty} \psi_k = \lim_{k \to \infty} \frac{1}{N_k} \ln Z_k = \lim_{N,V \to \infty} N^{-1} \ln Z(N,V,T) < \beta B + 1 + \ln\left(\frac{V}{N\lambda_T^3}\right). \quad (4.97)$$

Because V/N is fixed (thermodynamic limit), the right side of (4.97) supplies an upper bound on ψ (see Exercise 4.33), which exists in the desired form, $\lim_{k \to \infty} \psi_k = \psi(n, T)$.

The thermodynamic limit therefore exists, which we can write in the form of Eq. (4.89),

$$\lim_{N \to \infty} \left(\frac{F}{N}\right) = -kT \lim_{N,V \to \infty} \left(\frac{1}{N} \ln Z(N,V,T)\right) = -kT\psi\left(\frac{V}{N}, T\right) \equiv f(n, T). \quad (4.98)$$

Our proof, however, is based on unphysical assumptions—hard-core potential at short distances and an attractive tail that vanishes at finite distance. It can be made rigorous.[42] Sufficient conditions on $\phi(r)$ for the thermodynamic limit to exist are that: 1) $\phi(r) \ge -a$; 2) $|\phi(r)| \le C/r^{d+\epsilon}$ as $r \to \infty$, where d is the physical dimension of the system; and 3) $\phi(r) \ge C'/r^{d+\epsilon}$ as $r \to 0$. As mentioned on page 84, the thermodynamic limit exists for systems featuring pure Coulomb forces (no hard core).[37] One must invoke quantum mechanics for such systems, otherwise inequality (4.95) will not hold.[42] At least one of the species of particles in the system must obey Fermi statistics, so that the Pauli exclusion principle provides the role of a hard core (see Section 5.5.3).

4.3 CONVEXITY AND STABILITY

Stability of the equilibrium state was treated in Sections 1.12 and 2.4 from the perspective of thermodynamics, that for entropy changes associated with fluctuations, $d^2 S < 0$, which implies that $C_V > 0, \beta_T > 0$. We now show that $C_V > 0, \beta_T > 0$ follow readily from the convexity properties of the free energy function, which can be established using the methods of statistical mechanics. From Eq. (P1.5),

$$C_V = -T\left(\frac{\partial^2 F}{\partial T^2}\right)_V. \quad (4.99)$$

Invoking $C_V > 0$ implies F is a concave function of T, $(\partial^2 F/\partial T^2)_V < 0$. If the convexity properties of $F(N, V, T)$ were independently known, we could use Eq. (4.99) to conclude immediately that $C_V > 0$. The partition function is log-convex, $\partial^2 \ln Z/\partial \beta^2 > 0$ (Eq. (4.17)), implying F is a concave function of T; see Exercise 4.24. By the very structure of statistical mechanics (that the partition function is the Laplace transform of the density of states function, Section 4.1.2.4), we know once and for all that F is a concave function of T.

It's a little more involved to show that $F(N, V, T)$ is a *convex function of N and V*, which has as a consequence that $\beta_T > 0$. Referring to Fig. 4.1, where the domain Ω_{k+1} is constructed from eight cubes Ω_k, consider that we place $N_k^{(1)}$ particles into each of four Ω_k cubes, and $N_k^{(2)}$ particles into the remaining four, so that $N_{k+1} = 4(N_k^{(1)} + N_k^{(2)})$. Keeping $n_1 \equiv N_k^{(1)}/V_k$ and $n_2 \equiv N_k^{(2)}/V_k$ fixed, we have, using the reasoning leading to (4.94), the inequality,

$$Z_{k+1}\left(4(N_k^{(1)} + N_k^{(2)}), \Omega_{k+1}, T\right) > \left[Z_k(N_k^{(1)}, \Omega_k, T)\right]^4 \left[Z_k(N_k^{(2)}, \Omega_k, T)\right]^4. \quad (4.100)$$

---

[42]Why doesn't an electron spiral into the nucleus of the hydrogen atom?

Define (instead of Eq. (4.91)) a quantity related to the free energy per volume,

$$g_k(n_i) \equiv V_k^{-1} \ln Z(N_k^{(i)}, \Omega_k, T) = -\beta (F_{N_k}/V_k) . \qquad (i = 1, 2) \qquad (4.101)$$

From Eq. (4.91), $g_k = (N_k/V_k)\psi_k$. Because $\lim_{k\to\infty} \psi_k$ exists (see Section 4.2), $\lim_{k\to\infty} g_k$ exists, and we can write (compare with Eq. (4.98))

$$\lim_{V\to\infty} \left(\frac{F}{V}\right) = -kTg(n, T) .$$

We now show that $g(n, T)$ is a concave function of n, implying that *the free energy per volume is a convex function of n.*

Using inequality (4.100) and Eq. (4.101), we infer the inequality

$$g_{k+1}\left(\frac{1}{2}(n_k^{(1)} + n_k^{(2)})\right) > \frac{1}{2}\left[g_k(n_k^{(1)}) + g(n_k^{(2)})\right] ,$$

where we've used $V_{k+1} = 8V_k$. In the limit $k \to \infty$ (the sequence $\{g_k\}$ is monotonically increasing), we have

$$g\left(\frac{1}{2}(n_1 + n_2)\right) > \frac{1}{2}[g(n_1) + g(n_2)] . \qquad (4.102)$$

Thus, $g(n)$ is a concave function (see Fig. 4.2), that the value of the function at the midpoint of

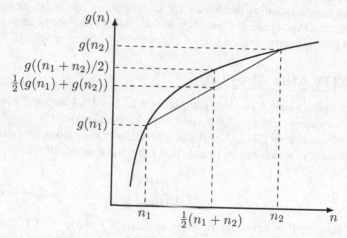

Figure 4.2: Concave function $g(n)$. The chord joining any two points on the graph of the function lies below the graph. More generally, for concave functions $g(\omega n) \geq \omega g(n)$ for $0 \leq \omega \leq 1$.

an arc is not less than the mean of the values of the function at the two end points of the arc. The defining property (4.102) generalizes (because $g(n)$ is bounded and hence continuous[43]) to[45]

$$g(\omega n) \geq \omega g(n) \qquad (0 \leq \omega \leq 1) \qquad (4.103)$$

if we agree to the reasonable request that $g(0) = 0$, which follows by demanding that the partition function of zero particles is unity, $Z(0, V, T) = 1$. The derivative dg/dn of a continuous concave function $g(n)$ is such that, an any point, the right derivative is not greater than the left derivative, with both derivatives non-increasing functions of n.

---

[43] A linear operator bounded on a given interval is continuous on that interval.[16, p47]

By definition, $g(n) = v^{-1}\psi(v)$, where $v \equiv V/N = n^{-1}$. Using (4.103), we have

$$\psi(v) \leq \psi(v'), \qquad \text{when } v \leq v' \tag{4.104}$$

i.e., $\psi(v)$ is a monotonic, nondecreasing function of v. Using Eq. (1.14) and $F = -NkT\psi$,

$$P = -\left(\frac{\partial F}{\partial V}\right)_{T,N} = kTN\left(\frac{\partial \psi}{\partial V}\right)_{T,N} = kT\left(\frac{\partial \psi}{\partial v}\right)_{T,N}.$$

Using the properties of the derivative of concave functions stated above, we conclude that *pressure is a monotonic nonincreasing function of the specific volume*, $(\partial P/\partial v)_{T,N} < 0$—the larger the volume per particle, the lower the pressure (at a fixed temperature). And, because $P(v)$ is monotonic, its derivative exists, and thus so does its inverse, the isothermal compressibility:

$$\beta_T \equiv -\frac{1}{V}\left(\frac{\partial V}{\partial P}\right)_{T,N} = -\frac{1}{V}\frac{1}{(\partial P/\partial V)_{T,N}} = -\frac{1}{v}\frac{1}{(\partial P/\partial v)_{T,N}} > 0. \tag{4.105}$$

Thus, $\beta_T > 0$ occurs naturally in the formalism of statistical mechanics. From Exercises 4.35 and 4.36, we can use $\beta_T > 0$ to establish the convexity properties of $F(N, V, T)$ and $G(N, P, T)$. To summarize:

- $F(N, V, T)$ is convex in N and V, but concave in T;

- $G(N, P, T)$ is convex in N, but concave in P and T.

4.4 QUANTUM ENSEMBLES: PROBABILITY DENSITY OPERATORS

Statistical mechanics requires modification when the quantum nature of matter is taken into account. The derivation of the canonical distribution in Section 4.1.2.7 invoked energy equipartition, Eq. (4.37), that the kinetic energy per particle $E/N = \frac{3}{2}kT$, implying that the specific heat is independent of T. The validity of that assumption breaks down at low temperature—specific heats vanish as $T \to 0$; we have to start over in devising a quantum statistical mechanics.

Consider a gas of diatomic molecules (which we treat in Section 5.4), such as HCl. The Hamiltonian for a *rigid* diatomic molecule consisting of structureless atoms has five additive quadratic degrees of freedom: three for the motion of the center of mass and two associated with angular momentum about the center of mass.[44] Assuming energy equipartition, we expect the specific heat to be $\frac{5}{2}k$, *and this is what is observed* (at room temperature). Shouldn't we expect, however, that the atoms vibrate with respect to each other? In that case there would be two additional quadratic degrees of freedom in the Hamiltonian (the kinetic and potential energies of vibration), implying that the specific heat should be $\frac{7}{2}k$. Experiment shows that as the temperature is raised (but before the molecule dissociates), the value of the specific heat rises towards $\frac{7}{2}k$. At room temperature, therefore, molecular vibrational degrees of freedom are *ineffective*—they don't become excited. Moreover, we know that atoms are not structureless and have internal, electronic degrees of freedom. Their failure to be represented in the value of the specific heat indicates that atomic degrees of freedom are ineffective. *Not all degrees of freedom participate equally in becoming excited* so as to be able to absorb energy at a given temperature (in contradistinction to the equipartition theorem). That specific heats vanish as $T \to 0$ suggests that *all* degrees of freedom become ineffective at low temperature, something beyond the ability of classical mechanics to describe.

The breakdown in our energy-equipartition-based formalism occurs when the mean energies it predicts become sufficiently small (be they related to translation, or rotation, or—whatever). It's here that quantum mechanics must replace classical dynamics, because the motions in question cannot be

---

[44] There are six degrees of freedom of two particles with one equation of constraint, the fixed distance between them.

even approximately described by means of classical dynamics. Under these circumstances, *what we have hitherto taken to be microstates is meaningless.* We can no longer use phase space to represent microstates. Move over Γ-space, here comes Hilbert space.

The *state* of a quantum system is denoted by the Zen-like symbol $|\psi\rangle$, the state vector. The *wavefunction*, $\psi(x)$, such that $|\psi(x)|^2$ is a probability density, is the abstract vector $|\psi\rangle$ expressed in the position basis: $\psi(x) \equiv \langle x|\psi\rangle$, where the angular brackets denote an inner product defined on the Hilbert space of all possible quantum states. The wavefunction is a function of the spatial coordinates of the particles of the system ("half" of phase space). There can also be internal degrees of freedom, so the wavefunction depends on a set of quantum numbers. For electrons, the wavefunction is written $\psi_{\sigma_1,\cdots,\sigma_N}(r_1,\cdots,r_N)$, where $\sigma_i = \pm\frac{1}{2}$. In order not to burden the notation, we won't write these indices.

Observable, i.e., measurable quantities, which in classical mechanics are phase-space functions $A(p,q)$, are in quantum mechanics represented as Hermitian *operators*, $\hat{A}$. We noted in Section 2.4 that fluctuations force us to interpret the measured values of macroscopic quantities as the mean of a large number of measurements. In quantum mechanics we have another source of indeterminacy: Measurements made on a system prepared in a known state occur with a *range* of values, and thus even for a *well defined* quantum system we must seek the mean of a number of measurements. Whereas in classical mechanics the trajectory of a particle subject to the same forces given the same initial conditions is reproducibly the same, in quantum mechanics measurements made on identically prepared systems do not produce the same values—the extra "layer" of indeterminism noted in Section 2.7. The *expectation value* $\overline{A}_\psi$ of a large number of measurements of $\hat{A}$ on systems prepared in state $|\psi\rangle$ is found from

$$\overline{A}_\psi = \int dx \psi^*(x)\hat{A}\psi(x) \equiv \langle\psi|\hat{A}\psi\rangle, \qquad (4.106)$$

when $|\psi\rangle$ is normalized, $\langle\psi|\psi\rangle = 1$. The act of measurement introduces an uncontrollable interaction with the system that modifies what we seek to measure. To see this, make use of the eigenfunctions of $\hat{A}$, a set of functions $\{\phi_n(x)\}_{n=0}^\infty$ such that $\hat{A}\phi_n = \lambda_n\phi_n$, where (because $\hat{A}$ is Hermitian) the eigenvalues λ_n are real numbers and the eigenfunctions are orthonormal, $\langle\phi_n|\phi_m\rangle = \delta_{nm}$. Eigenfunctions of Hermitian operators are also *complete sets*, whereby an arbitrary square-integrable function $\psi(x)$ can be represented by an infinite linear combination,[45]

$$\psi(x) = \sum_{n=0}^\infty a_n\phi_n(x), \qquad (4.107)$$

where the expansion coefficients (*probability amplitudes*) $a_n = \langle\phi_n|\psi\rangle$. Because $|\psi\rangle$ is normalized,

$$\sum_{n=0}^\infty |a_n|^2 = 1. \qquad (4.108)$$

Combining Eqs. (4.106) and (4.107),

$$\langle\psi|\hat{A}\psi\rangle = \sum_{n=0}^\infty |a_n|^2\lambda_n. \qquad (4.109)$$

Thus, the expectation value of $\hat{A}$ on systems in state $|\psi\rangle$ is an average of the eigenvalues of $\hat{A}$, which occur with a probability distribution $|a_n|^2 = |\langle\phi_n|\psi\rangle|^2$ (see Eq. (4.108)), and, as such, is an averaging process related to the probabilistic aspect of wavefunctions. To what extent can the

---

[45] See for example [16, Chapter 2].

quantum states of a system be controlled? As we now show, we must introduce a *second* kind of average over the quantum states of a system—the ensemble average.

Statistical mechanics employs wavefunctions of many particle systems, which often entail *identical* particles, and here we run into a new piece of physics, the *indistinguishability of identical particles*. Wavefunctions of an assembly of identical particles are either symmetric or antisymmetric under a permutation of any two particles (see Appendix D). Under the interchange of particles at positions r_j and r_k (leaving all other particles unchanged), the basic principles of quantum mechanics require that (for identical particles)

$$\psi(\cdots, r_j, \cdots, r_k, \cdots) = \theta \psi(\cdots, r_k, \cdots, r_j, \cdots), \tag{4.110}$$

where

$$\theta = \begin{cases} +1 & \text{for bosons} \\ -1 & \text{for fermions}. \end{cases}$$

In either case, the probability density is invariant under the interchange of particles, $|\psi(\cdots, r_k, \cdots, r_l, \cdots)|^2 = |\psi(\cdots, r_l, \cdots, r_k, \cdots)|^2$, implying that *identical particles can't be labeled*.[46] To construct an N-particle wavefunction, we use as basis functions, products of N single-particle wavefunctions,[47]

$$\psi(r_1, \cdots, r_N) = \sum_{m_1} \cdots \sum_{m_N} c(m_1, \cdots, m_N) \phi_{m_1}(r_1) \phi_{m_2}(r_2) \cdots \phi_{m_N}(r_N), \tag{4.111}$$

where the functions $\{\phi_k\}$ are a complete set (which could be plane-wave functions or atomic wave functions—whatever is most convenient for the problem at hand), and where $c(m_1, \cdots, m_N)$ are expansion coefficients. Equation (4.111) is to many particle wavefunctions what Eq. (4.107) is to single particle wavefunctions. The expansion coefficients $c(m_1, \cdots, m_N)$ "do the work" of providing the proper exchange symmetries (because the basis functions do not), i.e.,

$$c(\cdots, m_k, \cdots, m_l, \cdots) = \theta c(\cdots, m_l, \cdots, m_k, \cdots).$$

Note that if $m_l = m_k = m$, antisymmetry requires that $c(\cdots, m, \cdots, m, \cdots) = 0$, the usual form of the Pauli principle for fermions.

To avoid cumbersome notation, let's express Eq. (4.111) as

$$\psi(x) = \sum_r c_r \phi_r(x), \tag{4.112}$$

so that r is a multi-index symbol. The expectation value of $\hat{A}$ is then (see Exercise 4.37; we're not using a basis of eigenfunctions),

$$\overline{A} = \langle \psi | \hat{A} \psi \rangle = \sum_r \sum_s c_r^* c_s \langle r | \hat{A} s \rangle \equiv \sum_r \sum_s c_r^* c_s A_{rs}, \tag{4.113}$$

where $A_{rs} = \langle r | \hat{A} s \rangle$ are the *matrix elements* of $\hat{A}$, in the basis set used in Eq. (4.111).[48]

---

[46]This conclusion applies to identical particles only. The probability of an electron at point x and a proton at point y is different from the probability of a proton at x and an electron at y. For identical particles, quantum mechanics is incapable of saying *which* particle is in a given state, only *how many* are in a given state. This affects how we count configurations of identical particles. Basing the theory on position-dependent wavefunctions (the way we've presented it), that $\psi = \psi(r_1, \cdots, r_N)$, is a red herring because Eq. (4.111) is redundant. A better way to describe quantum states of identical particles is the *occupation number representation* discussed in Appendix D.

[47]The functions $\{\phi_n(r)\}$ form a complete set on the space of position coordinates for a single particle. *Products* of such functions form a complete set in the union of the domains of the spatial coordinates for each particle (see Section D.5).

[48]The action of an *abstract* operator as a mapping from one vector (element of a vector space) to another, is made concrete when we know its action on the basis vectors of the space. The collection of matrix elements (in a given basis) is the *matrix representation* of an operator. There are well known rules for how the matrix elements transform under a change of basis.

A system known to be in state $|\psi\rangle$ is said to be in a *pure state*. It's almost always the case that we have incomplete knowledge of the microscopic state of the system. And what do we do in the face of incomplete knowledge? I hope you're saying: "Resort to probability." Let γ_i denote the probability that a randomly selected member of the ensemble is in state $|\psi^{(i)}\rangle$, such that

$$\sum_i \gamma_i = 1 . \tag{4.114}$$

The wavefunction $\psi^{(i)}(x)$ of each possible state[49] can be expressed as in Eq. (4.112):

$$\psi^{(i)}(x) = \sum_r c_r^{(i)} \phi_r(x) .$$

The expectation value of $\hat{A}$ in state $|\psi^{(i)}\rangle$ is, from Eq. (4.113),

$$\overline{A}^{(i)} = \sum_r \sum_s c_r^{(i)*} c_s^{(i)} A_{rs} .$$

The ensemble average of a measurable quantity $\hat{A}$ is defined as

$$\langle A \rangle \equiv \sum_i \gamma_i \overline{A}^{(i)} = \sum_i \gamma_i \sum_r \sum_s c_r^{(i)*} c_s^{(i)} A_{rs} . \tag{4.115}$$

There are two averaging processes here: the quantum expectation value $\overline{A}^{(i)}$ of $\hat{A}$ in state $|\psi^{(i)}\rangle$, together with a thermal average over the states $|\psi^{(i)}\rangle$, which occur with probabilities γ_i.

Based on Eq. (4.115), define a matrix, the *density matrix* ρ (which represents the ensemble), with elements (note the order of the indices)

$$\rho_{sr} \equiv \sum_i \gamma_i c_s^{(i)} c_r^{(i)*} . \tag{4.116}$$

Combining Eq. (4.116) with Eq. (4.115),

$$\langle A \rangle = \sum_r \sum_s \rho_{sr} A_{rs} = \sum_s (\rho A)_{ss} = \text{Tr}\, \rho A ,$$

where we've used the rules of matrix multiplication, and where the trace of a matrix is the sum of its diagonal elements: $\text{Tr}\, M \equiv \sum_s M_{ss}$. Prior experience with quantum mechanics indicates that we should seek an *operator* to take the place of the phase-space probability *functions* of classical statistical mechanics. We define the *density operator* (or *statistical operator*) $\hat{\rho}$, which has for matrix elements those of the density matrix ρ_{sr},

$$\rho_{sr} \equiv \int \phi_s^*(x) \hat{\rho} \phi_r(x) \mathrm{d}x = \langle s | \hat{\rho} r \rangle .$$

The ensemble average can then be written

$$\langle A \rangle = \sum_s \sum_r \rho_{sr} A_{rs} = \text{Tr}\, (\rho A) = \text{Tr}\, \hat{\rho}\hat{A} = \text{Tr}\, \hat{A}\hat{\rho} . \tag{4.117}$$

Now, the trace is *independent of the basis in which the matrix elements of an operator are defined*.[16] Thus, we can associate the trace *with the operator itself*—hence the right side of Eq.

---

[49] The basis functions are a complete set; they can be used to represent arbitrary wavefunctions.

(4.117). We've also used the *cyclic invariance of the trace*, $\text{Tr } AB = \text{Tr } BA$; see Exercise 4.38. Note that

$$\text{Tr } \hat{\rho} = \text{Tr}(\rho)_{ss} = \sum_s \sum_i \gamma_i c_s^{(i)} c_s^{(i)*} = \sum_i \gamma_i \sum_s |c_s^{(i)}|^2 = \sum_i \gamma_i = 1, \qquad (4.118)$$

where we've used Eq. (4.108), true for normalized wavefunctions, and Eq. (4.114).

You may wish to refer to Section 2.5.1 to see how ensemble averages were defined for classical systems; Eqs. (4.117) and (4.118) are the *quantum analogs* of Eqs. (2.53) and (2.52). We've replaced the phase-space averages of classical statistical mechanics, $\langle A \rangle = \int_\Gamma A(p,q)\rho(p,q)d\Gamma$, involving probability density *functions* $\rho(p,q)$, with the trace of the density operator $\hat{\rho}$ acting on the operator $\hat{A}$ representing the observable, $\langle A \rangle = \text{Tr } \hat{\rho}\hat{A}$.

New features appear in quantum statistical mechanics that have no classical counterpart. Write Eq. (4.117), separating the diagonal from the off-diagonal matrix elements,

$$\langle A \rangle = \sum_s \rho_{ss} A_{ss} + \sum_s \sum_{r \neq s} \rho_{sr} A_{rs}. \qquad (4.119)$$

From Eq. (4.116), $\rho_{ss} = \sum_i \gamma_i |c_s^{(i)}|^2 \geq 0$, and from Eq. (4.118), $\sum_s \rho_{ss} = 1$: Thus, *the diagonal element ρ_{ss} is the probability of finding the system in state s.* The off-diagonal elements, however, have no definite sign and cannot be interpreted as probabilities.[50] These terms arise from the wave nature of matter and are without classical counterpart; they're linked to the interference effects observed in experiments such as electron diffraction. If in a given basis the density matrix is diagonal ($\rho_{rs} = 0$ for $r \neq s$), the definition of $\langle A \rangle$ in Eq. (4.119) is the same as in the classical theory when restricted to discrete probabilities. While the density matrix might be diagonal in a special basis, the diagonal character of a matrix is not basis-independent.

The density operator can be treated in a way that's independent of the basis used to define its matrix elements. It can be shown[17, Chapter 5] that $\hat{\rho}(t)$, generalized to include time,[51] satisfies a differential equation, the *von Neumann equation*,

$$i\hbar \frac{\partial}{\partial t}\hat{\rho}(t) = \left[\hat{H}, \hat{\rho}(t)\right], \qquad (4.120)$$

where $[\hat{H}, \hat{\rho}]$ denotes the commutator, $[A, B] \equiv AB - BA$. The von Neumann equation is the analog of the Liouville equation, Eq. (2.55); it's the fundamental equation of quantum statistical mechanics. Quantum statistical mechanics may be said to be the study of the solutions of Eq. (4.120).

Ensembles of equilibrium systems are *stationary*, $\partial\hat{\rho}/\partial t = 0$, implying that $\hat{\rho}$ commutes with the Hamiltonian,[52] $[\hat{H}, \hat{\rho}] = 0$. Commuting operators have a common set of eigenfunctions.[53] Using the eigenfunctions of $\hat{H}$ as a basis (the *energy representation*), the density matrix is *diagonal* (see Exercise 4.40),

$$\rho_{rs} = \rho_r \delta_{rs}. \qquad (4.121)$$

In what follows, we work with the density matrix in the energy representation.

---

[50]The density matrix as a whole is *positive definite*; see Exercise 4.39a.
[51]Equation (4.120) can be derived in the *Heisenberg representation* of quantum mechanics. While the von Neumann equation isn't difficult to derive, it would take us too far afield, and it's not used anywhere else in this book.
[52]It also implies that the Hamiltonian is independent of time.
[53]See most any text on quantum mechanics.

4.4.1 Microcanonical ensemble

Consider an ensemble of isolated systems having a fixed number of particles in a fixed volume. The density matrix is diagonal, as in Eq. (4.121), with

$$
\rho_r = \begin{cases} \dfrac{1}{\Omega} & \text{if } E = E_r \\ 0 & \text{otherwise}, \end{cases}
$$

where Ω is the number of states available to the system at energy E_r.

4.4.2 Canonical ensemble

Consider an ensemble of systems with well defined values of N, V, T. The probability that a randomly chosen system has energy E_n is proportional to the Boltzmann factor. The density matrix is diagonal as in Eq. (4.121) with

$$
\rho_n = \frac{1}{Z(\beta)} g_n e^{-\beta E_n} ,
\tag{4.122}
$$

where $Z(\beta)$ is defined in terms of the discrete energies

$$
Z(\beta) = \sum_n g_n e^{-\beta E_n} ,
\tag{4.123}
$$

where g_n is the *degeneracy* of the nth energy level, the number of linearly independent eigenfunctions each having eigenvalue E_n. The density operator in the canonical ensemble may be evaluated using Eq. (P4.14),

$$
\hat{\rho} = \sum_n \rho_n |\phi_n\rangle\langle\phi_n| = \frac{1}{Z} \sum_n g_n e^{-\beta E_n} |\phi_n\rangle\langle\phi_n| = \frac{1}{Z} e^{-\beta\hat{H}} \sum_n |\phi_n\rangle\langle\phi_n| = \frac{1}{Z(\beta)} e^{-\beta\hat{H}} ,
\tag{4.124}
$$

where we've used that $e^{-\beta\hat{H}}\phi_n = e^{-\beta E_n}\phi_n$, Eq. (P4.16), and the completeness relation, Eq. (P4.15). The ensemble average of an observable quantity A is therefore

$$
\langle A \rangle = \text{Tr}\left(\hat{\rho}\hat{A}\right) = \frac{1}{Z} \text{Tr} \, e^{-\beta\hat{H}} \hat{A} = \frac{1}{Z(\beta)} \sum_n g_n e^{-\beta E_n} \langle\phi_n|\hat{A}\phi_n\rangle .
\tag{4.125}
$$

4.4.3 Grand canonical ensemble

The density operator in this case must commute with the Hamiltonian and the *number operator* $\hat{N}$ whose eigenvalues are $0, 1, 2, \ldots$ (see Appendix D). The density operator is

$$
\hat{\rho} = \frac{1}{Z_G} e^{-\beta(\hat{H}-\mu\hat{N})} ,
\tag{4.126}
$$

where

$$
Z_G = \text{Tr} \, e^{-\beta(\hat{H}-\mu\hat{N})} = \sum_{N=0}^{\infty} e^{\beta\mu N} Z_N(\beta)
\tag{4.127}
$$

is the grand partition function, with Z_N the canonical partition function for an N-particle system. Ensemble averages are then given by

$$
\langle A \rangle = \frac{1}{Z_G} \sum_{N=0}^{\infty} e^{\beta\mu N} Z_N(\beta)\langle A\rangle_N ,
\tag{4.128}
$$

where $\langle A \rangle_N$ denotes the canonical ensemble average for an N-particle system.

SUMMARY

> This fundamental law [the Boltzmann probability distribution] is the summit of statis-
> tical mechanics, and the entire subject is either the slide-down from the summit, as the
> principle is applied to various cases, or the climb-up to where the fundamental law is
> derived and the concepts of thermal equilibrium and temperature T clarified.
> —R.P. Feynman[46, p1]

In this chapter, we developed the ensemble formalism of statistical mechanics (a huge undertak-
ing, by the way; Chapter 4 is foundational and is no doubt the most difficult in this book). In doing
so, it concludes the first part of this book, Chapters 1–4. The aforementioned quote by R.P. Feynman
aptly pertains to the organization of this book: Part I covers theoretical foundations, culminating in
the derivation of the Boltzmann distribution, with Part II devoted to applications.

- The phase-space probability density for isolated classical systems (fixed N, V, E) is

$$\rho(p,q) = \frac{1}{\Omega(E)}\delta\left(E - H(p,q)\right),$$

where $\Omega(E)$ is the density of states function for the system (which depends on the external
variables V and N). The density of states function plays a *fundamental* role in the theory.
There is a tendency in teaching statistical mechanics to impart the impression that the density
of states function is a *convenient* quantity that allows us to convert integrations over Γ-space
(or k-space) to integrations over the energy. While true, we've seen that $\Omega(E)$ makes an
early entrance into the foundations of the subject. The density of states function isn't simply
convenient, it's essential. Expressions for the density of states of free, nonrelativistic particles
and photons were derived in Section 2.1.5. The density of states of free, relativistic, massive
particles is developed in Section 5.7.1.

- The phase-space probability density for closed classical systems (fixed N, V, T) is

$$\rho(p,q) = \frac{1}{N!Z_{\text{can}}}e^{-\beta H(p,q)} \implies \rho(E) = \frac{1}{N!Z_{\text{can}}}e^{-\beta E}\Omega(E),$$

where $\beta = 1/(kT)$ and the normalizing factor, the partition function, is given by

$$Z_{\text{can}}(N,V,T) = \frac{1}{N!}\int e^{-\beta H(p,q)}d\Gamma = \frac{1}{N!}\int e^{-\beta E}\Omega(E)dE,$$

where the Hamiltonian is a function of the external parameters V and N. The presence
of $N!$ is required to prevent overcounting configurations of identical particles (which are
invariant under permutations of the particles). Except for calculations of entropy ($S =
-k\beta^2\partial\left(\beta^{-1}\ln Z_{\text{can}}\right)/\partial\beta$, Eq. (4.56)), the factor of $N!$ is unnecessary in calculations involv-
ing logarithmic derivatives of the partition function, such as for the average internal energy.

The partition function is the Laplace transform of the density of states function,

$$Z(\beta) = \int_0^\infty e^{-\beta E}\Omega(E)dE,$$

which shows why it serves as the normalization factor on the probability density. The Boltzmann factor $e^{-\beta E}$ is the extent to which the energy states of the system are populated at
temperature T, and $\Omega(E)dE$ is the number of energy states between E and $E + dE$. Thus
$Z = \int_0^\infty e^{-\beta E}\Omega(E)dE$ is *the total number of possible states accessible to the system at temperature T*. The probability $P(E)dE = e^{-\beta E}\Omega(E)dE/Z$ is the ratio of two numbers: the
number of states available to the system at energy E when it has temperature T, to the total
number of states available to the system at temperature T.

- For open classical systems (fixed μ, V, T), we can't write down a simple formula for the phase-space probability density, because ensemble averages involve integrations over phase spaces of varying dimensionality. The ensemble average of a phase-space function $A(p, q, N)$ (that depends on the number of particles) is, by combining Eqs. (4.73) and (4.75),

$$\langle A \rangle = \frac{1}{Z_G} \sum_{N=0}^{\infty} \frac{e^{\beta N \mu}}{N!} \int_{\Gamma_N} A(p, q, N) e^{-\beta H_N(p,q)} \mathrm{d}\Gamma_N = \frac{1}{Z_G} \sum_{N=0}^{\infty} e^{\beta \mu N} Z_{\text{can},N} \langle A \rangle_N \,,$$

where the grand partition function Z_G is, from Eq (4.77),

$$Z_G(\mu, V, T) = \sum_{N=0}^{\infty} \frac{1}{N!} e^{\beta \mu N} \int_{\Gamma_N} e^{-\beta H_N(p,q)} \mathrm{d}\Gamma_N = \sum_{N=0}^{\infty} e^{\beta \mu N} Z_{\text{can}}(N, V, T) \,.$$

- Ensembles are specified by *given* thermodynamic variables, whereas other thermodynamic quantities are obtained from derivatives of Z. In the canonical ensemble there is a connection between Z and the Helmholtz energy $F(N, V, T)$, from Eq. (4.57),

$$Z_{\text{can}} = e^{-\beta F} \,.$$

In the grand canonical ensemble the partition function is related to the grand potential $\Phi(\mu, V, T)$, from Eq. (4.76),

$$Z_G = e^{-\beta \Phi} \,.$$

The thermodynamic variables associated with the ensembles are summarized in Table 4.1. Each ensemble involves seven thermodynamic variables, those connected by the Euler relation, Eq. (1.53). The normalizing factor in the microcanonical ensemble, $\Omega(E)$, is related to the entropy through the Boltzmann relation, $\Omega = \exp(S/k)$.

Table 4.1: Statistical ensembles and their associated thermodynamic variables

Ensemble	Fundamental variables	Thermodynamic potential	Implied variables
Microcanonical	U, V, N	$S(U, V, N)$	T, P, μ
Canonical	T, V, N	$F(T, V, N)$	S, P, μ, U
Grand canonical	T, V, μ	$\Phi(T, V, \mu)$	S, P, N, U

- From first derivatives of Z one can calculate various ensemble averages. In the canonical ensemble (variable energy), the average energy is, from Eq. (4.40),

$$U = \langle H \rangle = - \left(\frac{\partial}{\partial \beta} \ln Z_{\text{can}} \right)_V \,.$$

The values of (μ, P, S) can be calculated using Eq. (4.58) and knowledge of the Hamiltonian. In the grand canonical ensemble (varying energy and particle number), the average number of particles is, from Eq. (4.78),

$$\langle N \rangle = \left(\frac{\partial}{\partial (\beta \mu)} \ln Z_G \right)_{T,V} \,.$$

One can calculate P and S using the other expressions in Eq. (4.78). The average energy is obtained from Eq. (4.82),

$$U = \langle H \rangle = - \left(\frac{\partial \ln Z_G}{\partial \beta} \right)_{z,V},$$

where $z \equiv e^{\beta \mu}$ is the fugacity. In the canonical ensemble, the chemical potential $\mu = \mu(N, V, T)$ is a function of particle number (from Eq. (4.45)), whereas in the grand canonical ensemble $N = N(\mu, V, T)$ is a function of the chemical potential (from Eq. (4.78)).

- From second derivatives of Z, one can calculate the values of fluctuations. In the canonical ensemble, from Eqs. (4.61) and (4.62), we have that the heat capacity is a measure of energy fluctuations:

$$\frac{\partial^2 \ln Z_{\text{can}}}{\partial \beta^2} = \langle H^2 \rangle - \langle H \rangle^2 = \left\langle (H - \langle H \rangle)^2 \right\rangle = kT^2 C_V .$$

In the grand canonical ensemble, we have that the isothermal compressibility β_T is a measure of fluctuations in particle number. From Eq. (4.81),

$$\langle N^2 \rangle - \langle N \rangle^2 = \frac{\partial^2 \ln Z_G}{\partial (\beta \mu)^2} = \frac{kTN^2}{V} \beta_T .$$

Energy fluctuations in the grand canonical ensemble are, from Eq. (4.87),

$$\langle H^2 \rangle - \langle H \rangle^2 = kT^2 C_V + \left\langle (N - \langle N \rangle)^2 \right\rangle \left(\frac{\partial U}{\partial N} \right)_{T,V}^2 .$$

The relative root-mean-square deviations

$$\frac{\sqrt{\left\langle (H - \langle H \rangle)^2 \right\rangle}}{\langle H \rangle} \qquad \frac{\sqrt{\left\langle (N - \langle N \rangle)^2 \right\rangle}}{\langle N \rangle}$$

are of order $N^{-1/2} \approx 10^{-12}$ for Avogadro's number.

- Thermodynamic relations derived in the different ensembles are not in general the same for finite-size systems. They become identical for macroscopic systems which we can model with the limit $N \to \infty$. The limiting process $N \to \infty$, $V \to \infty$, but with N/V fixed, is known as the thermodynamic limit. We presented an outline of the proof of the existence of the thermodynamic limit. A nodding acquaintance with the thermodynamic limit will be required in the subsequent developments in this book.

- The thermodynamic theory of the stability of the equilibrium state (discussed in Sections 1.12 and 2.4) shows that $C_V > 0$ and $\beta_T > 0$ from an analysis of entropy changes produced by fluctuations. The same conclusions follow in statistical mechanics from the convexity properties of the Helmholtz energy, $F(N, V, T)$; see Section 4.3.

- In quantum statistical mechanics (Section 4.4), we transition from phase space descriptions to a picture that focuses on the energy levels of systems, or, more generally, on the eigenvalues of Hermitian operators that represent physical observables. We also transition away from probability density *functions* (in classical statistical mechanics) to a probability density *operator*, $\hat{\rho}$, with ensemble averages of an observable A given by Eq. (4.117), $\langle A \rangle = \text{Tr} \, \hat{\rho} \hat{A}$, where, to make use of the trace operation, we must declare a basis in which to evaluate the matrix elements of the operators $\hat{\rho}$ and $\hat{A}$. Fortunately, the trace is independent of basis. In a basis of the eigenstates of the Hamiltonian, the density matrix is diagonal. In that case, averages in the canonical and grand canonical ensemble are given by Eqs. (4.125) and (4.128).

- Having now developed the basic formalism of statistical mechanics, it's instructive to return to the beginning of this chapter and consider where it comes from. Equation (4.1) is entirely from the theory of probability, with little input from physics, other than the Hamiltonian defines a constant-energy hypersurface in Γ-space, $H(p, q) = E$. The Hamiltonian is in fact all the physics it *can* have. The equilibrium probability distribution is stationary, $\partial \rho / \partial t = 0$, implying from Liouville's theorem that ρ is a function of H. The Hamiltonian is the "place" where microscopic information about the components of the system enters the theory. Most results in the theory of probability are based on the assumption of equal probabilities of elementary events, implying that if we draw on the methods of probability, *we've signed off on that assumption*. Lacking precise information to the contrary, we have no basis on which *not* to assume that microscopic events occur with equal likelihood. Ensembles are our answer to the problem that we as humans lack the ability to control the microscopic events underlying macroscopic systems; ensembles are abstract collections of identical copies of a system that embody all possible initial conditions consistent with known macroscopic information. Some books take the microcanonical distribution function, Eq. (4.3), as a *postulate* because it effectively embodies the ergodic hypothesis (phase trajectories uniformly cover a constant-energy surface). We argued in Chapter 2 that the ergodic hypothesis is not necessary for the construction of statistical mechanics—we're free to adopt whatever assumptions we require so long as they're sufficient to guarantee that the laws of thermodynamics emerge from the statistical theory. There are books that take the canonical distribution function as a postulate as well, Eq. (4.32). The derivation of Eq. (4.13) relied on the *geometry* of Γ-space and requires only that the density of states function exists; it has almost no physics in it other than the assumption of weakly interacting subsystems (energy of interaction between subsystems negligible in relation to bulk energies, a picture justified by the existence of the thermodynamic limit). The subsequent derivation from Eq. (4.13) to Eq. (4.32) invoked the central limit theorem which relies on the large number of components systems have; therein lies the universality of statistical mechanics—it applies to any macroscopic system. As an alternative (acknowledging that a fundamental law of nature—the Boltzmann distribution—cannot be derived from something more fundamental), we derived the canonical distribution starting from the microcanonical distribution of an isolated compound system by eliminating the environmental degrees of freedom (in reducing a joint probability density of the compound system to the probability density of a single system that interacts with its environment). In that approach we encountered the assumption of energy equipartition, implying the temperature-independence of the heat capacity—true of many materials—but which ultimately can't be justified at low temperature, signaling the need for a theory based on quantum mechanics. Statistical mechanics inherits quite a bit of structure from the theory of thermodynamics (which underscores why it's really helpful to understand thermodynamics in its own right). Once it's been shown that statistical mechanics is consistent with the second law of thermodynamics, Eq. (4.45), we're free to make use of relations derived in thermodynamics, knowing that points of contact occur between the partition function and the relevant thermodynamic potential associated with the types of systems comprising the ensemble. Out of this melange of ideas, can we pinpoint the essential ideas that the edifice of statistical mechanics rests on—equal a priori probabilities, weakly interacting systems, energy equipartition? The thermodynamic limit, on which extensivity is based, requires a system to possess a minimum-potential-energy-configuration. Perhaps the essential idea is simply *randomness*—that which is not under our control we can only assume occurs randomly. And at that point, the theory of probability takes over.

EXERCISES

4.1 Derive Eq. (2.19).

4.2 Consider a system of free particles confined to two dimensions. There are several ways, for example, to confine electrons in semiconductor devices to two-dimensional regions (because motion in the third dimension is precluded by a confining potential). Repeat the calculations from Eqs. (2.15) to (2.17) for particles confined to an area A to derive an expression for the density of single-particle states. A: $g_d Am/(2\pi\hbar^2)$. Hint: What is the size of the unit cell for a two-dimensional k-space?

4.3 Derive the density of states of a free relativistic particle of rest mass m in three spatial dimensions of volume V. Show that

$$g(T) = g_d \frac{V}{2\pi^2(\hbar c)^3} \sqrt{T^2 + 2Tmc^2}\,(T + mc^2)\ ,$$

where T indicates the *kinetic energy* of the particle, with $E = \sqrt{(pc)^2 + (mc^2)^2} = T + mc^2$. If this seems like a difficult problem (and it kind of *is* a difficult problem), look ahead to Section 5.7.

4.4 Verify Eq. (4.8). It suffices to show that

$$\frac{E^{(n+m)/2-1}}{\Gamma(\frac{n+m}{2})} = \frac{1}{\Gamma(\frac{n}{2})\Gamma(\frac{m}{2})} \int_0^E y^{n/2-1}(E-y)^{m/2-1}dy\ .$$

Hint: Use the beta function defined in Appendix B.

4.5 Show that the nth moment of $\Omega(x)$ can be found by taking derivatives of the partition function,

$$\langle x^n \rangle = \frac{1}{Z(\alpha)}(-1)^n \frac{d^n}{d\alpha^n} Z(\alpha)\ .$$

The partition function can then be referred to as the generating function for the density of states function; see Section 3.5.3.

4.6 Derive Eq. (4.16) starting from Eq. (4.9). Keep in mind that $\Omega(x) = 0$ for $x \le 0$.

4.7 Derive Eq. (4.17).

4.8 Derive Eqs. (4.20) and (4.21) using Eq. (4.19).

4.9 Derive the results in Eq. (4.24) by substituting Eq. (4.23) into Eqs. (4.20) and (4.21).

4.10 Consider the function $f(x) = e^{\alpha x}e^{-\beta x^2}$ $(\alpha, \beta > 0)$, the product of an exponential function and a Gaussian function centered about the origin. Find the location $x = x^*$ where $f(x)$ has a maximum (A: $x^* = \alpha/2\beta$). Show that $f(x)$ is a Gaussian centered about a new mean:

$$f(x) = e^{\alpha^2/4\beta} e^{-\beta(x-x^*)^2}\ .$$

Hint: Complete the square. Thus, multiplying a Gaussian by an exponential function results in another Gaussian that's centered about a new location and that's been scaled in magnitude, but which retains the width of the original Gaussian, $\approx 1/\sqrt{\beta}$.

4.11 Derive Eq. (4.35). Hint: Use Eq. (B.14) and the properties of the gamma function.

4.12 Work through the steps in Eq. (4.36).

4.13 What's the Hamiltonian for a system of N structureless particles of mass m contained in a rigid container of volume V? Hint: This *is* a trick question. Ordinarily one would say that H is the kinetic energy, $H = (1/(2m)) \sum_{i=1}^{3N} p_i^2$. To take into account the finite volume, one should add to H a function, call it $\theta(q)$, such that $\theta(q) = 0$ or ∞ depending on whether the values of q_k correspond to points inside or outside the container; $\theta(q)$ is formally the potential associated with an external force that confines the particles to a finite volume.

4.14 Show that

$$C_V = k\beta^2 \left(\frac{\partial^2}{\partial \beta^2} \ln Z \right)_V . \tag{P4.1}$$

4.15 Derive Eqs. (4.41) and (4.42).

4.16 Show the identities (which hold for any smooth function Z, i.e., these identities are obtained using calculus)

a. $\beta \dfrac{\partial}{\partial \beta} \left(\beta^{-1} \ln Z \right) = \dfrac{\partial \ln Z}{\partial \beta} - \dfrac{1}{\beta} \ln Z$.

b. $\dfrac{\partial}{\partial \beta} \left(\beta^2 \dfrac{\partial}{\partial \beta} \left(\beta^{-1} \ln Z \right) \right) = \beta \dfrac{\partial^2 \ln Z}{\partial \beta^2}$.

4.17 Fill in the steps leading to Eq. (4.45).

4.18 For the canonical ensemble with $\rho = e^{-\beta H}/Z$ and $Z = \int e^{-\beta H} d\Gamma$:

a. Show that variations in ρ, $\delta\rho$, are given by (for the variation symbol δ defined in Eq. (4.44)),

$$\delta\rho = \underbrace{-\rho(H - \langle H \rangle)\delta\beta}_{\text{thermal variations}} - \underbrace{\beta\rho \left(\delta_m H - \langle \delta_m H \rangle \right)}_{\text{mechanical variations}},$$

where δ_m refers to mechanical variations in external parameters, $\delta_m H \equiv \sum_i (\partial H/\partial X_i) \, \delta X_i$.

b. Show that $\int \delta\rho \, d\Gamma = 0$, i.e., variations of the normalization condition $\int \rho \, d\Gamma = 1$ vanish.

c. Equation (4.43) defines the heat transferred to a system as the difference between variations in internal energy (produced by any means) and adiabatic work. Show that

$$\delta Q = \int H \delta\rho \, d\Gamma = - \left(\langle H^2 \rangle - \langle H \rangle^2 \right) \delta\beta - \beta \left(\langle H \delta_m H \rangle - \langle H \rangle \langle \delta_m H \rangle \right) .$$

Note that δQ is not an ensemble average in the usual sense.

4.19 Derive Eq. (4.46) from Eq. (4.45), taking into account that the thermodynamic definition of entropy is $(dQ)_{\text{rev}}/T = dS$. Hint: $\partial/\partial\beta = (\partial T/\partial\beta)\partial/\partial T = (-1/(k\beta^2))\partial/\partial T$.

4.20 Verify Eq. (4.47) for an ideal gas contained in volume V in equilibrium with its environment at absolute temperature T.

4.21 Is Eq. (4.55) satisfied by partition function of the ideal gas?

4.22 Consider a gas of relativistic particles, for which the density of states for a single relativistic particle was derived in Exercise 4.4. What is the equation of state of such a system? A: $PV = NkT$. Hint: The partition function for N free particles is $Z = (Z_1)^N$, where Z_1 is the partition function of a single particle. Show that Z_1 has the form $Z_1 = V f(T)$, where T is the temperature. Using Eq. (4.58), you don't have to know the explicit form of $f(T)$.

4.23 Verify all parts of Eq. (4.61).

4.24 Show in the canonical ensemble that, as an exercise in the chain rule,

$$\frac{\partial^2 \ln Z}{\partial \beta^2} = -kT^3 \frac{\partial^2 F}{\partial T^2} \, .$$

Argue therefore that

$$\frac{\partial^2 F}{\partial T^2} < 0 \, ,$$

and hence that $F(N,V,T)$ is a concave function of T. Hint: $\left\langle (H - \langle H \rangle)^2 \right\rangle \geq 0$ from Eq. (4.61), or see Eq. (4.17).

4.25 Consider a system that interacts with an external magnetic field, H. Which ensemble should we use to predict its thermodynamic behavior? What are the relevant variables for such systems? Start with the first law of thermodynamics. The work done on a system in reversibly changing its magnetization by dM is $dW = \mu_0 H dM = B dM$. The first law for such a system is

$$dU = TdS - PdV + \mu_0 H dM + \mu dN \, .$$

Assume in experiments on such systems that PdV work is negligible. The first law simplifies: $dU = TdS + BdM + \mu dN$. Thus, $U = U(S, M, N)$. Define the magnetic Gibbs energy,

$$G_m \equiv U - TS - BM = F - BM \, . \tag{P4.2}$$

Form its differential,

$$dG_m = -SdT - MdB + \mu dN \, , \tag{P4.3}$$

and therefore $G_m = G_m(T, B, N)$. Work in the T, B, N ensemble.

a. Show that Eq. (P4.3) implies the thermodynamic identities,

$$S = -\left(\frac{\partial G_m}{\partial T}\right)_{B,N} \qquad M = -\left(\frac{\partial G_m}{\partial B}\right)_{T,N} \qquad \mu = \left(\frac{\partial G_m}{\partial N}\right)_{T,B} , \tag{P4.4}$$

and the Maxwell relation

$$\left(\frac{\partial M}{\partial T}\right)_{B,N} = \left(\frac{\partial S}{\partial B}\right)_{T,N} \, . \tag{P4.5}$$

b. The magnetization M is the sum of the individual magnetic dipole moments of the particles of the system, $M \equiv \sum_{i=1}^{N} m_i$, with the energy of interaction with the field given by $E[m_1, m_2, \cdots, m_N] = -B \sum_{i=1}^{N} m_i$. (The notation $E[m_1, m_2, \cdots, m_N]$ indicates that E depends on the configuration of all the magnetic moments.) The partition function is then given by (we're ignoring translational kinetic energy of the dipoles—they could be located on the sites of a crystalline lattice):

$$Z(T, B, N) = \sum_{\{m_i\}} \exp\left(-\beta E[m_1, m_2, \cdots, m_N]\right) = \sum_{\{m_i\}} \exp\left(\beta B \sum_{i=1}^{N} m_i\right) ,$$

where $\sum_{\{m_i\}}$ indicates a sum over all configurations of the dipoles. (We're being a bit obscure here; we get more precise in Section 5.2.) Show that

$$G_m(T, B, N) = -kT \ln Z(T, B, N) - BM(T, B, N) ,$$

where

$$M(T, B, N) = \frac{\partial}{\partial(\beta B)} \ln Z(T, B, N) .$$

Hint: Use Eqs. (4.40) and (4.46), and make use of $\partial/\partial T = -(\beta/T)\partial/\partial\beta$. Thus we have

$$Z(T, B, N) = \exp\left(-\beta[G_m(T, B, N) + BM(T, B, N)]\right) \equiv e^{-\beta F(T,B,N)} ,$$

where $G_m + BM$ effectively defines a Helmholtz energy as a function of (T, B, N).

c. Referring to the magnetic susceptibility $\chi \equiv (\partial M/\partial B)_T$, Eq. (1.50), show that

$$\chi = \beta\left[\langle M^2\rangle - \langle M\rangle^2\right] = \beta\langle(M - \langle M\rangle)^2\rangle . \tag{P4.6}$$

The susceptibility is thus a measure of fluctuations in the magnetization, just as C_V is a measure of energy fluctuations, Eq. (4.62), and the compressibility β_T is a measure of particle fluctuations, Eq. (4.81).

4.26 Derive Eq. (4.70). Start with Eq. (4.67) and re-express it as a relation for dS. Note that $S = S(U, X_i, N_j)$, that S is naturally a function of the extensive variables (U, X_i, N_j). Invoke extensivity, $S(\lambda U, \lambda X_i, \lambda N_j) = \lambda S(U, X_i, N_j)$, and reach for Euler's theorem.

4.27 Show, by combining Eqs. (4.71) and (4.69), that we have the most general form of the Gibbs-Duhem equation:

$$\sum_j N_j d\mu_j = -SdT - \sum_i X_i dY_i . \tag{P4.7}$$

Compare with Eq. (P1.1) for a single-component system with V the only external parameter.

4.28 From Eq. (4.76), $Z_G = e^{-\beta\Phi}$, and for a single component system with V the only relevant external parameter, $\Phi = -PV$. Thus, in this case, $Z_G = e^{\beta PV}$. For the ideal gas, therefore, $Z_G = e^{\langle N\rangle}$. Show this is consistent with the expression derived in Eq. (4.79). In the grand canonical ensemble, N is not a known quantity, but rather $N = N(\mu, T, V)$.

4.29 Derive Eq. (4.77). Show that for Eq. (4.75) to satisfy Eq. (4.74), Eq. (4.77) ensues.

4.30 a. Show that

$$\frac{\partial}{\partial\beta} \ln Z_G\bigg|_\mu = \mu\langle N\rangle - U ,$$

where $U \equiv \sum_{N=0}^\infty \int_{\Gamma_N} H_N \rho_N d\Gamma_N$.

b. Show, for a single-component system having only one relevant external parameter V, that $\ln Z_G = V (\partial \ln Z_G/\partial V)$, implying that $\ln Z_G$ varies linearly with V.

c. Show for the ideal gas that

$$\frac{\partial}{\partial\beta} \ln Z_G\bigg|_\mu = \langle N\rangle \left(\mu - \frac{3}{2}kT\right) .$$

Use Eq. (4.79). The internal energy of the ideal gas in the grand canonical ensemble is therefore given by the expression $U = \frac{3}{2}\langle N\rangle kT$.

d. Show that the second and third relations in Eq. (4.78) lead to the known expressions for the equation of state and entropy of the ideal gas.

4.31 a. Show, for $Z_G = Z_G(\mu, V, T)$, that $\left.\dfrac{\partial^2 \ln Z_G}{\partial \mu^2}\right|_{T,V} = \left.\dfrac{1}{Z_G}\dfrac{\partial^2 Z_G}{\partial \mu^2}\right|_{T,V} - \left(\left.\dfrac{\partial \ln Z_G}{\partial \mu}\right|_{T,V}\right)^2.$

There is no physics in this formula, just calculus.

b. Show that $\left.\dfrac{1}{Z_G}\dfrac{\partial^2 Z_G}{\partial \mu^2}\right|_{T,V} = \beta^2 \langle N^2 \rangle$. Hint: Start from Eq. (4.77) and recognize that

$$\langle N^2 \rangle = \sum_{N=0}^{\infty} N^2 P_N = \frac{1}{Z_G} \sum_{N=0}^{\infty} N^2 e^{\beta \mu N} Z_{\text{can}}(N, V, T).$$

c. Using Eq. (4.78), conclude that

$$\langle N^2 \rangle - \langle N \rangle^2 = (kT)^2 \left.\frac{\partial^2 \ln Z_G}{\partial \mu^2}\right|_{T,V} = kT \left.\frac{\partial \langle N \rangle}{\partial \mu}\right|_{T,V}. \tag{P4.8}$$

4.32 Guided exercise to establish the identity in Eq. (4.81),

$$\left(\frac{\partial N}{\partial \mu}\right)_{T,V} = \frac{N^2}{V} \beta_T. \tag{P4.9}$$

We do this in steps.

a. The fundamental variables in the grand canonical ensemble are μ, T, V, from which the average number of particles can be calculated using Eq. (4.78). For simplicity, we use N here to denote $\langle N \rangle$. Thus, $N = N(\mu, T, V)$, and implicitly, $V = V(\mu, T, N)$. We seek to develop an identity for $(\partial N/\partial \mu)_{T,V}$ (look ahead to Eq. (P4.10)). You might want (right now) to consult Table 1.3 and Eq. (P1.8) to see if this derivative occurs in a Maxwell relation. It doesn't, but it's always worth a try! From calculus

$$dN = \left(\frac{\partial N}{\partial \mu}\right)_{T,V} d\mu + \left(\frac{\partial N}{\partial T}\right)_{\mu,V} dT + \left(\frac{\partial N}{\partial V}\right)_{\mu,T} dV.$$

We can also expand dV in terms of differentials of μ, T, N. Thus,

$$dN = \left(\frac{\partial N}{\partial \mu}\right)_{T,V} d\mu + \left(\frac{\partial N}{\partial T}\right)_{\mu,V} dT +$$

$$\left(\frac{\partial N}{\partial V}\right)_{\mu,T} \left[\left(\frac{\partial V}{\partial \mu}\right)_{T,N} d\mu + \left(\frac{\partial V}{\partial T}\right)_{\mu,N} dT + \left(\frac{\partial V}{\partial N}\right)_{\mu,T} dN\right].$$

Using Eq. (1.30), we find

$$0 = \left[\left(\frac{\partial N}{\partial \mu}\right)_{T,V} + \left(\frac{\partial N}{\partial V}\right)_{\mu,T}\left(\frac{\partial V}{\partial \mu}\right)_{T,N}\right] d\mu + \left[\left(\frac{\partial N}{\partial T}\right)_{\mu,V} + \left(\frac{\partial N}{\partial V}\right)_{\mu,T}\left(\frac{\partial V}{\partial T}\right)_{\mu,N}\right]$$

implying

$$\left(\frac{\partial N}{\partial T}\right)_{\mu,V} = -\left(\frac{\partial N}{\partial V}\right)_{\mu,T}\left(\frac{\partial V}{\partial T}\right)_{\mu,N}$$

$$\left(\frac{\partial N}{\partial \mu}\right)_{T,V} = -\left(\frac{\partial N}{\partial V}\right)_{\mu,T}\left(\frac{\partial V}{\partial \mu}\right)_{T,N}. \tag{P4.10}$$

These expressions generalize the cyclic relation Eq. (1.31) to four variables, N, μ, T, V.

b. What do we know about $(\partial V/\partial \mu)_{T,N}$? This derivative is part of a Maxwell relation (see Table 1.3), but I don't see how it helps us.[54] We can get at this derivative through the

---

[54]Maxwell relations can work wonders in streamlining calculations, but not always.

Gibbs-Duhem equation[55] for a single-component system where V is the only relevant external variable, Eq. (P1.1), $N d\mu = V dP - S dT$. We have in the grand canonical ensemble that $N = N(\mu, V, T)$ (implying $\mu = \mu(N, V, T)$) and $P = P(\mu, V, T)$, implying that $P = P(N, V, T)$. Expand P assuming it's a function of N, V, T:

$$N d\mu = V \left(\frac{\partial P}{\partial N}\right)_{T,V} dN + V \left(\frac{\partial P}{\partial V}\right)_{N,T} dV + \left[V \left(\frac{\partial P}{\partial T}\right)_{N,V} - S\right] dT .$$

Holding T, N fixed,

$$\left(\frac{\partial \mu}{\partial V}\right)_{T,N} = \frac{V}{N} \left(\frac{\partial P}{\partial V}\right)_{N,T} = -\frac{1}{N \beta_T} . \tag{P4.11}$$

Equation (P4.11) is not a Maxwell relation.

c. Note, however, that Eq. (P4.11) implies

$$\left(\frac{\partial \mu}{\partial P}\right)_{T,N} = \frac{V}{N} . \tag{P4.12}$$

Using a Maxwell relation, Eq. (P1.8), Eq. (P4.12) implies that

$$\left(\frac{\partial N}{\partial V}\right)_{\mu,T} = \frac{N}{V} . \tag{P4.13}$$

Combining Eqs. (P4.10), (P4.11), and (P4.13), we obtain Eq. (P4.9).

4.33 Calculate the quantity $\psi \equiv \ln Z(N, V, T)/N$ for the ideal gas. (A: $1 + \ln(V/N\lambda_T^3)$.) Note that the same group of terms appears on the right side of the inequality (4.97).

4.34 Show that Eq. (4.104) implies Eq. (4.103). Hint: Let $n' = \omega n$.

4.35 Show for the Helmholtz energy $F(N, V, T)$ that

a. $\left(\frac{\partial^2 F}{\partial N^2}\right)_{T,V} = \frac{V}{N^2 \beta_T}$. Hint: Use the result derived in Exercise 4.32.

b. $\left(\frac{\partial^2 F}{\partial V^2}\right)_{T,N} = \frac{1}{\beta_T V}$.

Conclude that $F(N, V, T)$ is a convex function of N and V. Hint: $\beta_T > 0$. We already established that F is a concave function of T.

4.36 Show for the Gibbs energy $G = G(N, P, T)$, that

$$\left(\frac{\partial^2 G}{\partial T^2}\right)_{P,N} = -\frac{C_P}{T} \qquad \left(\frac{\partial^2 G}{\partial P^2}\right)_{T,N} = -\beta_T V \qquad \left(\frac{\partial^2 G}{\partial N^2}\right)_{T,P} = \left(\frac{\partial \mu}{\partial N}\right)_{T,P} .$$

Argue that $G(N, P, T)$ is a concave function of T and P, but a convex function of N. Hint: For the third relation, $(\partial \mu/\partial N)_{T,P} > 0$.[3, p47]

4.37 a. Derive Eq. (4.109); supply all the steps.

---

[55] The Gibbs-Duhem equation is the first law of thermodynamics cast in terms of the differentials of intensive variables.

b. Suppose we use a complete orthonormal set of functions $\{f_n(x)\}$ that are *not* the eigen-functions of $\hat{A}$, so that $\psi(x) = \sum_n b_n f_n(x)$, where $b_n \equiv \langle f_n | \psi \rangle$. Show that

$$\langle \psi | \hat{A} \psi \rangle = \sum_l \sum_n b_l^* b_n \langle f_l | \hat{A} f_n \rangle \, .$$

4.38 Show that for square matrices A, B, and C that $\operatorname{Tr} ABC = \operatorname{Tr} BCA = \operatorname{Tr} CAB$. The trace of the product of any number of square matrices is invariant under cyclic permutations of the matrices. Note, however, that $\operatorname{Tr} ABC \neq \operatorname{Tr} ACB$.

4.39 Show:

a. The density matrix is *positive definite*. Starting from the definition in Eq. (4.116), show that for arbitrary complex numbers z_r:

$$\sum_r \sum_s \rho_{sr} z_r z_s^* \geq 0 \, .$$

This is the closest analog to the requirement that the classical phase-space probability distribution $\rho(p, q) \geq 0$.

b. The density operator can be written as a weighted sum of *projection operators*:

$$\hat{\rho} = \sum_i \gamma_i \left| \psi^{(i)} \right\rangle \left\langle \psi^{(i)} \right| \, . \tag{P4.14}$$

Hint: $\hat{\rho}$ has matrix elements ρ_{rs}, where $c_s^{(i)} = \langle \phi_s | \psi^{(i)} \rangle$.

c. As a reminder of what's implied by Dirac notation, go through the steps, starting from the assumption of a complete orthonormal set of states $|\phi_n\rangle$, that for an arbitrary state $|\psi\rangle$

$$|\psi\rangle = \sum_n c_n |\phi_n\rangle = \sum_n \langle \phi_n | \psi \rangle |\phi_n\rangle \equiv \left(\sum_n |\phi_n\rangle\langle\phi_n| \right) |\psi\rangle \, ,$$

and thus

$$\sum_n |\phi_n\rangle\langle\phi_n| = I \, , \tag{P4.15}$$

where I is the identity operator. Equation (P4.15) is the completeness relation $\sum_n \phi_n^*(x)\phi_n(y) = \delta(x - y)$ written in Dirac notation.

d. For an operator A which has an eigenfunction f with $Af = \lambda f$, that

$$e^A f = e^\lambda f \, , \tag{P4.16}$$

where e^A is defined by the Taylor series, $e^A = \sum_{n=0}^{\infty}(A)^n/n!$.

4.40 Show that if the density operator commutes with the Hamiltonian, $[\hat{H}, \hat{\rho}] = 0$, then the density matrix is diagonal in a basis of the eigenfunctions of $\hat{H}$. Hint: $[\hat{H}, \hat{\rho}] = 0$ implies $\langle \phi_r | [\hat{H}, \hat{\rho}] | \phi_s \rangle = 0$ in any basis set. Let the functions $\{\phi_r\}$ be eigenfunctions of $\hat{H}$, with $\hat{H}\phi_r = \lambda_r \phi_r$.

II

Applications of Equilibrium Statistical Mechanics

CHAPTER 5

Ideal systems

N OW that we've developed the basic formalism of statistical mechanics (Chapter 4), we can proceed to applications. Measurable quantities can be calculated once the partition function is known, which requires a specification of 1) the Hamiltonian[1] and 2) the defining macrovariables associated with the type of ensemble, TVN, etc. In this chapter, we consider *ideal* systems composed of *noninteracting* constituents. Statistical mechanics naturally incorporates interactions through the potential energy part of the Hamiltonian, although as we'll see in Chapter 6, evaluating the partition function for systems of interacting particles is a challenging problem.

5.1 THE MAXWELL SPEED DISTRIBUTION

The Hamiltonian of a gas of N noninteracting particles is $H = \sum_{i=1}^{N} p_i^2/(2m)$. The partition function for this system (volume V, temperature T) is found from Eqs. (4.47) and (4.53),

$$
Z_{\text{can}}(N, V, T) = \frac{1}{N!} \left(\frac{V}{\lambda_T^3} \right)^N \equiv \frac{1}{N!} Z(N, V, T) \,, \tag{5.1}
$$

where λ_T is the thermal wavelength, Eq. (1.65), which results from integrating over the momentum variables. With Z_{can} one can calculate the equation of state and the entropy using Eq. (4.58) (Exercise 5.1). The phase-space probability density is, from Eq. (4.54),

$$
\rho(p, q) = \frac{1}{Z} \exp\left(-\beta \sum_{i=1}^{N} p_i^2/(2m) \right) = \prod_{i=1}^{N} \left(\frac{\lambda_T^3}{V} e^{-\beta p_i^2/(2m)} \right) \equiv \prod_{i=1}^{N} \rho_i \,, \tag{5.2}
$$

where ρ_i is a *one-particle* distribution function. Because the Hamiltonian is separable, the N-particle distribution occurs as the product of N, single-particle distributions, i.e., the particles are *independently distributed*.[2] Note that ρ_i is normalized on a one-particle phase space:

$$
\int \rho_i \mathrm{d}\Gamma_i \equiv \frac{\lambda_T^3}{h^3 V} \int_V \mathrm{d}x \mathrm{d}y \mathrm{d}z \int_{-\infty}^{\infty} \mathrm{d}p_x \mathrm{d}p_y \mathrm{d}p_z e^{-\beta(p_x^2 + p_y^2 + p_z^2)/(2m)} = 1 \,. \tag{5.3}
$$

Another way to calculate the entropy is through the distribution function, Eq. (4.60). One can show that Eq. (4.60) yields the Sackur-Tetrode formula when combined with Eq. (5.2) (see Exercise 5.3).

---

[1] All microscopic information about a system enters the theory of statistical mechanics through the Hamiltonian.

[2] Noninteracting particles are of necessity independently distributed—the motion of each particle is not influenced by that of the other particles.

We can express ρ_1 (the index denotes a single-particle distribution) as a probability density of the *speeds* of the particle. Start from the normalization integral, Eq. (5.3):

$$\frac{1}{h^3}\int \rho_1 d^3r d^3p = 1 = \frac{1}{h^3}\frac{\lambda_T^3}{V}\left(\int d^3r\right)\int d^3p\, e^{-\beta p^2/(2m)} = \frac{1}{(2\pi mkT)^{3/2}}\int_0^\infty 4\pi p^2 e^{-\beta p^2/(2m)}dp$$

(5.4)

$$= \left(\frac{2}{\pi}\right)^{1/2}\left(\frac{m}{kT}\right)^{3/2}\int_0^\infty v^2 e^{-mv^2/(2kT)}dv \equiv \int_0^\infty f(v)dv\,,$$

where

$$f(v) = \left(\frac{2}{\pi}\right)^{1/2}\left(\frac{m}{kT}\right)^{3/2}v^2 e^{-mv^2/(2kT)}$$

(5.5)

is the *Maxwell speed distribution*[3,4,5] for $v \geq 0$, the probability $f(v)dv$ of finding a particle with speed between v and $v + dv$. We cited the Maxwell distribution as an example of a probability density function in Eq. (3.27). Equation (5.5) shows there is a *distribution* in molecular speeds in a gas in equilibrium at temperature T—a big conceptual discovery by Maxwell.[6] A gas in thermal equilibrium—seemingly a quiescent system—actually consists of a collection of molecules having a range of speeds: a few slow ones, a few fast ones, with most having speeds near the mean.[7] The shape of the distribution is shown in Fig. 5.1. The speed distribution confirms our physical expectation that

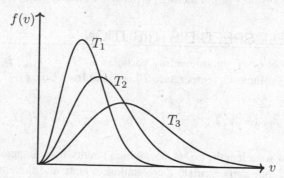

Figure 5.1: Maxwell speed distribution $f(v)$ for $T_1 < T_2 < T_3$. The area under the curve is the same for each temperature.

as the temperature is lowered, progressively more of the molecules have slower speeds. We'll see that such an expectation can fail when quantum mechanics is brought into account.

---

[3] Also known as the Maxwell-Boltzmann distribution. Maxwell's 1860 derivation (not involving statistical mechanics) can only be considered heuristic. Boltzmann in the 1870's derived it starting from more general considerations.

[4] Note what's happened here. We integrated over the spatial coordinates of a phase-space distribution $\rho(p, q)$ to produce a probability density involving only the momentum variables, converted into a distribution over speeds, $f(v)$, through a change in variables $p = mv$. Such a procedure is trivial in this case because $\rho(p, q)$ is independent of the spatial coordinates.

[5] Equation (5.5) applies for a gas in thermal equilibrium, which to maintain, requires collisions between particles, a piece of physics not contained in the starting Hamiltonian. Understanding the role of collisions in establishing and maintaining equilibrium is the province of non-equilibrium statistical mechanics, outside the scope of this book.

[6] Maxwell's theoretical advance has been verified experimentally using velocity selectors—spinning wheels with holes in them from which the gas molecules effuse.

[7] There's a fascinating thought problem known as *Maxwell's demon* afforded by the fact that a system in thermal equilibrium has a spread in the speeds of its particles. A "demon," a hypothesized intelligent being with the ability to discern the fast from the slow moving molecules and which can open a trap door without performing work, could separate fast from slow and thereby effect a violation of the second law (transfer heat from cold to hot without the performance of work). The history of Maxwell's demon is nicely recounted in Rex and Leff.[47] See also [3, Chapter 12]. It took approximately 100 years to conclusively demonstrate that Maxwell's demon could not exist within the confines of the laws of physics.

We can calculate the mean speed using the rules of probability,

$$\bar{v} = \int_0^\infty v f(v) dv = \sqrt{\frac{8kT}{m\pi}}. \tag{5.6}$$

There are other ways, however, to characterize the speed of atoms in a gas. What's the "rms" (root-mean-square) speed? By definition,

$$v_{\text{rms}} = \sqrt{\overline{v^2}} = \left[\int_0^\infty v^2 f(v) dv\right]^{1/2} = \sqrt{\frac{3kT}{m}}. \tag{5.7}$$

Note that we don't actually need to do the integral in Eq. (5.7); it follows from the equipartition theorem (Section 4.1.2.11), $\langle mv^2/2 \rangle = (3/2)kT$—why the factor of 3? We can also ask for the *most probable* speed v_{mp} at which the distribution has a maximum. This is readily found to have the value

$$v_{\text{mp}} = \sqrt{\frac{2kT}{m}}. \tag{5.8}$$

These speeds are shown in Fig. 5.2.

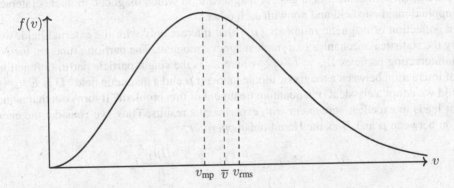

Figure 5.2: Characteristic molecular speeds of a gas in thermal equilibrium: $v_{\text{mp}}, \bar{v}, v_{\text{rms}}$.

Example. Nitrogen is the largest component of air (approximately 78%, with oxygen comprising 21%).[8] What is the mean speed $\bar{v}$ of a nitrogen molecule at room temperature, $T = 293$ K, given that nitrogen occurs as a diatomic molecule N_2 at this temperature? To apply Eq. (5.6), we need the molecular mass. The mass number of a nitrogen *atom* is approximately 14 grams/mole (consult a periodic table of the elements, 14.007 when isotopic variances are taken into account). The mass of the molecule is therefore 28 grams/mole; Avogadro's number of N_2 molecules has a mass of 28 grams. The mass of one molecule is therefore $m = 28\text{g}/(6.02 \times 10^{23}) = 4.65 \times 10^{-23}\text{g} = 4.65 \times 10^{-26}$ kg. Using Eq. (5.6), we find $\bar{v} = 471$ m/s. That's *fast*! The speed of a bullet fired from a gun is ≈ 500 m/s. Keep in mind that Eq. (5.5) is a distribution of *speeds*, not velocities. In equilibrium, the molecules of a gas have their velocities directed at random, implying the net velocity is zero. Table 5.1 lists the mean speed $\bar{v}$ for various gases at room temperature.

---

[8]We tend of think of air as a single entity, but it consists of a mixture of gases.

Table 5.1: Mean speed $\bar{v}$ of selected gases at $T = 293$ K

Element	Molar mass (g mol^{-1})	$\bar{v}$ (m s^{-1})
H_2	2	1754
He^4	4	1245
H_2O	18	585
N_2	28	471
Ar	40	394
CO_2	44	375

5.2 PARAMAGNETS

Some of the most successful applications of statistical mechanics involve the magnetic properties of materials. Under the general banner of magnetism there are different types of magnetic phenomena: ferromagnetism, antiferromagnetism, paramagnetism, diamagnetism, and others. In the limited space of this book we can only offer a cursory treatment of the subject. Ferro- and antiferromagnetism are *cooperative effects* produced by interactions among the magnetic dipoles of the atoms in a solid. Paramagnetism is the "ideal gas" of magnetism, in which magnetic moments interact only with an applied magnetic field and not with each other.

For a collection of magnetic moments $\{\mu_i\}$ that interact only with the external field, we need treat only the statistical mechanics of a single magnetic moment. The partition function for N identical, noninteracting particles $Z_N = (Z_1)^N$, where Z_1 is the single-particle partition function. The energy of interaction between a magnetic dipole moment μ and a magnetic field[9] B is $E = -\mu \cdot B$.

Should we adopt a classical or a quantum treatment of this problem? It turns out that a quantum treatment leads to excellent agreement with experimental results. Thus, we consider the energy of interaction between μ and B as the Hamiltonian *operator*,

$$\hat{H} = -B \cdot \hat{\mu} = \frac{g\mu_B}{\hbar} B \cdot \hat{J} = \frac{gB\mu_B}{\hbar} \hat{J}_z , \qquad (5.9)$$

where we've used Eq. (E.4), $\mu = -g\mu_B J/\hbar$, where $\mu_B \equiv e\hbar/(2m)$ is the Bohr magneton, g is the Landé g-factor (see Appendix E), and the operator $\hat{J}_z$ is the z-component of the total angular momentum (the B-field defines the z-direction). To use Eqs. (4.123) or (4.125) (quantum statistical mechanics in the canonical ensemble), we require the eigenfunctions and eigenvalues of the Hamiltonian operator, which in this case is proportional to $\hat{J}_z$ (Eq. (5.9)). As is well known, $\hat{J}^2$ and $\hat{J}_z$ have a common set of eigenfunctions $|J, m\rangle$ (a complete orthonormal set), such that

$$\hat{J}^2 |J, m\rangle = J(J+1)\hbar^2 |J, m\rangle$$
$$\hat{J}_z |J, m\rangle = m\hbar |J, m\rangle ,$$

where the quantum number J has the values $J = 0, 1, 2, \cdots$ or $J = \frac{1}{2}, \frac{3}{2}, \frac{5}{2}, \cdots$, and $m = -J, -J+1, \cdots, J-1, J$ so that there are $(2J+1)$ values of m. The energy eigenvalues are therefore $E_m = g\mu_B m B$. From Eq. (4.123),[10]

$$Z_1 = \sum_{m=-J}^{J} e^{-\beta m \mu_B g B} = \frac{\sinh\left(y(J+\frac{1}{2})\right)}{\sinh(y/2)} , \qquad (5.10)$$

where $y \equiv \beta\mu_B g B$. The summation in Eq. (5.10) is simple because it's a finite geometric series.

---

[9]See the remark in Section 1.9, Footnote 62 about the distinction between B and H.

[10]There are *three* minus signs at work here: one in the definition of the Hamiltonian, $\hat{H} = -\mu \cdot B$, one from the magnetic moment, $\mu = -g\mu_B J/\hbar$, and another from the Boltzmann factor, $e^{-\beta\hat{H}}$.

To calculate the average value of[11] μ_z, we use Eq. (4.125), (where $|m\rangle \equiv |J,m\rangle$)

$$\langle \mu_z \rangle = \frac{1}{Z_1} \sum_{m=-J}^{J} e^{-\beta E_m} \langle m|\hat{\mu}_z|m\rangle = -\frac{g\mu_B}{\hbar} \frac{1}{Z_1} \sum_{m=-J}^{J} e^{-\beta E_m} \langle m|\hat{J}_z|m\rangle = -\frac{g\mu_B}{Z_1} \sum_{m=-J}^{J} e^{-\beta E_m} m \tag{5.11}$$

$$= \frac{1}{\beta} \frac{\partial}{\partial B} \ln Z_1 .$$

By evaluating the derivative indicated in Eq. (5.11), we find, after some algebra,

$$\langle \mu_z \rangle = \mu_B g J B_J(\beta \mu_B g B J) , \tag{5.12}$$

where $B_J(x)$ is the *Brillouin function* of order J, defined as

$$B_J(x) \equiv \frac{2J+1}{2J} \coth\left(\frac{2J+1}{2J}x\right) - \frac{1}{2J}\coth\left(\frac{1}{2J}x\right) . \tag{5.13}$$

Graphs of these functions are shown in Fig. 5.3. They demonstrate the characteristic feature of

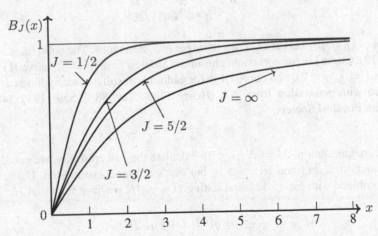

Figure 5.3: Brillouin functions $B_J(x)$ for $J = 1/2, 3/2, 5/2$, and $J = \infty$.

saturation, that $\lim_{x\to\infty} B_J(x) = 1$ for all J. At a fixed temperature, for increasing values of the magnetic field, μ becomes (on average) progressively more aligned with the direction of the field. For strong enough fields, the moments are ostensibly all aligned with the field; increasing the field further can only keep the moments at their maximum alignment with the field—saturation.

Figure 5.4 shows measured values of $\langle \mu_z \rangle$ as a function of B/T at several values of T for three paramagnetic salts which contain an ion for which $g = 2$, but for which the values of J are different, $J = \frac{3}{2}, \frac{5}{2}, \frac{7}{2}$. The data are presented as $\langle \mu_z \rangle/\mu_B$, which, from Eq. (5.12), because $g = 2$, saturate at the values of 3, 5, 7, precisely what is found experimentally. Furthermore, the data fall almost perfectly on the Brillouin functions, validating the predictions of Eq. (5.12).

Another measured quantity is the magnetic susceptibility,

$$\chi \equiv \left(\frac{\partial M}{\partial H}\right)\bigg|_{H=0} , \tag{5.14}$$

---

[11]The dipole moment μ precesses around the direction of B (see Appendix E); the components of μ orthogonal to B average to zero—we want the component of μ along the direction of B.

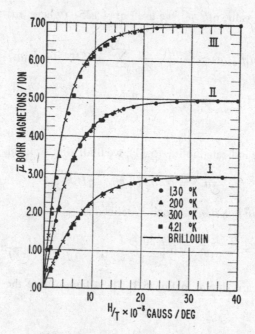

Figure 5.4: Plot of $\langle \mu_z \rangle / \mu_B$ versus B/T for three paramagnetic ions. The solid lines are the predictions of Eq. (5.12). Curve I is for potassium chromium alum ($J = \frac{3}{2}, g = 2$), curve II is for iron ammonium alum ($J = \frac{5}{2}, g = 2$), and curve III is for gadolinium sulfate octahydrate ($J = \frac{7}{2}, g = 2$). Reprinted figure with permission from W.E. Henry, Phys. Rev. **88**, p.559 (1952).[48] Copyright (2020) American Physical Society.

where M is the magnetization, $M = N\langle \mu_z \rangle$. To calculate the susceptibility, we could differentiate the Brillouin function $B_J(x)$ and let $x \to 0$, but that's an unnecessary step. Using the result of Exercise 5.9, combined with Eq. (5.12), and setting $B = \mu_0 H$, we have for small H,

$$M \stackrel{H \to 0}{\sim} \frac{N}{3kT}(\mu_B g)^2 J(J+1)\mu_0 H \equiv C\frac{H}{T} , \tag{5.15}$$

where

$$C = \frac{N}{3k}\mu_0 (\mu_B g)^2 J(J+1) . \tag{5.16}$$

Equation (5.15), *Curie's law*, is the equation of state of a paramagnet—the magnetization is linear with H for small M. The constant C is the *Curie constant*, the value of which is material specific.[12] From Eq. (5.15), as $H \to 0$, $M \to 0$, the hallmark of paramagnetism—the system acquires a magnetization in an applied magnetic field, which vanishes in zero field.[13] For small fields, the larger the susceptibility, the larger is the magnetization obtained for the same field strength.

Paramagnetism can be treated classically if we consider the magnetic moment a vector (not an operator): $\mu = \mu\hat{e}$, where $\hat{e}$ is a unit vector that can point in any direction. The energy $E = -\mu B \cos\theta$, where θ is the angle between B and μ. We use Eq. (4.15) for the partition function, $Z = \int e^{-\beta E}\Omega(E)dE$. The density of states function $\Omega(E)$ is found by differentiating the formula $E(\theta) = -\mu B \cos\theta$, implying $\Omega(\theta) = \sin\theta$. (We leave off the factor of μB to keep the density of

---

[12]The Curie constant C is therefore not a universal constant, as is the gas constant. There isn't just one Curie constant; every material has its own. C has the base units K·m³; C must scale with the size of the system—the form of Curie's law, Eq. (5.15), relates M, an extensive quantity, to the ratio of intensive quantities, H/T.

[13]Ferromagnets develop a spontaneous magnetization in the absence of an applied field at sufficiently low temperature.

states dimensionless, the *number* of states in the range $[\theta, \theta + d\theta]$). Thus,

$$Z = \int_0^\pi e^{\beta\mu B \cos\theta} \sin\theta d\theta = \frac{2}{\beta\mu B} \sinh(\beta\mu B) . \qquad (5.17)$$

To obtain the average value of μ_z, we use Eq. (5.11),

$$\langle \mu_z \rangle = \frac{\partial}{\partial(\beta B)} \ln Z = \mu L(x) , \qquad (5.18)$$

where $x \equiv \beta\mu B$ and $L(x)$ is the *Langevin function*,

$$L(x) \equiv \coth x - \frac{1}{x} . \qquad (5.19)$$

From Exercise 5.9, $L(x)$ is the limiting form of $B_J(x)$ as $J \to \infty$.

5.3 HARMONIC OSCILLATORS, QUANTUM AND CLASSICAL

No physics book can be complete without treating the harmonic oscillator, as it's among the few exactly solved problems in physics. We're fortunate this problem *can* be solved exactly, because it occurs widely in physics. For a potential energy function $V(r)$ that has a minimum at $r = r_0$, its Taylor series about r_0 is $V(r) = V(r_0) + (r - r_0)dV/dr|_{r_0} + \frac{1}{2}(r - r_0)^2 d^2V/d^2r|_{r_0} + \cdots$. If the system is in equilibrium at $r = r_0$ (no force acting), $dV/dr|_{r_0} = 0$, and assuming $d^2V/dr^2|_{r_0} > 0$ (stable equilibrium), then small excursions about $r = r_0$ map onto the harmonic oscillator with $m\omega^2 \equiv d^2V/dr^2|_{r_0}$. In what follows, assume we have a harmonic oscillator of mass m and angular frequency ω in equilibrium with a heat bath at temperature T. We consider the problem from the quantum and the classical perspectives.

5.3.1 Quantum treatment

Harmonic oscillators have quantized energy levels[14] $E_n = (n + \frac{1}{2})\hbar\omega$, $n = 0, 1, 2, \cdots$. The energy associated with $n = 0$, $\frac{1}{2}\hbar\omega$, is the *zero-point energy*, the lowest possible energy that a quantum system may have (which, we note, is not zero).[15] The canonical partition function for a single oscillator is, from Eq. (4.123),[16]

$$Z_1(\beta) = \sum_{n=0}^\infty e^{-\beta(n+\frac{1}{2})\hbar\omega} = \frac{1}{2\sinh(\beta\hbar\omega/2)} . \qquad (5.20)$$

The partition function specifies the number of states a system has available to it at temperature T. As $\beta \to 0$ (high temperature), we have from Eq. (5.20),

$$Z_1(\beta) \overset{\beta\to 0}{\sim} \frac{1}{\beta\hbar\omega} , \qquad (5.21)$$

that *all* of the infinite number of energy states of the harmonic oscillator become thermally accessible, that Z diverges as we (formally) allow $T \to \infty$. Compare with the $\beta \to 0$ limit of the partition function for a paramagnetic ion, Eq. (5.17), $Z(\beta \to 0) = 2$. In that case there are only two states available to the system: aligned or antialigned with the direction of the magnetic field. Consider the other limit of Eq. (5.20),

$$Z_1(\beta) \overset{\beta\to\infty}{\sim} e^{-\beta\hbar\omega/2} . \qquad (5.22)$$

---

[14] Shown in any text on quantum mechanics.
[15] Zero-point energy is a distinctively quantum mechanical effect related to the Heisenberg uncertainty principle.
[16] The sum in Eq. (5.20) is a geometric series; it converges because $e^{-\beta\hbar\omega} < 1$ for all T.

For temperatures such that $kT \lesssim \hbar\omega/2$, $Z_1 \ll 1$; the number of states available to the system is exponentially smaller than unity. As $T \to 0$ there are no states available to the system: $Z \to 0$.

Applying Eq. (5.20) to Eq. (4.40), we have the average energy of the oscillator,

$$\langle E \rangle = \frac{\hbar\omega}{2} \coth\left(\frac{1}{2}\beta\hbar\omega\right) = \hbar\omega\left(\frac{1}{e^{\beta\hbar\omega} - 1} + \frac{1}{2}\right) \equiv \hbar\omega\left(\langle n \rangle + \frac{1}{2}\right). \tag{5.23}$$

Let's look at the limiting forms of Eq. (5.23):

$$\langle E \rangle = \frac{\hbar\omega}{2} \qquad (T \to 0)$$
$$\langle E \rangle = kT. \qquad (T \to \infty) \tag{5.24}$$

At low temperature, the system occupies its ground state, with energy $E = \hbar\omega/2$. At sufficiently high temperatures, the system behaves classically, with energy given by the equipartition theorem. We've written Eq. (5.23) in the form $\langle E \rangle = (\langle n \rangle + \frac{1}{2})\hbar\omega$, where $\langle n \rangle$ denotes the *occupation number* specifying the effective average state that the system is "in" (or occupies) in thermal equilibrium,

$$\langle n \rangle = \frac{1}{e^{\beta\hbar\omega} - 1}. \tag{5.25}$$

The occupation number $\langle n \rangle$ is not an integer. The system (oscillator) is continually exchanging energy with its environment, causing it to momentarily occupy the allowed states of the system labeled by integer n. The occupation number is the average value in thermal equilibrium of the quantum number n. We can either say that the system has energy $\langle E \rangle = (\langle n \rangle + \frac{1}{2})\hbar\omega$, or, equivalently (because the energy of an ideal system is the sum of the energies of its components), that the system consists of $\langle n \rangle$ *quanta*[17] each of energy $\hbar\omega$.

We note that Eq. (5.25) is the Bose-Einstein distribution function for bosons with $\mu = 0$ (see Eq. (5.61)), which is also the Planck distribution for the number of thermally excited photons of energy $E = \hbar\omega$ (see Section 5.8.2). Is there a connection between bosons and harmonic oscillators? There is indeed, as we now show.[18]

For a system of N independent oscillators, the partition function $Z_N(\beta) = (Z_1(\beta))^N$. Thus, using Eq. (5.20),

$$Z_N(\beta) = [2\sinh(\beta\hbar\omega/2)]^{-N} = \frac{e^{-N\beta\hbar\omega/2}}{[1 - e^{-\beta\hbar\omega}]^N}. \tag{5.26}$$

Equation (5.26) is a closed-form expression for the partition function of N independent harmonic oscillators, from which one could calculate the heat capacity—see Eq. (5.41). We can, however, express Eq. (5.26) in another way. Apply the binomial theorem Eq. (3.10) to Eq. (5.26):

$$Z_N(\beta) = e^{-N\beta\hbar\omega/2} \sum_{k=0}^{\infty} \binom{N+k-1}{k} e^{-k\beta\hbar\omega}. \tag{5.27}$$

Writing $Z_N(\beta)$ in the form of Eq. (4.15) (the Laplace transform of the density of states), $Z_N(\beta) = \int_0^\infty \Omega_N(E)e^{-\beta E}dE$, we infer from Eq. (5.27) that the density of states for N independent harmonic oscillators has the form

$$\Omega_N(E) = \sum_{k=0}^{\infty} \binom{k+N-1}{k} \delta\left(E - (k+N/2)\hbar\omega\right), \tag{5.28}$$

---

[17] *Quanta* is the plural of *quantum*.

[18] We quote from the writings of P.A.M. Dirac, one of the founders of quantum mechanics and quantum electrodynamics: "...*the dynamical system consisting of an assembly of similar bosons is equivalent to the dynamical system consisting of a set of oscillators—the two systems are just the same system looked at from two different points of view. There is one oscillator associated with each independent boson state. We have here one of the most fundamental results of quantum mechanics, which enables a unification of the wave and corpuscular theories of light to be effected.*"[49, p229]

where $\delta(x)$ is the Dirac delta function. What combinatorial problem does the binomial coefficient $\binom{N+k-1}{k}$ pertain to? Consider, starting from N oscillators each in its ground state (so $E = N\hbar\omega/2$ is the zero-point energy of the system), how many ways can k quanta, each of energy $\hbar\omega$, be added to N oscillators so that the energy of the system is $E = (k+N/2)\hbar\omega$? Quanta are *indistinguishable*; we can't say *which* quantum of energy is added to an oscillator, all we can say is that k quanta have been added to the system. The combinatorial problem is therefore how many distinct ways can k indistinguishable quanta be added to N *distinguishable* oscillators? Figure 5.5 shows 17 "dots,"

Figure 5.5: One of the $\binom{k+N-1}{k}$ ways of distributing k indistinguishable energy quanta (shown here as 17 identical dots) among N oscillators (shown here as 7 boxes, delineated by 6 vertical lines). In this example, a box in the middle has no quanta.

where the dots represent energy quanta, distributed among 7 oscillators, where the oscillators have been conceptually arranged in a line, separated by vertical lines. N oscillators are delineated by $(N-1)$ vertical lines. There would be $(N+k-1)!$ permutations of the $(N+k-1)$ symbols in Fig. 5.5, but that would overcount the number of distinct configurations of the system. We should divide $(N+k-1)!$ by $k!$ for permutations of the k dots, and by $(N-1)!$ for permutations of the oscillators. Thus, the number of distinct permutations of k dots among the $N-1$ lines is $\binom{k+N-1}{k}$.

Example. There are $\binom{3}{2} = 3$ ways of adding $k = 2$ quanta to $N = 2$ oscillators; see Fig. 5.6.

Figure 5.6: The three distinct configurations of two quanta added to two oscillators.

5.3.2 Classical treatment

The Hamiltonian function for the harmonic oscillator is $H = p^2/(2m) + \frac{1}{2}m\omega^2 x^2$, implying

$$Z_1(\beta) = \frac{1}{h}\int_{-\infty}^{\infty} dx \int_{-\infty}^{\infty} dp e^{-\beta[p^2/(2m)+\frac{1}{2}m\omega^2 x^2]} = \frac{1}{\beta\hbar\omega}. \tag{5.29}$$

Planck's constant enters the evaluation of Z in *classical* statistical mechanics (see Section 2.3). The partition function for the quantum harmonic oscillator in the high-temperature limit is the same as that for the classical oscillator.

5.4 DIATOMIC GASES

In Section 5.1, we treated the ideal gas of structureless molecules, the most salient feature of which is the translational kinetic energy of its point particles. Translational motion is present in any gas. The constituents of real gases have *internal* motions that we have yet to take into account. We consider the ideal diatomic gas,[19] i.e., we ignore inter-particle interactions and we assume the conditions for

---

[19]The noble gases (helium, neon, argon, krypton, xenon, and radon) are monatomic. There are *homonuclear* diatomic gases such as hydrogen (H_2), nitrogen (N_2), oxygen (O_2), and chlorine (Cl_2), and *heteronuclear* diatomic gases such as carbon monoxide (CO), nitric oxide (NO), and hydrogen chloride (HCl).

classical behavior apply, $n\lambda_T^3 \ll 1$ (see Section 1.11). It should be clear that internal motions must be treated using quantum mechanics—classical statistical mechanics brings with it the equipartition theorem, which we know is insufficient to explain the heat capacity of real gases.

Energies of translational degrees of freedom and those of internal motions can be written in terms of individual Hamiltonians,

$$H = H_{\text{trans}} + H_{\text{rot}} + H_{\text{vib}} + H_{\text{elec}} + \cdots , \tag{5.30}$$

where H_{trans} is the Hamiltonian associated with translational motion, H_{rot} is that for rotations, H_{vib} for vibrations, H_{elec} for electronic degrees of freedom, and so on. Writing H in this form assumes the degrees of freedom underlying each Hamiltonian are noninteracting, that the various modes of excitation occur independently—an approximation that's not always true. When H can be written in separable form, the partition function occurs as the product of the partition functions associated with each part of the Hamiltonian:[20]

$$Z(T,V) = \frac{1}{N!} \left(Z_{\text{trans}}(T,V) \cdot Z_{\text{rot}}(T) \cdot Z_{\text{vib}}(T) \cdot \ldots \right) . \tag{5.31}$$

In that case, the heat capacity occurs as the sum of the heat capacities for each of the modes of excitation (because C_V is related to $\ln Z$; see Exercise 4.14):

$$C_V(T) = C_{\text{trans}} + C_{\text{rot}}(T) + C_{\text{vib}}(T) + \cdots . \tag{5.32}$$

For an ideal gas, $C_{\text{trans}} = \frac{3}{2}Nk$ is independent of temperature.[21] In this section, we calculate the heat capacities for the rotational and vibrational degrees of freedom of the ideal diatomic gas.

5.4.1 Rotatonal motion

The *rigid rotor* problem treats the two atoms of a diatomic molecule as having a fixed separation distance r_0. The allowed rotational energies depend on the moment of inertia $I = \mu r_0^2$, where μ is the reduced mass of the two atomic masses, $\mu = m_1 m_2 / (m_1 + m_2)$. The rotational state is determined by the angular momentum operator, $\hat{\boldsymbol{L}}$. $\hat{L}^2$ and $\hat{L}_z$ have a common set of eigenfunctions,

$$\hat{L}^2 |l,m\rangle = l(l+1)\hbar^2 |l,m\rangle$$
$$\hat{L}_z |l,m\rangle = m\hbar |l,m\rangle ,$$

where $l = 0,1,2,\cdots$ and $m = -l, -l+1, \cdots, l-1, l$ so that there are $2l+1$ values of m. The Hamiltonian for rotational motion about the center of mass is $\hat{H}_{\text{rot}} = \boldsymbol{L}^2/(2I)$, and thus the rotational energy eigenvalues are $E_l = \hbar^2 l(l+1)/(2I)$. Because E_l is independent of the quantum number m, each state is $(2l+1)$-fold degenerate. The partition function is, using Eq. (4.123),[22]

$$Z_{1,\text{rot}}(T) = \sum_{l=0}^{\infty} (2l+1)e^{-\beta E_l} . \tag{5.33}$$

---

[20]Partition functions for internal degrees of freedom are independent of the volume of the system. Internal variables are restricted to domains associated with isolated molecules, implying internal partition functions depend on temperature alone.

[21]Not to flog a dead horse, the gaseous state of matter ceases to exist at sufficiently low temperatures where phase transitions occur from the gaseous to the liquid to the solid state.

[22]Equation (5.33) is correct for heteronuclear diatomic molecules. The treatment of homonuclear diatomic molecules requires more care due to the indistinguishability of the two atoms, the wave function of which must be symmetric or antisymmetric under interchange of the atoms. For example, in the *parahydrogen* form of H_2, the spins of the two hydrogen nuclei (protons) combine to form a singlet state—antisymmetric under exchange of the nuclear spins—implying that the spatial part of the wave function is even under interchange, which has as a consequence that terms associated with $l = $ odd are excluded from the sum in Eq. (5.33). For the *orthohydrogen* form of H_2, the nuclei combine in a triplet state—even under spin exchange—implying the spatial part of the wave function is antisymmetric under exchange, which excludes from the sum in Eq. (5.33) terms associated with $l = $ even. Unless the H_2 gas is specifically prepared in either the ortho- or para-forms, Eq. (5.33) must be replaced with $Z = \sum_{l=0,2,4,\ldots}^{\infty}(2l+1)e^{-\beta E_l} + 3\sum_{l=1,3,5,\ldots}^{\infty}(2l+1)e^{-\beta E_l}$. The factor of 3 results from the 3 spin wave functions associated with the triplet state. We ignore these subtleties in analyzing Eq. (5.33).

The sum in Eq. (5.33) cannot be evaluated in closed analytic form, and we must introduce approximations. We examine the high and low-temperature limits.

5.4.1.1 High-temperature form

As $\beta \to 0$ there are contributions to Eq. (5.33) from large values of the quantum number l, which suggests we approximate the sum in Eq. (5.33) with an integral, using the form of Z in Eq. (4.15). That route requires the density-of-states function, $\Omega(E)$, the derivative with respect to energy of the total number of energy states up to and including E. Energy at a specified value E implies a maximum value of l determined by $E = \hbar^2 l_{max}(l_{max} + 1)/(2I) \approx \hbar^2 l_{max}^2/(2I)$ because $l_{max} \gg 1$. How many states are there for $0 \le l \le l_{max}$? It can be shown that

$$\sum_{l=0}^{l_{max}} (2l + 1) = (l_{max} + 1)^2 \approx l_{max}^2 \approx \frac{2I}{\hbar^2} E .\tag{5.34}$$

The density of states is therefore $\Omega(E) = 2I/\hbar^2$. Thus, we can approximate Eq. (5.33),

$$Z_{1,rot}(T) = \frac{2I}{\hbar^2} \cdot \int_0^\infty e^{-\beta E} dE = \frac{2I}{\beta \hbar^2} \equiv \frac{T}{\Theta_r} , \qquad (T \gg \Theta_r) \tag{5.35}$$

where $\Theta_r = \hbar^2/(2Ik)$ sets a characteristic temperature for rotational motions.[23] Using equations that we've now used several times (Eqs. (4.40) and (P4.1)), with $Z = (Z_1)^N$,

$$\langle E \rangle_{rot} = NkT$$
$$(C_V)_{rot} = Nk , \qquad (T \to \infty) \tag{5.36}$$

the same as what we obtain from the equipartition theorem.

A more accurate high-temperature form can be obtained using the result of Exercise 5.11:

$$Z_{1,rot}(T) = \frac{T}{\Theta_r} + \frac{1}{3} + \frac{1}{15}\frac{\Theta_r}{T} + \frac{4}{315}\left(\frac{\Theta_r}{T}\right)^2 + \cdots . \qquad (T \gg \Theta_r) \tag{5.37}$$

From Eq. (5.37) we obtain an expression for the heat capacity more general than Eq. (5.36) (see Exercise 5.12),

$$(C_V(T))_{rot} = Nk \left[1 + \frac{1}{45}\left(\frac{\Theta_r}{T}\right)^2 + \frac{16}{945}\left(\frac{\Theta_r}{T}\right)^3 + \cdots \right] . \qquad (T \gg \Theta_r) \tag{5.38}$$

We see that $(C_V(T))_{rot}$ exceeds the classical value Nk, a value that it tends to as $T \to \infty$.

5.4.1.2 Low-temperature form

In the low-temperature regime, $T \ll \Theta_r$, we have, from Eq. (5.33),

$$Z(T)_{1,rot} = 1 + 3e^{-2\Theta_r/T} + 5e^{-6\Theta_r/T} + \cdots . \tag{5.39}$$

In this case, the variable $e^{-\Theta_r/T}$ is exponentially small as $T \to 0$. From Eq. (5.39), we find to lowest order

$$(C_V(T))_{rot} \approx 12Nk \left(\frac{\Theta_r}{T}\right)^2 e^{-2\Theta_r/T} . \qquad (T \ll \Theta_r) \tag{5.40}$$

---

[23]What's a typical value of Θ_r? Experimental data is tabulated in terms of *rotation constants*, $B \equiv \hbar/(4\pi Ic)$, where c is the speed of light and B has the units cm^{-1}. One can convert from B to Θ_r through $k\Theta_r = hcB$. For HCl, B has the value 10.6 cm^{-1}, implying $\Theta_r \approx 15$ K.

As $T \to 0$, $(C_V(T))_{\text{rot}}$ drops to zero exponentially fast; rotational degrees of freedom can't be excited at sufficiently low temperature—they become "frozen out."

The two equations, (5.38) and (5.40), are limiting forms of $(C_V(T))_{\text{rot}}$ in the high- and low-temperature regimes. They each show that the heat capacity is temperature dependent. To obtain the complete temperature dependence of $(C_V(T))_{\text{rot}}$ requires the use of a computer to evaluate the sum in Eq. (5.33) at each temperature. A detailed analysis shows there is a maximum value of $(C_V(T))_{\text{rot}} \approx 1.1Nk$ at $T \approx 0.81\Theta_r$. Given that $\Theta_r \approx 10$ K, measurements of C_V on diatomic gases at room temperature are consistent with the prediction of the equipartition theorem.

5.4.2 Vibrational motion

The partition function for the quantum harmonic oscillator is given in Eq. (5.20), from which may be derived an expression for the heat capacity,

$$(C_V(T))_{\text{vib}} = Nk \left(\frac{\Theta_v}{T}\right)^2 \frac{e^{\Theta_v/T}}{\left(e^{\Theta_v/T} - 1\right)^2}, \qquad (5.41)$$

where $\Theta_v \equiv \hbar\omega/k$ is a characteristic temperature associated with vibrational energies. For HCl, $\Theta_v \approx 4300$ K, and for H_2, $\Theta_v \approx 6300$ K. Thus, only for temperatures on the order of 10^4 K would we expect vibrational modes to be sufficiently excited that they contribute to C_V at the level required by the equipartition theorem. From Eq. (5.41) we have the high and low-temperature limits:

$$(C_V)_{\text{vib}} = Nk \qquad\qquad (T \gg \Theta_v)$$

$$(C_V)_{\text{vib}} \sim Nk \left(\frac{\Theta_v}{T}\right)^2 e^{-\Theta_v/T}. \qquad (T \ll \Theta_v) \qquad (5.42)$$

Given that $\Theta_v \sim 10^3$ K, vibrational modes are frozen out at room temperature and don't appreciably contribute to C_V.

5.5 IDENTICAL FERMIONS AND BOSONS

We now take into account the indistinguishability of identical particles required by quantum mechanics. We start with the Hamiltonian for a system of N noninteracting identical particles, $\hat{H} = \sum_{n=1}^{N} \hat{h}_n$, where $\hat{h}_n$ is a function of the coordinates and momenta associated with an isolated atom or molecule, which can include the translational motion of the center of mass or the degrees of freedom associated with rotation, vibration, intra-atomic electronic structure, and the intrinsic spin, S. The energy of an ideal system is the sum of the energies of its constituents, $E = \sum_{n=1}^{N} E^{(n)}$, where $E^{(n)}$ is the energy of the nth particle, which is any one of the eigenvalues belonging to $\hat{h}$. (We don't have to label $\hat{h}$ with an index—it's the same function for all particles.) The quantity $E^{(n)}$ is a function of the quantum numbers associated with the aforementioned degrees of freedom, such as the wavevector k introduced in Section 2.1.5, rotational or vibrational quantum numbers, or the z-component of the spin. We denote the *collection* of relevant quantum numbers associated with the nth particle as m_n. Thus, $E^{(n)} = E^{(n)}(m_n)$.

5.5.1 Partition function

We might assume that the canonical partition function can be written in the form

$$Z_N = \sum_{m_1} \cdots \sum_{m_N} \exp\left(-\beta \sum_{n=1}^{N} E^{(n)}\right). \qquad \text{(wrong!)} \qquad (5.43)$$

Equation (5.43) is incorrect because it overcounts the allowed states of identical particles. The *occupation number* n_k is the number of particles in the system having the eigenstate associated with eigenvalue E_k (see Appendix D). The energy of the system can therefore be written not as a sum over *particles*, as in $E = \sum_{n=1}^{N} E^{(n)}$, but as a sum over energy levels,

$$E = \sum_m n_m E_m \,. \tag{5.44}$$

Equation (5.44) is an important step in setting up the statistical mechanics of identical particles. The occupation numbers must satisfy the constraint

$$\sum_m n_m = N \,. \tag{5.45}$$

Equations (5.44) and (5.45) are *unrestricted* sums over all possible energy states.[24]

How many ways can the energy E be partitioned over the particles of the system? For a *given set* of occupation numbers $\{n_k\}$ satisfying Eq. (5.45), there are, using Eq. (3.12), $N!/\prod_k (n_k!)$ ways of permuting the particles, which, by the indistinguishability of identical particles, are equivalent and *have to be treated as a single state*. To correct for overcounting, Eq. (5.43) should be written

$$Z_N = \frac{1}{N!} \sum_{m_1} \cdots \sum_{m_N} \prod_k (n_k!) \exp\left(-\beta \sum_{n=1}^{N} E^{(n)}\right) \,. \tag{5.46}$$

Equation (5.46) presents a challenging combinatorial problem because it connects the energy levels of individual particles, $E^{(n)}$, to the occupation numbers n_k, which apply to the entire system. To apply Eq. (5.46), we must know the occupation numbers associated with a system in thermal equilibrium, which is what we're trying to solve for! At high temperature ($\beta \to 0$) the number of energy levels that can make a significant contribution to the sum in Eq. (5.46) becomes quite large. One would expect, for fixed values of N and E, that in this limit the occupation numbers will be predominately 0 or 1 and thus $(n_k)! = 1$ for most configurations. If we set $(n_k)! = 1$ in Eq. (5.46), we have an approximate expression for the partition function (which factorizes)

$$Z \approx \frac{1}{N!} \sum_{m_1} \cdots \sum_{m_N} e^{-\beta \sum_{n=1}^{N} E^{(n)}} = \frac{1}{N!} \prod_{k=1}^{N} \left(\sum_{m_k} e^{-\beta E^{(k)}}\right) = \frac{1}{N!} (Z_1)^N \,, \tag{5.47}$$

where in the last step we've used that all particles are identical, where Z_1 is the partition function for a single particle. Equation (5.47) is the high-temperature limit of the partition function for bosons or fermions.

It was noted in Section 1.11 that for combinations of temperature and density such that $n\lambda_T^3 \gtrsim 1$, the boson or fermion character of the particles of the system must be taken into account. We'll see in Section 5.5.3 how this criterion emerges from a calculation of the equation of state.[25] We consider an ideal gas of point particles in the absence of an external magnetic field, so that the relevant quantum numbers[26] are the wavevector k associated with the particle's kinetic energy (see Eq. (2.14)) and its

---

[24]There may be many states for which $n_m = 0$.

[25]The criterion $n\lambda_T^3 \gtrsim 1$ for the necessity of a quantum treatment was suggested by Eq. (1.64), where, for $n\lambda_T^3 \ll 1$, the entropy of the system is 1) positive and 2) offers the interpretation that $1/(n\lambda_T^3)$ is the number of ways a particle, associated with the intrinsic quantum volume λ_T^3, can be arranged in the volume per particle, V/N. For $n\lambda_T^3 \ll 1$, particles behave independently of each other. For $n\lambda_T^3 \gtrsim 1$, the density has become sufficiently large that the particles can no longer be treated as independent. In this regime, new physics can emerge.

[26]There's no new physics to be understood by including internal degrees of freedom (rotation and vibration) in our treatment here, which are uninvolved in the spatial permutations considered in applications of the Pauli principle. Internal degrees of freedom are unaffected by the Fermi or Bose statistics.

spin quantum number σ. The energy eigenvalues in this case[27] are

$$E_{k,\sigma} = E_k = \frac{\hbar^2}{2m} k^2 , \qquad (5.48)$$

where the eigenvalues are g-fold degenerate, with $g = 2S + 1$. It's shown in Appendix D how, in creating wavefunctions displaying the proper symmetries under permutations of particles, that information about particle identity is lost. We can't say *which* particle has a given energy, because of the indistinguishability of identical particles. The most we can say is how many particles have a given energy (the occupation number), and not which particles have that energy. We can therefore write the N-particle partition function, not as in Eq. (5.46) (a sum over particles), but in the form

$$Z_N = \sum_{\substack{\{n_{k,\sigma}\} \\ (\sum_{k,\sigma} n_{k,\sigma}=N)}} e^{-\beta \sum_{k,\sigma} n_{k,\sigma} E_{k,\sigma}} , \qquad (5.49)$$

a sum over energy levels, where $\sum_{\{n_{k,\sigma}\}}$ denotes a summation over all quantum numbers (k,σ), $\sum_{\{n_{k,\sigma}\}} \equiv \prod_{k,\sigma} \sum_{k,\sigma}$. Equation (5.49) indicates, for each allowed (k,σ), to sum over the occupation numbers $n_{k,\sigma}$, subject to the constraint $\sum_{k,\sigma} n_{k,\sigma} = N$.

Equation (5.49) is in general impossible to evaluate, because of the constraint of a fixed number of particles, Eq. (5.45). One has a formidable combinatorial problem of summing over all sets of occupation numbers consistent with the constraint on the number of particles. Without the constraint, Eq. (5.49) would simply factorize. Surprisingly, this problem *simplifies* in the grand canonical ensemble where, by allowing N to vary, the constraint is eliminated. Referring to Eq. (4.127),

$$Z_G = \sum_{N=0}^{\infty} \sum_{\substack{\{n_{k,\sigma}\} \\ (\sum_{k,\sigma} n_{k,\sigma}=N)}} \exp\left(\beta\mu \sum_{k,\sigma} n_{k,\sigma}\right) \exp\left(-\beta \sum_{k,\sigma} E_k n_{k,\sigma}\right) . \qquad (5.50)$$

As we'll explain, the sums in Eq. (5.50) can be rearranged so that

$$Z_G = \sum_{\{n_{k,\sigma}\}} \exp\left(\beta \sum_{k,\sigma} (\mu - E_k) n_{k,\sigma}\right) = \prod_{k,\sigma} \sum_{n_{k,\sigma}=0}^{\infty} \exp\left(\beta n_{k,\sigma}[\mu - E_k]\right) , \qquad (5.51)$$

where now the sums over occupation numbers are *unrestricted*. The transition from Eq. (5.50) to Eq. (5.51) can be shown by the method of staring at it long enough. Consider a system that has only two energy levels, E_1 and E_2. In that case, we would have from Eq. (5.50)

$$Z_G = \sum_{N=0}^{\infty} \sum_{n_1} \sum_{\substack{n_2 \\ (n_1+n_2=N)}} e^{\beta(\mu-E_1)n_1 + \beta(\mu-E_2)n_2} \qquad (5.52)$$

Let $a \equiv e^{\beta(\mu-E_1)}$ and $b \equiv e^{\beta(\mu-E_2)}$, so that Eq. (5.52) can be expressed

$$Z_G = \sum_{N=0}^{\infty} \sum_{n_1=0}^{N} \sum_{n_2=0}^{N-n_1} a^{n_1} b^{n_2} .$$

---

[27]We show in Eq. (5.48) the form of the energy eigenvalues of the Schrödinger equation as an example; nothing in the argument relies on the specific form of the eigenvalues—all we require is that there *are* a set of allowed energies.

Write out the first few terms in the series: For $N = 0$, $a^0 b^0 = 1$; for $N = 1$, $a^0 b^1 + a^1 b^0 = a + b$; for $N = 2$, $a^2 + ab + b^2$; and for $N = 3$, $a^3 + a^2 b + ab^2 + b^3$. Add these terms (up to and including $N = 3$):

$$Z_G = (1 + b + b^2 + b^3) + a(1 + b + b^2) + a^2(1 + b) + a^3(1) .$$

We see the pattern: As $N \to \infty$, we have the unrestricted sums,

$$Z_G = \sum_{n_1=0}^{\infty} \sum_{n_2=0}^{\infty} a^{n_1} b^{n_2} = \left(\sum_{n_1=0}^{\infty} a^{n_1} \right) \left(\sum_{n_2=0}^{\infty} b^{n_2} \right) .$$

Generalize to a set of k energy levels $E_1, E_2, \cdots, E_k$, with $a_i \equiv e^{\beta(\mu - E_i)}$, $i = 1, \cdots, k$. Then, from Eq. (5.50),

$$Z_G = \sum_{N=0}^{\infty} \sum_{n_1=0}^{N} \sum_{n_2=0}^{N} \cdots \sum_{n_{k-1}=0}^{N} \sum_{n_k=0}^{N - \sum_{l=1}^{k-1} n_l} (a_1)^{n_1} (a_2)^{n_2} \cdots (a_{k-1})^{n_{k-1}} (a_k)^{n_k} .$$

The sums become unrestricted for $N \to \infty$, with

$$Z_G = \prod_{i=1}^{k} \sum_{n_i=0}^{\infty} (a_i)^{n_i} ,$$

the same as Eq. (5.51) when we let $k \to \infty$. With Eq. (5.51) established, we can quickly complete the calculation.

The occupation numbers of fermions are restricted to the values $n = 0, 1$ for any state (the Pauli principle). In that case, we have from Eq. (5.51)

$$Z_G(\beta, \mu) = \prod_{k,\sigma} \left(1 + e^{\beta(\mu - E_{k,\sigma})} \right) . \qquad (F) \qquad (5.53)$$

For bosons there are no restrictions on the occupation numbers. The sum in Eq. (5.51) can be evaluated explicitly as a geometric series, with the result

$$Z_G(\beta, \mu) = \prod_{k,\sigma} \frac{1}{1 - e^{\beta(\mu - E_{k,\sigma})}} . \qquad (B) \qquad (5.54)$$

For the infinite series to converge requires that $\mu \leq 0$ for bosons. There is no restriction on μ for fermions. These formulas can be combined into a common expression. Let $\theta = +1$ for bosons and $\theta = -1$ for fermions (the same factor of θ introduced in Eq. (4.110)). Equations (5.53) and (5.54) can then be written as a common expression

$$Z_G(\beta, \mu) = \prod_{k,\sigma} \left(1 - \theta e^{\beta(\mu - E_{k,\sigma})} \right)^{-\theta} . \qquad (\theta = \pm 1) \qquad (5.55)$$

The partition function for the ideal quantum gas thus factorizes, but the factors don't pertain to individual particles—they refer to individual energy levels.

5.5.2 The Fermi-Dirac and Bose-Einstein distributions

The average number of particles a system has (associated with given values of μ, T, V—we're using the grand canonical ensemble) can be found from the derivative (see Eq. (4.78))

$$\langle N \rangle = kT \left(\frac{\partial \ln Z_G}{\partial \mu} \right)_{T,V} .$$

Using Eq. (5.55) for Z_G,

$$\frac{\partial \ln Z_G}{\partial \mu}\bigg|_{T,V} = g\beta \sum_{k} \frac{1}{e^{\beta(E_k-\mu)} - \theta}, \tag{5.56}$$

where $g = 2S + 1$ is the spin degeneracy (see Eq. (5.48)). Thus,

$$\langle N \rangle = g \sum_{k} \frac{1}{e^{\beta(E_k-\mu)} - \theta}. \tag{5.57}$$

The sum in Eq. (5.57) can be converted to an integral over k-space, $\sum_{k} \rightarrow (V/(8\pi^3)) \int d^3k$ (see Section 2.1.5),

$$\langle N \rangle = g\frac{V}{8\pi^3} \int d^3k \frac{1}{e^{\beta(E_k-\mu)} - \theta} = g\frac{V}{8\pi^3} \int_0^\infty 4\pi k^2 dk \frac{1}{e^{\beta(E_k-\mu)} - \theta}, \tag{5.58}$$

where, because for free particles E_k is isotropic in k-space (see Eq. (5.48)), we work in spherical coordinates. Change variables[28] in Eq. (5.58); let $E = \hbar^2 k^2/(2m)$. Then,

$$\langle N \rangle = g\frac{V}{4\pi^2}\left(\frac{2m}{\hbar^2}\right)^{3/2} \int_0^\infty \frac{\sqrt{E}dE}{e^{\beta(E-\mu)} - \theta} = \int_0^\infty \frac{g(E)dE}{e^{\beta(E-\mu)} - \theta} \equiv \int_0^\infty g(E)\langle n \rangle_\theta dE, \tag{5.59}$$

where we've used the free-particle density of states, $g(E)$, Eq. (2.18), and where we've introduced the average occupation numbers,

$$\langle n \rangle_\theta = \frac{1}{e^{\beta(E-\mu)} - \theta}. \qquad (\theta = \pm 1) \tag{5.60}$$

The functions implied by Eq. (5.60) for $\theta = \pm 1$ are called the *Bose-Einstein distribution* function ($\theta = 1$),

$$\langle n \rangle_{BE} = \frac{1}{e^{\beta(E-\mu)} - 1}, \tag{5.61}$$

and the *Fermi-Dirac distribution* function ($\theta = -1$),

$$\langle n \rangle_{FD} = \frac{1}{e^{\beta(E-\mu)} + 1}. \tag{5.62}$$

For fermions, $0 < \langle n \rangle_{FD} \leq 1$, which (as we'll discuss) reflects the requirements of the Pauli principle. There is no restriction on $\langle n \rangle_{BE}$ (other than $\langle n \rangle_{BE} > 0$) because $\mu \leq 0$ for bosons. These functions (Eqs. (5.61) and (5.62)) are fundamental to any discussion of identical bosons or fermions. We've derived them from the partition function in the grand canonical ensemble, the appropriate ensemble if the chemical potential is involved. There's another method, however, for deriving these functions that's often presented in textbooks, the method of the most probable distribution, shown in Appendix F.

The structure of Eq. (5.59) *is a generic result in statistical mechanics*:

$$\langle N \rangle = \int_0^\infty g(E)n(E)dE. \tag{5.63}$$

The average number of particles is obtained from a sum over energy levels, with the density of states $g(E)$ telling us the number of allowed states per energy range, multiplied by $n(E)$, the average number of particles occupying those states in thermal equilibrium. One is from quantum mechanics—the allowed states of the system, the other is from statistical physics, the number of particles actually occupying those states in equilibrium at temperature T and chemical potential μ.

---

[28]We use the energy levels of the Schrödinger equation only to make contact with the density of states, Eq. (2.18). The Bose-Einstein and Fermi-Dirac distributions for occupation numbers are independent of the nature of the allowed energies, and, for example, can be used in relativistic applications of quantum statistics. Only the density of states function $g(E)$ requires modification in relativistic calculations.

5.5.3 Equation of state, fugacity expansions

From Eq. (4.76), $\Phi = -kT \ln Z_G$. For a system with V as the only relevant external parameter, $\Phi = -PV$, Eq. (4.69). For such a system, $PV = kT \ln Z_G$. Using Eq. (5.55) for Z_G, we have the equation of state:

$$PV = -\theta g k T \sum_k \ln\left(1 - \theta e^{\beta(\mu - E_k)}\right) . \tag{5.64}$$

The sum in Eq. (5.64) can be converted to an integral, $\sum_k \longrightarrow (V/(8\pi^3)) \int d^3 k$ (see Section 2.1.5). Thus,

$$PV = -\theta g k T \frac{V}{8\pi^3} \int d^3 k \ln\left(1 - \theta e^{\beta(\mu - E_k)}\right) . \tag{5.65}$$

Because E_k is isotropic in k-space, work with spherical coordinates:

$$PV = -g\theta k T \frac{V}{8\pi^3} \int_0^\infty 4\pi k^2 dk \ln\left(1 - \theta e^{\beta[\mu - \hbar^2 k^2/(2m)]}\right) .$$

Change variables: Let $x \equiv \beta \hbar^2 k^2/(2m)$, a dimensionless variable. Then,

$$P = -\frac{g\theta k T}{\lambda_T^3} \frac{2}{\sqrt{\pi}} \int_0^\infty dx \sqrt{x} \ln\left(1 - \theta e^{\beta\mu - x}\right) .$$

Integrate by parts:

$$P(T, \mu) = g \frac{kT}{\lambda_T^3} \frac{1}{\Gamma(5/2)} \int_0^\infty \frac{x^{3/2}}{e^{x - \beta\mu} - \theta} dx . \tag{5.66}$$

Note that Eq. (5.66) provides an expression for P that's *intensive* in character; $P = P(T, \mu)$ is independent of V. Pressure P, which has the dimension of energy density, is equal to kT, an energy, divided by λ_T^3, a volume, multiplied by a dimensionless function of $e^{\beta\mu}$ (the integral in Eq. (5.66)).

The two equations implied by Eq. (5.66) (one for each of $\theta = \pm 1$) involve a class of integrals known as *Bose-Einstein integrals* for $\theta = 1$, $G_n(z)$, defined in Eq. (B.18), and *Fermi-Dirac integrals* for $\theta = -1$, $F_n(z)$, defined in Eq. (B.29), where $z = e^{\beta\mu}$. Written in terms of these functions, we have from Eq. (5.66),

$$P(T, \mu) = g \frac{kT}{\lambda_T^3} \begin{cases} G_{5/2}(z) & \theta = +1 \\ F_{5/2}(z) & \theta = -1 . \end{cases} \tag{5.67}$$

Because for bosons $0 < z \le 1$, an expansion in powers of z can be developed for $G_n(z)$, a *fugacity expansion* (see Eq. (B.21)). For fermions there is no restriction on z (only that $z > 0$). A fugacity expansion can be developed for $F_n(z)$ for $0 < z \le 1$ (see Eq. (B.32)); for $z \gg 1$, one has to rely on asymptotic expansions (see Eq. (B.40)), or numerical integration.

Using Eqs. (B.21) and (B.32) for the small-z forms of $G_n(z), F_n(z)$, we have from Eq. (5.67),

$$P = g \frac{kT}{\lambda_T^3} z \left[1 + \theta \frac{z}{2^{5/2}} + \frac{z^2}{3^{5/2}} + \theta \frac{z^3}{4^{5/2}} + \cdots\right] . \qquad (\theta = \pm 1) \tag{5.68}$$

For the classical ideal gas, the chemical potential is such that $z = e^{\beta\mu} = n\lambda_T^3$ (see Eq. (P1.3)). In the classical regime, $n\lambda_T^3 \ll 1$, implying *classical behavior is associated with $z \ll 1$*. Ignoring the terms in square brackets in Eq. (5.68), we recover the equation of state for the classical ideal gas, $P = nkT$ (the degeneracy factor g "goes away," for reasons we explain shortly). Fugacity is a proxy for pressure (see page 100), and we see from Eq. (5.68) that $z \propto P$ in the classical limit.

From Eqs. (5.66) or (5.68) we have the pressure in terms of μ and T. The chemical potential, however, is not easily measured, and P is more conveniently expressed in terms of the density $n \equiv N/V$. It can be shown using Eqs. (5.59), (B.18), and (B.29), that

$$N(T, V, \mu) = \frac{gV}{\lambda_T^3} \begin{cases} G_{3/2}(z) & \theta = +1 \\ F_{3/2}(z) & \theta = -1 . \end{cases} \tag{5.69}$$

From Eq. (5.69) and Eqs. (B.21) and (B.32), we find a fugacity expansion for the density:

$$\frac{\lambda_T^3}{g} n = z \left(1 + \theta \frac{z}{2^{3/2}} + \frac{z^2}{3^{3/2}} + \theta \frac{z^3}{4^{3/2}} + \cdots \right).$$ (5.70)

We can *invert* the series in Eq. (5.70) to obtain z in terms of a *density expansion*.[29] Working consistently to terms of third order, we find

$$z = \frac{\lambda_T^3 n}{g} \left[1 - \frac{\theta}{2^{3/2}} \left(\frac{\lambda_T^3 n}{g} \right) + \left(\frac{1}{4} - \frac{1}{3^{3/2}} \right) \left(\frac{\lambda_T^3 n}{g} \right)^2 + \cdots \right].$$ (5.71)

When $n\lambda_T^3 \ll 1$, the terms in square brackets in Eq. (5.71) can be ignored, leaving us (in this limit) with $z = n\lambda_T^3/g$, *a generalization of what we found from thermodynamics*, Eq. (P1.3), to include spin degeneracy. Thus, from Eq. (5.68), $P = nkT$ in the classical limit.

Substitute Eq. (5.71) into the fugacity expansion for the pressure, Eq. (5.68). We find, working to third order,

$$P = nkT \left[1 + a_1 \theta \left(\frac{\lambda_T^3 n}{g} \right) + a_2 \left(\frac{\lambda_T^3 n}{g} \right)^2 + \cdots \right]$$ (5.72)

where

$$a_1 = -\frac{1}{2^{3/2}} = -0.1768 \qquad a_2 = -\left(\frac{2}{3^{5/2}} - \frac{1}{8} \right) = -0.0033.$$

A density expansion of P (as in Eq. (5.72)) is called a virial expansion. For classical gases, the terms in a virial expansion result from interactions between particles (see Section 6.2). The terms in square brackets in Eq. (5.72) are a result of the quantum nature of identical particles and not from "real" interactions. That is, nominally noninteracting quantum particles act in such a way as to be effectively interacting. The requirements of permutation symmetry imply a type of interaction between particles—because of Eq. (4.110), correlations exist among nominally noninteracting particles, *a state of matter not encountered in classical physics*. We see from Eq. (5.72) that the pressure of a dilute gas of fermions (bosons) is greater (lesser) than the pressure of the classical ideal gas. The Pauli principle effectively introduces a repulsive force between fermions: Configurations of particles in identical states occupying the same spatial position do not occur in the theory, *as if* there was a repulsive force between particles. For bosons, it's as if there's an attractive force between particles.

5.5.4 Thermodynamics

We can now consider the other thermodynamic properties of ideal quantum gases. From Eq. (4.82),

$$U = -\left(\frac{\partial \ln Z_G}{\partial \beta} \right)_{z,V} = g \sum_k \frac{E_k}{e^{\beta(E_k - \mu)} - \theta} \longrightarrow g \frac{V}{8\pi^3} \int d^3k \frac{E_k}{e^{\beta(E_k - \mu)} - \theta}$$ (5.73)

where we've used Eq. (5.55). Using the same steps as in previous sections, we find

$$U(T, V, \mu) = \frac{3}{2} \frac{gV}{\lambda_T^3} kT \begin{cases} G_{5/2}(z) & \theta = +1 \\ F_{5/2}(z) & \theta = -1. \end{cases}$$ (5.74)

Equation (5.74) is quite similar to Eq. (5.67), which we may use to conclude that

$$U = \frac{3}{2} PV \implies P = \frac{2}{3} \frac{U}{V}.$$ (5.75)

---

[29] A given series $y = a_1 x + a_2 x^2 + a_3 x^3 + \cdots$ can be inverted to develop a series for $x = A_1 y + A_2 y^2 + A_3 y^3 + \cdots$, where $A_1 = a_1^{-1}$, $A_2 = -a_1^{-3} a_2$, $A_3 = a_1^{-5} \left(2a_2^2 - a_1 a_3 \right)$, etc. See example [50, p16].

Equation (5.75) holds for *all ideal gases*, which is noteworthy because the equations of states of the classical and two quantum ideal gases are different.[30]

To calculate the heat capacity, it's useful[31] to divide Eq. (5.74) by Eq. (5.69):

$$\frac{U}{N} = \frac{3}{2}kT \begin{cases} \dfrac{G_{5/2}(z)}{G_{3/2}(z)} & \theta = +1 \\[2ex] \dfrac{F_{5/2}(z)}{F_{3/2}(z)} & \theta = -1 \end{cases} \equiv \frac{3}{2}kT\frac{A_{5/2}(z)}{A_{3/2}(z)}, \tag{5.76}$$

where, to save writing, $A_n(z)$ denotes either $G_n(z)$ or $F_n(z)$. Then, using Eq. (5.76),

$$\begin{aligned} C_V = \left(\frac{\partial U}{\partial T}\right)_{V,N} &= \frac{3}{2}NkT\frac{\partial}{\partial T}\left(\frac{A_{5/2}(z)}{A_{3/2}(z)}\right)\bigg|_n + \frac{3}{2}N\frac{A_{5/2}(z)}{A_{3/2}(z)} \\ &= \frac{3}{2}NkT\left[\frac{A'_{5/2}}{A_{3/2}} - \frac{A_{5/2}A'_{3/2}}{A^2_{3/2}}\right]\left(\frac{\partial z}{\partial T}\right)\bigg|_n + \frac{3}{2}N\frac{A_{5/2}(z)}{A_{3/2}(z)}, \end{aligned} \tag{5.77}$$

where primes indicate a derivative with respect to z, and where (shown in Exercise 5.14)

$$\left(\frac{\partial z}{\partial T}\right)\bigg|_n = -\frac{3}{2}\frac{z}{T}\frac{A_{3/2}(z)}{A_{1/2}(z)}. \tag{5.78}$$

Combining Eqs. (5.78) and (5.77), and making use of Eqs. (B.20) and (B.31), we find

$$\frac{C_V}{Nk} = \frac{15}{4}\frac{A_{5/2}(z)}{A_{3/2}(z)} - \frac{9}{4}\frac{A_{3/2}(z)}{A_{1/2}(z)}. \tag{5.79}$$

Equation (5.79) applies for fermions and bosons. It's easy to show, using either Eq. (B.21) or (B.32) that as $z \to 0$, $C_V \to Nk(\frac{15}{4} - \frac{9}{4}) = \frac{3}{2}Nk$, the classical value. We'll examine the low-temperature properties in upcoming sections.

An efficient way to calculate the entropy is to use the Euler relation, Eq. (1.53),

$$\begin{aligned} S &= \frac{1}{T}(U + PV - N\mu) = \frac{1}{T}\left(\frac{5}{3}U - N\mu\right) \\ &= Nk\left(\frac{5}{2}\frac{A_{5/2}(z)}{A_{3/2}(z)} - \ln z\right), \end{aligned} \tag{5.80}$$

where we've used Eqs. (5.75) and (5.76) and the definition of $z = e^{\beta\mu}$. (See Exercise 5.15.) The form of Eq. (5.80) for small z is, using Eqs. (B.21) and (B.32) and Eq. (5.71) at lowest order,

$$S = Nk\left[\frac{5}{2} - \ln\left(\frac{\lambda^3_T n}{g}\right)\right], \qquad (z \ll 1) \tag{}$$

a generalization of the Sackur-Tetrode formula to include spin degeneracy (see Eq. (1.64)).

Equation (5.75) allows an easy way to calculate the heat capacity in the regime $n\lambda^3_T \ll 1$, without the heavy machinery of Eq. (5.79). From

$$C_V = \left(\frac{\partial U}{\partial T}\right)_V = \frac{3}{2}\left(\frac{\partial(PV)}{\partial T}\right)_V = \frac{3}{2}Nk\left[1 - \frac{1}{2}a_1\theta\left(\frac{n\lambda^3_T}{g}\right) - 2a_2\left(\frac{n\lambda^3_T}{g}\right)^2\right] \tag{5.81}$$

---

[30]This is a good place to interject the comment that pressure has the dimension of energy per volume, energy density, as we see in Eq. (5.75). In certain fields of study, energy densities are quoted in pascals, often gigapascals (GPa).

[31]Keep in mind we're working in the grand canonical ensemble, where N is not a fixed quantity.

where we've used Eq. (5.72). Putting in numerical values,

$$C_V = \frac{3}{2} N k \left[1 + 0.0884 \theta \left(\frac{n \lambda_T^3}{g} \right) + 0.0066 \left(\frac{n \lambda_T^3}{g} \right)^2 \cdots \right]. \tag{5.82}$$

As $T \to \infty$, C_V approaches its classical value, $\frac{3}{2} N k$. For bosons ($\theta = +1$) the value of C_V for large but finite temperatures, is larger than the classical value, implying that the *slope* of C_V versus T is negative, which a calculation of $\partial C_V / \partial T$ from Eq. (5.82) shows. We know that heat capacities must vanish as $T \to 0$, a piece of physics not contained in Eq. (5.82), a high-temperature result.

As Eqs. (5.72) and (5.82) show, there are minor differences between the two types of quantum gases in the regime $n \lambda_T^3 \ll 1$, with the upshot that the two cases can be treated in tandem (as we've done in Sections 5.5.3 and 5.5.4). We now consider the opposite regime $(n \lambda_T^3)/g \gtrsim 1$ in which quantum effects are more clearly exhibited, in which Fermi and Bose gases behave differently, and which must be treated separately.

5.6 DEGENERATE FERMI GAS: $T = 0$ AND $0 < T \ll T_F$

We start with the Fermi gas, which is simpler. At a given temperature, particles distribute themselves so as to minimize the energy.[32] Our expectation is that as the temperature is lowered, progressively more of the lower-energy states become occupied (such as we see in Fig. 5.1). Because of the Pauli principle, however, fermions cannot accumulate in any energy level, which, as a consequence, implies that at low temperature the lowest-energy configuration of a system of noninteracting fermions is to have one particle in every energy level, starting at the lowest possible energy, with all energy levels filled until the supply of particles is exhausted. The energy of the top-most-occupied state (technically at $T = 0$) is the *Fermi energy*, E_F, the zero-point energy of a collection of noninteracting fermions. *This is the most important difference between Fermi and Bose gases.* The state of matter in which all the lowest energy states of the system are occupied is called *degenerate*. Particles of a degenerate Fermi gas occupy states of high kinetic energy at low temperature.[33] The *degeneracy parameter*

$$\delta \equiv \frac{n \lambda_T^3}{g} = \frac{n}{g} \frac{h^3}{(2 \pi m k T)^{3/2}} \tag{5.83}$$

distinguishes systems in which quantum effects are small ($\delta \ll 1$) from those in which there are strong quantum effects ($\delta \gtrsim 1$).[34]

5.6.1 Identical fermions at $T = 0$

In the limit $T \to 0$, the Fermi factor becomes a step function,[35]

$$n(E) = \frac{1}{e^{\beta(E - \mu)} + 1} \xrightarrow{T \to 0} \begin{cases} 1 & E < E_F \\ 0 & E > E_F \end{cases} \equiv \theta(E_F - E), \tag{5.84}$$

---

[32] Recall the Gibbs criteria for equilibrium (Section 1.12), that for a system of fixed entropy, the equilibrium state is one of minimum energy, and for fixed energy, the equilibrium state is one of maximum entropy. In either case, the equilibrium state is one of minimum Helmholtz energy $F = U - TS$ in the canonical ensemble (see Table 1.2) or, for the grand canonical ensemble, minimum grand potential $\Phi = U - TS - N\mu$ (see Exercise 1.11).

[33] We treat the degenerate Bose gas in Section 5.9.

[34] As noted in Section 1.11, $\delta \ll 1$ is a regime in which particles can be treated as independent of each other, for which we expect classical behavior of noninteracting particles. In the regime $\delta > 1$, however, even nominally noninteracting particles are not independent of one another and become effectively coupled through the wave nature of matter.

[35] Up to now, we have consistently referred to Eq. (5.62) as the Fermi-Dirac distribution function, which becomes a bit of a mouthful after a while. Equation (5.62) is also known simply as the *Fermi factor*.

where $E_F \equiv \mu(T = 0)$ is the zero-temperature limit[36] of the chemical potential, and $\theta(x)$ is the

Heaviside step function, $\theta(x) \equiv \begin{cases} 0 & x < 0 \\ 1 & x > 0 \end{cases}$, the graph of which is shown in Fig. 5.7.

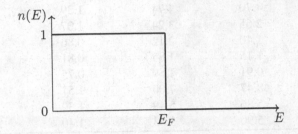

Figure 5.7: The Fermi-Dirac distribution at $T = 0$. States for which $E < E_F$ are all occupied ($n = 1$) and states for which $E > E_F$ are all unoccupied ($n = 0$).

The Fermi energy is easy to calculate. Combining Eq. (5.84) with Eq. (5.59),

$$N = \int_0^\infty g(E)n(E)\mathrm{d}E = \int_0^{E_F} g(E)\mathrm{d}E = g\frac{V}{4\pi^2}\left(\frac{2m}{\hbar^2}\right)^{3/2} \int_0^{E_F} \sqrt{E}\mathrm{d}E$$

$$= g\frac{V}{6\pi^2}\left(\frac{2m}{\hbar^2}\right)^{3/2} E_F^{3/2} \tag{5.85}$$

implying

$$E_F = \frac{\hbar^2}{2m}\left(\frac{6\pi^2}{g}n\right)^{2/3}. \tag{5.86}$$

The energy of the "last" particle added to the system, E_F, is therefore the chemical potential of the N-particle system—the energy required to add one more particle. From Eq. (5.86), we see that the higher the density, the larger the Fermi energy—adding more particles to the same volume requires progressively higher and higher energy levels to be filled. Equation (5.86) implies an equivalent temperature associated with particles at the Fermi energy, the *Fermi temperature*, T_F defined so that $kT_F = E_F$,

$$T_F = \frac{\hbar^2}{2mk}\left(\frac{6\pi^2}{g}n\right)^{2/3}. \tag{5.87}$$

It should not be construed that particles actually have the Fermi temperature (the temperature here is $T = 0$); T_F is a convenient way to characterize the Fermi energy.

The Fermi energy can be quite large relative to other energies. Table 5.2 lists the density n of *free electrons* (those available to conduct electricity[37]) for selected elements in their metallic states, together with the value of E_F as calculated from Eq. (5.86) using $g = 2$ for electrons ($S = \frac{1}{2}$). We can associate with E_F an equivalent speed known as the *Fermi velocity*, $v_F \equiv \sqrt{2E_F/m}$. The Fermi velocity is the *speed* of an electron having energy E_F. Compare such speeds ($\approx 10^6$ m/s at metallic densities) with thermal speeds in gases at room temperature, ≈ 500 m/s (see Table 5.1). That v_F can be so large (at *zero* temperature) is a purely quantum effect—as more particles are added to the same volume, they must occupy progressively higher energy levels because of the Pauli

---

[36]We can never achieve a state of matter associated with zero Kelvin (third law of thermodynamics; see [3, Chapter 8]). As we'll soon see, there is another temperature associated with a collection of fermions, $T_F \equiv E_F/k$, the *Fermi temperature* that can be quite large (depending on the density). Even at room temperature T, it's common to have $T \ll T_F$ and thus it's often the case we can invoke the *approximation* of $T = 0$ for systems of fermions in which $T \neq 0$.

[37]The density of charge carriers n can be measured from the Hall effect.

Table 5.2: Electron densities, Fermi energies, velocities, and temperatures of selected metallic elements. Source: N.W. Ashcroft and N.D. Mermin, *Solid State Physics*[18].

Element	n (10^{22} cm^{-3})	E_F (eV)	v_F (10^6 m s^{-1})	T_F (10^4 K)
Li	4.70	4.74	1.29	5.51
Na	2.65	3.24	1.07	3.77
K	1.40	2.12	0.86	2.46
Rb	1.15	1.85	0.81	2.15
Cs	0.91	1.59	0.75	1.84
Cu	8.47	7.00	1.57	8.16
Ag	5.86	5.49	1.39	6.38
Au	5.90	5.53	1.40	6.42

principle, which at metallic densities are ≈ 5 eV. The thermal equivalent of 1 eV is 11,600 K (show this), implying $T_F \approx 50,000$ K. For sufficiently large densities—such as occur in astrophysical applications—fermions would have to be treated relativistically (see Section 5.7). Electron densities in metals ($n \approx 10^{22}$ cm^{-3}) should be compared with the densities of ordinary gases. At *standard temperature and pressure* (STP), $T = 273.15$ K and $P = 1$ atm, the density of the ideal gas is $n \approx 2.7 \times 10^{19}$ cm^{-3}.

Another way to calculate E_F is to work in k-space (see Section 2.1.5). Associated with the Fermi energy is a wavevector, the (you guessed it) *Fermi wavevector* k_F such that

$$\frac{\hbar^2}{2m} k_F^2 \equiv E_F .$$
(5.88)

Between Eqs. (5.88) and (5.86),

$$k_F = \left(\frac{6\pi^2}{g} n \right)^{1/3} .$$
(5.89)

The Fermi wavevector therefore probes a distance $d \sim n^{-1/3} = (V/N)^{1/3}$, the distance between particles. As the density increases, the distance between particles decreases, and k_F increases.

The Fermi wavevector defines a surface in k-space, the *Fermi surface*, a sphere of radius k_F (the *Fermi sphere*), that separates filled from unfilled energy states, depicted in Fig. 5.8. Each allowed

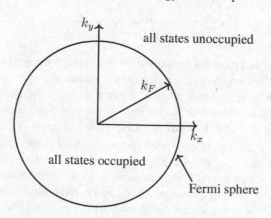

Figure 5.8: The Fermi sphere separates states occupied for $k \leq k_F$ and unoccupied for $k > k_F$. Third dimension not shown.

k-vector is uniquely associated with a small volume of k-space, $8\pi^3/V$ (Section 2.1.5). Each k-vector (representing solutions of the free-particle Schrödinger equation satisfying periodic boundary conditions) is associated with $g = 2S + 1$ energy states, when spin degeneracy is accounted for. The number of particles N in the lowest-energy state,[38] those with k-vectors lying within the Fermi sphere, can be found from the number of allowed k-vectors interior to the Fermi surface, multiplied by g:

$$N = g\frac{(4\pi/3)k_F^3}{(8\pi^3/V)} = g\frac{V}{6\pi^2}k_F^3 , \tag{5.90}$$

a calculation that reproduces Eq. (5.89).

What is the *energy* of the ground-state configuration? Is it E_F? From Eq. (5.73) it can be shown that

$$U = \int_0^\infty Eg(E)n(E)\mathrm{d}E . \tag{5.91}$$

Equation (5.91) has a similar structure to Eq. (5.63): The average energy of the system is given by an integral over the density of states $g(E)$ (number of allowed states per energy range), multiplied by the average number of particles occupying those states in thermal equilibrium $n(E)$ (either the Fermi-Dirac or Bose-Einstein distributions), multiplied by the energy, the quantity we're trying to find the average of. Combining Eq. (5.84) with Eq. (5.91),

$$U = \int_0^{E_F} Eg(E)\mathrm{d}E = g\frac{V}{4\pi^2}\left(\frac{2m}{\hbar^2}\right)^{3/2}\int_0^{E_F} E^{3/2}\mathrm{d}E = \frac{2}{5}g\frac{V}{4\pi^2}\left(\frac{2m}{\hbar^2}\right)^{3/2}E_F^{5/2} . \tag{5.92}$$

Divide Eq. (5.92) by Eq. (5.87),

$$\frac{U}{N} = \frac{3}{5}E_F . \tag{5.93}$$

Thus, the average energy per particle (at $T = 0$) is $\frac{3}{5}$ of the Fermi energy. That implies the mean speed per particle is $\sqrt{3/5}v_F \approx 0.77v_F$. The pressure of the Fermi gas can be found by combining Eq. (5.75) with Eq. (5.93):

$$P = \frac{2}{3}\frac{U}{V} = \frac{2}{3}\frac{U}{N}\frac{N}{V} = \frac{2}{5}nE_F = \frac{\hbar^2}{5m}\left(\frac{6\pi^2}{g}\right)^{2/3}n^{5/3} . \tag{5.94}$$

Because of its large zero-point energy, the Fermi gas has a considerable pressure[39] at $T = 0$. Equation (5.94) specifies the *degeneracy pressure* of the Fermi gas at $T = 0$. Can the degeneracy pressure be measured? The *bulk modulus* B is the inverse of the isothermal compressibility,

$$B \equiv -V\left(\frac{\partial P}{\partial V}\right)_T = \frac{2}{5}nE_F = \frac{5}{3}P , \tag{5.95}$$

where we've used Eq. (5.94). (Note that for the classical ideal gas, $B = P$.) Table 5.3 lists values of B as calculated from Eq. (5.95) and measured values. The agreement is satisfactory, but even when the predictions are considerably off, it gets you on the same page as the data. The degeneracy pressure alone can't explain the bulk modulus of real materials, but it's an effect at least as important as other physical effects. The degeneracy pressure underscores that the Pauli principle in effect introduces a repulsive force between identical fermions, a rather strong one. Note that nowhere have we brought in the electron *charge*—the predictions of this section would apply equally as well for a collection of neutrons. *Why* the Fermi gas model works so well for electrons in solids was something of a puzzle when the theory was developed by Sommerfeld in 1928. Coulomb interactions between electrons would seemingly invalidate the assumptions of an ideal gas. The answer came only later,

---

[38]Appreciate that N particles are *all* associated with the lowest-energy state of the system, a collective state of matter.
[39]For Bose gases, $P \to 0$ as $T \to 0$—the molecules of a Bose gas have zero momentum at $T = 0$, whereas fermions have considerable momentum at $T = 0$, $p_F = mv_F$.

Table 5.3: Bulk modulus (Gpa) of selected elements in their metallic states. Source: N.W. Ashcroft and N.D. Mermin, *Solid State Physics*[18].

Element	B (Eq. (5.95))	B (expt)
Li	23.9	11.5
Na	9.23	6.42
K	3.19	2.81
Rb	2.28	1.92
Cs	1.54	1.43
Cu	63.8	134.3
Ag	34.5	99.9
Al	228	76.0

in the development of *Fermi liquid theory* (beyond the scope of this book). In solids, electrons *screen* the positive charges of the ions that remain on the lattice sites of a crystalline solid, after each atom donates a few valence electrons to the lattice. As a result, electrons in metals act as almost-independent "quasiparticles," a topic we won't develop in this book.

5.6.2 Finite-temperature Fermi gas, $0 < T \ll T_F$

For $T \neq 0$, we can no longer use the "easy" equations (5.87) and (5.92). Instead we must use Eqs. (5.69) and (5.74), which, for arbitrary temperatures, must be handled numerically. At sufficiently low temperatures, however, deviations from the results at $T = 0$ are small, and we can approximate Eqs. (5.63) and (5.91) appropriately. Of course, we have to quantify what constitutes low temperature, which, as we'll see, are temperatures $T \ll T_F$. Because T_F can be rather large (for metallic densities), one often has systems for which room temperature can be considered "low temperature."

Figure 5.9 shows the Fermi-Dirac distribution at finite temperature. We see a softening of the

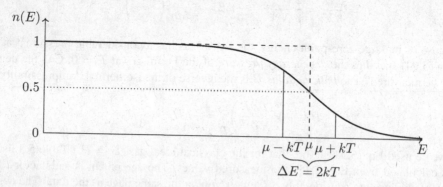

Figure 5.9: The Fermi-Dirac distribution at finite temperature (solid line). Dashed line corresponds to $T = 0$. Temperature was chosen so that $kT = 0.1\mu$.

characteristic sharp edge between occupied and unoccupied states that occurs for $T = 0$ (see Fig. 5.7). States at $E = \mu - kT$ are not all occupied ($n(E) < 1$), and states at $E = \mu + kT$ are not all unoccupied ($n(E) > 0$). The temperature used to make Fig. 5.9 is such that $kT = 0.1\mu$, which is actually quite large (for systems of metallic density). Such a temperature was chosen so that the features of the Fermi distribution could be easily discerned in a figure such as Fig. 5.9. For a more realistic temperature such as $kT = 0.01\mu$, the transition region of approximate energy width $\Delta E \approx 2kT$ would occur over a smaller energy range and be more difficult to display.

Referring to Fig. 5.9, the unoccupied states for energy $\mu - \alpha kT$ (where α is a number) occur as a result of transitions induced by thermal energies $2\alpha kT$ to occupy states of energy $\mu + \alpha kT$. We expect that the heat capacity of a system of fermions (the ability to absorb energy) would therefore be controlled by states occurring within an approximate energy range kT of μ. We'll see that's indeed the case (see Eq. (5.106)). For $n(E)$ the average number of states occupied at energy E, its complement $(1 - n(E))$ is the average number of unoccupied states at energy E. Unoccupied states are referred to as *holes*. The product of the two, $n(E)(1 - n(E))$, is largest for energies at which particles and holes are both prevalent. As is readily shown:

$$n(E)(1 - n(E)) = \frac{e^{\beta(E-\mu)}}{\left(e^{\beta(E-\mu)} + 1\right)^2}. \tag{5.96}$$

Note that the right side of Eq. (5.96) (plotted in Fig. 5.10) is an even function[40] of $E - \mu$.

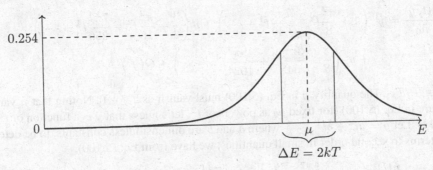

Figure 5.10: Product $n(E)(1 - n(E))$ of the number of occupied states and the number of holes (unoccupied states) versus E. Note change in vertical scale from Fig. 5.9. $kT = 0.1\mu$.

5.6.2.1 *Chemical potential*

Referring to Eq. (5.63), at low temperatures, $kT \ll \mu$, i.e., $\beta\mu \gg 1$, the Fermi distribution transitions from 1 to 0 over a narrow energy range about $E = \mu$. Under these circumstances there exists a method for accurately approximating integrals such as we have in Eq. (5.63), the *Sommerfeld expansion*, Eq. (B.49). Using Eq. (5.69),

$$
\begin{aligned}
N(T, V, \mu) &= \frac{gV}{\lambda_T^3} F_{3/2}(\beta\mu) = \frac{gV}{\lambda_T^3} \frac{1}{\Gamma(3/2)} \int_0^\infty \frac{\sqrt{x}}{e^{x - \beta\mu} + 1} dx \\
&= \frac{gV}{\lambda_T^3} \frac{1}{\Gamma(3/2)} \left[\int_0^{\beta\mu} \sqrt{x} dx + \frac{\pi^2}{12\sqrt{\beta\mu}} + \frac{7\pi^4}{960(\beta\mu)^{5/2}} + \cdots \right] \\
&= \frac{gV}{\lambda_T^3} \frac{1}{\Gamma(3/2)} \frac{2}{3}(\beta\mu)^{3/2} \left[1 + \frac{\pi^2}{8} \frac{1}{(\beta\mu)^2} + \frac{7\pi^4}{640} \frac{1}{(\beta\mu)^4} + \cdots \right],
\end{aligned} \tag{5.97}
$$

where in the second line we've applied the Sommerfeld expansion to second order in small quantities. Equation (5.97) implies, for the density,

$$\left(\frac{\hbar^2}{2m}\right)^{3/2} \frac{6\pi^2}{g} n(T, \mu) = \mu^{3/2} \left[1 + \frac{\pi^2}{8(\beta\mu)^2} + \frac{7\pi^4}{640(\beta\mu)^4} + \cdots \right].$$

---

[40]The right side of Eq. (5.96) had *better* be symmetric about $E = \mu$; in thermal equilibrium, there are just as many positive energy transfers to the system as there are negative energy transfers to the environment.

This equation simplifies upon dividing by $E_F^{3/2}$:

$$\frac{n(T,\mu)}{n_0} = \left(\frac{\mu}{E_F}\right)^{3/2}\left[1 + \frac{\pi^2}{8(\beta\mu)^2} + \frac{7\pi^4}{640(\beta\mu)^4} + \cdots\right], \tag{5.98}$$

where n_0 denotes the density at $T = 0$. Our goal is to invert Eq. (5.98),[41] to obtain an implicit expression for $\mu(n,T)$ for temperatures $T \ll T_F$.

Equation (5.98) reduces to an identity at $T = 0$: $1 = 1$. Let's write in Eq. (5.98), $\mu \equiv E_F(1+y)$, where y is a dimensionless function that vanishes as $T \to 0$:

$$\frac{n(T,y)}{n_0} = (1+y)^{3/2}\left[1 + \frac{\pi^2(T/T_F)^2}{8(1+y)^2} + \frac{7\pi^4(T/T_F)^4}{640(1+y)^4} + \cdots\right]. \tag{5.99}$$

Expand the right side of Eq. (5.99) to second order in y. We find:

$$\frac{n(t,y)}{n_0} = y^0\left(1 + \frac{\pi^2}{8}t^2 + \frac{7\pi^4}{640}t^4 + \cdots\right) + y\left(\frac{3}{2} - \frac{\pi^2}{16}t^2 - \frac{7\pi^4}{256}t^4 + \cdots\right)$$
$$+ y^2\left(\frac{3}{8} + \frac{3\pi^2}{64}t^2 + \frac{49\pi^4}{1024}t^4 + \cdots\right) + O(y^3), \tag{5.100}$$

where $t \equiv T/T_F$. The quantity y in Eq. (5.100) must vanish as $T \to 0$. Noting that n varies with temperature in Eq. (5.100) (for fixed μ) as powers of t^2, let's guess that y is a function of t^2 as well (for small t). Let $y = at^2 + bt^4 + \cdots$ where a and b are dimensionless constants, to be determined. Keeping terms to second order in small quantities, we have from Eq. (5.100),

$$\frac{n(t)}{n_0} = 1 + t^2\left[\frac{\pi^2}{8} + \frac{3}{2}a\right] + t^4\left[\frac{7\pi^4}{640} - \frac{\pi^2}{16}a + \frac{3}{2}b + \frac{3}{8}a^2\right] + \cdots. \tag{5.101}$$

Let's find solutions of Eq. (5.101) at *fixed density*, which we take to be the zero-temperature value, n_0 (which asserts itself through the value of E_F, or T_F). With that assumption, setting $n(t) = n_0$, the terms in square brackets in Eq. (5.101) must vanish, implying $a = -\pi^2/12$ and $b = -\pi^4/80$. Thus,

$$\mu(n,T) = E_F\left[1 - \frac{\pi^2}{12}\left(\frac{T}{T_F}\right)^2 - \frac{\pi^4}{80}\left(\frac{T}{T_F}\right)^4 + \cdots\right], \tag{5.102}$$

where the density dependence is implicit through the value of T_F. For $T \ll T_F$, we can treat systems *as if* the temperature is $T = 0$.

The chemical potential for a system of fermions *decreases* for $T > 0$. As the temperature is raised from $T = 0$, particles are promoted to states of energies $E > E_F$. We note that μ is, from the definition of the Fermi-Dirac distribution, the energy of the state for which the average occupation number $n(E = \mu) = 0.5$ if $\mu > 0$. As particles are promoted to higher-lying states, μ must shift downward—the state for which $n(E) = 0.5$ occurs at a smaller energy. As the temperature is increased so that $T \lesssim T_F$, a detailed analysis shows that $\mu \to 0$, implying that the state for which $n(E) = 0.5$ occurs at $E = 0$. At even higher temperatures, μ becomes negative. To see this, consider the value of $n(E = 0) = (e^{-\beta\mu} + 1)^{-1}$. One has, for any value of $\beta\mu$, $n(E = 0) < 1$, and we know that for $T \ll T_F$, $n(E = 0)$ is almost unity, $(e^{-\beta E_F} + 1)^{-1}$. One can show that if $\mu > 0$, then $n(E = 0) > \frac{1}{2}$. To have $n(E = 0) < \frac{1}{2}$, it must be the case that $\mu < 0$, at which point μ loses its interpretation as the energy for which $n(E) = 0.5$. For $\mu < 0$, we have that the limiting form of the Fermi factor for $T \to \infty$ is $n(E) = \frac{1}{2}$ for all E—the average value of $n(E) = 0.5$ for every state; every state is either occupied or unoccupied with equal probability in the extreme high-temperature limit.

---

[41] Keep in mind that in the grand canonical ensemble, (μ, V, T) are givens, with $N = N(T, V, \mu)$ an implied quantity.

5.6.2.2 Internal energy

Starting from Eq. (5.74) and using the Sommerfeld expansion, Eq. (B.49), we find

$$U(T,V,\mu) = \frac{gV}{10\pi^2}\left(\frac{2m}{\hbar^2}\right)^{3/2}\mu^{5/2}\left[1+\frac{5\pi^2}{8}\frac{1}{(\beta\mu)^2}-\frac{7\pi^4}{384}\frac{1}{(\beta\mu)^4}+\cdots\right]. \qquad (5.103)$$

Divide Eq. (5.103) by Eq. (5.97); we find, working to second order,

$$\frac{U}{N} = \frac{3}{5}\mu\left[1+\frac{\pi^2}{2}\frac{1}{(\beta\mu)^2}-\frac{11\pi^4}{120}\frac{1}{(\beta\mu)^4}+\cdots\right]. \qquad (5.104)$$

By substituting Eq. (5.102) into Eq. (5.104) and working consistently to second order, we find

$$\frac{U}{N} = \frac{3}{5}E_F\left[1+\frac{5\pi^2}{12}\left(\frac{T}{T_F}\right)^2-\frac{5\pi^4}{80}\left(\frac{T}{T_F}\right)^4+\cdots\right]. \qquad (5.105)$$

Equation (5.105) is the finite-temperature generalization of Eq. (5.93).

5.6.2.3 Heat capacity

Using Eq. (5.104) we can calculate the heat capacity,

$$C_V = \left(\frac{\partial U}{\partial T}\right)_{V,N} = Nk\left[\frac{\pi^2}{2}\frac{T}{T_F}-\frac{3\pi^4}{20}\left(\frac{T}{T_F}\right)^3+\cdots\right]. \qquad (T\ll T_F) \qquad (5.106)$$

Clearly, C_V vanishes as $T\to 0$ (compare with Eq. (5.82)). We also see a characteristic feature of the Fermi gas: C_V is *linear* with T at low temperature.

One can understand $C_V \propto T$ at low temperature qualitatively.[42] As per the considerations leading to Eq. (5.96), the energy states that can participate in energy transfers between the system and its environment are those roughly within kT of μ, which at low temperature is ostensibly the Fermi energy, E_F. How many of those states are there? The number of energy states per energy range is the density of states, $g(E)$. Thus, $g(E_F)\times kT$ is approximately the number of states near E_F available to participate in energy exchanges, and is therefore the number of states that contribute to the heat capacity.[43] The excitation energy for each energy transfer is approximately kT. Thus, we can estimate the energy of the Fermi gas at low temperatures as

$$U \approx U_0 + [g(E_F)\times kT]\times kT,$$

where $U_0 = \frac{3}{5}NE_F$, Eq. (5.93). As shown in Exercise 5.19, $g(E_F)E_F = \frac{3}{2}N$. The heat capacity based on this line of reasoning would then be

$$C_V \approx 2k^2g(E_F)T = 3Nk\frac{T}{T_F}.$$

Such an argument gets you on the same page with the exact result, Eq. (5.106); the two differ by a multiplicative factor of order unity, $\pi^2/6$.

---

[42] Pay attention to this argument: It can make you look smart on a PhD oral exam.
[43] It's readily shown that $g(E_F)kT = \frac{3}{2}N(T/T_F)$, the number one would expect from the classical equipartition theorem, multiplied by T/T_F. The severe reduction in the number of states available to participate in energy exchanges with the environment is a direct consequence of the Pauli exclusion principle.

5.7 DEGENERACY PRESSURE IN THE LIFE OF STARS

Degeneracy pressure, associated with the large zero-point energy of collections of identical fermions, accounts reasonably well for the bulk modulus of metals (Section 5.6)—that compressing an electron gas meets with a significant resisting force associated with the Pauli exclusion principle. In this section we consider an astrophysical application of degeneracy pressure.

Stars generate energy through nuclear fusion, "burning" nuclei in processes that are fairly well understood. Stars convert hydrogen to helium by a series of fusion reactions: $H^1 + H^1 \rightarrow H^2 + e^+ + \nu$, $H^2 + H^1 \rightarrow He^3 + \gamma$, $He^3 + He^3 \rightarrow He^4 + H^1 + H^1$. When all hydrogen has been converted to helium, this phase of the burning process stops. Gravitational contraction then compresses the helium until the temperature rises sufficiently that a new sequence of reactions can take place, $He^4 + He^4 \rightarrow Be^8 + \gamma$, $Be^8 + He^4 \rightarrow C^{12} + \gamma$. As helium is exhausted, gravitational contraction resumes, heating the star until new burning processes are initiated. The variety of nuclear processes gets larger as new rounds of nuclear burning commence (the subject of stellar astrophysics). Eventually burning stops when the star consists of iron, silicon, and other elements. The force of gravity, however, is ever present. Can gravitational collapse can be forestalled? We now show that the degeneracy pressure of electrons in stars is enough to balance the force of gravity under certain circumstances.

We assume temperatures are sufficiently high that all atoms in stars are completely ionized, i.e., stars consist of fully-ionized plasmas, gases of electrons and positively charged species. Let there be N electrons (of mass m) and assume conditions of electrical neutrality, that $N = N_p$, the number of protons (of mass m_p). Assume that the number of neutrons N_n (locked up in nuclei) is the same as the number of protons, an assumption valid only for elements up to $Z \approx 20$; for heavier nuclei there might be ≈ 1.5 neutrons per proton. We take $N_n = N_p$ to simplify the analysis. The neutron mass is nearly equal to the proton mass. The mass M of the star is then, approximately (because $m_p \gg m$),

$$M \approx N(m + 2m_p) \approx 2Nm_p .\tag{5.107}$$

We make another simplifying assumption that the mass density $\rho = M/V$ is uniform[44] throughout a star of volume V. With these assumptions, the average electron number density $n \equiv N/V$ is

$$n = \frac{N}{V} = \frac{M/(2m_p)}{M/\rho} = \frac{\rho}{2m_p} .\tag{5.108}$$

White dwarfs are a class of stars thought to be in the final evolutionary state wherein nuclear burning has ceased and the star has become considerably reduced in size through gravitational contraction, where a star of solar mass $M_\odot$ might have been compressed into the volume of Earth. Sirius B, for example, is a white dwarf of mass $1.018 M_\odot$ and radius $8.4 \times 10^{-3} R_\odot$, implying an average electron density 7.2×10^{29} cm^{-3}, some seven orders of magnitude larger than the electron density in metals (see Table 5.2). We can take $n = 10^{30}$ cm^{-3} as characteristic of white dwarf stars. *Such high densities necessitate a relativistic treatment of the electrons* (see Exercise 5.20).

Before delving into a relativistic treatment of the electron gas, let's calculate the gravitational pressure, for which we use classical physics. Assuming the mass density of a spherical star of radius R is uniform, the contribution to the gravitational potential energy from a spherical shell of matter between r and $r + dr$ is (do you see Gauss's law at work here?),

$$dV_g = -G\frac{(4\pi r^2 dr \rho)(4\pi r^3 \rho/3)}{r} = -G\frac{16\pi^2}{3}\rho^2 r^4 dr .$$

---

[44] As an example, the sun has a mass $M_\odot = 1.989 \times 10^{30}$ kg and a radius $R_\odot = 6.957 \times 10^8$ m, implying an average mass density 1.4×10^3 kg m^{-3}. The number of protons is therefore $N_p \approx M/(2m_p) \approx 5 \times 10^{56}$, implying an electron number density $n \approx 4 \times 10^{29}$ m$^{-3} = 4 \times 10^{23}$ cm^{-3}, approximately an order of magnitude larger than the electron densities in metals; see Table 5.2. In actual stars, the mass density is inhomogeneous and varies with position.

Thus, the gravitational potential energy of a uniform, spherical mass distribution of radius R is

$$V_g = -G\frac{16\pi^2}{15}\rho^2 R^5 = -\frac{3}{5}G\frac{M^2}{R} = -\frac{12}{5}\left(\frac{4\pi}{3}\right)^{1/3}Gm_p^2 N^2 V^{-1/3} ,$$

where it's a good habit in these calculations to display the dependence on the number of electrons N and the volume V. The gravitational pressure, P_g, is found from

$$P_g \equiv -\left(\frac{\partial V_g}{\partial V}\right)_N = -\frac{4}{5}\left(\frac{4\pi}{3}\right)^{1/3}Gm_p^2 N^2 V^{-4/3} . \tag{5.109}$$

We have a negative pressure; gravity is an attractive force.

The question is: Under what conditions can the degeneracy pressure balance the gravitational pressure? Can we reach for Eq. (5.94), the previously-derived expression for the degeneracy pressure? We can't—Eq. (5.94) was derived using the density of states for solutions of the Schrödinger equation (see Section 2.1.5). For the electron densities of white dwarf stars, the formulas derived in Section 5.6.1 imply Fermi velocities in excess of the speed of light; a relativistic treatment is called for. If you examine the arguments in Sections 5.5.1 and 5.5.2, in the derivation of the Fermi-Dirac and Bose-Einstein distributions, the specific form of the energy levels is not invoked, i.e., whether or not the energy levels are solutions of the Schrödinger equation. *The distribution functions for occupation numbers require no modification for use in relativistic calculations.* The density of states function, however, must be based on the allowed energy levels of relativistic electrons, which are found from the solutions of the Dirac equation. It would take us too far afield to discuss the Dirac equation, but we can say the following. The solutions of the free-particle, time-independent Schrödinger equation are in the form of plane waves, $\psi \sim e^{i\mathbf{k}\cdot\mathbf{r}}$, with the wave vector related to the energy through $E = \hbar^2 k^2/(2m)$. The solutions of the Dirac equation (which pertains to non-interacting particles) are in the form of traveling plane waves,[45] $\psi \sim e^{i(\mathbf{k}\cdot\mathbf{r}-\omega t)}$ where $\omega \equiv E/\hbar$ and $\mathbf{k} = \mathbf{p}/\hbar$ are connected[46] through the relativistic energy-momentum relation

$$E^2 = (pc)^2 + (mc^2)^2 . \tag{5.110}$$

We've repeatedly invoked the replacement $\Sigma_{\mathbf{k}} \to (V/(8\pi)^3)\int d^3k$ in converting sums over allowed energy levels (parameterized by their association with the points of k-space permitted by periodic boundary conditions) to integrals over k-space. Does it hold for relativistic energy levels? Yes, periodic boundary conditions can be imposed on the solutions of the Dirac equation and thus the enumeration of states in k-space proceeds as before, with one modification: The energy associated with $k = 0$ is mc^2 and not simply zero as in the nonrelativistic case. Moreover, there are two linearly independent solutions of the Dirac equation for wave vectors $\mathbf{k}$ satisfying Eq. (5.110),[47] and thus $g = 2$. We can calculate k_F as previously, $N = 2\sum_{|\mathbf{k}|\le k_F} = 2\frac{V}{8\pi^3}\int_0^{k_F}4\pi k^2 dk = \frac{V}{3\pi^2}k_F^3$, implying

$$k_F = \left(3\pi^2 n\right)^{1/3} , \tag{5.111}$$

equivalent to Eq. (5.90) with $g = 2$. To find the Fermi energy, we would seemingly proceed as in Section 5.6.1: Substitute Eq. (5.111) into Eq. (5.110),

$$E_F = mc^2\sqrt{1 + \left(\frac{\hbar}{mc}(3\pi^2 n)^{1/3}\right)^2} . \qquad \text{(wrong!)} \tag{5.112}$$

---

[45] A relativistic formulation must treat space and time on equal footing; the Schrödinger equation is nonrelativistic from the outset. The plane wave can be written in covariant form, $e^{ik_\mu x^\mu}$, where $k_\mu \equiv (-\omega/c, \mathbf{k})$ denotes the four-wavevector.

[46] Note that we make use of both the Planck and the de Broglie relations, $E = \hbar\omega$ and $\mathbf{p} = \hbar\mathbf{k}$.

[47] See, for example, Chandrasekhar[51, p366].

Equation (5.112) implies, for $n = 10^{30}$ cm^{-3}, $E_F = 0.796$ MeV, quite a large energy. In relativity theory, the zero of energy is the rest energy, mc^2, as we see from Eq. (5.110). In what follows, we'll use the *kinetic energy* $T \equiv E - mc^2$ (not the temperature), the difference between the total energy and the rest energy. It's the kinetic energy that's involved in finding the pressure. We define

$$E_F \equiv mc^2 \left[\sqrt{1 + \left(\frac{\hbar}{mc} (3\pi^2 n)^{1/3} \right)^2} - 1 \right] . \tag{5.113}$$

The dimensionless group of terms in Eq. (5.113), $(\hbar/(mc))(3\pi^2 n)^{1/3}$, occurs frequently in the theory we're about to develop; let's give it a name, $x \equiv (\hbar/(mc))(3\pi^2 n)^{1/3}$. For $n = 10^{30}$ cm^{-3}, $x = 1.194$, and Eq. (5.113) predicts $E_F = 0.285$ MeV. The temperature equivalent of 0.285 MeV (the Fermi temperature) is 3.3×10^9 K. In comparison, the temperature of white dwarf stars is $10,000 - 20,000$ K. We're justified therefore in treating the electron gas as if it was at zero temperature! Equation (5.113) reduces to the nonrelativistic Fermi energy, Eq. (5.88), for $x \ll 1$:

$$E_F = \frac{\hbar^2}{2m} (3\pi^2 n)^{2/3} \left[1 + O\left((\lambda_C/\lambda_F)^2\right) \right] , \tag{5.114}$$

where $\lambda_C \equiv h/(mc)$ is the *Compton wavelength*, and $\lambda_F \equiv 2\pi/k_F$ is the Fermi wavelength.[48] The low-density limit is equivalent to $\lambda_F \gg \lambda_C$. It can be shown that the inverse of Eq. (5.113) is

$$3\pi^2 n = \frac{1}{(\hbar c)^3} \left(E_F^2 + 2E_F mc^2 \right)^{3/2} . \tag{5.115}$$

5.7.1 Relativistic degenerate electron gas

We must develop the density of kinetic-energy levels for relativistic electrons. From Eq. (5.110),

$$\hbar c k = \sqrt{T^2 + 2Tmc^2} . \tag{5.116}$$

With E_F defined as in Eq. (5.113), we have

$$N = 2 \int_0^{E_F} \frac{dN}{dT} dT = 2 \int_0^{E_F} \frac{dN}{dk} \frac{dk}{dT} dT = 2 \cdot \frac{V}{2\pi^2} \cdot \frac{1}{\hbar c} \int_0^{E_F} k^2 \cdot \frac{T + mc^2}{\sqrt{T^2 + 2Tmc^2}} dT$$

$$= \frac{V}{\pi^2 (\hbar c)^3} \int_0^{E_F} (T + mc^2) \sqrt{T^2 + 2Tmc^2} dT \equiv \int_0^{E_F} g(T) dT , \tag{5.117}$$

where we've used Eq. (2.16) and Eq. (5.116). The density of states is therefore

$$g(T) = \frac{V}{\pi^2 (\hbar c)^3} (T + mc^2) \sqrt{T^2 + 2Tmc^2} . \tag{5.118}$$

Equation (5.117) simplifies with a substitution. Let

$$\sinh \theta \equiv (1/(mc^2)) \sqrt{T^2 + 2Tmc^2} , \tag{5.119}$$

under which it's readily shown that

$$g(T) dT = 8\pi \frac{V}{\lambda_C^3} \cosh \theta \sinh^2 \theta d\theta . \tag{5.120}$$

---

[48] We've seen that the degeneracy parameter $n\lambda_T^3$ distinguishes classical ($n\lambda_T^3 \ll 1$) from quantum regimes ($n\lambda_T^3 \gtrsim 1$), where λ_T is the thermal wavelength associated with temperature T. We now have another discriminator, that $n\lambda_C^3 \ll 1$ distinguishes the nonrelativistic regime at zero temperature. A photon at the Compton wavelength has energy $E = mc^2$, almost enough to cause pair production (which requires $2mc^2$). The opposite limit of $\lambda_F \ll \lambda_C$ is associated with a new range of physical phenomena involving pair production, beyond the scope of this book.

The total kinetic energy of the gas is, using[49] $T = mc^2(\cosh\theta - 1)$ and Eq. (5.120),

$$U = \int_0^{E_F} T g(T) dT = 8\pi \frac{V}{\lambda_C^3} mc^2 \int_0^{\theta_F} (\cosh\theta - 1)\cosh\theta \sinh^2\theta d\theta , \qquad (5.121)$$

where $\sinh\theta_F \equiv \sqrt{E_F^2 + 2E_F mc^2}/(mc^2) = (\hbar/(mc))(3\pi^2 n)^{1/3} \equiv x$ (where we've used Eq. (5.115)). The nonrelativistic limit $E_F \ll 2mc^2$ is therefore $x \ll 1$. The integral in Eq. (5.121) takes some algebra to evaluate. We find[50]

$$U = \pi \frac{V}{\lambda_C^3} mc^2 \left[x\sqrt{1+x^2}(1+2x^2) - \sinh^{-1} x - \frac{8}{3}x^3 \right] \equiv \pi \frac{V}{\lambda_C^3} mc^2 f(x) . \qquad (5.122)$$

Compare the structure of Eq. (5.122) with Eq. (5.74); U is extensive (scales with V) but it's also an energy. We see the ratio of V to λ_C^3, the cube of the Compton wavelength, multiplied by an energy, mc^2, whereas in Eq. (5.74) we have the ratio V/λ_T^3 multiplied by kT.

To calculate the pressure, return to Eq. (5.65),

$$PV = kT \frac{V}{\pi^2} \int_0^\infty k^2 \ln\left(1 + e^{\beta(\mu - E_k)}\right) dk ,$$

and integrate by parts:

$$P = \frac{kT}{\pi^2} \left[\frac{k^3}{3} \ln\left(1 + e^{\beta(\mu - E_k)}\right) \Big|_0^\infty + \frac{\beta}{3} \int_0^\infty k^3 \frac{1}{e^{\beta(E_k - \mu)} + 1} \frac{dE_k}{dk} dk \right] .$$

The integrated part vanishes. In the remaining integral, let the temperature go zero, and let the Fermi factor provide the cutoff at $k = k_F$:

$$P = \frac{1}{3\pi^2} \int_0^{k_F} k^3 \frac{dE_k}{dk} dk .$$

Change variables: Let $E_k = T + mc^2$ and use Eq. (5.116),

$$P = \frac{1}{3\pi^2} \frac{1}{(\hbar c)^3} \int_0^{E_F} (T^2 + 2Tmc^2)^{3/2} dT .$$

Change variables again, using Eq. (5.119): We have, for the degeneracy pressure,

$$P = \frac{8\pi}{3} \frac{mc^2}{\lambda_C^3} \int_0^{\theta_F} \sinh^4\theta d\theta . \qquad (5.123)$$

Once again we see that pressure is an energy density, mc^2/λ_C^3; compare with Eq. (5.66), where the pressure is related to another energy density, kT/λ_T^3. Equation (5.123) presents us with another tough integral.[51] We find (where $x = \sinh\theta_F = (\hbar/(mc))(3\pi^2 n)^{1/3}$),

$$P = \frac{\pi}{3} \frac{mc^2}{\lambda_C^3} \left[x(2x^2 - 3)\sqrt{1+x^2} + 3\sinh^{-1} x \right] \equiv \frac{\pi}{3} \frac{mc^2}{\lambda_C^3} g(x) . \qquad (5.124)$$

Equation (5.124) is the equation of state for the degenerate electron gas as a general function of $x = (\hbar/(mc))(3\pi^2 n)^{1/3}$. We can identify the two limiting forms (see Exercise 5.40),

$$P = \begin{cases} \dfrac{\hbar^2}{5m}(3\pi^2)^{2/3} n^{5/3} & x \ll 1 \quad \text{(nonrelativistic)} \\[2mm] \dfrac{\hbar c}{4}(3\pi^2)^{1/3} n^{4/3} & x \gg 1 . \quad \text{(extreme relativistic)} \end{cases} \qquad (5.125)$$

---

[49]For those familiar with relativity theory, $\cosh\theta$ plays the role of the Lorentz factor, γ.

[50]After the substitution $u = \sinh\theta$, $\int_0^x u^2(\sqrt{1+u^2} - 1)du = \frac{1}{8}\left[x(2x^2 + 1)\sqrt{1+x^2} - \sinh^{-1} x - \frac{8}{3}x^3 \right]$.

[51]After the substitution $u = \sinh\theta$, $\int_0^x (u^4/\sqrt{1+u^2})du = \frac{1}{8}\left[x(2x^2 - 3)\sqrt{1+x^2} + 3\sinh^{-1} x \right]$.

We can use Eqs. (5.122) and (5.124) to form the ratio $U/(PV)$ (see Exercises 5.39 and 5.40):

$$\frac{U}{PV} = 3\frac{f(x)}{g(x)} = \begin{cases} 3/2 & x \ll 1 \\ 3 & x \gg 1. \end{cases} \tag{5.126}$$

5.7.2 White dwarf stars

We've now developed the machinery to address the question we posed earlier: Can the degeneracy pressure of electrons P balance the gravitational pressure, P_g? Under what conditions is $P \geq |P_g|$? Using Eq. (5.109) and the strong relativistic form of Eq. (5.125), we require

$$\frac{\hbar c}{4}(3\pi^2)^{1/3}N^{4/3}V^{-4/3} \geq \frac{4}{5}\left(\frac{4\pi}{3}\right)^{1/3}Gm_p^2N^2V^{-4/3}.$$

The volume dependence is the same on both sides of the inequality; we therefore have a criterion on the number of electrons involving fundamental constants of nature:

$$N \leq \left[\frac{5}{16}\left(\frac{9\pi}{4}\right)^{1/3}\frac{\hbar c}{Gm_p^2}\right]^{3/2} = 1.024 \times 10^{57}.$$

A star has the same number of protons (electric neutrality), and by assumption the same number of neutrons. The mass of a star with $N = 1.024 \times 10^{57}$ electrons is, from Eq. (5.107),

$$M = 2m_pN = 2(1.673 \times 10^{-27})(1.024 \times 10^{57}) \text{ kg} = 3.426 \times 10^{30} \text{ kg} = 1.72M_\odot. \tag{5.127}$$

Equation (5.127) would specify the theoretical maximum mass of a star that can hold off gravitational collapse if the assumptions we've made are all accurate. We've used the strong relativistic form of Eq. (5.125) for $x \gg 1$, whereas for stars with electron concentrations of 10^{30} cm^{-3}, $x = 1.194$. A more drastic assumption is that the star consists of a fully-ionized plasma. The basic theory of the upper mass of white dwarf stars was developed by S. Chandrasekhar in the 1930s.[52] When the degree of ionization is properly taken into account, the upper mass in Eq. (5.127) is revised downward. Detailed investigations by Chandrasekhar led to $1.44M_\odot$ as the upper bound, the *Chandrasekhar limit*.

5.7.3 Neutron stars

A star with mass in excess of the Chandrasekhar limit therefore cannot hold off gravitational collapse, right? Yes and no: Yes, the electron gas has a degeneracy pressure insufficient to balance the force of gravity, and no, because a new fusion reaction sets in. At sufficiently high pressures, the reaction $e^- + p \rightarrow n + \nu$ takes place, fusing electrons and protons into neutrons.[53] The neutrinos escape; degenerate matter is transparent to neutrinos, leaving a *neutron star*. Neutrons are fermions and therefore have a degeneracy pressure. Neutrons, however, because they are so much more massive than electrons, can be treated nonrelativistically. The question becomes, under what conditions can the *neutron* degeneracy pressure exceed the gravitational pressure. Using the nonrelativistic form of P from Eq. (5.125), we have the inequality

$$\frac{\hbar^2}{5m_n}(3\pi^2)^{2/3}N^{5/3}V^{-5/3} \geq \frac{1}{5}\left(\frac{4\pi}{3}\right)^{1/3}Gm_n^2N^2V^{-4/3},$$

---

[52]Chandrasekhar received the 1983 Nobel Prize in Physics for work on the physics of stars.

[53]The opposite reaction, $n \rightarrow e^- + p + \bar{\nu}$, takes place spontaneously (*exothermic*) because $m_n > m_p$; free neutrons decay into protons with a half-life of 15 minutes. The reaction we're talking about, $e^- + p \rightarrow n + \nu$, is *endothermic*—it absorbs energy of the reactants.

where N refers to the number of neutrons, and we've used the neutron mass, m_n. This inequality is equivalent to

$$R \leq \frac{\hbar^2}{m_n^3 G} \left(\frac{9\pi}{4}\right)^{2/3} N^{-1/3} = 1.31 \times 10^{23} N^{-1/3} \text{ m} , \qquad (5.128)$$

where $V^{1/3} = (4\pi/3)^{1/3} R$ specifies a radius R. For a neutron star of two solar masses, $N = 2.4 \times 10^{57}$, in which case the critical radius $R \approx 10$ km! Nonrelativistic neutrons therefore have a degeneracy pressure sufficient to hold off gravitational collapse. If, however, as gravity compresses the star the neutrons are heated to such an extent that they "go relativistic," there is no counterbalance to the force of gravity, and a *black hole* forms.

5.8 CAVITY RADIATION

Photons are special particles: spin-1 bosons of zero mass that travel at the speed of light[54] and have two spin states.[55] They have another property that's directly relevant to our purposes: Photons in thermal equilibrium have zero chemical potential, $\mu = 0$. Electromagnetic radiation is not ordinarily in equilibrium with its environment,[56] implying it can't ordinarily be described by equilibrium statistical mechanics. *Cavity radiation*, however, is the singularly important problem of electromagnetic energy contained within a hollow enclosure bounded by thick opaque walls maintained at a uniform temperature, and thus *is* in equilibrium with its environment. In this section, we apply the methods of statistical physics to cavity radiation.

5.8.1 Why is $\mu = 0$ for cavity radiation?

The assignment of $\mu = 0$ to cavity radiation can be established in thermodynamics. Consider the electromagnetic energy U contained within a cavity of volume V that's surrounded by matter at temperature T. Denote the energy density as $u(T) \equiv U(T)/V$. Cavity radiation is independent of the *specifics* of the cavity—the size and shape of the cavity or the material composition of the walls—and depends only on the temperature of the walls. That conclusion follows from thermodynamics: Cavity radiation *not* independent of the specifics of the cavity would imply a violation of the second law[3, p68]. In particular, "second-law arguments" show that the *density* of electromagnetic energy is the same for any type of cavity and depends only on the temperature. Thus, we have "line one" for cavity radiation:

$$\left(\frac{\partial u}{\partial V}\right)_T = 0 . \qquad (5.129)$$

Equation (5.129) is the analog of Joule's law for the ideal gas, Eq. (1.44), $((\partial U/\partial V)_T = 0)$. Cavity radiation is *not* an ideal gas (see Table 5.4), even though both systems are collections of non-interacting entities. For the ideal gas, $\Delta U = 0$ in an isothermal expansion, i.e., U is independent of V. For cavity radiation, the energy *density* is independent of volume, with $\Delta U = u \Delta V$ in an

---

[54]Various experiments place an upper bound on the photon mass ranging from 10^{-17} eV/c^2 to 10^{-27} eV/c^2, i.e., at least 20 orders of magnitude smaller than the electron mass. Standard practice is to treat the photon as having zero mass. There are theoretical reasons for assuming the photon has no mass. Einstein's second postulate for the theory of special relativity is actually an assertion that the photon has zero mass.[13, p50] From the relativistic energy-momentum relation, $E = \sqrt{(pc)^2 + (mc^2)^2}$, if $m = 0$, then $|p| = E/c$, whereas for massive particles, $p = Ev/c^2$. These formulas are compatible only if $|v| = c$ for $m = 0$. Photons have momentum because they have energy, even though they have no mass. That photons act as particles carrying energy and momentum is verified in Compton scattering experiments.

[55]The photon is a spin-1 particle, but, because it's massless, has only two spin states, $S_z = \pm \hbar$. Thus, $g = 2$ for photons. Actually, massless particles don't have spin, they have *helicity*, the value of the projection of the spin operator onto the momentum operator—a topic beyond the scope of this book. "Up" and "down" can't be defined in a relativistically invariant way, but projections onto the direction of propagation can.

[56]Shine a flashlight at night toward the stars—does that light ever come back?

isothermal expansion.[57] The number of atoms in a gas is fixed, and heat absorbed from a reservoir in an isothermal expansion keeps the temperature of the particles constant. In an isothermal expansion of cavity radiation, heat is absorbed from a reservoir, but goes into creating new photons to keep the energy density fixed.

The equation of state for cavity radiation[58] is[3, p70]

$$P = \frac{1}{3}u(T) . \tag{5.130}$$

Compare with the equation of state for ideal gases, $P = \frac{2}{3}u$, Eq. (5.75). Equation (5.130) follows from an analysis of the momentum imparted to the walls of the cavity based on the *isotropy* of cavity radiation (another conclusion from second-law arguments[59]), that photons travel at the same speed, independent of direction, and that the momentum density of the electromagnetic field[60] is u/c. With Eq. (5.130) established, using the methods of thermodynamics it can be shown that[3, p70]

$$u(T) = aT^4 , \tag{5.131}$$

where a is a constant, the *radiation constant*, the value of which cannot be obtained from thermodynamics. We'll derive the value of this constant using statistical mechanics; see Eq. (5.140).

Once we know Eq. (5.131), we know a great deal. Combining Eq. (5.131) with Eq. (5.130),

$$P = \frac{a}{3}T^4 . \tag{5.132}$$

The radiant energy in a cavity of volume V at temperature T is, from Eqs. (5.131) and (5.129),

$$U = aVT^4 . \tag{5.133}$$

The heat capacity is therefore

$$C_V = \left(\frac{\partial U}{\partial T} \right)_V = 4aVT^3 . \tag{5.134}$$

Note that $C_V(T) \to 0$ as $T \to 0$, as required by the third law of thermodynamics. From Eq. (5.134), we can calculate the entropy:

$$\left(\frac{\partial S}{\partial T} \right)_V = \frac{1}{T}C_V \implies S = \frac{4}{3}aVT^3 . \tag{5.135}$$

With Eqs. (5.132), (5.133), and (5.135) (those for P, U, and S), we have the ingredients to construct the thermodynamic potentials:

$$U = aVT^4 = \frac{3}{4}TS = 3PV \qquad F = U - TS = -\frac{1}{3}aVT^4$$

$$H = U + PV = TS = \frac{4}{3}aVT^4 \qquad G = H - TS = 0 . \tag{5.136}$$

The Gibbs energy is identically zero, *a result obtained strictly from thermodynamics*. In general, $G = N\mu$, Eq. (1.54). Because $G = 0$ for cavity radiation for any N, we infer $\mu = 0$, implying it costs no *extra* energy to add photons to cavity radiation (but of course it costs energy to *make* photons). Photons are a special case in that creating them is the same as adding them to the system.[61]

---

[57]While the energy density of cavity radiation is independent of V under isothermal conditions, it's not independent of V for isentropic processes; see Eq. (P5.6).

[58]Equation (5.130) was derived by Ludwig Boltzmann in 1884.

[59]The Michelson-Morley experiment also shows that the speed of light is independent of direction.

[60]The momentum density of the electromagnetic field follows from Maxwell's equations.

[61]We can't get into the rest frame of a photon: A photon held in your hand is a destroyed (absorbed) photon.

Now that we've brought it up, however, what *is* the number of photons in a cavity? Thermodynamics has no way to calculate that number. Can we use the formalism already developed, say Eq. (5.69), to calculate the average number of photons? Not directly: The thermal wavelength $\lambda_T = h/\sqrt{2\pi mkT}$ depends on the mass of the particle; moreover the formulas derived in Section 5.5 for bosons utilize the density of states of nonrelativistic particles, Eq. (2.17). To calculate the number of photons, we must start over with Eq. (5.63) using the density of states for relativistic particles; see Eq. (2.20). Thus, with $g = 2$ and $\mu = 0$,

$$N = \frac{V}{\pi^2(\hbar c)^3} \int_0^\infty \frac{E^2}{e^{\beta E} - 1} dE = \frac{V}{\pi^2(\hbar c\beta)^3} \int_0^\infty \frac{x^2}{e^x - 1} dx = \frac{2V}{\pi^2} \left(\frac{kT'}{\hbar c}\right)^3 \zeta(3). \quad (5.137)$$

Between Eqs. (5.133), (5.135), and (5.137), we have

$$\frac{S}{N} = \frac{2a\pi^2}{3\zeta(3)} \left(\frac{\hbar c}{k}\right)^3 \qquad \frac{U}{N} = \frac{a\pi^2}{2\zeta(3)} \left(\frac{\hbar c}{k}\right)^3 T. \quad (5.138)$$

For cavity radiation, therefore, S is *strictly* proportional to N, $S \propto N$. For the ideal gas (see Eq. (1.64)), $S \sim N$, but is not strictly proportional to N. From the definition of the chemical potential, Eq. (1.22),

$$\mu \equiv \left(\frac{\partial U}{\partial N}\right)_{S,V},$$

we see that one can't take a derivative of U with respect to N holding S fixed, because holding S fixed is to hold N fixed (what is $(\partial f/\partial x)_x$?). *It's not possible to change the number of photons keeping entropy fixed*, and thus chemical potential is not well defined for cavity radiation.

We arrived at that conclusion, however, assuming $\mu = 0$. Is the assignment of $\mu = 0$ for cavity radiation consistent with general thermodynamics? From Table 1.2, $[\Delta F]_{T,V} = W'$, the amount of "other work." Using Eq. (5.136), $[\Delta F]_{T,V} = 0$. The maximum work in *any* form is $[\Delta F]_T$ (see Section 1.5), and from Eq. (5.136), $[\Delta F]_T = -\frac{1}{3}aT^4 \Delta V = -P\Delta V$, where we've used Eq. (5.132). Thus, *no forms of work other than PdV work are available to cavity radiation*, which is consistent with $\mu = 0$. Thermodynamic equilibrium is achieved when the intensive variables *conjugate to conserved quantities* (energy, volume, particle number) equalize between system and surroundings (see Section 1.12). *Photons are not conserved quantities*. Photons are *created and destroyed* in the exchange of energy between the cavity walls and the radiation in the cavity. There's no population of photons external to the cavity for which those in the cavity can come to equilibrium with. The natural variables to describe the thermodynamics of cavity radiation are T, V, S, P, or U, but not N. Cavity radiation should not be considered an open system, but a closed system that exchanges energy with its surroundings. The confusion here is that photons are particles of energy, a quintessential quantum concept. Table 5.4 summarizes the thermodynamics of the ideal gas and the photon gas.

5.8.2 The Planck distribution

With $\mu = 0$ established, it remains to ascertain the radiation constant a before we can say the thermodynamics of the photon gas is completely understood (see Table 5.4). Without statistical mechanics, the radiation constant would have been considered a fundamental constant of nature, akin to the gas constant, R. A perfect marriage of thermodynamics and statistical mechanics, we can calculate the internal energy U (combining Eqs. (5.91) and (2.20), with $g = 2$ and $\mu = 0$) and

Table 5.4: Thermodynamics of the ideal gas and the photon gas

	Ideal gas	Photon gas
Internal energy	$U = \frac{3}{2}NkT$	$U = aVT^4$
Volume dependence of U	$\left(\frac{\partial U}{\partial V}\right)_T = 0$	$\left(\frac{\partial u}{\partial V}\right)_T = 0$
Equation of state	$P = NkT/V = \frac{2}{3}u$	$P = \frac{1}{3}aT^4 = \frac{1}{3}u$
Heat capacity	$C_V = \frac{3}{2}Nk$	$C_V = 4aVT^3$
Entropy	$S = Nk\left[\frac{5}{2} + \ln\left(\frac{V}{N\lambda_T^3}\right)\right]$	$S = \frac{4}{3}aVT^3$
Chemical potential	$\mu = -kT\ln\left(\frac{V}{N\lambda_T^3}\right)$	$\mu = 0$
Adiabatic process	$TV^{\gamma-1} = \text{constant}$	$TV^{1/3} = \text{constant}$

compare with Eq. (5.133):

$$U = \int_0^\infty Eg(E)n(E)dE = \frac{V}{\pi^2(\hbar c)^3}\int_0^\infty \frac{E^3}{e^{\beta E}-1}dE = \frac{V}{\pi^2(\hbar c)^3}\frac{1}{\beta^4}\int_0^\infty \frac{x^3}{e^x-1}dx$$
$$= \frac{V(kT)^4}{\pi^2(\hbar c)^3}\Gamma(4)\zeta(4) = \frac{\pi^2 V(kT)^4}{15(c\hbar)^3} \equiv aVT^4 , \tag{5.139}$$

implying

$$a = \frac{\pi^2 k^4}{15(c\hbar)^3} = 7.5657 \times 10^{-16} \text{ J m}^{-3}\text{ K}^{-4} . \tag{5.140}$$

Equation (5.140) is one of the triumphs of statistical mechanics. Note that we can't take the classical limit of the radiation constant by formally letting $\hbar \to 0$. *There is no classical antecedent of cavity radiation*; it's an intrinsically quantum problem from the outset.[62]

Let $u(\lambda, T)$ denote the *energy spectral density*, the energy per volume contained in the wavelength range $(\lambda, \lambda + d\lambda)$ in equilibrium at temperature T. (Thus, $u(\lambda, T)$ has the units J m^{-4}; do you see why?) Starting from Eq. (5.139) and changing variables, $E = hc/\lambda$, we find

$$U = 8\pi hcV\int_0^\infty \frac{d\lambda}{\lambda^5}\frac{1}{\exp(hc/(\lambda kT))-1} \equiv V\int_0^\infty u(\lambda, T)d\lambda ,$$

and therefore we identify

$$u(\lambda, T) = 8\pi hc\frac{1}{\lambda^5}\frac{1}{\exp(hc/(\lambda kT))-1} . \tag{5.141}$$

---

[62] Also, the radiation constant does not contain a mass or a charge. Photons carry no charge, and have zero mass. The radiation constant does, however, contain Boltzmann's constant, which appears in physics through the average kinetic energy per particle of an ideal gas (particles that have mass) at temperature T. As discussed earlier (Section 1.10), measuring absolute temperature in degrees Kelvin artificially introduces a conversion factor between energy and Kelvin. Absolute temperature "should" have the dimension of energy, in which case Boltzmann's constant would be unnecessary.

Equation (5.141) is called (among other names) the *Planck distribution law*.[63] By changing variables, $\lambda = c/\nu$, we have the frequency distribution (see Exercise 5.32),

$$u(\nu, T) = \frac{8\pi h}{c^3} \frac{\nu^3}{e^{\beta h \nu} - 1}.$$ (5.142)

5.8.3 The Wien displacement law

We've written the energy spectral density function as $u(\lambda, T)$ and $u(\nu, T)$ in Eqs. (5.141) and (5.142), indicating they are functions of two variables, (λ, T) or (ν, T). This isn't correct, however. Wilhelm Wien showed, in 1893, that $u(\lambda, T)$, presumed a function of two variables, is actually a function of a *single variable*, λT. He showed that the spectral density function must occur in the form

$$u(\nu, T) = \nu^3 \psi \left(\frac{\nu}{T} \right) \quad \text{or} \quad u(\lambda, T) = \frac{1}{\lambda^5} f(\lambda T),$$ (5.143)

where ψ and f are functions of a single variable, *precisely* in the forms of Eqs. (5.141) and (5.142). Wien didn't derive the Planck distribution—Planck's constant had not yet been discovered—but his work placed an important constraint on its possible form.[64]

Wien's central result is a partial differential equation for u as a function of ν and V (see [3, pp73–78]):

$$V \left(\frac{\partial u(\nu)}{\partial V} \right)_S = \frac{\nu}{3} \frac{\partial u(\nu)}{\partial \nu} - u(\nu).$$ (5.144)

Equation (5.144) follows from an analysis of the means by which u can change in a reversible, adiabatic process. As can readily be verified, the solution of Eq. (5.144) is in the form

$$u(\nu) = \nu^3 \phi(V \nu^3),$$ (5.145)

where ϕ is *any function* of a single variable. The functional form of ϕ cannot be established by this means of analysis.[65]

In a reversible, adiabatic process ($\mathrm{d}S = 0$), we have for cavity radiation that $VT^3 = $ constant (see Table 5.4). Equation (5.145) can therefore be written (for cavity radiation)

$$u(\nu, T) = \nu^3 \phi \left(\frac{\nu^3}{T^3} \right) \equiv \nu^3 \psi \left(\frac{\nu}{T} \right),$$ (5.146)

where ψ is a function of a single variable. From Eq. (5.146), we have under the change of variables $\nu = c/\lambda$, using energy conservation $u(\nu)\mathrm{d}\nu = u(\lambda)\mathrm{d}\lambda$,

$$u(\lambda, T) = \left| \frac{\mathrm{d}\nu}{\mathrm{d}\lambda} \right| u(\nu) = \frac{c^4}{\lambda^5} \psi \left(\frac{c}{\lambda T} \right) \equiv \lambda^{-5} f(\lambda T).$$ (5.147)

If every dimension of a cavity is expanded uniformly, the wavelength of every mode of electromagnetic oscillation would increase in proportion—that is, there would be a *redshift*. For a length $L \equiv V^{1/3}$ associated with the cavity, every wavelength λ scales with L, $\lambda \sim L = V^{1/3}$, and, because $TV^{1/3}$ is constant in an isentropic process, $\lambda T = $ constant, or, equivalently, $\nu/T = $ constant.

---

[63]The Planck distribution is also called the *black body* spectral distribution. A *black body* is one that absorbs (i.e., does not reflect) all incident radiation, and then emits it with the characteristic Planck spectrum. The *emissivity* $E(T)$ of a black body (energy radiated per unit time per unit surface area) is related to the energy density of cavity radiation through the simple formula $E(T) = \frac{c}{4} u(T)$ [3, p69]. *Stefan's constant* $\sigma \equiv ca/4 = 5.670 \times 10^{-8}$ W m^{-2} K^{-4}.

[64]Wien received the 1911 Nobel Prize in Physics for work on heat radiation.

[65]A "problem" with Eq. (5.144) is that it's too general—it applies to the spectral distribution of *any* system of radiant energy, not just thermal radiation. Equation (5.146) applies to cavity radiation.

We know that $\int_0^\infty u(\nu, T)d\nu = U/V = aT^4$. Assume that in an isentropic expansion the radiation temperature changes from $T_1 \to T_2$. Then, we have the equality

$$\frac{1}{T_1^4} \int_0^\infty u(\nu, T_1)d\nu = \frac{1}{T_2^4} \int_0^\infty u(\nu', T_2)d\nu' . \tag{5.148}$$

Applying Wien's result, Eq. (5.146), to Eq. (5.148),

$$\int_0^\infty \left(\frac{\nu}{T_1}\right)^3 \psi\left(\frac{\nu}{T_1}\right) d\left(\frac{\nu}{T_1}\right) = \int_0^\infty \left(\frac{\nu'}{T_2}\right)^3 \psi\left(\frac{\nu'}{T_2}\right) d\left(\frac{\nu'}{T_2}\right) . \tag{5.149}$$

Equation (5.149) implies the *Wien displacement law*: Under the expansion, for every frequency ν for the system at temperature T_1, there's an associated ("displaced") frequency $\nu' = (T_2/T_1)\nu$ at temperature T_2 (change variables in Eq. (5.149), $\nu/T_1 = \nu'/T_2$). Wien's law therefore implies that *the form of the Planck distribution remains invariant in an isentropic expansion*; one has to "displace" the frequencies $\nu \to \nu' = (T_2/T_1)\nu$. This finds applications in the subject of cosmology, where the expansion of the universe is modeled as an isentropic process.[66]

It's clear that $u(\nu) \to 0$ as $\nu \to 0$ (as $\nu \to 0$, $\lambda \to \infty$, and a cavity of finite size cannot support an infinite-wavelength mode of electromagnetic vibration). It's also true that $u(\nu) \to 0$ as $\nu \to \infty$ (see Exercise 5.35). Mathematically, the energy spectral density $u(\lambda T)$ must have a maximum (Rolle's theorem). Suppose the function $f(\lambda T)$ in Eq. (5.147) is known (which it is, from the Planck distribution). Its maximum with respect to λ is obtained by satisfying the condition $xf'(x) = 5f(x)$, where $x \equiv \lambda T$ (show this). The wavelength where maximum energy density occurs is

$$\lambda_{\text{max}} = \frac{1}{T}\left(\frac{5f}{f'}\right)_{\text{max}} \equiv \frac{b}{T} , \tag{5.150}$$

where the quantity b is known as *Wien's constant*; it has the value $b = 2.898 \times 10^{-3}$ m K.

5.9 DEGENERATE BOSE GAS, BOSE-EINSTEIN CONDENSATION

In Section 5.5 we looked at the thermodynamic properties of ideal gases of fermions or bosons, Eqs. (5.67), (5.69), (5.74), and (5.80), those[67] for P, N, U, and S. These formulas require modification at low temperature for bosons. At low temperature, bosons occupy low-lying energy states with no limit on the number of particles that can occupy a given state.[68] As $T \to 0$, nothing prevents all particles from occupying (*condensing into*) the lowest energy state, which we've taken to be at

---

[66]It's not out of place to discuss the cosmic microwave background (CMB). In 1964, it was found that the universe is permeated by electromagnetic radiation that's not associated with any star or galaxy, that's unpolarized and highly isotropic, what we'd expect of cavity radiation. (The 1978 Nobel Prize in Physics was award to A.A. Penzias and R.W. Wilson for the discovery of the CMB.) Is the universe one big "cavity"? The measured spectral energy density of the CMB is found to occur as a Planck distribution associated with absolute temperature $T = 2.7260 \pm 0.0013$ K. The universe thus contains a population of photons in thermal equilibrium at $T = 2.726$ K. Why?

The CMB is taken to be a thermal relic of a universe that was at one time considerably hotter than $T = 2.73$ K. By the *Big Bang* theory, with a universe initially in a hot, dense state, photons are scattered by charged particles (Thomson scattering), particularly free electrons, providing an efficient mechanism for establishing thermal equilibrium (with free charges about, the universe has "opaque walls," just what we need for cavity radiation!) because the expansion rate is less than the speed of light, $\dot{R} < c$ (R is the scale length of the universe). If hotter conditions prevail at one region of the early universe than in another, photons act to equalize temperatures spatially. In an adiabatic expansion of a photon gas (the universe is a closed system, right?), $VT^3 = $ constant, Eq. (5.135), and thus the temperature of the photon population decreases as the universe expands, $T \propto R^{-1}(t)$. When the temperature cools to approximately 3000 K, neutral atoms can form and stay in that state, stable against disruption by photons. At that point, photons are said to *decouple* from matter: Neutral species are significantly less effective in scattering photons. By photons having been in thermal equilibrium at the time of decoupling, they maintain their black-body spectrum as the universe keeps adiabatically expanding.

[67]In the grand canonical ensemble, T, V, μ are givens, with U, S, P, N derived quantities. Sometimes we can invert $N = N(\mu, T, V)$ to obtain $\mu = \mu(n, T)$, as in Eq. (5.102).

[68]That is, there is no Pauli principle at work for bosons.

$E = 0$. For bosons, $E = 0$ is "where the action is" at low temperature.[69] The density of states function, however, $g(E)$ *vanishes* as $E \to 0$ (see Eq. (2.18)). Because $g(E)$ pertains to single-particle states, one might question the assumption of noninteracting particles as $T \to 0$. Can we even speak of a gaseous state at absolute zero temperature? As we'll see, if we can "fix up" the treatment of Bose gases to properly capture the special state associated with $E = 0$, it leads to a fascinating piece of physics, the *Bose-Einstein condensation*, a new state of matter predicted in 1924 but not observed until 1995, which was the basis for the 2001 Nobel Prize in Physics.[70]

To see what the issue is, consider Eqs. (5.67) and (5.69) for bosons ($\theta = 1$), which involve the Bose-Einstein functions $G_{5/2}(z)$ and $G_{3/2}(z)$. From Eq. (B.21),

$$G_{5/2}(z) = \sum_{m=1}^{\infty} \frac{z^m}{m^{5/2}} \qquad G_{3/2}(z) = \sum_{m=1}^{\infty} \frac{z^m}{m^{3/2}} \cdot$$

Power-series representations are valid provided the series converge. The interval of convergence for both series is $0 < z \le 1$; they diverge for $z > 1$. The function $G_{3/2}$ is plotted in Fig. 5.11 for $0 \le z \le 1$; that for $G_{5/2}$ has a similar shape. The integral representation of $G_{3/2}$ is (see Eq. (B.18))

$$G_{3/2}(z) = \frac{1}{\Gamma(3/2)} \int_0^{\infty} \frac{\sqrt{x}}{z^{-1}e^x - 1} dx .$$

The denominator of the integrand vanishes when $e^x = z$. For $z > 1$, there exists a value of $x > 0$ (i.e., within the range $0 < x < \infty$) at which the integral diverges. *The functions $G_{3/2}(z)$ and $G_{5/2}(z)$ do not exist for $z > 1$.* Mathematically, the fact that these functions have maximum values as $z \to 1^-$ lies at the root of the theory of Bose-Einstein condensation.[71]

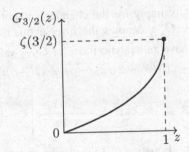

Figure 5.11: Bose-Einstein function $G_{3/2}(z)$. The slope becomes infinite as $z \to 1^-$.

From Eq. (5.69) we have (where for simplicity we take $g = 1$), $n\lambda_T^3 = G_{3/2}(z)$. For a fixed density, at sufficiently high temperatures $n\lambda_T^3$ is small, and one has a solution of Eq. (5.69) for small[72] z. As the temperature is lowered (for fixed density) $n\lambda_T^3$ increases in value, for which, referring to Fig. 5.11, there will be solutions of Eq. (5.69) for $z < 1$. At a *critical temperature*, however, the "last" solution of Eq. (5.69) occurs for $z = 1$, where[73] $n\lambda_T^3 = \zeta(3/2)$, at the temperature

$$T_B \equiv \frac{\hbar^2}{2mk} \left(\frac{8\pi^3}{\zeta(3/2)} n \right)^{2/3} . \tag{5.151}$$

---

[69] For fermions, the low-temperature ($T \ll T_F$) action is at the Fermi energy E_F.

[70] Awarded to E.A. Cornell, W. Ketterle, and C.E. Wieman for observation of the Bose-Einstein condensation.

[71] The Fermi-Dirac functions are unbounded as $z \to \infty$, growing like $F_n(z) \sim (\ln z)^n$ for large z (see Eq. (B.40)).

[72] As noted previously, $z \to 0$ corresponds to classical behavior. At sufficiently high temperature, we have the classical ideal-gas solution $z = e^{\beta\mu} = n\lambda_T^3$.

[73] The maximum value of the Bose-Einstein functions occurs at $z = 1$, with $G_n(1) = \zeta(n)$, where $\zeta(n)$ is the Riemann zeta function (see Appendix B). Select values of $\zeta(n)$ are given in Table B.1.

Equation (5.151) defines the *Bose temperature*, T_B. Aside from numerical factors of order unity, the Bose temperature is similar to the Fermi temperature, Eq. (5.87): $\hbar^2/2m$, which from the Schrödinger equation has the dimension of energy-(length)^2, multiplied by $n^{2/3}$, which provides $1/(\text{length}^2)$. The length in question, $n^{-1/3}$, is the average distance between particles. Whereas the Fermi temperature associated with electrons at metallic densities is quite large (see Table 5.2), the Bose temperature is quite small because the mass m refers to that of *atoms*.[74] There is no solution of Eq. (5.69) at fixed density for $T < T_B$. Equation (5.151) specifies a density-dependent temperature at which $n\lambda_{T_B}^3 = \zeta(3/2)$. The *critical density*, n_c is such that $n_c\lambda_T^3 \equiv \zeta(3/2)$,

$$n_c = \zeta(3/2)\left(\frac{2\pi mkT}{h^2}\right)^{3/2}. \tag{5.152}$$

At a given temperature, there is no solution to Eq. (5.69) for $n > n_c$.

Example. Consider a gas of 6×10^{22} He^4 atoms (one-tenth of a mole) confined to a volume of one liter. What is its Bose temperature, T_B? He^4 is a boson: Its nucleus consists of two protons and two neutrons. By the rules of angular momentum addition in quantum mechanics, the spin of a composite object of two spin-$\frac{1}{2}$ particles is either 1 or 0, and is therefore a boson. Nature seeks the lowest energy configuration, which would be for two identical fermions to pair such that the net spin is zero (no Pauli exclusion effect and hence lower energy). The two electrons in He^4 also pair such that their spins are oppositely directed: He^4 is a boson of spin zero. (By this reasoning, is He^3 a boson?) What's the mass of a He^4 atom? I hope you're reaching for Avogadro's number: 6.64×10^{-27} kg. Using Eq. (5.151), we find $T_B = 0.059$ K. In everyday experience that's an exceedingly small temperature, but one that's routinely achievable in modern laboratories.[75]

What does a maximum density imply for the chemical potential? We have, from Eq. (5.69), $z = e^{\beta\mu} = G_{3/2}^{-1}(n\lambda_T^3)$, where $G_{3/2}^{-1}$ denotes the inverse function of $G_{3/2}$. We don't have an expression for $G_{3/2}^{-1}$, but we do have an analytic form for $G_{3/2}$ valid for $z \lesssim 1$ that can easily be inverted, Eq. (B.23): $G_{3/2}(\beta\mu) = \zeta(3/2) - 2\sqrt{\pi}\sqrt{-\beta\mu} + O(-\beta\mu)$, implying for $T \gtrsim T_B$,

$$\beta\mu \approx -\frac{1}{4\pi}\left[\zeta(3/2) - n\lambda_T^3\right]^2 = -\frac{\zeta(3/2)^2}{4\pi}\left[1 - \left(\frac{T_B}{T}\right)^{3/2}\right]^2, \tag{5.153}$$

where we've used $\zeta(3/2) = n\lambda_{T_B}^3$. Equation (5.153) can be written, to lowest order,

$$\frac{\mu}{kT_B} \approx -C\left(\frac{T - T_B}{T_B}\right)^2, \qquad (T \gtrsim T_B) \tag{5.154}$$

where $C \equiv 9\zeta(3/2)^2/(16\pi) \approx 1.22$. Thus, μ *vanishes* as $T \to T_B^+$, as shown in Fig. 5.12.

The upshot is that Eq. (5.69) is valid only for $n\lambda_T^3 \leq \zeta(3/2)$, implying $\mu \leq 0$ for $T \geq T_B$ only.[76] It's *as if* at low temperatures, in approaching the maximum density n_c, the extra particles

---

[74] Are there massive bosons (i.e., possessing mass) among the elementary particles? I'll leave that for you to look up. Do any have a mass as small as that of the electron? Are they stable? Could one form an assembly of such bosons long enough and with sufficient density to measure the Bose temperature?

[75] Students sometimes say, "The Bose temperature is so small, who cares about $T < T_B$—it's not a practical reality." It may indeed be impractical, but that's not why we do science. Leave aside questions of practicality when investigating fundamental physics. Instead of "is it practical," ask "is it *impossible*?" Applications may come later; don't preclude yourself from what is possible because you can't foresee any applications. Just the beauty of the manifestations of nature is enough for scientific work. Does every discovery of science have to result in an application?

[76] The chemical potential of bosons must be nonpositive ($\mu \leq 0$) for the grand partition function to exist—see discussion around Eq. (5.54). Physical reasons why the chemical potential of noninteracting particles is negative are given in [3, p39]. For the statistical mechanics of noninteracting bosons to be internally consistent, we require $\mu \leq 0$ for all temperatures, not just for $T \geq T_B$.

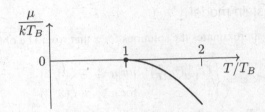

Figure 5.12: Chemical potential of the ideal Bose gas for $T \gtrsim T_B$ (from Eq. (5.154)).

we're adding to the system (at fixed volume) simply "disappear." Where did we go wrong? We went wrong in the transition from equations such as (5.57) and (5.64) to Eqs. (5.58) and (5.66), in converting sums over allowed states to integrals over k-space, in making the replacement $\sum_k \rightarrow (V/8\pi^3) \int d^3 k$. The state associated with $k = 0$ is given zero weight in the resulting integral, yet that state dominates the physics of bosons at low temperature. Let's split off the $k = 0$ state from the sum in Eq. (5.57), and *then* let $\sum_k \rightarrow (V/8\pi^3) \int d^3 k$:

$$N = \underbrace{g \frac{1}{e^{-\beta\mu} - 1}}_{N_0, \text{number of particles in the ground state}} + \underbrace{g \sum_{k \neq 0} \frac{1}{e^{\beta(E_k - \mu)} - 1}}_{N_{\text{ex}}, \text{number of particles in excited states}}$$

$$= g \frac{z}{1 - z} + g \frac{V}{8\pi^3} \int \frac{d^3 k}{e^{\beta(E_k - \mu)} - 1} = g \frac{z}{1 - z} + g \frac{V}{\lambda_T^3} G_{3/2}(z)$$

$$\equiv N_0 + N_{\text{ex}} . \tag{5.155}$$

In Eq. (5.155), N_0 denotes the number of particles in the ground state, with N_{ex} the number of particles in all other, excited states of the system. The "missing" particles are in the ground state! This is the Bose-Einstein *condensation*: Starting at $T = T_B$ and continuing to $T = 0$, particles begin occupying the ground state of the system with significant (macroscopic) numbers, until, at $T = 0$, all particles of the system occupy the ground state.[77] Equation (5.155) can be written (compare with Eq. (5.69)),

$$\frac{n\lambda_T^3}{g} = \frac{\lambda_T^3}{V} \frac{z}{1 - z} + G_{3/2}(z) . \tag{5.156}$$

Henceforth we will set $g = 1$ in formulas related to the Bose gas; experiments on Bose-Einstein condensation utilize atoms for which $g = 1$.

As long as $z \ll 1$, the new term in Eq. (5.156) is negligibly small because of the factor of V^{-1}, with the value of z implied by Eq. (5.156) not appreciably different from what we obtained from Eq.(5.69). When, however, z has a value infinitesimally close to $z = 1$, which we can characterize as $z \approx 1 - O(N^{-1})$ (for macroscopic N), then the factor of $z/(1 - z)$ in Eq. (5.156) cannot be neglected. As $z \rightarrow 1^-$, this term saves us from the apparent missing-particle "trap" set by Eq. (5.69)—there is no longer a maximum density of Bose gases. Indeed, we can find a solution for z from Eq. (5.156) for every value of $n\lambda_T^3$. As $z \rightarrow 1^-$, the factor of $z/(1 - z)$ is a steeply rising curve that goes to infinity. This term is added to $G_{3/2}(z)$, which has a finite value as $z \rightarrow 1^-$.

---

[77] It's clear that the Pauli exclusion effect prevents such a condensation from occurring for a collection of fermions.

5.9.1 The Bose-Einstein model

The Bose-Einstein model approximates the solutions for z that would be obtained numerically from Eq. (5.156) as

$$z = \begin{cases} G_{3/2}^{-1}(n\lambda_T^3) & \text{for } n\lambda_T^3 \leq \zeta(3/2) \\ 1 & \text{for } n\lambda_T^3 > \zeta(3/2) . \end{cases} \tag{5.157}$$

We therefore set $\mu = 0$ for $T < T_B$, that it costs no energy to add particles to the ground state. By definition (Eq. (5.155)), we have for the number of particles in the ground state $(g = 1)$

$$N_0 = \frac{z}{1-z} \implies z = \frac{1}{1+N_0^{-1}} \approx 1 - \frac{1}{N_0} . \qquad (N_0 \gg 1)$$

The model consists of the assignment $z = 1$ for $T < T_B$; $z \approx 1 - 1/N_0$ is consistent with that for macroscopic values[78] of N_0. The number of particles in excited states is therefore, for $T < T_B$,

$$N_{ex} = \frac{V}{\lambda_T^3}G_{3/2}(1) = \frac{V}{\lambda_T^3}\zeta(3/2) = Vn\frac{\lambda_{T_B}^3}{\lambda_T^3} = N\left(\frac{T}{T_B}\right)^{3/2} . \qquad (T \leq T_B) \tag{5.158}$$

The number of particles in the ground state is, from $N_0 = N - N_{ex}$,

$$N_0 = N\left[1 - \left(\frac{T}{T_B}\right)^{3/2}\right] . \qquad (T \leq T_B) \tag{5.159}$$

At $T = T_B$, *all* particles are in excited states and none are in the ground state. The temperature dependence of N_0 and N_{ex} is shown in Fig. 5.13. For $T \geq T_B$, we can apply the formulas previously

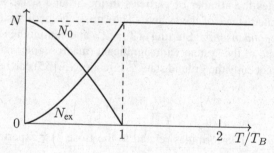

Figure 5.13: Variation with temperature of N_0 and N_{ex} for the ideal Bose gas.

obtained (Eqs. (5.67), (5.69), (5.74), and (5.80)), but for $T < T_B$ we must use Eqs. (5.156) for the density and Eq. (5.157) for the fugacity. The fraction of particles in the ground state, N_0/N, is nonzero for $T < T_B$ and vanishes at $T = T_B$. Such behavior is characteristic of the *order parameter* associated with phase transitions (Chapter 7). Phase transitions ordinarily connote a change of phase in real space. The Bose-Einstein condensation occurs in *momentum space*—particles condense into the state of zero momentum (from the de Broglie relation $p = \hbar k$). As $k \to 0$, the de Broglie wavelength becomes infinite, implying that nominally noninteracting particles are correlated in a new state of matter having no classical counterpart.

---

[78]Of course, $N_0 \to 0$ as $T \to T_B^-$, and thus there's an inconsistency in the model. To do a better job, one has to solve Eq. (5.156) for z numerically over the narrow range of temperatures for which $N_0 \gg 1$. Equation (5.157) has proven quite successful in treating real systems.

5.9.2 Equation of state

For the equation of state, we return to Eq. (5.64) and split off the term associated with $k = 0$:

$$P = kT \left(-\frac{1}{V} \ln(1 - z) + \frac{1}{\lambda_T^3} G_{5/2}(z) \right) . \tag{5.160}$$

For $z \ll 1$, the first term in Eq. (5.160) is negligible because of the factor of V^{-1}. For $z \to 1^-$, which we can characterize as $z \approx 1 - O(1/N)$, the first term behaves like $(\ln N)/V$, which tends to zero in the thermodynamic limit. The equation of state in the Bose-Einstein model is

$$P = \frac{kT}{\lambda_T^3} \begin{cases} G_{5/2}(z) & v > v_c \\ G_{5/2}(1) & v < v_c , \end{cases} \tag{5.161}$$

where $v \equiv V/N = 1/n$ is the *specific volume*, the volume per particle, and where $z = G_{3/2}^{-1}(n\lambda_T^3)$ from Eq. (5.157). The *isotherms*[79] of the ideal Bose gas therefore have a flat section for $0 < v < v_c$ (associated with densities $n > n_c$). As the system is compressed (increasing the density), particles condense into the ground state, leaving the pressure unchanged. Particles in the ground state, which have zero momentum, do not exert a pressure.

5.9.3 Internal energy

For the average internal energy U, there is no need to split off the ground state from the sum in Eq. (5.73), because the state with $E = 0$ does not contribute to the energy, and we have from Eq. (5.74), $U = (3gVkT/(2\lambda_T^3))G_{5/2}(z)$. We must still distinguish, however, the two cases for $T > T_B$ and $T < T_B$ because of the fugacity, Eq. (5.157). Thus, with $g = 1$,

$$U(T, V, \mu) = \frac{3}{2} \frac{V}{\lambda_T^3} kT \begin{cases} G_{5/2}(z) & T > T_B \\ G_{5/2}(1) & T < T_B , \end{cases} \tag{5.162}$$

where $z = G_{3/2}^{-1}(n\lambda_T^3)$ for $T \geq T_B$. Note that P as given by Eq. (5.161) and U/V from Eq. (5.162) satisfy the $P = \frac{2}{3}U/V$ relationship established in Eq. (5.75). Equation (5.162) can be written

$$U = \frac{3}{2} NkT \begin{cases} \dfrac{G_{5/2}(z)}{G_{3/2}(z)} & T > T_B \\ \left(\dfrac{T}{T_B}\right)^{3/2} \dfrac{\zeta(5/2)}{\zeta(3/2)} & T < T_B , \end{cases} \tag{5.163}$$

where we've used $n\lambda_T^3 = G_{3/2}(z)$ for $T > T_B$ and $n\lambda_{T_B}^3 = \zeta(3/2)$ for $T \leq T_B$. Note that for $T > T_B$, the energy is that of the classical ideal gas, $\frac{3}{2}NkT$, multiplied by $G_{5/2}(z)/G_{3/2}(z)$. For $T < T_B$, the energy is in the form $U = \frac{3}{2}N_{ex}kT\zeta(5/2)/\zeta(3/2)$, where N_{ex} is defined in Eq. (5.158). Particles condensed in the ground state have zero energy; only particles in excited states contribute to the energy of the gas for $T < T_B$.

We can use Eq. (B.28) (an expansion of $G_{5/2}$ in terms of $G_{3/2}$) to eliminate reference to the fugacity in Eq. (5.163),

$$\frac{G_{5/2}(z)}{G_{3/2}(z)} = 1 - 0.1768 G_{3/2}(z) - 0.0033 G_{3/2}^2(z) - 0.00011 G_{3/2}^3(z) - \cdots , \tag{5.164}$$

---

[79] An isotherm is the locus of points in thermodynamic state space associated with a fixed temperature.[3, p12]

and thus, using $G_{3/2}(z) = \zeta(3/2)(T_B/T)^{3/2}$, we have from Eq. (5.163),

$$U = \frac{3}{2}NkT \begin{cases} \left[1 - 0.4618\left(\dfrac{T_B}{T}\right)^{3/2} - 0.0225\left(\dfrac{T_B}{T}\right)^{3} - 0.0020\left(\dfrac{T_B}{T}\right)^{9/2} - \cdots\right] & T > T_B \\[2ex] \dfrac{\zeta(5/2)}{\zeta(3/2)}\left(\dfrac{T}{T_B}\right)^{3/2} & T < T_B \end{cases}$$

$$(5.165)$$

5.9.4 Heat capacity

The heat capacity is found from the derivative of U in Eq. (5.165):

$$C_V = \left(\frac{\partial U}{\partial T}\right)_{V,N}$$

$$= \frac{3}{2}Nk \begin{cases} \left[1 + 0.231\left(\dfrac{T_B}{T}\right)^{3/2} + 0.045\left(\dfrac{T_B}{T}\right)^{3} + 0.007\left(\dfrac{T_B}{T}\right)^{9/2} + \cdots\right] & T > T_B \\[2ex] \dfrac{5}{2}\dfrac{\zeta(5/2)}{\zeta(3/2)}\left(\dfrac{T}{T_B}\right)^{3/2} & T < T_B. \end{cases}$$

$$(5.166)$$

Figure 5.14 shows the variation of $C_V(T)$. Clearly it approaches the classical value for $T \gg T_B$.

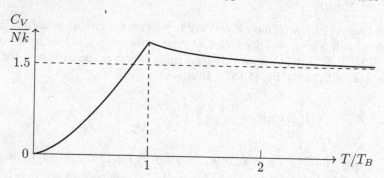

Figure 5.14: Heat capacity of the ideal Bose gas as function of temperature.

Starting at $T = 0$, C_V grows like $T^{3/2}$, i.e., $C_V \sim N_{\text{ex}}$. Whereas for the low-temperature Fermi gas, $C_V \sim T$, because the number of particles available to participate in energy exchanges is restricted to those within kT of E_F, for the low-temperature Bose gas, $C_V \sim T^{3/2}$, because of a different restriction—only particles in excited states participate.

In Eq. (5.79) we derived a general expression for C_V not taking into account the Bose-Einstein condensation:

$$C_V = \frac{3}{2}Nk\left(\frac{5}{2}\frac{G_{5/2}(z)}{G_{3/2}(z)} - \frac{3}{2}\frac{G_{3/2}(z)}{G_{1/2}(z)}\right). \qquad (T > T_B)$$

We now know that such an equation applies only for $T \geq T_B$. In the limit $T \to T_B^{+}$ ($z \to 1^{-}$), $G_{1/2}$ diverges (see Eq. (B.23)), implying

$$\lim_{T \to T_B^{+}} \frac{C_V}{Nk} = \frac{15}{4}\frac{\zeta(5/2)}{\zeta(3/2)} = \lim_{T \to T_B^{-}} \frac{C_V}{Nk},$$

and thus we have on general grounds (i.e., not based on the expansion in Eq. (5.166)) that C_V is continuous at $T = T_B$ in the Bose-Einstein model. Its derivative, however, $\partial C_V/\partial T$ is discontinuous at $T = T_B$, as we see from the cusp in Fig. 5.14.

5.9.5 Entropy

We can use the general formula for the entropy of the ideal Bose gas in Eq. (5.80) for temperatures $T \geq T_B$:

$$\frac{S}{Nk} = \frac{5}{2}\frac{G_{5/2}(z)}{G_{3/2}(z)} - \ln z. \qquad (T \geq T_B) \qquad (5.167)$$

For $T < T_B$, we can calculate S using Eq. (5.166) for C_V, and the thermodynamic relation

$$S = \int_0^T \frac{C_V(T)}{T} dT = \frac{5}{2}Nk \left(\frac{T}{T_B}\right)^{3/2} \frac{\zeta(5/2)}{\zeta(3/2)} = \frac{5}{2}\frac{\zeta(5/2)}{\zeta(3/2)} N_{ex} k. \qquad (T < T_B) \quad (5.168)$$

Equations (5.167) and (5.168) become equal at $T = T_B$ where $z = 1$, and thus entropy is continuous at $T = T_B$. We see (for $T < T_B$), that $S \propto N_{ex}$ (which we should expect on general grounds—entropy is extensive), except S is proportional to the number of particles in excited states, implying that particles in the "condensate," those in the ground state, *carry no entropy*. Each particle in an excited state contributes an entropy $\frac{5}{2}k\zeta(5/2)/\zeta(3/2) = 1.283k$, implying that when a particle makes a transition to the ground state, it gives up a *latent heat* of $T\Delta S = \frac{5}{2}kT\zeta(5/2)/\zeta(3/2)$. If we use the branch of the function C_V for $T > T_B$ (see Eq. (5.166)) in Eq. (5.168), we find the expansion for $T > T_B$,

$$S = \frac{3}{2}Nk \left[\ln\left(\frac{T}{T_B}\right) + 1.0265 - 0.1537\left(\frac{T_B}{T}\right)^{3/2} - 0.0150\left(\frac{T_B}{T}\right)^3 - 0.0015\left(\frac{T_B}{T}\right)^{9/2} - \cdots\right].$$

$$(5.169)$$

5.9.6 Thermodynamic potentials

The Helmholtz energy, $F = U - TS$, can be calculated by combining Eqs. (5.163), (5.167), and (5.168),

$$F = -NkT \begin{cases} \dfrac{G_{5/2}(z)}{G_{3/2}(z)} - \ln z & T > T_B \\[2mm] \left(\dfrac{T}{T_B}\right)^{3/2} \dfrac{\zeta(5/2)}{\zeta(3/2)} & T < T_B \end{cases}. \qquad (5.170)$$

The enthalpy $H = U + PV$ is particularly simple because $PV = \frac{2}{3}U$ for ideal gases, and thus $H = \frac{5}{3}U$, from which, using Eq. (5.163),

$$H = \frac{5}{2}NkT \begin{cases} \dfrac{G_{5/2}(z)}{G_{3/2}(z)} & T > T_B \\[2mm] \left(\dfrac{T}{T_B}\right)^{3/2} \dfrac{\zeta(5/2)}{\zeta(3/2)} & T < T_B \end{cases}. \qquad (5.171)$$

The Gibbs energy, $G = H - TS$, is found by combining Eqs. (5.171), (5.167), and (5.168):

$$G = \begin{cases} NkT\ln z = N\mu & T > T_B \\ 0 & T < T_B \end{cases}. \qquad (5.172)$$

For $T > T_B$, Eq. (5.172) reproduces what have from general thermodynamics, Eq. (1.54), that $G = N\mu$. For $T < T_B$, $G = 0$ because $\mu = 0$ for the Bose condensate.

SUMMARY

This chapter considered applications of equilibrium statistical mechanics to systems of noninteracting particles. Despite its length, it can be summarized fairly succinctly.

- We started with the classical ideal gas of N structureless particles in a volume V at temperature T (and so, the canonical ensemble). The partition function—the number of states accessible to the system at temperature T—is quite simple: $Z_{can} = (V/\lambda_T)^N/N!$, Eq. (5.1). By considering the associated phase-space distribution function, Eq. (5.2), we obtained the Maxwell speed distribution function, $f(v)$, Eq. (5.5), the probability $f(v)dv$ that a particle selected at random has a speed between v and $v + dv$. From this probability density, one can calculate the most probable speed, the average speed, and the root-mean-square speed. Referring to Fig. 5.1, there is a maximum in the Maxwell distribution, a result of the product of the Boltzmann factor, $\exp(-\beta mv^2/2)$, a decreasing function of the speed, and the volume of phase space, which grows like v^2.

- We considered paramagnets—the "ideal gas" of magnetism—systems of magnetic moments that don't interact with each other, but interact with an external magnetic field. We treated the problem from the quantum and classical points of view. The quantum treatment leads to the Brillouin functions, B_J, Eq. (5.13), where J is the total angular momentum, while the classical treatment leads to the Langevin function, L, Eq. (5.19), which is the $J \to \infty$ limit of the Brillouin functions. As shown in Fig. 5.4, there is excellent agreement between theory and experiment for paramagnets.

- We considered the statistical mechanics of harmonic oscillators. The partition function for a single, quantum-mechanical harmonic oscillator, $Z_1(\beta) = (2\sinh(\beta\hbar\omega/2))^{-1}$, Eq. (5.20), becomes, in the high temperature limit, $Z_1(\beta) \sim (\beta\hbar\omega)^{-1}$, i.e., *all* of the infinite number of energy states of the harmonic oscillator become thermally accessible as $T \to \infty$. At low temperature, $\beta \to \infty$, $Z_1(\beta) \sim e^{-\beta\hbar\omega/2}$, and the number of states accessible to the system is exponentially small. As $T \to 0$, there are no states available to the system, $Z \to 0$. We introduced the idea of occupation number, $\langle n \rangle = (e^{\beta\hbar\omega} - 1)^{-1}$, the average value of the quantum number n at temperture T. We noted the connection between harmonic oscillators and bosons, that the number of ways quanta can be distributed over oscillators obeys Bose-Einstein statistics (see Eq. (5.61)).

- We treated the diatomic ideal gas—molecules that don't interact with each other, but which have internal degrees of freedom. Perhaps the most important lesson here is that if the Hamiltonian can be written in separable form, Eq. (5.30), then the degrees of freedom underlying each part of the Hamiltonian are noninteracting, and the partition function occurs in the form of Eq. (5.31), as the product of partition functions associated with each part of the Hamiltonian. In particular, partition functions associated with internal degrees of freedom are independent of the volume of the system. That in turn implies that the heat capacity of the system is the sum of heat capacities for each of the modes of excitation, Eq. (5.32). We treated rotational degrees of freedom and found that the characteristic temperature at which rotation becomes activated, Θ_r, is quite low, ≈ 10 K. At room temperature, rotational degrees of freedom contribute an extra factor of Nk to the heat capacity, Eq. (5.36), explaining why C_V has the value $\frac{5}{2}Nk$ for diatomic gases (a problem noted by Gibbs before the advent of quantum mechanics). The characteristic temperature at which vibrations become activated is much greater than room temperature, $\Theta_v \sim 4000$ K.

- We then took up the ideal quantum gases, the *bulk* of Chapter 5.

  - We derived the partition function for ideal quantum gases, Eq. (5.55), which leads to the Bose-Einstein and Fermi-Dirac distribution functions for occupation numbers, Eqs.

(5.61) and (5.62). We derived the equation of state, Eq. (5.66), which introduced the Bose-Einstein and Fermi-Dirac integral functions, $G_n(z)$ and $F_n(z)$, the properties of which are developed in Appendix B. We worked out general thermodynamic expressions for N, U, C_V, and S, Eqs. (5.69), (5.74), (5.79), and (5.80), and we developed high-temperature expansions in powers of $n\lambda_T^3$ (for the regime $n\lambda_T^3 \ll 1$). At high temperatures there are minor differences between the two types of quantum gas.

- In the opposite regime, $n\lambda_T^3 \gtrsim 1$, quantum effects are markedly exhibited, in which Fermi and Bose gases behave differently. A feature common to both is the concept of degenerate matter, in which all the lowest energy states of the system are occupied. For fermions, there is a large zero-point energy (Fermi energy) because by the Pauli principle. For bosons, at a sufficiently low temperature, all particles condense into the ground state, the Bose-Einstein condensation; Section 5.9.

- We included two important applications: degeneracy pressure in stars, Section 5.7, and cavity radiation, Section 5.8.

EXERCISES

5.1 Use the partition function for the ideal gas, Eq. (5.1), together with the expressions in Eq. (4.58) to derive the equation of state for the ideal gas and the Sackur-Tetrode formula for the entropy, Eq. (1.64). Use the Stirling approximation.

5.2 Derive an expression for the entropy of an ideal gas of N identical particles confined to a d-dimensional volume $V = L^d$. A: $S/(Nk) = 1 + \dfrac{d}{2} + \ln\left(V/(N\lambda_T^d)\right)$.

5.3 Use Eq. (4.60) to calculate the entropy of the ideal gas using the distribution function ρ in Eq. (5.2). Let's do this in steps.

a. Show that Eq. (4.60) is equivalent to $S = -k\left[\ln N! + \int_\Gamma \rho \ln \rho d\Gamma\right]$.

b. Show, when using the factorized distribution function $\rho = \prod_{i=1}^N \rho_i$ (as in Eq. (5.2)) that

$$\int_\Gamma \rho \ln \rho d\Gamma = \sum_{i=1}^N \int_{\Gamma_i} \rho_i \ln \rho_i d\Gamma_i \,.$$

c. Show, using the form for ρ_i in Eq. (5.2) that

$$\int_{\Gamma_i} \rho_i \ln \rho_i d\Gamma_i = \ln\left(\lambda_T^3/V\right) - \frac{\beta}{2m}\int_{\Gamma_i} p^2 \rho_i d\Gamma_i$$

d. Show that

$$\int_{\Gamma_i} p^2 \rho_i d\Gamma_i = \frac{3}{2}\frac{\lambda_T^3}{h^3}\pi^{3/2}\left(\frac{2m}{\beta}\right)^{5/2} = \frac{3}{2}\left(\frac{2m}{\beta}\right)$$

Use an integral from Appendix B.

e. Combine the results of these steps, and make use of the Stirling approximation. You should obtain the Sackur-Tetrode formula, Eq. (1.64).

5.4 Consider the momentum version of the Maxwell speed distribution. Based on Eq. (5.4), we can "read off" the normalized momentum probability density function $f(p)$:

$$f(p) = \frac{4\pi}{(2\pi mkT)^{3/2}}p^2 e^{-\beta p^2/(2m)} \,.$$

a. Show that the mean momentum of magnitude p

$$\bar{p} = \int_0^\infty p f(p) \mathrm{d}p = \sqrt{\frac{8mkT}{\pi}} \; .$$

b. Show that the mean *inverse* momentum $\overline{p^{-1}}$ has the value

$$\overline{p^{-1}} = \int_0^\infty \frac{1}{p} f(p) \mathrm{d}p = \frac{2}{\sqrt{2\pi mkT}} \; .$$

c. Is $(\bar{p})^{-1} = \overline{p^{-1}}$?

d. Show that $h \left\langle \dfrac{1}{p} \right\rangle = 2\lambda_T$ and that $\dfrac{h}{\langle p \rangle} = \dfrac{\pi}{2}\lambda_T$. The thermal wavelength is *almost* the average of the de Broglie wavelengths of the particles of a gas in equilibrium at temperature T, but not exactly. Give a qualitative argument why we should expect $\langle p^{-1} \rangle > (\langle p \rangle)^{-1}$.

5.5 Verify the results in Eqs. (5.6)–(5.8). Give a qualitative argument why $\bar{v} < v_{rms}$.

5.6 Diatomic nitrogen (N_2) is the largest component of the air we breathe. What is mean speed of an N_2 molecule at $T = 300$ K? (A: 476 m s^{-1}.)

5.7 Equation (5.5) is the Maxwell speed distribution. Derive a similar expression for the distribution of the energies of the particles, through the substitution $E = \frac{1}{2}mv^2$. Hint: You want to obtain a normalized distribution, so that $\int_0^\infty f_E(E)\mathrm{d}E = 1$.

5.8 Derive Eq. (5.12), including the form of the Brillouin function in Eq. (5.13), starting from the derivative of the partition function as indicated in Eq. (5.11).

5.9 Properties of the Brillouin functions.

a. Show for small x that the Langevin function has the series expansion

$$L(x) = \frac{x}{3} - \frac{x^3}{45} + O(x^5) \; . \tag{P5.1}$$

b. Show that

$$\lim_{x \to \infty} B_J(x) = 1 \qquad \lim_{J \to \infty} B_J(x) = \coth x - \frac{1}{x} \qquad B_J(x) \overset{x \to 0}{\sim} \frac{1}{3}\left(1 + \frac{1}{J}\right)x \; .$$

c. Show that $B_{\frac{1}{2}}(x) = \tanh(x)$.

5.10 Referring to curve I in Fig. 5.4 associated with $J = \frac{3}{2}$ and $g = 2$, the data point for $T = 2$ K occurs close to the value of 20×10^{-3} Gauss/Kelvin (a Gauss, a non-SI unit, is 10^{-4} Tesla). What value of the dimensionless variable $x = \beta\mu_B g J B$ does this point correspond to? (A: $x \approx 4$.) Referring to Fig. 5.3, does the value of $B_{\frac{3}{2}}(4)$ look about right in relation to the saturation value?

5.11 The purpose of this exercise is to derive the high-temperature form of the partition function for rotational motions of the diatomic gas, Eq. (5.37). The Euler-Maclaurin summation formula[50, p806] is an expression for the difference between an integral and a closely related sum:

$$\sum_{n=0}^\infty f(n) = \int_0^\infty f(x)\mathrm{d}x + \frac{1}{2}f(0) + \sum_{k=1}^\infty \frac{B_{2k}}{(2k)!} f^{(2k-1)}(0) \; , \tag{P5.2}$$

where B_{2k} are the *Bernoulli numbers* ($B_2 = \frac{1}{6}$, $B_4 = -\frac{1}{30}$, $B_6 = \frac{1}{42}$, etc.), $f^{(n)}$ denotes the nth derivative of $f(x)$, and where the form of the formula in Eq. (P5.2) assumes that $f(\infty) = 0$, as well as all the derivatives, $f^{(n)}(\infty) = 0$. Apply the Euler-Maclaurin formula to the summation in Eq. (5.33) for the function $f(x) = (2x+1)e^{-\Theta_r x(x+1)/T}$ through second order in Θ_r/T, which requires you to evaluate the term for $k = 3$ in Eq. (P5.2) (that is, you have to find the *fifth* derivative of $f(x)$). Use Eq. (5.35) for the integral in Eq. (P5.2).

5.12 Derive Eq. (5.38) starting from Eq. (5.37). Hint: It helps to separate the big from the small as $\beta \to 0$. Write Z in the form $Z_{1,\text{rot}} = \frac{1}{\alpha\beta}\left(1 + a\beta + b\beta^2 + c\beta^3 + \cdots\right)$, where $\alpha \equiv \hbar^2/(2I)$. Thus, $\ln Z_{1,\text{rot}} = -\ln(\alpha\beta) + \ln\left(1 + a\beta + b\beta^2 + c\beta^3 + \cdots\right)$. Use the Taylor expansion for $\ln(1+x)$, and work consistently to order β^3. The point of this exercise is to show that it's easier to develop an approximation for $\ln Z$ instead of one for Z.

5.13 Calculate the entropy of an ideal gas of diatomic molecules at room temperature. Show that

$$S = Nk\left[\frac{7}{2} + \ln\left(\frac{V}{N\lambda_T^3}\frac{T}{2\Theta_r}\right)\right].$$

The entropy per particle of the diatomic gas is therefore larger than that for the monatomic gas; the internal degrees of freedom per particle (rotational, vibrational, etc.) add to the entropy of the system. Hint: In Section 5.4 we noted that the partition function Z that ensues from a separable Hamiltonian H, Eq. (5.30), occurs as a product of the partition functions associated with the degrees of freedom of each part of H, Eq. (5.31). That implies, from $F = -kT \ln Z$, that the free energy occurs as a sum of free energies associated with the degrees of freedom of each part of H, and hence, using Eq. (4.58), that the entropy also occurs as such a sum. You already know the form of the entropy of an ideal gas of structureless particles; you need to find the entropy associated with the relevant internal degrees of freedom. Make suitable approximations based on the fact that at room temperature, $\Theta_r \ll T \ll \Theta_v$.

5.14 Write Eq. (5.69) (for the density of fermions or bosons) in the form

$$y \equiv \frac{n\lambda_T^3}{g} = A_{3/2}(z),$$

where $A_m(z)$ represents *either* $G_m(z)$ or $F_m(z)$. We can therefore write for the fugacity,

$$z = A_{3/2}^{-1}(y),$$

where $A_{3/2}^{-1}$ is the inverse function to $A_{3/2}$. Show that

$$\left(\frac{\partial z}{\partial T}\right)_n = -\frac{3}{2}\frac{z}{T}\frac{A_{3/2}(z)}{A_{1/2}(z)}.$$

Hint: Make use of the inverse function theorem of differential calculus, and Eqs. (B.20) and (B.31).

5.15 We calculated the entropy of the ideal quantum gases in Eq. (5.80) using the Euler relation, Eq. (1.53). Show that the Euler relation is obtained using Eq. (4.78),

$$S = k\frac{\partial}{\partial T}(T\ln Z_G)\Big|_{V,\mu},$$

the formula for S in the grand canonical ensemble, when Eq. (5.55) is used for Z_G. Hint: Use Eqs. (5.57), (5.73), and (5.64).

5.16 Show that the Helmholtz free energy for the ideal quantum gases is given by the expression

$$F = NkT \left(\ln z - \frac{A_{5/2}(z)}{A_{3/2}(z)} \right),$$

where the notation $A_{5/2}$ and so on is explained in Section 5.5.4. Hint: $F = -PV + N\mu$.

5.17 The heat capacity C_V can be obtained either from a derivative of the internal energy, Eq. (1.58) ($C_V = (\partial U/\partial T)_{V,N}$), or from a derivative of the entropy, Eq. (P1.5),

$$C_V = T \left(\frac{\partial S}{\partial T} \right)_{V,N}. \tag{P5.3}$$

Show that Eq. (5.79) for C_V follows by differentiating Eq. (5.80) for S.

5.18 Show, using the string of definitions in Section 5.6.1, that $v_F = \hbar k_F/m$. What is the interpretation of this result?

5.19 Show that $g(E_F)E_F = \frac{3}{2}N$.

5.20 What would the density of electrons ($g = 2$) have to be so that from Eq. (5.86) we would find a Fermi velocity equal to the speed of light? What would the Fermi energy be in this case? A: 5.9×10^{29} cm^{-3}, $\frac{1}{2}mc^2 \approx 0.256$ MeV. Of course, we can't use Eq. (5.86) in this case; we would have to find a dedicated formula for the Fermi energy of a relativistic gas.

5.21 Derive Eq. (5.91) from Eq. (5.73).

5.22 Derive Eq. (5.95) for the bulk modulus of the degenerate Fermi gas.

5.23 Compare the right side of Eq. (5.96) with the derivative of the Fermi-Dirac distribution function with respect to energy. You should find they are the same up to a minus sign.

5.24 Starting from the exact result Eq. (5.79),

$$\frac{C_V}{Nk} = \frac{15}{4} \frac{F_{5/2}}{F_{3/2}} - \frac{9}{4} \frac{F_{3/2}}{F_{1/2}},$$

derive an expression for the heat capacity of the ideal Fermi gas at low temperatures. You should find that at leading order Eq. (5.79) reduces to the results of Eq. (5.106) for low temperatures. Hint: Use Eq. (B.40).

5.25 Use the exact result Eq. (5.80) for the entropy of the Fermi gas,

$$\frac{S}{Nk} = \frac{5}{2} \frac{F_{5/2}}{F_{3/2}} - \ln z,$$

to derive an expression for the entropy at low temperatures. You should find the same result as for the low-temperature heat capacity,

$$\frac{S}{Nk} \approx \frac{\pi^2}{2} \frac{1}{\beta\mu} = \frac{\pi^2}{2} \left(\frac{T}{T_F} \right).$$

Does it make sense that for a heat capacity that's linear with T at low temperature, the entropy would have exactly the same form? See Eq. (P5.3). Both C_V and S have the base units of Joule per Kelvin; both have Boltzmann's constant as a prefactor.

5.26 Show that the pressure of the ideal Fermi gas is, at low temperatures, given by

$$P = \frac{gV}{15\pi^2}\mu^{5/2}\left[1 + \frac{5\pi^2}{8}\frac{1}{(\beta\mu)^2} - \frac{7\pi^4}{384}\frac{1}{(\beta\mu)^4} + \cdots\right].$$

Hint: This is actually a one-liner. Equation (5.75) holds for all temperatures and densities.

5.27 Assume that a Bose-Einstein condensation experiment uses 10^7 rubidium-87 atoms confined to a volume of 10^{-15} m^3 at a temperature of 200 nK. Calculate the Bose temperature, T_B. What fraction of the particles are in the ground state?

5.28 Derive Eq. (5.154).

5.29 Derive the following expression for the heat capacity at constant pressure, C_P, of the ideal Bose gas for $T > T_B$:

$$C_P = \frac{5}{2}Nk\left[\frac{5}{2}\frac{G_{5/2}^2(z)G_{1/2}(z)}{G_{3/2}^3(z)} - \frac{3}{2}\frac{G_{5/2}(z)}{G_{3/2}(z)}\right]. \tag{P5.4}$$

We'll take this in steps.

a. Start from (see, for example, [3, p65])

$$C_P = \left(\frac{\partial H}{\partial T}\right)_P.$$

Use Eq. (5.171) to show that

$$\left(\frac{\partial H}{\partial T}\right)_P = \frac{5}{2}Nk\left[\frac{T}{z}\left(\frac{\partial z}{\partial T}\right)_P\left(1 - \frac{G_{5/2}(z)G_{1/2}(z)}{G_{3/2}^2(z)}\right) + \frac{G_{5/2}(z)}{G_{3/2}(z)}\right],$$

where we've used Eq. (B.20).

b. Let's work on the derivative $(\partial z/\partial T)_P$. Write out Eq. (5.161) explicitly:

$$P = \frac{kT}{h^3}(2\pi mkT)^{3/2}G_{5/2}(z).$$

Form the differential of this relation, which will be in the form $dP = AdT + Bdz$, where you need to evaluate the quantities A and B. The derivative $(\partial z/\partial T)_P$ therefore equals $(-A/B)$. Show that

$$\frac{T}{z}\left(\frac{\partial z}{\partial T}\right)_P = -\frac{5}{2}\frac{G_{5/2}(z)}{G_{3/2}(z)}.$$

Compare this result with the result of Exercise 5.14. Equation (P5.4) then follows. Show this.

c. Conclude that $C_P \to \infty$ as $T \to T_B^+$. The isobars[80] of the ideal Bose gas are vertical (infinite slope) for $T \le T_B$ (the converse of the level isotherms noted in Eq. (5.161)), and thus the thermal expansivity $\alpha = \frac{1}{V}(\partial V/\partial T)_P$ diverges as $T \to T_B^+$. From Eq. (1.42), $C_P = C_V + \alpha^2 TV/\beta_T$, where $\beta_T > 0$ is the isothermal compressibility.

---

[80] An isobar is the locus of all points in thermodynamic state space associated with constant pressure.

5.30 Show that for cavity radiation, the entropy per photon $S/N \approx 3.6k$ and the energy per photon $U/N \approx 2.7kT$. Compare numerically with the entropy per particle of the ideal gas of monatomic particles, the ideal gas of diatomic molecules (at room temperature), and the Bose-Einstein condensate. A: $2.5k$, $3.5k$, and $1.28k$. The entropy of thermal radiation is large compared with other systems. These differences in entropy per particle may seem minor, but actually they're *huge*. A difference in entropy per particle $\Delta S = k$ implies from Eq. (1.62) an increase in the number of ways W by a factor of e^N.

5.31 Referring to Table 5.4, you sometimes see the statement that $\gamma \equiv C_P/C_V$ (the ratio of heat capacities) for cavity radiation is $\gamma = 4/3$, which is simply incorrect. It's true that for adiabatic processes, $TV^{1/3} = $ constant for the photon gas, but one can't make the comparison with the ideal gas to conclude that $\gamma = \frac{4}{3}$—the photon gas is not the ideal gas! Make an argument that C_P is not defined for the photon gas. Hint: $C_P = (\partial H/\partial T)_P$—can one take the derivative with respect to T, holding P fixed? See Eq. (5.132).

5.32 Derive Eq. (5.142). Hint: By energy conservation, $|u(\lambda, T)d\lambda| = |u(\nu, T)d\nu|$.

5.33 Show that Eq. (5.145) follows from Eq. (5.144).

5.34 Is there a conflict between Eq. (5.144) and Eq. (5.129)? The following relation can be established as a general result of thermodynamics (see [3, p52]):

$$\left(\frac{\partial U}{\partial V}\right)_S = \left(\frac{\partial U}{\partial V}\right)_T - T\left(\frac{\partial P}{\partial T}\right)_V, \qquad (P5.5)$$

Show, using $U = uV$ for cavity radiation, that (a result *specific to cavity radiation*)

$$\left(\frac{\partial u}{\partial V}\right)_S = -\frac{4}{3}\frac{u}{V}. \qquad (P5.6)$$

Hint: Use Eqs. (5.129), (5.132), and (5.133). Note that you're not asked to derive Eq. (P5.5). Thus, while the energy density of cavity radiation is independent of volume under *isothermal* conditions, *it's not for adiabatic processes*—temperature changes in adiabatic processes.

5.35 Show, by integrating Wien's equation, (5.144), over all frequencies ($u = \int_0^\infty u(\nu)d\nu$),

$$\left(\frac{\partial u}{\partial V}\right)_S = \frac{1}{3V}\left[\nu u(\nu)\right]\Big|_0^\infty - \frac{4}{3}\frac{u}{V}. \qquad (P5.7)$$

Hint: Integrate by parts. Conclude that consistency between Eqs. (P5.7) and (P5.6) requires

$$\lim_{\nu \to \infty} \nu u(\nu) = 0,$$

a consequence of *classical* physics.

5.36 Estimate the number of thermally excited photons in the universe. Hint: From Eq. (5.137), we need to know the volume V and temperature T. Assume the universe is a spherical cavity of radius $R \approx 10^{26}$ m (or whatever your favorite number is for the size of the universe). The universe is thought to be approximately 14 billion years old; how far can light travel in that time? For the temperature, measurements of the *cosmic microwave background* radiation fall on the Planck curve with high accuracy for a temperature of 2.725 K.

5.37 Derive Eq. (5.114) from Eq. (5.113).

5.38 Show that Eq. (5.118), the density of states for relativistic electrons, reduces to the nonrelativistic expression, Eq. (2.18), in the limit $mc^2 \gg T$ (where T is kinetic energy).

5.39 a. Show that the function $f(x)$ defined in Eq. (5.122) has the form for $x \to 0$:

$$f(x) \overset{x\to 0}{\sim} \frac{4}{5}x^5 - \frac{1}{7}x^7 + O(x^9) .$$

Show that, at lowest order in x, we recover from Eq. (5.122) the nonrelativistic expression, Eq. (5.92),

$$U = \frac{V}{5\pi^2} \left(\frac{2m}{\hbar^2}\right)^{3/2} E_F^{5/2} ,$$

where $E_F = (\hbar^2/(2m))(3\pi^2 n)^{2/3}$.

b. Consider the opposite limit of $x \gg 1$. Show that $f(x)$ defined in Eq. (5.122) has the form for large x:

$$f(x) \overset{x\to\infty}{\sim} 2x^4 \left[1 - \frac{4}{3x} + O(1/x^2)\right] .$$

Show that in the extreme relativistic limit,

$$U = \frac{V}{4\pi^2} \hbar c (3\pi^2 n)^{4/3} .$$

5.40 a. Show that the function $g(x)$ defined in Eq. (5.124) has the form for small x:

$$g(x) \overset{x\to 0}{\sim} \frac{8}{5}x^5 - \frac{4}{7}x^7 + O(x^9) .$$

Show that Eq. (5.124) reduces to Eq. (5.94) in this limit.

b. Show that $g(x)$ in Eq. (5.124) has the form for large x:

$$g(x) \overset{x\to\infty}{\sim} 2x^4 \left[1 - \frac{1}{x^2} + O(1/x^4)\right] .$$

Derive the form of the degeneracy pressure shown in Eq. (5.125).

CHAPTER 6

Interacting systems

T HE systems treated in Chapter 5 are valuable examples of systems for which all mathematical steps can be implemented and the premises of statistical mechanics tested. Instructive (and relevant) as they are, these system lack an important detail: interactions between particles. In this chapter, we step up our game and consider systems featuring inter-particle interactions. Statistical mechanics can treat interacting systems, but no one said it would be easy.

6.1 THE MAYER CLUSTER EXPANSION

Consider N identical particles of mass m that interact through *two-body interactions*, v_{ij}, with Hamiltonian

$$H = \frac{1}{2m} \sum_i p_i^2 + \sum_{j>i} v_{ij} , \qquad (i, j = 1, 2, \cdots, N) \qquad (6.1)$$

where $v_{ij} \equiv v(r_i - r_j)$ denotes the potential energy associated with the interaction between particles at positions r_i, r_j, and $\sum_{j>i}$ indicates a sum over $\binom{N}{2}$ pairs of particles.[1] For *central forces*, v_{ij} depends only on the magnitude of the distance between particles, $v_{ij} = v(|r_i - r_j|)$, which we assume for simplicity. The methods developed here can be extended to quantum systems, but the analysis becomes more complicated; we won't consider interacting quantum gases.[2]

For our purposes, the precise nature of the interactions underlying the potential energy function $v(r)$ is not important as long as there is a long-range attractive component together with a short-range repulsive force. To be definite, we mention the *Lennard-Jones potential* (shown in Fig. 6.1) for the interaction between closed-shell atoms, which has the parameterized form,

$$v(r) = 4\epsilon \left[\left(\frac{\sigma}{r} \right)^{12} - \left(\frac{\sigma}{r} \right)^6 \right] , \qquad (6.2)$$

where ϵ is the depth of the potential well, and σ is the distance at which the potential is zero. The r^{-6} term describes an attractive interaction between neutral molecules that arises from the energy of interaction between fluctuating multipoles of the molecular charge distributions.[3] The r^{-12} term models the repulsive force at short distances that arises from the Pauli exclusion effect of overlapping electronic orbitals. There's no science behind the r^{-12} form; it's analytically convenient, and it provides a good approximation of the interactions between atoms. For the noble gases, ϵ ranges from 0.003 eV for Ne to 0.02 eV for Xe [18, p398]. The parameter σ is approximately 0.3 nm.

---

[1] For $N = 3$, $\sum_{j>i} v_{ij} = v_{12} + v_{13} + v_{23}$. Each pair of particles is counted only once.

[2] Reference [52] is an extensive review of the statistical mechanics of interacting quantum systems. See also [53].

[3] Such forces are variously referred to as *van der Waals forces* or *London dispersion forces*.

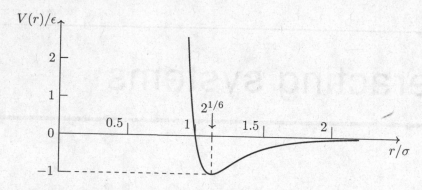

Figure 6.1: The Lennard-Jones inter-atomic potential as a function of r/σ.

To embark on the statistical-mechanical-road, we have in the canonical ensemble[4]

$$Z(T,V,N) = \frac{1}{N!} \int d\Gamma_N e^{-\beta\left(\sum_i p_i^2/2m + \sum_{i>j} v_{ij}\right)} \equiv \frac{1}{N!\lambda_T^{3N}} Q(T,V,N), \qquad (6.3)$$

where λ_T occurs from integrating the momentum variables and Q is the *configuration integral*, the part of the partition function associated with the potential energy of particles,

$$Q(T,V,N) = \int d^N r \exp\left(-\beta \sum_{i>j} v_{ij}\right) = \int d^N r \prod_{i>j} e^{-\beta v_{ij}} \equiv \int d^N r \prod_{i>j} (1 + f_{ij}), \quad (6.4)$$

where

$$f_{ij} = f(r_{ij}) \equiv e^{-\beta v(r_{ij})} - 1 \qquad (6.5)$$

is the *Mayer function*,[5] shown in Fig. 6.2. In the absence of interactions, $f_{ij} = 0$, $Q = V^N$, and we

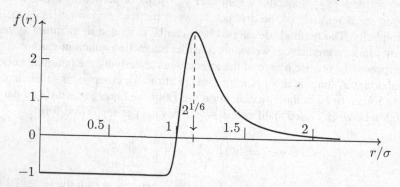

Figure 6.2: Mayer function $f(r)$ associated with the Lennard-Jones potential. $\beta\epsilon = 1.33$.

recover Eq. (5.1). With interactions, $f(r)$ is bounded between -1 and $\left(e^{-\beta V_{\min}} - 1\right)$, where $V_{\min}$ is the minimum value of the interaction potential; $f(r)$ is small for inter-particle separations in excess of the effective range of the potential. Mayer functions allow us to circumvent problems associated with potential functions that diverge[6] as $r \to 0$. At sufficiently high temperatures, $|f(r)| \ll 1$, which provides a way of approximately treating the non-ideal gas.

---

[4]Where is the factor of h^{3N} in Eq. (6.3)? Recall that we defined $d\Gamma$ as dimensionless in Eq. (2.51).
[5]After Joseph Mayer, who developed this method in the 1930s. See [54, Chapter 13].
[6]We can't apply methods of perturbation theory if the potential contains a divergent term—it can't be considered a "small correction"; the Lennard-Jones potential is not suitable for perturbation theory, but the Mayer function is.

By expanding out the product in Eq. (6.4), a $3N$-fold integration is converted into a sum of lower-dimensional integrals known as *cluster integrals*. The product expands into a sum of terms each involving products of Mayer functions, from zero to all $\binom{N}{2} = N(N-1)/2$ Mayer functions,

$$\prod_{j>i}(1+f_{ij}) = 1 + \sum_{\text{pairs}}' f_{ij} + \sum_{\text{triples}}' [f_{ij}f_{jk} + f_{ij}f_{jk}f_{ik}] + \sum_{\text{quadruples}}' [f_{ij}f_{kl} + \cdots] + \cdots, \quad (6.6)$$

where pairs, triples, quadruples, etc., refer to configurations of two, three, four, and so on, particles known as *clusters*.[7] The primes on the summation signs indicate restrictions so that we never encounter terms such as f_{mm} (particles don't interact with themselves) or $f_{12}f_{12}$; *interactions are between distinct pairs of particles*, counted only once. For $N = 3$ particles, Eq. (6.6) generates $2^{\binom{3}{2}} = 2^3 = 8$ terms:

$$(1 + f_{12})(1 + f_{13})(1 + f_{23}) = 1 + f_{12} + f_{13} + f_{23} + f_{12}f_{13} + f_{12}f_{23} + f_{13}f_{23} + f_{12}f_{13}f_{23}.$$

As N increases, the number of terms rises rapidly. For $N = 4$, there are $2^{\binom{4}{2}} = 2^6 = 64$ terms generated by the product; 64 integrals that contribute to Eq. (6.4). Fortunately, many of them are the same; our task is to learn how to characterize and count the different types of integrals. For $N = 5$, there are 1024 terms—we *better* figure out how to systematically count the relevant contributions to the partition function if we have any hope of treating macroscopic values of N.

A productive strategy is to draw a picture, or *diagram*, representing each term in Eq. (6.6). Figure 6.3 shows two circles with letters in them denoting particles at positions r_i and r_j. The

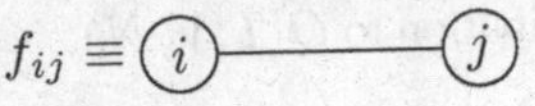

$$f_{ij} \equiv \text{(i)} \longrightarrow \text{(j)}$$

Figure 6.3: Cluster diagram representing the Mayer function f_{ij} between particles i, j.

circles represent particles, and the line between them represents the Mayer function f_{ij}. Note that the physical distance between i and j is taken into account through the value of the Mayer function; the line in Fig. 6.3 indicates an interaction between particles, regardless of the distance between them. If one were to imagine a numbered circle for each of the N particles of the system, with a line drawn between circles i and j for every occurrence of f_{ij} in the terms of Eq. (6.6), every term would be represented by a diagram.[8] Let the drawing of diagrams begin!

We'll do that in short order, but let's first consider what we do with diagrams, a process known as "calculating the diagram." Starting with two-particle diagrams (Fig. 6.3), for each term associated with $\sum_{\text{pairs}}' f_{ij}$ in Eq. (6.6), there is a corresponding contribution to Eq. (6.4):

$$\int d^N r f_{ij} = V^{N-2} \int dr_i dr_j f(|r_i - r_j|) \equiv V^{N-2}(2b_2(T)V). \quad (6.7)$$

The integrations in Eq. (6.7) over the spatial coordinates *not* associated with particles i and j leave us with V^{N-2} on the right side. The cluster integral $b_2(T)$ is, by definition,

$$b_2(T) \equiv \frac{1}{2V} \int dr_i dr_j f(r_{ij}) = \frac{1}{2} \int dr f(r) = \frac{1}{2} \int_0^\infty 4\pi r^2 f(r) dr. \quad (6.8)$$

The second equality in Eq. (6.8) is a step we'll take frequently. Define a new coordinate system centered at the position specified by r_i. With $r_j - r_i \equiv r_{ji}$ as a new integration variable, we're

---

[7]There is no question whether the expansion in Eq. (6.6) converges—it has a finite number of terms.
[8]The use of diagrams to represent the terms generated by $\prod_{j>i}(1 + f_{ij})$ was the first systematic use of diagrams in theoretical physics, starting in the 1930s. *Feynman diagrams* were introduced in the late 1940s for similar purposes of representing mathematical expressions with pictures.

free to integrate over r_i. We'll denote this step as $dr_i dr_j \rightarrow dr_i dr_{ji}$. The quantity b_2 probes the effective range of the two-body interaction at a given temperature (the second moment of the Mayer function) and has the dimension of (length)^3. The cluster expansion method works best when the volume per particle $V/N \equiv 1/n$ is large relative to the volume of interaction, i.e., when $nb_2 \ll 1$. The identity of particles is lost in Eq. (6.8). Thus, the $\binom{N}{2} = N(N-1)/2$ terms in $\sum'_{\text{pairs}} f_{ij}$ all contribute the same value to the configuration integral. Through first order in Eq. (6.6), we have

$$Q(N, T, V) = V^N \left(1 + N(N-1) \frac{b_2(T)}{V} + \cdots \right) .$$

The partition function (number of states available to the system) is modified (relative to noninteracting particles) by pairs of particles that bring with them an effective volume of interaction, $2b_2$.

The next term in Eq. (6.6), a sum over three-particle clusters, involves products of two and three Mayer functions. Figure 6.4 shows the diagrams associated with three distinct particles (ijk) joined

Figure 6.4: Linked clusters of three particles.

by two or three lines. For their contribution to $Q(T, V, N)$,

$$\int d^N r \left[f_{ik} f_{kj} + f_{ik} f_{ij} + f_{ij} f_{jk} + f_{ij} f_{jk} f_{ik} \right] \equiv V^{N-3} (3! V b_3(T)) , \tag{6.9}$$

where, by definition,

$$\begin{aligned}
b_3(T) &\equiv \frac{1}{3!V} \left[3 \int dr_i dr_j dk_k f_{ik} f_{kj} + \int dr_i dr_j dr_k f_{ik} f_{kj} f_{ji} \right] \\
&\rightarrow \frac{1}{3!V} \left[3 \int dr_k dr_{ik} dr_{jk} f(r_{ik}) f(r_{jk}) + \int dr_j dr_{ij} dr_{kj} f(r_{ik}) f(r_{kj}) f(r_{ij}) \right] \\
&= \frac{1}{3!} \left[3 \left(\int dr f(r) \right)^2 + \int dr_{ij} dr_{kj} f(r_{ik}) f(r_{kj}) f(r_{ij}) \right] \\
&= 2b_2^2(T) + \frac{1}{3!} \int dr_{ij} dr_{kj} f(r_{ik}) f(r_{kj}) f(r_{ij}) .
\end{aligned} \tag{6.10}$$

The third line of Eq. (6.10) follows because $r_{ik} = r_{ij} - r_{kj}$ (and thus the integral is completely determined by r_{ij} and r_{kj}; see Exercise 6.2), and we've used Eq. (6.8) in the final line. The factor of 3! is included in the definition to take into account permutations of i, j, k, and thus b_3 is *independent of how we label the vertices of the diagram.*[9] The factor of 3 inside the square brackets comes from the equivalence of the three diagrams in Fig. 6.4 under cyclic permutation, $i \rightarrow j \rightarrow k \rightarrow i$. The quantity b_3 has the dimension of (volume)^2.

---

[9]*Graph theory* is a branch of mathematics devoted to the study of graphs, and the use of its terminology is helpful. A *linear graph* is a collection of vertices (circles), between some pairs of which there are lines (bonds). Two vertices are *adjacent* is there is a bond joining them directly. A *path* is a sequence of adjacent vertices. A graph is *connected* (linked) if there exists at least one path between any pair of vertices in the graph. Otherwise the graph is *disconnected*. In *labeled graphs*, each vertex is distinguished by some index. In a *free* (or *topological*) graph, the vertices are regarded as indistinguishable, and are unlabeled. The most important characteristic of graphs is not their geometry (shape and size, position, and orientation), but rather their *topology*—the scheme of connections among its vertices and bonds.

With Eq. (6.10), we've evaluated (formally) the contribution to $Q(N, T, V)$ of a given set of three particles (ijk) that are coupled through pairwise interactions. How many ways can we choose triples? Clearly, $\binom{N}{3} = N(N-1)(N-2)/3!$. Through second order in Eq. (6.6), we have for the configuration integral

$$Q(T, V, N) = V^N \left(1 + N(N-1)\frac{b_2(T)}{V} + N(N-1)(N-2)\frac{b_3(T)}{V^2} + \cdots\right).$$

6.1.1 Disconnected diagrams and the linked-cluster theorem

It might seem we've discerned the pattern now and we could start generalizing. With the next term in Eq. (6.6) ("quadruples"), we encounter a qualitatively new type of diagram. Figure 6.5 shows

Figure 6.5: Diagram involving four distinct particles interacting through two pairwise interactions.

the diagram associated with the product of two Mayer functions with four distinct indices. The contribution of this diagram to the configuration integral is

$$\int d^N \mathbf{r} f_{ji} f_{lk} = V^{N-4} \int d\mathbf{r}_i d\mathbf{r}_j d\mathbf{r}_k d\mathbf{r}_l f(r_{ji}) f(r_{lk}) \to V^{N-4} \int d\mathbf{r}_i d\mathbf{r}_{ji} d\mathbf{r}_k d\mathbf{r}_{lk} f(r_{ji}) f(r_{lk})$$

(6.11)

$$= V^{N-2} \left(\int d\mathbf{r} f(r)\right)^2 = 4V^{N-2} b_2^2,$$

where we've used Eq. (6.8). What is the *multiplicity* of this diagram? There are $\binom{N}{2} \times \binom{N-2}{2} \times \frac{1}{2} = \frac{1}{8}N(N-1)(N-2)(N-3)$ distinct ways the cluster in Fig. 6.5 can be synthesized out of N particles. Figure 6.6 shows the three equivalent diagrams for $N = 4$ particles coupled through two

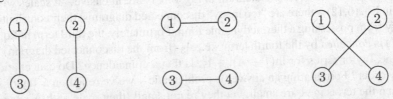

Figure 6.6: Disconnected diagrams of four particles and two pairwise interactions.

pairwise interactions. Including the cluster integral Eq. (6.11) together with its multiplicity,

$$Q(T, V, N) = V^N \left(1 + N(N-1)\frac{b_2(T)}{V} + N(N-1)(N-2)\frac{b_3(T)}{V^2}\right.$$
$$\left. + \frac{1}{2}N(N-1)(N-2)(N-3)\frac{b_2^2(T)}{V^2} + \cdots\right).$$

(6.12)

It's not apparent yet, but the term we just added to Eq. (6.12) is bad news.

Let's think about what we're trying to do. We seek the partition function for a system of interacting particles. But what do we do with that once we find it? All thermodynamic information can be obtained from the free energy, $\ln Z$ (Eq. (4.58) or (4.78)). Write Eq. (6.3), $Z(N, T, V) = Z_{tr}Q(N, T, V)/V^N$, where Z_{tr} is the partition function for the translational degrees of freedom, Eq. (5.1). Then, $\ln Z = \ln Z_{tr} + \ln(Q/V^N) \equiv \ln Z_{tr} + \ln(1 + A)$, where

$$A \equiv \left[\frac{1}{V^N} \int d^N r \prod_{j>i} (1 + f_{ij})\right] - 1 .$$

Make the assumption (to be verified) that A is small compared with unity. Using Eq. (6.12), we can write $A = A_1 + A_2 + A_3 + \cdots$. Apply the Taylor series,[10]

$$\ln(1 + A) \approx A_1 + A_2 + A_3 + \cdots - \frac{1}{2}\left(A_1^2 + 2A_1 A_2 + 2A_1 A_3 + \cdots\right) + \frac{1}{3}\left(A_1^3 + \cdots\right) + \cdots .$$

Put in the explicit expressions for A_1, A_2, A_3 from Eq. (6.12), keeping terms proportional to b_2^2:

$$\ln(1 + A) = N(N-1)\frac{b_2}{V} + N(N-1)(N-2)\frac{1}{V^2}\left(2b_2^2\right) + \frac{1}{2}N(N-1)(N-2)(N-3)\frac{b_2^2}{V^2}$$

$$- \frac{1}{2}N^2(N-1)^2\frac{b_2^2}{V^2} + \cdots , \tag{6.13}$$

where we've used Eq. (6.10). We know that the free energy is extensive in the thermodynamic limit (see Eq. (4.89)),

$$\lim_{\substack{N,V \to \infty \\ n=N/V \text{ fixed}}} N^{-1} F(N, V, T) = f(n, T) ,$$

and thus we expect that $\ln(1 + A) \sim N f(n, T)$ as $N \to \infty$. Examine Eq. (6.13) for $N \gg 1$:

$$\ln(1 + A) = N\left(nb_2 + 2n^2 b_2^2 + \underbrace{\frac{1}{2}Nn^2 b_2^2}_{\substack{\text{Trouble:} \\ \text{Diverges with } N}} - \underbrace{\frac{1}{2}Nn^2 b_2^2}_{\substack{\text{Tragedy averted:} \\ \text{Divergent term removed}}} + \cdots\right). \tag{6.14}$$

The first two terms in Eq. (6.14) are indeed intensive quantities that depend on n and T. The third term, however, which comes from the diagram in Fig. 6.5, is not intensive—it scales with N. That's the bad news in Eq. (6.12)—there are "too many" disconnected diagrams; their contributions prevent the free energy from possessing a thermodynamic limit. Fortunately, the third term in Eq. (6.14) (that scales with N) is *cancelled* by the fourth term, i.e., A_3 (from the disconnected diagram) is cancelled by a term in the Taylor series for $\ln(1 + A)$, $\frac{1}{2}A_1^2$. Is that a coincidence? Do cancellations like that occur at every order? Formulating an answer is problematic—we've relied on a Taylor series that's valid only when the terms in A are small, yet they're not small (they scale with N), but at the same time they seem to miraculously disappear from the expansion. Something deeper is at work.

The *linked-cluster theorem* is a fundamental theorem in graph theory, that *only connected diagrams contribute to the free energy*. Before stating the theorem (which we won't prove), let's try to "psyche out" what the issue is with disconnected diagrams. The diagrams in Fig. 6.4 are *linked clusters*, where each vertex of a diagram is connected to at least one line. The diagram in Fig. 6.5 is a *disconnected* cluster—there is not a path between *any* vertices of the graph. Figure 6.7 shows graphs involving $N = 3, 4, 5, 6$ particles interacting by three lines, where we've left the vertices unlabeled (free graphs). The first three are linked clusters, the remaining two are disconnected. Suppose one of the particles, k, in the linked clusters of Fig. 6.4 is far removed from particles i and j; in that case, the Mayer function f_{ik} or f_{jk} vanishes, implying that the contribution of the diagram

---

[10]For $x \ll 1$, $\ln(1 + x) \approx x - \frac{1}{2}x^2 + \frac{1}{3}x^3 - \frac{1}{4}x^4 + \cdots$.

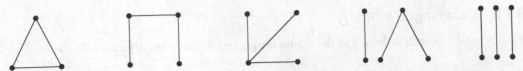

Figure 6.7: Diagrams composed of three lines. The first three are linked, the last two are not.

to the configuration integral vanishes. When particles are within an interaction distance, there is a distinct type of energy configuration that's counted in the partition function. As the particles become sufficiently separated, leaving no interaction among them, such contributions vanish. Now consider the disconnected diagrams, such as in Fig. 6.5 or 6.7: One can freely separate the disjoint parts of the diagram (which are not in interaction with each other), placing them anywhere in the system, in which case *the interactions represented by the disconnected parts have already been counted in the partition function*. The expansion we started with in Eq. (6.6) generates disconnected diagrams, which overcount various configurations. The linked-cluster theorem tells us that only connected diagrams contribute to the free energy (and thus to all thermodynamic information). *We need evaluate the partition function taking into account connected diagrams only.*

The precise form of the linked-cluster theorem depends on whether we're in the classical or quantum realm, Fermi or Bose, but the central idea remains the same. We present a version given by Uhlenbeck and Ford[55, p40]. Consider a quantity F_N that's a weighted sum over the graphs G_N (connected or disconnected) of N labeled points, $F_N \equiv \sum_{G_N} W(G_N)$, where $W(G_N)$, the weight given to a graph, is in our application the product of the multiplicity and the cluster integral associated with that type of graph. The N-particle configuration integral $Q(N, T, V)$ is just such a function as F_N. Define another quantity f_l as a weighted sum over connected graphs, $f_l \equiv \sum_{C_l} W(C_l)$, where the sum is over the connected graphs C_l (of the set G_N) with labeled points. The theorem states that

$$1 + F(x) = e^{f(x)} , \tag{6.15}$$

where $F(x)$ and $f(x)$ are *generating functions*[11] of the quantities F_N and f_l:

$$F(x) \equiv \sum_{N=1}^{\infty} F_N \frac{x^N}{N!} \qquad f(x) \equiv \sum_{l=1}^{\infty} f_l \frac{x^l}{l!} . \tag{6.16}$$

So far, we've considered the case of fixed N, yet the generating functions in Eq. (6.15) apply for an unlimited number of particles. That finds a perfect application in the grand canonical ensemble, which is where we're heading (see Section 6.1.3). The linked-cluster theorem can be remembered as the equality

$$\sum(\text{all diagrams}) = \exp\left(\sum(\text{all connected diagrams})\right) .$$

---

[11] A generating function $G(\{a_n\}, x) \equiv \sum_{n=0}^{\infty} a_n x^n$ is way of encoding an infinite sequence $\{a_n\}|_{n=0}^{\infty}$ by treating them as the expansion coefficients of a power series. An example is the generating function of Legendre polynomials, $G(x, y) = \sum_{n=0}^{\infty} P_n(x) y^n$. In that case, $G(x, y) = (1 - 2xy + y^2)^{-1/2}$ has a closed-form expression, but generating functions need not be in closed form. To quote G. Polya, "A generating function is a device somewhat similar to a bag. Instead of carrying many little objects detachedly, which could be embarrassing, we put them all in a bag, and then we have only one object to carry, the bag."[56, p101]. We previously encountered the moment generating function, Eq. (3.35), and the cumulant generating function, Eq. (3.59). Note that the generating functions in Eq. (6.16) start at $N = 1$ and $l = 1$. For Eq. (6.15) to be valid at $x = 0$, $F(x)$ and $f(x)$ must vanish as $x \to 0$. Mathematically speaking, the variable x need not have a physical interpretation, but in our applications x will have a well-defined physical meaning.

6.1.2 Obtaining $Z(N, T, V)$

We now give a general definition of the cluster integral associated with n-particle diagrams:

$$b_n(T) \equiv \frac{1}{n!V} \sum_{\substack{\text{all connected} \\ n\text{-particle diagrams}}} \int \left(\prod_{\substack{lk \,\in\, \text{the set of bonds in a} \\ \text{connected } n\text{-particle diagram}}} f_{lk} \right) \mathrm{d}\boldsymbol{r}_1 \cdots \mathrm{d}\boldsymbol{r}_n . \tag{6.17}$$

Equation (6.17) is consistent with Eqs. (6.8) and (6.10) for $n = 2, 3$. (By definition, $b_1 = 1$.) The purpose of the factor of $n!$ is to make the value of b_n independent of how we've labeled the n vertices of the diagram (required by the linked-cluster theorem), and the factor of $1/V$ cancels the factor of V that always occurs in evaluating cluster integrals of connected diagrams—we're free to take one vertex of the graph and place it anywhere in the volume V. The quantity b_n has the dimension of $(\text{volume})^{n-1}$: For an n-particle diagram, we integrate over the coordinates of $n - 1$ particles relative to the position of the nth particle. The cluster integral b_n is therefore independent of the volume of the system as long as V is not too small.

There are many ways that a *given set* of particles can be associated with clusters. Suppose K particles are partitioned into m_2 two-particle clusters, m_3 three-particle clusters, and so on. The integral over $\mathrm{d}\boldsymbol{r}_1 \cdots \mathrm{d}\boldsymbol{r}_N$ of this collection of clusters *factorizes* (because each cluster is connected)

$$(1!Vb_1)^{m_1} (2!Vb_2)^{m_2} \cdots (j!Vb_j)^{m_j} \cdots = \prod_{j=1}^{N} (j!Vb_j)^{m_j} , \tag{6.18}$$

where, to systematize certain formulas we're going to derive, we've introduced the *unit cluster*, b_1, which is not a particle; such terms contribute the factors of V^{N-K} seen in Eqs. (6.7) and (6.9). Associated with any given placement of N particles into a cluster is a constraint,

$$\sum_{l=1}^{N} l m_l = N . \tag{6.19}$$

Figure 6.8 shows a set of diagrams for $N = 4$ particles in which we show the unit clusters. You should verify that Eq. (6.19) holds for each of the diagrams in Fig. 6.8.

Figure 6.8: Some of the diagrams for a system of $N = 4$ particles, including the unit clusters. In the first $m_2 = 1$ and $m_1 = 2$, in the second $m_3 = 1$ and $m_1 = 1$, and in the third, $m_4 = 1$ and $m_1 = 0$.

How many distinct ways can N distinguishable particles[12] be partitioned into m_1 unit clusters, m_2 two-particle clusters, $\cdots$, m_j clusters of j-particles, and so on? The number of ways of dividing N distinguishable objects among labeled boxes so that there is one object in each of m_1 boxes, two objects in each of m_2 boxes, etc., is given by the multinomial coefficient, Eq. (3.12),

$$\frac{N!}{(1!)^{m_1} (2!)^{m_2} \cdots (j!)^{m_j} \cdots} .$$

We don't want to count as separate, however, configurations that differ by permutations among clusters *of the same kind*. To prevent overcounting, we have to divide the multinomial coefficient by

---

[12]The indistinguishability of identical particles is taken into account by the factor of $N!$ in Eq. (6.4).

$m_1! m_2! \cdots m_j! \cdots$. The combinatorial factor is therefore

$$\frac{N!}{\prod_{j=1}^{N}(j!)^{m_j} m_j!}. \tag{6.20}$$

The contribution to the configuration integral of the collection of clusters characterized by the particular set of integers $\{m_j\}$ is therefore the product of the expressions in (6.18) and (6.20):

$$N! \prod_j \frac{(j! V b_j)^{m_j}}{(j!)^{m_j} m_j!} = N! \prod_j \frac{(V b_j)^{m_j}}{m_j!}. \tag{6.21}$$

Note that we don't need to indicate the range of the index j—clusters for which $m_j = 0$ don't affect the value of the product.

There will be a contribution to the configuration integral for *each set* of the numbers $\{m_j\}$,

$$Q(N, T, V) = N! \sum_{\substack{\{m_j\} \\ (\sum_{j=1}^{N} j m_j = N)}} \prod_{j=1}^{N} \frac{(V b_j)^{m_j}}{m_j!}, \tag{6.22}$$

where $\sum_{\{m_j\}}$ indicates to sum over all conceivable sets of the numbers m_j that are consistent with Eq. (6.19). For the noninteracting system, there are no clusters: $m_1 = N$ with $m_{j \neq 1} = 0$, for which Eq. (6.22) reduces to V^N. The partition function for N particles is therefore (see Eq. (6.3)):

$$Z(N, T, V) = \frac{1}{\lambda_T^{3N}} \sum_{\substack{\{m_j\} \\ (\sum_{j=1}^{N} j m_j = N)}} \prod_{j=1}^{N} \frac{(V b_j)^{m_j}}{m_j!}. \tag{6.23}$$

6.1.3 Grand canonical ensemble, $Z_G(\mu, T, V)$

Equation (6.23) is similar to Eq. (5.49) (the partition function of ideal quantum gases). For quantum systems, we have sums over occupation numbers, which satisfy a constraint, $\sum_{k,\sigma} n_{k,\sigma} = N$. Here we have a constrained sum over m_j, the number of j-particle clusters.[13] And just as with Eq. (5.49), Eq. (6.23) is impossible to evaluate because of the combinatorial problem of finding all sets of numbers $\{m_j\}$ that satisfy $\sum_j j m_j = N$. But, just as with Eq. (5.49), Eq. (6.23) simplifies in the grand canonical ensemble, where the constraint of a fixed number of particles is removed.

Combining Eq. (6.23) with Eq. (4.77) (where $z = e^{\beta \mu}$ is the fugacity), we have the grand partition function (generating function for the quantities $\{Z_N\}$)

$$Z_G(\mu, T, V) = \sum_{N=0}^{\infty} Z(N, T, V) z^N = \sum_{N=0}^{\infty} \left(\frac{e^{\beta \mu}}{\lambda_T^3} \right)^N \sum_{\{m_j\}_N} \prod_{j=1}^{N} \frac{(V b_j)^{m_j}}{m_j!}$$

$$= \prod_{j=1}^{\infty} \sum_{m_j=0}^{\infty} \frac{(V b_j)^{m_j}}{m_j!} \left(\frac{e^{\beta \mu}}{\lambda_T^3} \right)^{j m_j} \equiv \prod_{j=1}^{\infty} \sum_{m_j=0}^{\infty} \frac{(V b_j \xi^j)^{m_j}}{m_j!}, \tag{6.24}$$

---

[13] What does a classical system of interacting particles have in common with an ideal quantum system? The energy of an ideal system is a sum of the energies of its constituents, which can be found from the occupation numbers $n_{k,\sigma}$ associated with each single-particle energy level $E_{k,\sigma}$ (which for fixed N must satisfy $\sum_{k,\sigma} n_{k,\sigma} = N$). By breaking up configurations of interacting particles into connected clusters, we have sums over *noninteracting* clusters; all we need to know is how many j-particle clusters we have, which must satisfy, for fixed N, the constraint $\sum_j j m_j = N$.

where the transition to the second line of Eq. (6.24) follows from the same reasoning used in the transition from Eq. (5.50) to Eq. (5.51), $\xi \equiv e^{\beta\mu}/\lambda_T^3$, and we've used Eq. (6.19) for N. We can then sum the infinite series[14] in Eq. (6.24), with the result

$$Z_G(\mu, T, V) = \prod_{j=1}^{\infty} \exp\left(V b_j \xi^j\right) = \exp\left(V \sum_{j=1}^{\infty} b_j \xi^j\right). \tag{6.25}$$

Equation (6.25) reduces to Eq. (4.79) in the noninteracting case. From Z_G, we have the grand potential (see Eq. (4.76))

$$\Phi(T, V, \mu) = -kT \ln Z_G(T, V, \mu) = -kTV \sum_{j=1}^{\infty} b_j \xi^j. \tag{6.26}$$

The thermodynamics of interacting gases is therefore reduced to evaluating the cluster integrals b_j.

6.2   VIRIAL EXPANSION, VAN DER WAALS EQUATION OF STATE

Once we have the partition function, we know a lot. Combining Eq. (6.25) with Eq. (4.78),[15]

$$P = kT \frac{\partial \ln Z_G}{\partial V}\bigg|_{T,\mu} = kT \sum_{j=1}^{\infty} b_j \xi^j = kT \left(\xi + b_2 \xi^2 + b_3 \xi^3 + \cdots\right)$$

$$N = z \frac{\partial \ln Z_G}{\partial z}\bigg|_{T,V} = V \sum_{j=1}^{\infty} j b_j \xi^j = V \left(\xi + 2b_2 \xi^2 + 3b_3 \xi^3 + \cdots\right). \tag{6.27}$$

These formulas are fugacity expansions ($\xi = z/\lambda_T^3$), such as we found for the ideal quantum gases (Section 5.5.3). It's preferable to express P in the form of a density expansion (density is more easily measured than chemical potential). We can invert the expansion for N to obtain a density expansion of the fugacity. Starting from $n = \sum_{j=1}^{\infty} j b_j \xi^j$ in Eq. (6.27), we find using standard series inversion methods, through third order in n:

$$\frac{e^{\beta\mu}}{\lambda_T^3} = \xi = n - 2b_2 n^2 + \left(8b_2^2 - 3b_3\right) n^3 + O(n^4). \tag{6.28}$$

Equation (6.28) should be compared with Eq. (5.71). Substituting Eq. (6.28) into the expression for P in Eq. (6.27), we find, through second order,

$$P = nkT \left[1 - b_2 n + \left(4b_2^2 - 2b_3\right) n^2 + O(n^3)\right]. \tag{6.29}$$

Equation (6.29), a density expansion of P, is known as the *virial expansion*. It reduces to the ideal gas equation of state in the case of no interactions.

The virial expansion was introduced (in 1901[16]) as a parameterized equation of state,

$$P = nkT \left[1 + B_2(T)n + B_3(T)n^2 + B_4(T)n^3 + \cdots\right],$$

---

[14]Because ξ has the dimension of (volume)$^{-1}$, the quantity $V\xi^j b_j$ is dimensionless for each value of j. The infinite series $V \sum_{j=1}^{\infty} \xi^j b_j$ converges if, for large j, $|b_{j+1}| \ll \xi^{-1}|b_j|$.

[15]We take the opportunity, again, to emphasize that in the grand canonical ensemble the number of particles in the system is not a fixed quantity, but is determined by the chemical potential, as we see in Eq. (6.27), i.e., $n = n(\mu, T)$. Equation (6.28) specifies a relation between the chemical potential and the density of particles, $\mu = \mu(n, T)$.

[16]By K. Onnes, winner of the 1913 Nobel Prize in Physics for the production of liquid helium. The virial expansion abandons the idea of a closed-form expression for the equation of state.

where the quantities $B_n(T)$ are the *virial coefficients*, which are tabulated for many gases.[17] Virial coefficients are not measured directly; they're determined from an analysis of PVT data. The most common practice is a least-squares fit of PV values along isotherms as a function of density. Statistical mechanics provides (from Eq. (6.29)) theoretical expressions for the virial coefficients:

$$B_2 = -b_2 \qquad B_3 = 4b_2^2 - 2b_3 \qquad B_4 = -20b_2^3 + 18b_2b_3 - 3b_4 , \qquad (6.30)$$

where the expression for B_4 is the result of Exercise 6.6. *The virial coefficients require inter-particle interactions for their existence.* For example, using Eq. (6.8),

$$B_2(T) = -\frac{1}{2} \int d\mathbf{r} f(r) . \qquad (6.31)$$

An equation of state proposed by van der Waals[18] in 1873 takes into account the finite size of atoms as well as their interactions. The volume available to gas atoms is reduced (from the volume V of the container) by the volume occupied by atoms. Van der Waals modified the ideal gas law, to

$$P = \frac{NkT}{V - Nb} ,$$

where $b > 0$ is an experimentally determined parameter for each type of gas. The greater the number of atoms, the greater is the *excluded volume*. Van der Waals further reasoned that the pressure would be lowered by attractive interactions between atoms. The decrease in pressure is proportional to the probability that two atoms interact, which, in turn, is proportional to the square of the particle density. In this way, van der Waals proposed the equation of state,

$$P = \frac{NkT}{V - Nb} - an^2 ,$$

where $a > 0$ is another material-specific parameter to be determined from experiment.[19] The *van der Waals equation of state* is usually written

$$(P + an^2)(V - Nb) = NkT . \qquad (6.32)$$

It's straightforward to show that Eq. (6.32) implies for the second virial coefficient,

$$B_2^{\text{vdw}}(T) = b - \frac{a}{kT} . \qquad (6.33)$$

The van der Waals equation of state provides a fairly successful model of the thermodynamic properties of gases. It doesn't predict all properties of gases, but it predicts enough of them for us to take the model seriously. It's the simplest model of an interacting gas we have. Can the *phenomenological* parameters of the model, (a, b), can be related to the properties of the inter-particle potential? Let's see if the second virial coefficient as predicted by statistical mechanics, $-b_2$, has the form of that in Eq. (6.33), i.e., can we establish the correspondence

$$b_2 \overset{?}{\leftrightarrow} -b + \frac{a}{kT}$$

for suitably defined quantities (a, b)? To do that, let's calculate b_2 for the Lennard-Jones potential. From Eq. (6.8),

$$\frac{1}{2\pi} b_2 = \int_0^\infty r^2 \left(e^{-\beta v(r)} - 1\right) dr \approx -\int_0^\sigma r^2 dr - \beta \int_\sigma^\infty v(r) r^2 dr , \qquad (6.34)$$

[17] Values of $B_2(T)$ are tabulated in the *CRC Handbook of Chemistry and Physics*[57, pp6-24–6-32], a valuable resource.
[18] Van der Waals received the 1910 Nobel Prize in Physics for work on the equation of state of gases and liquids.
[19] Values of the van der Waals parameters a and b are tabulated in [57, p6–33].

where, referring to Fig. 6.2, for $0 \leq r \leq \sigma$ we've taken the repulsive part of the potential as infinite (*hard core* potential) so that the Mayer function is equal to -1, and for $r \geq \sigma$ we've approximated the Mayer function with its high-temperature form, $-\beta v(r)$. With these approximations, we find

$$b_2 = -\frac{2\pi}{3}\sigma^3 + \frac{16\pi}{9}\frac{\epsilon}{kT}\sigma^3 . \tag{6.35}$$

The correspondence therefore holds: $b \sim \sigma^3$ is an excluded volume provided by the short-range repulsive part of the inter-particle potential, and the parameter $a \sim \epsilon\sigma^3$ is an energy-volume associated with the attractive part of the potential. *Statistical mechanics validates the assumptions underlying the van der Waals model*, illustrating the role of microscopic theories in deriving phenomenological theories. Moreover, as we now show, with some more analysis we can provide a physical interpretation of the types of interactions that give rise to the virial coefficients.

6.3 CUMULANT EXPANSION OF THE FREE ENERGY

We now derive the virial expansion in another way, one that features additional techniques of diagrammatic analyses.[20] Consider $B_3 = 2(2b_2^2 - b_3)$; Eq. (6.30). Using Eq. (6.10) (for b_3), we see that a cancellation occurs among the terms contributing to B_3, leaving us with one integral:

$$B_3 = -\frac{1}{3}\int d\mathbf{r}_{ij} d\mathbf{r}_{kj} f(r_{ik}) f(r_{kj}) f(r_{ij}) . \tag{6.36}$$

The diagram corresponding to this integral is shown in the left part of Fig. 6.9. What about the

Figure 6.9: Irreducible (left) and reducible (right) connected clusters of three particles.

diagrams that *don't* contribute to B_3? Time for another property of diagrams. A graph is *irreducible* when each vertex is connected by a bond to at least two other vertices, as in the left part of Fig. 6.9.[21] A *reducible* graph has certain points, *articulation points*, where it can be cut into two or more disconnected parts, as in the right part of Fig. 6.9. Graphs can have more than one articulation point; see Fig. 6.10. *A linked graph having no articulation points is irreducible*. Figure 6.11 shows the

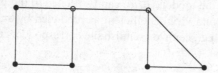

Figure 6.10: Example of a graph with two articulation points (open circles)

---

[20]Experience shows that the more ways you have of looking at something, the better. As noted by Richard Feynman: "It always seems odd to me that the fundamental laws of physics, when discovered, can appear in so many different forms that are not apparently identical at first, but, with a little mathematical fiddling you can show the relationship. ... There is always another way to say the same thing that doesn't look at all like the way you said it before. I don't know what the reason for this is. I think it is somehow a representation of the simplicity of nature. ... Perhaps a thing is simple if you can describe it fully in several different ways without immediately knowing that you are describing the same thing."[58]

[21]As an exception, the graph associated with b_2 is included in the class of irreducible clusters.

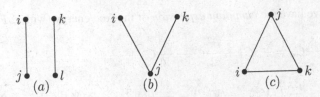

Figure 6.11: Three types of graphs: (a) Unlinked (not all vertices connected by bonds); (b) reducibly linked (every vertex connected by a least one bond); (c) irreducibly linked (every vertex connected by at least two bonds).

three types of diagrams: Unlinked, reducible, and irreducible. As we now show, *only irreducible diagrams contribute to the virial coefficients.*

Rewrite Eq. (6.3) (as we did in Section 6.1.1), $Z(T, V, N) = Z_{tr}Q(T, V, N)/V^N$, where Z_{tr} is the partition function for the ideal gas, Eq. (5.1). We now write Q/V^N in a new way:

$$\frac{1}{V^N}Q(T, V, N) = \frac{1}{V^N}\int d^N r e^{-\beta V(r_1,\cdots,r_N)} \equiv \left\langle e^{-\beta V(r_1,\cdots,r_N)} \right\rangle_0 , \qquad (6.37)$$

where $V(r_1, \cdots, r_N) \equiv \sum_{j>i} v(r_{ij})$ is the total potential energy of particles having the instantaneous positions $r_1, \cdots, r_N$, and we've introduced a new average symbol,

$$\langle\langle\cdots\rangle\rangle_0 \equiv \frac{1}{V^N}\int d^N r(\cdots) .$$

Equation (6.37) interprets the configuration integral as the expectation value of $e^{-\beta V(r_1,\cdots,r_N)}$ with respect to a *non-thermal* (in fact, geometric) probability distribution[22] where the variables $r_1, \cdots, r_N$ have a uniform probability density $(1/V)^N$ inside a container of volume V. Using Eq. (4.57), we obtain

$$-\beta(F - F_{ideal}) = \ln\left\langle e^{-\beta V(r_1,\cdots,r_N)} \right\rangle_0 , \qquad (6.38)$$

where $F_{ideal} \equiv -kT \ln Z_{tr}$ is the free energy of the ideal gas. The right side of Eq. (6.38) is the contribution to the free energy *arising solely from inter-particle interactions.*

We encountered just such a quantity in Eq. (3.35), the moment generating function $\left\langle e^{\theta x} \right\rangle = \sum_{n=0}^{\infty} \theta^n \langle x^n \rangle /n!$, where, we stress, the average symbols $\langle\ \rangle$ are associated with a *given* (not necessarily thermal) probability distribution. The logarithm of $\left\langle e^{\theta x} \right\rangle$ defines the cumulant generating function, Eq. (3.59),[23]

$$\ln\left\langle e^{\theta x} \right\rangle = \sum_{n=1}^{\infty} \frac{\theta^n}{n!}C_n , \qquad (6.39)$$

where each quantity C_n (*cumulant*) contains combinations[24] of the moments $\langle x^k \rangle$, $1 \leq k \leq n$. Explicit expressions for the first few cumulants are listed in Eq. (3.62). To apply Eq. (6.39) to Eq. (6.38), set $\theta = -\beta$ and associate the random variable x with the total potential energy, $V =$

---

[22]We easily see that a probability density function $P(r_1, r_2, \cdots, r_N) = (1/V)^N$ is properly normalized: $\int P(r_1, r_2, \cdots, r_N)d^N r = (1/V^N)\int d^N r = 1$.

[23]Set $\omega = -i\theta$ in Eq. (3.59).

[24]It's instructive to point out parallels with the generating function for Legendre polynomials, $\left(1 - 2xy + y^2\right)^{-1/2} = \sum_{n=0}^{\infty} y^n P_n(x)$, a power series in y, with coefficients $P_n(x)$ that are n^{th}-degree polynomials in the variable x. For the cumulant generating function, Eq. (6.39), the cumulants C_n are simple algebraic functions of the moments of x (up to order n), calculated with respect to the averaging procedure associated with the brackets $\langle\ \rangle$.

$\sum_{j>i} v(r_{ij})$. Thus we have the *cumulant expansion* of the free energy (with $\Delta F \equiv F - F_{\text{ideal}}$):

$$-\beta \Delta F = \sum_{n=1}^{\infty} \frac{(-\beta)^n}{n!} C_n = -\beta C_1 + \frac{\beta^2}{2} C_2 - \frac{\beta^3}{3!} C_3 + \cdots \tag{6.40}$$

$$= -\beta \langle V \rangle_0 + \frac{\beta^2}{2} \left(\langle V^2 \rangle_0 - \langle V \rangle_0^2 \right) - \frac{\beta^3}{3!} \left(\langle V^3 \rangle_0 - 3 \langle V^2 \rangle_0 \langle V \rangle_0 + 2 \langle V \rangle_0^3 \right) + O(\beta^4),$$

where we've used Eq. (3.62) for C_1, C_2, C_3. We examine the first few terms of Eq. (6.40).
    The first cumulant C_1 is the *average potential energy*:

$$C_1 = \langle V(r_1, \cdots, r_N) \rangle_0 = \sum_{j>i} \langle v(r_{ij}) \rangle_0 = \sum_{j>i} \frac{1}{V^N} \int d^N r v(r_{ij}) \to \binom{N}{2} \frac{1}{V} \int d r v(r),$$

$$\tag{6.41}$$

where we've used Eq. (3.66), that the cumulant associated with a sum of independent random variables is the sum of cumulants associated with each variable. We're interested in the thermodynamic limit. Making use of Eq. (3.44), $\binom{N}{n} \overset{N \to \infty}{\sim} N^n/n!$, in Eq. (6.41),

$$\lim_{\substack{N, V \to \infty \\ N/V = n}} \left(\frac{C_1}{N} \right) = \frac{1}{2} n \int v(r) d r. \tag{6.42}$$

Cumulants must be extensive, so that the free energy as calculated with Eq. (6.40) is extensive.[25] We associate the integral in Eq. (6.42) with the graph in Fig. 6.12, which is nominally the same as Fig. 6.3, with an important exception—Fig. 6.3 represents the Mayer function f_{ij} between particles i, j, whereas Fig. 6.12 represents their direct interaction, v_{ij}.

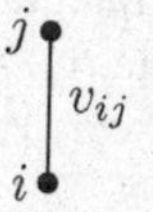

Figure 6.12: The one diagram contributing to C_1.

The cumulant C_2 is the fluctuation in potential energy:[26]

$$C_2 = \langle V^2(r_1, \cdots, r_N) \rangle_0 - (\langle V(r_1, \cdots, r_N) \rangle_0)^2 = \sum_{j>i} \sum_{l>k} \left[\langle v_{ij} v_{kl} \rangle_0 - \langle v_{ij} \rangle_0 \langle v_{kl} \rangle_0 \right], \tag{6.43}$$

where $v_{ij} \equiv v(r_{ij})$. Let's analyze the structure of the indices in Eq. (6.43), because that's the key to this method. There are three and only three possibilities in this case:

(a) *No indices in common; unlinked terms.* Consider $\langle v_{12} v_{34} \rangle_0$. Because we're averaging with respect to a probability distribution in which r_{12} and r_{34} can be varied independently,

$$\langle v_{12} v_{34} \rangle_0 = \langle v_{12} \rangle_0 \langle v_{34} \rangle_0. \tag{6.44}$$

The averages of products of potential functions with distinct indices *factor*, and as a result C_2 vanishes identically—a fortunate development: There are $\frac{1}{2} \binom{N}{2} \binom{N-2}{2} \overset{N \to \infty}{\sim} N^4/8$ ways to choose, out of N indices, two sets of pairs having no elements in common. We'll call terms that scale with N too rapidly to let the free energy have a thermodynamic limit, *super extensive*. The important point is that every unlinked term arising from $\langle v_{ij} v_{kl} \rangle_0$ (which we're calling super extensive) is cancelled by a counterpart, $\langle v_{ij} \rangle_0 \langle v_{kl} \rangle_0$. The diagrams representing $\langle v_{ij} v_{kl} \rangle_0$ for no indices in common are those in Fig. 6.11(a).

---

[25] Or, better said, we'll keep only the parts of cumulants that properly scale with N and survive the thermodynamic limit.
[26] The potential energy fluctuates because the atoms of a gas move around.

(b) *One index in common; reducibly-linked terms.* Consider $\langle v_{12} v_{23} \rangle_0$:

$$\langle v_{12} v_{23} \rangle_0 = \frac{1}{V^3} \int d\mathbf{r}_1 d\mathbf{r}_2 d\mathbf{r}_3 v_{12} v_{23} \rightarrow \frac{1}{V^2} \int d\mathbf{r}_{12} d\mathbf{r}_{32} v_{12} v_{23} = \langle v_{12} \rangle_0 \langle v_{23} \rangle_0 . \quad (6.45)$$

Because of the averaging procedure $\langle \; \rangle_0$, $\langle v_{12} v_{23} \rangle_0$ factorizes, implying that C_2 vanishes for these terms. Such terms correspond to reducibly linked diagrams (see Fig. 6.11(b))—the common index represents an articulation point where the graph can be cut and separated into disconnected parts. *Reducible graphs make no contribution to the free energy.*

(c) *Both pairs of indices in common; irreducibly-linked graphs.* Consider the case, from Eq. (6.43), where $k = i$ and $l = j$,

$$C_2 = \sum_{j>i} \left[\langle v_{ij}^2 \rangle_0 - \left(\langle v_{ij} \rangle_0 \right)^2 \right] . \quad (6.46)$$

The first term in Eq. (6.46), the average of the square of the inter-particle potential,

$$\langle v_{ij}^2 \rangle_0 = \frac{1}{V^2} \int d\mathbf{r}_i d\mathbf{r}_j v_{ij}^2 \rightarrow \frac{1}{V} \int d\mathbf{r}_{ij} v_{ij}^2 \equiv \frac{1}{V} \int d\mathbf{r} v^2(r) \sim \frac{1}{V} , \quad (6.47)$$

whereas

$$\left(\langle v_{ij} \rangle_0 \right)^2 = \left(\frac{1}{V} \int d\mathbf{r} v(r) \right)^2 \sim \frac{1}{V^2} . \quad (6.48)$$

We're assuming the integrals $\int d\mathbf{r} v(r)$ and $\int d\mathbf{r} v^2(r)$ exist. Noting how the terms scale with volume in Eqs. (6.47), (6.48), only the first term survives the thermodynamic limit,

$$\lim_{\substack{N,V \to \infty \\ N/V = n}} \left(\frac{C_2}{N} \right) = \frac{1}{2} n \int v^2(r) d\mathbf{r} , \quad (6.49)$$

where we've used $\sum_{j>i} = \binom{N}{2}$. Figure 6.13 shows the diagram representing the integral in Eq.

Figure 6.13: The one diagram contributing to C_2.

(6.49). This is a new kind of diagram where v_{ij}^2 is represented by two bonds between particles i, j; it has no counterpart in the Mayer cluster expansion (we never see terms like $f_{12} f_{12}$).

Most of the complexity associated with higher-order cumulants is already present in C_3, so let's examine it in detail.

$$C_3 = \sum_{j>i} \sum_{l>k} \sum_{m>n} \left[\langle v_{ij} v_{kl} v_{mn} \rangle_0 - \langle v_{ij} v_{kl} \rangle_0 \langle v_{mn} \rangle_0 - \langle v_{mn} v_{ij} \rangle_0 \langle v_{kl} \rangle_0 - \langle v_{kl} v_{mn} \rangle_0 \langle v_{ij} \rangle_0 \right.$$
$$\left. + 2 \langle v_{ij} \rangle_0 \langle v_{kl} \rangle_0 \langle v_{mn} \rangle_0 \right] .$$

For all indices distinct, $i \neq j \neq k \neq l \neq m \neq n$, the average of three potential functions factors,

$$\langle v_{ij} v_{kl} v_{mn} \rangle_0 = \langle v_{ij} v_{kl} \rangle_0 \langle v_{mn} \rangle_0 = \langle v_{ij} \rangle_0 \langle v_{kl} \rangle_0 \langle v_{mn} \rangle_0 ,$$

and $C_3 = 0$. Disconnected diagrams (Fig. 6.14(a)) make no contribution. There are other kinds of unlinked diagrams, however. For one index in common between a pair of potential functions ($l = j$, for example), with no overlap of indices from the third function, we have (see Fig. 6.14(b))

$$\langle v_{ij} v_{jk} v_{mn} \rangle_0 = \langle v_{ij} v_{jk} \rangle_0 \langle v_{mn} \rangle_0 , \qquad (m, n \neq i, j, k)$$

and C_3 vanishes. If we set $k = i$ (with $l = j$ and $m, n \neq i, j$), $C_3 = 0$ for the diagram in Fig. 6.14(c). *All unlinked diagrams arising from $\langle v_{ij} v_{kl} v_{mn} \rangle_0$ are cancelled by the terms of C_3.*

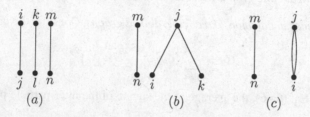

Figure 6.14: The unlinked diagrams associated with $\langle v_{ij} v_{kl} v_{mn} \rangle_0$: (a) $i \neq j \neq k \neq l \neq m \neq n$; (b) $l = j$ and $m, n \neq i, j, k$; (c) $l = j$ and $k = i$, $m, n \neq i, j$. $C_3 = 0$ for these diagrams.

The reducible diagrams associated with $\langle v_{ij} v_{kl} v_{mn} \rangle_0$ (see Fig. 6.15) make no contribution to

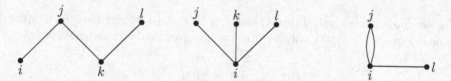

Figure 6.15: The reducible diagrams associated with $\langle v_{ij} v_{kl} v_{mn} \rangle_0$. $C_3 = 0$ for these diagrams.

C_3. For the left diagram in Fig. 6.15, $\langle v_{ij} v_{jk} v_{kl} \rangle_0$ factors into $\langle v_{ij} \rangle_0 \langle v_{jk} \rangle_0 \langle v_{kl} \rangle_0$ on passing to relative coordinates. For the middle diagram, $\langle v_{ij} v_{ik} v_{il} \rangle_0 = \langle v_{ij} \rangle_0 \langle v_{ik} \rangle_0 \langle v_{il} \rangle_0$, and for that on the right, $\langle v_{ij}^2 v_{il} \rangle_0 = \langle v_{ij}^2 \rangle_0 \langle v_{il} \rangle_0$. In all cases, $C_3 = 0$ for reducible diagrams.

That leaves the irreducible diagrams, for which $C_3 \neq 0$. The two (and only two) irreducible diagrams associated with $\langle v_{ij} v_{kl} v_{mn} \rangle_0$ are shown in Fig. 6.16. There are $\binom{N}{2}$ contributions of the

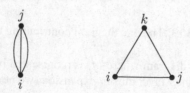

Figure 6.16: The two (irreducible) diagrams having nonzero contributions to C_3.

diagram on the left, and $N(N-1)(N-2)$ contributions of the diagram on the right. In the thermodynamic limit,

$$\lim_{\substack{N,V \to \infty \\ N/V=n}} \left(\frac{C_3}{N} \right) = \frac{1}{2} n \int v^3(r) \mathrm{d}\mathbf{r} + n^2 \int v_{12} v_{23} v_{31} \mathrm{d}\mathbf{r}_{12} \mathrm{d}\mathbf{r}_{23} . \qquad (6.50)$$

These examples show that only irreducible diagrams contribute to the cumulant expansion of the free energy. Unlinked and reducible parts of cumulants either cancel identically, or don't survive the thermodynamic limit. Whatever are the unlinked or reducible graphs in a cumulant, factorizations occur such as in Eq. (6.45), causing it to vanish. The *nonzero* contributions to C_n consist of the irreducible clusters arising from the leading term,[27] $\langle V^n \rangle_0$. Our task therefore reduces to finding all irreducible diagrams at each order. Figure 6.17 shows the three (and only three) irreducible diagrams

Figure 6.17: Irreducible diagrams associated with $n = 4$ bonds and $m = 2, 3, 4$ vertices.

at order $n = 4$. The graphs associated with $\langle V^n \rangle_0$ have n bonds because each of the n copies of the potential function v_{ij} represents a bond. Such graphs have m vertices, $2 \leq m \leq n$. One can have n powers of a single term v_{ij}, in which case $m = 2$, up to the case of $m = n$ vertices which occurs for a diagram like $\langle v_{12}v_{23}v_{34}v_{41} \rangle_0$ (a *cycle* in graph-theoretic terms). Let $D(n,m)$ denote an irreducible diagram having n bonds and m vertices. Table 6.1 lists the irreducible diagrams

Table 6.1: Irreducible graphs $D(n,m)$ classified by number of bonds n and vertices $2 \leq m \leq n$.

bonds, $n \downarrow$		vertices, $m \rightarrow$ 2	3	4
1	(β)	$D(1,2)$		
2	(β^2)	$D(2,2)$		
3	(β^3)	$D(3,2)$	$D(3,3)$	
4	(β^4)	$D(4,2)$	$D(4,3)$	$D(4,4)$

classified by the number of bonds n ($n \leq 4$) and the number of vertices, $2 \leq m \leq n$. The diagram $D(1,2)$ is shown in Fig. 6.12, $D(2,2)$ is shown in Fig. 6.13, $D(3,2)$ and $D(3,3)$ are shown in Fig. 6.16, and $D(4,2)$, $D(4,3)$, and $D(4,4)$ are shown in Fig. 6.17. Each cumulant $C_{n>1}$ can be written, in the thermodynamic limit, as a sum of irreducible diagrams, $C_n = \sum_{m=2}^{n} D(n,m)$. The case of C_1 is special: $C_1 = D(1,2)$.

Thus, the free energy is determined by irreducible diagrams, but we haven't shown (as advertised) that the virial coeficients are so determined. Returning to Eq. (6.40), we can write

$$-\beta \Delta F = \sum_{n=1}^{\infty} \frac{(-\beta)^n}{n!} C_n = \sum_{n=1}^{\infty} \frac{(-\beta)^n}{n!} \sum_{m=2}^{n} D(n,m) ,$$

where we've in essence summed "across" the entries in Table 6.1 for each n (which is natural—that's how the equation is written). We can reverse the order of summation, however, and sum over columns,

$$-\beta \Delta F = \sum_{m=2}^{\infty} \sum_{n=1}^{\infty} \frac{(-\beta)^n}{n!} D(n,m) = \sum_{n=1}^{\infty} \frac{(-\beta)^n}{n!} D(n,2) + \sum_{n=1}^{\infty} \frac{(-\beta)^n}{n!} D(n,3) + \cdots . \quad (6.51)$$

Let's see how this works for two-vertex diagrams. Their contribution to $-\beta \Delta F$ is, from Eq. (6.51) with[28] $m = 2$, for large N,

$$\frac{N}{2} n \sum_{k=1}^{\infty} \frac{(-\beta)^k}{k!} \int v^k(r) \mathrm{d}\boldsymbol{r} = \frac{N}{2} n \int \left[\sum_{k=1}^{\infty} \frac{1}{k!} (-\beta v(r))^k \right] \mathrm{d}\boldsymbol{r} = \frac{N}{2} n \int \left(e^{-\beta v(r)} - 1 \right) \mathrm{d}\boldsymbol{r} = N n b_2$$
$$(6.52)$$

---

[27] We noted in Section 3.7 that cumulant C_n has moments of nth order and less.
[28] For each diagram in the class $D(n,2)$, there are $\binom{N}{2}$ contributions to ΔF of the integral $(1/V) \int v^n(r) \mathrm{d}\boldsymbol{r}$.

where we've used Eq. (6.8). Thus, the sum over the class of irreducible diagrams $D(n, 2)$ has reproduced (up to multiplicative factors) the cluster integral b_2. It can then be shown from (6.52) that (see Exercise 6.12)

$$P = nkT \left(1 + B_2 n + \cdots \right), \tag{6.53}$$

where $B_2 = -b_2$, Eq. (6.30). We'll stop now, but the process works the same way for the other virial coefficients—summing over irreducibly linked, topologically distinct diagrams associated with the same number of vertices, one arrives at *irreducible* cluster integrals for each virial coefficient, such as we've seen already for B_2 and B_3, Eqs. (6.31) and (6.36).

6.4   THE TONKS AND TAKAHASHI GASES

Consider a system of N identical particles confined to a one-dimensional region $0 \le x \le L$ that interact through an inter-particle potential energy function $\phi(x)$. In the *Tonks gas*[59], $\phi(x)$ has the form

$$\phi(x) = \begin{cases} \infty & |x| < a \\ 0 & |x| > a, \end{cases} \tag{6.54}$$

which is therefore a collection of *hard rods* of length a. The *Takahashi gas*[29] is a generalization with

$$\phi(x) = \begin{cases} \infty & |x| < a \\ v(x - a) & a < |x| < 2a \\ 0 & |x| > 2a, \end{cases} \tag{6.55}$$

which is a collection of hard rods with *nearest-neighbor interactions*.[30]
The canonical partition function is

$$
\begin{aligned}
Z(L, N, T) &= \frac{1}{h^N} \frac{1}{N!} \int \mathrm{d}^N p \exp\left(-\beta \sum_{i=1}^N p_i^2/(2m) \right) \int \mathrm{d}^N x \exp\left(-\beta \sum_{1 \le i \le j \le N} \phi(|x_i - x_j|) \right) \\
&= \frac{1}{N!} \frac{1}{\lambda_T^N} \int_0^L \mathrm{d}x_1 \cdots \int_0^L \mathrm{d}x_N \exp\left(-\beta \sum_{1 \le i \le j \le N} \phi(|x_i - x_j|) \right).
\end{aligned}
\tag{6.56}
$$

Note the factor of h^N and not h^{3N}—we're in one spatial dimension with N particles. The integrand is a totally symmetric function of its arguments $(x_1, \cdots, x_N)$, which can be permuted in $N!$ ways. One can show[31] for $f(x_1, \cdots, x_N)$ a symmetric function, the multiple integral $\int_0^L \cdots \int_0^L f \mathrm{d}x_1 \cdots \mathrm{d}x_N$ over an N-dimensional cube of length L is equal to $N!$ times the nested integrals: $N! \int_0^L \mathrm{d}x_N \int_0^{x_N} \mathrm{d}x_{N-1} \cdots \int_0^{x_2} f \mathrm{d}x_1$. This *ordering* of the integration variables such that $0 \le x_1 \cdots \le x_N \le L$ allows us to evaluate the configuration integral exactly.
For the Tonks gas, which has only the feature of a hard core, the Boltzmann factors are either zero or one:

$$\exp\left(-\beta \sum_{1 \le i \le j \le N} \phi(|x_i - x_j|) \right) = \prod_{1 \le i \le j \le N} S(|x_i - x_j|), \tag{6.57}$$

where

$$S(x) \equiv \begin{cases} 1 & |x| > a \\ 0 & |x| < a. \end{cases}$$

---

[29]Originally published in 1942[60]. An English translation is available in Lieb and Mattis[61, p25].
[30]The Takahashi potential is a one-dimensional version of the Van Hove potential, Eq. (4.92).
[31]A rigorous proof of this statement can be found in [62].

The region of integration for the configuration integral is therefore specified by $|x_i - x_j| > a$ for all i, j. It should be noted that in the sum $\sum_{j>i} \phi(|x_i - x_j|)$, every particle in principle interacts with every other particle, even though in this example the interactions between particles are zero except when they're close enough to encounter the hard core of the potential. Given the aforementioned result on symmetric functions, we can set $j = i + 1$ in Eq. (6.57), with the region of integration specified by the inequalities $x_{i+1} > x_i + a$, $i = 1, \cdots, N - 1$. Referring to Fig. 6.18, we see that

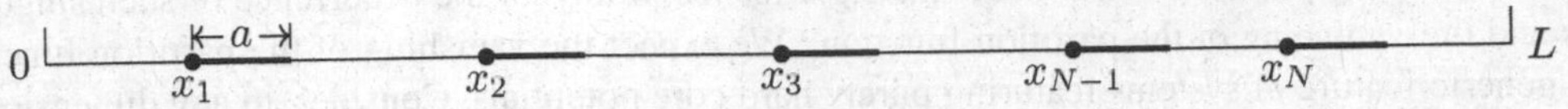

Figure 6.18: Region of integration for the partition function of a one-dimensional gas of hard rods.

$0 \le x_1 < x_2 - a, \, a < x_2 < x_3 - a, \, \cdots, \, (i-1)a < x_i < x_{i+1} - a, \, \cdots, \, (N-1)a < x_N \le L$.
With a change of variables, $y_i = x_i - (i-1)a$, $i = 1, \cdots, N$,

$$Z(L, N, T) = \frac{1}{\lambda_T^N} \int_0^l dy_N \int_0^{y_N} dy_{N-1} \cdots \int_0^{y_3} dy_2 \int_0^{y_2} dy_1 = \frac{1}{N!} \left(\frac{l}{\lambda_T} \right)^N, \tag{6.58}$$

where $l \equiv L - (N-1)a$ is the *free* length available to the collection of rods; $(N-1)a$ is the excluded volume (length). We see the "return" of the factor of $N!$.

With $Z(L, N, T)$ in hand, we can calculate the free energy and the associated thermodynamic quantities μ, P, S (see Eq. (4.58)). Starting from Eq. (4.57), and using Eq. (6.58),

$$-\beta F(L, N, T) = \ln Z(L, N, T) = \ln \left(\frac{1}{N!} \left(\frac{l}{\lambda_T} \right)^N \right) \overset{N \gg 1}{\Rightarrow} N \left[1 + \ln \left(\frac{L - Na}{N \lambda_T} \right) \right], \tag{6.59}$$

where we've used Stirling's approximation for $N \gg 1$. The pressure is obtained from

$$P \equiv -\left(\frac{\partial F}{\partial L} \right)_{N, T} = \frac{NkT}{L - Na}. \tag{6.60}$$

Pressure in one dimension is an energy per length—an energy density—just as pressure in three dimensions is an energy density, energy per volume, or force per area. One-dimensional pressure is simply a force—an effective force experienced by hard rods. We see the excluded volume effect in Eq. (6.60), just as in the van der Waals equation of state; there is clearly no counterpart to the other parameter of the van der Waals model. The entropy and chemical potential are found from the appropriate derivatives of the free energy, with the results:

$$S = Nk \left[\frac{3}{2} + \ln \left(\frac{L - Na}{N \lambda_T} \right) \right]; \tag{6.61}$$

$$\beta \mu = \frac{Na}{L - Na} - \ln \left(\frac{L - Na}{N \lambda_T} \right). \tag{6.62}$$

Equation (6.61) becomes, in the limit $a \to 0$, the formula for the entropy of a one-dimensional ideal gas (see Exercise 5.2). We see from Eq. (6.62) a positive contribution to the chemical potential from the repulsive effect of the hard core potential (for $L > Na$).

In the thermodynamic limit ($N \to \infty$, $L \to \infty$, $v \equiv L/N$ fixed), these quantities reduce to the expressions (from Eqs. (6.59)–(6.62)),

$$\lim_{\substack{N, L \to \infty \\ v \equiv L/N \text{ fixed}}} \left(\frac{-\beta F}{N} \right) = 1 + \ln \left(\frac{v - a}{\lambda_T} \right) \qquad P = \frac{kT}{v - a}$$

$$\lim_{\substack{N, L \to \infty \\ v \equiv L/N \text{ fixed}}} \left(\frac{S}{Nk} \right) = \frac{3}{2} + \ln \left(\frac{v - a}{\lambda_T} \right) \qquad \beta \mu = \frac{a}{v - a} - \ln \left(\frac{v - a}{\lambda_T} \right). \tag{6.63}$$

The thermodynamic limit therefore exists (something already known from Section 4.2): Extensive quantities (F, S) are extensive, and intensive quantities (P, μ) are intensive. The free energy (from which the other quantities are found by taking derivatives) is an analytic function of v for $v > a$, where $v = L/N$ is the average length per particle. For $v < a$, rods overlap, which from Eq. (6.56) implies that $Z(L, N, T) = 0$ when $N/L > a^{-1}$, where a^{-1} is the *close packing density*.

The thermodynamic quantities associated with this one-dimensional system become singular at the close-packing point, $v = a$. What is the significance, if any, of the occurrence of such singularities and the vanishing of the partition function? We expect the vanishing of the partition function as a generic feature of systems featuring purely hard core potentials. Consider, in any dimension d, the canonical partition function

$$Z(V, N, T) = \frac{1}{\lambda_T^{dN}} \frac{1}{N!} \int d^N r \prod_{1 \leq i \leq j \leq N} S(|r_i - r_j|), \qquad (6.64)$$

where the quantities r_i are d-dimensional vectors, and we've used Eq. (6.57). The partition function vanishes for densities exceeding the close packing density $n_0 \equiv N_{max}/V$, where N_{max} is the maximum number of hard spheres of radius a that can be contained in the volume V. It might be thought that the vanishing of Z at the close packing density signals a phase transition from a gaseous to a solid phase.[32] It was conjectured in 1939 that a system of particles with hard-core potentials would undergo a phase transition at a density well below the close packing density[63]; evidence was first found in computer simulations[64]. A phase transition associated with purely repulsive interactions is known as the *Kirkwood-Alder transition*. Evidence for this transition has been found experimentally in colloidal suspensions [65].

For the Takahashi gas, we have, instead of Eq. (6.58),

$$Z(L, N, T) = \frac{1}{\lambda_T^N} \int_0^l dy_N \int_0^{y_N} dy_{N-1} \cdots \int_0^{y_3} dy_2 \int_0^{y_2} dy_1 \exp\left(-\beta \sum_{i=1}^{N-1} v(y_{i+1} - y_i)\right).$$
$$(6.65)$$

Equation (6.65) is in the form of an iterated Laplace convolution, which suggests a method of solution. Take the Laplace transform of Z as a function of l. It can be shown from Eq. (6.65) that

$$\widetilde{Z}(s) \equiv \int_0^\infty e^{-ls} Z(L, N, T) dl = \frac{1}{\lambda_T^N} \frac{1}{s^2} [K(s)]^{N-1}, \qquad (6.66)$$

where

$$K(s) \equiv \int_0^\infty e^{-sx - \beta v(x)} dx. \qquad (6.67)$$

The partition function then follows by finding the inverse Laplace transform of $\widetilde{Z}(s)$. Without specifying the form of the potential function $v(x)$, this is about as far as we can go. The method cannot be extended to systems having more than nearest-neighbor interactions. Thus, the partition function of a one-dimensional collection of particles having hard core potentials can be solved exactly for nearest-neighbor interactions only.

6.5   THE ONE-DIMENSIONAL ISING MODEL

The Ising model, conceived in 1924 as a model of magnetism, has come to occupy a special place in theoretical physics with an enormous literature.[33] Consider a one-dimensional crystalline lattice— a set of uniformly spaced points (*lattice sites*) separated by a distance a. Referring to Fig. 6.19, at each lattice site assign the value of a variable that can take one of two values, conventionally

---

[32]Note that the vanishing of the configuration integral occurs independently of the temperature; it's a geometric effect.

[33]The history of the Ising model up to 1967 is reviewed by Brush[66]. The Ising model—which seems to provide an unending wellspring of ideas for modeling physical systems—has in recent years found use in the field of *quantum computation*[67], providing yet another reason for knowing about the Ising model.

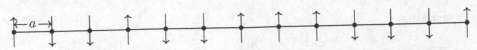

Figure 6.19: One-dimensional system of Ising spins with lattice constant a.

denoted $\sigma_i = \pm 1$, $i = 1, \cdots, N$, where N is the number of lattice sites. The variables σ_i can be visualized as vertical arrows, up or down (as in Fig. 6.19), and for that reason are known as *Ising spins*. Real spin-$\frac{1}{2}$ particles have two values of the projection of their spin vectors S onto a pre-selected z-axis, $S_z = \pm \frac{1}{2}\hbar$, and thus S_z can be written $S_z = \frac{1}{2}\hbar\sigma$, but that is the extent to which Ising spins have any relation to quantum spins. Ising spins are two-valued classical variables. In this section we consider one-dimensional systems of Ising spins, which can be solved exactly. Ising spins on two-dimensional lattices can also be solved exactly, but the mathematics is more difficult. We touch on the two-dimensional Ising model in Chapter 7; what we learn here will help.

Paramagnetism was treated in Chapter 5 in which independent magnetic moments interact with an externally applied magnetic field. Many other types of magnetic phenomena occur as a result of *interactions* between moments located on lattice sites of crystals. In ferromagnets, moments at widely separated sites become aligned, spontaneously producing (at sufficiently low temperatures) a magnetized sample *in the absence of an applied field*. In antiferromagnets, moments become anti-aligned at different sites, in which there is a spontaneous ordering of the individual moments, even though there is no net magnetization. The *Heisenberg spin Hamiltonian* models the coupling of spins S_i, S_j located at lattice sites (i, j) in the form $H = -\sum_{ij} J_{ij} S_i \cdot S_j$, where the coefficients J_{ij} are the *exchange coupling constants*.[34] Positive (negative) exchange coefficients promote ferromagnetic (antiferromagnetic) ordering. The microscopic underpinnings of the interaction J_{ij} is a complicated business we won't venture into.[35] In statistical mechanics, the coupling coefficients are taken as given parameters. The symbol S_i strictly speaking refers to a quantum-mechanical operator, but in many cases is approximated as a classical vector. The Ising model replaces S_i with its z-component, normalized to unit magnitude.

We take as the Hamiltonian for a one-dimensional system of Ising spins having nearest-neighbor interactions,[36]

$$H(\sigma_1, \cdots, \sigma_N) = -J \sum_{i=1}^{N-1} \sigma_i \sigma_{i+1} - b \sum_{i=1}^{N} \sigma_i . \tag{6.68}$$

The magnetic field parameter b is an energy (Ising spins are dimensionless) and thus $b = \mu_B \overline{B}$, where μ_B is the Bohr magneton and $\overline{B}$ is the "real" magnetic field. We're going to reserve B (a traditional symbol for magnetic field) for a dimensionless field strength, $B \equiv \beta b$. We'll also define a dimensionless coupling constant $K \equiv \beta J$. It's obvious, but worth stating: Equation (6.68) specifies the magnetic energy of a *given* assignment of spin values $(\sigma_1, \cdots, \sigma_N)$. The recipe of statistical mechanics is to sum over all 2^N possible configurations in obtaining the partition function.

6.5.1 Zero external field, free boundaries

The canonical partition function for N Ising spins in one dimension having nearest-neighbor interactions in the absence of an applied magnetic field and for *free boundary conditions* is obtained

---

[34]The form of the exchange interaction $-JS_1 \cdot S_2$ was derived by Dirac [68]. Heisenberg[69] was the first to realize that the exchange energy offers an explanation of ferromagnetism. In the older literature, $H = -\sum_{ij} J_{ij} S_i \cdot S_j$ is known as the Heisenberg-Dirac Hamiltonian. Magnetic moments are proportional to the total angular momentum J; see Eq. (E.4). For simplicity we take $L = 0$ and thus $J = S$.

[35]See for example [70].

[36]While it's common to refer to H in Eq. (6.68) as a Hamiltonian (a practice we'll adhere to), it's not strictly speaking a Hamiltonian, but rather an energy function. Ising spins are not canonical variables; they have no natural dynamics.

from the summation:[37]

$$Z_N(K) = \sum_{\{\sigma\}} e^{-\beta H(\sigma_1,\cdots,\sigma_N)} \equiv \sum_{\sigma_1=-1}^{1} \cdots \sum_{\sigma_N=-1}^{1} e^{-\beta H(\sigma_1,\cdots,\sigma_N)} = \sum_{\sigma_1=-1}^{1} \cdots \sum_{\sigma_N=-1}^{1} e^{K\sum_{i=1}^{N-1} \sigma_i \sigma_{i+1}}$$

(6.69)

$$= \sum_{\sigma_1=-1}^{1} \cdots \sum_{\sigma_N=-1}^{1} e^{K\sigma_1\sigma_2} e^{K\sigma_2\sigma_3} \cdots e^{K\sigma_{N-1}\sigma_N} = (2\cosh K) Z_{N-1}(K) ,$$

where we've summed over the last spin, σ_N,

$$\sum_{\sigma_N=-1}^{1} e^{K\sigma_{N-1}\sigma_N} = e^{K\sigma_{N-1}} + e^{-K\sigma_{N-1}} = 2\cosh(K\sigma_{N-1}) = 2\cosh K ,$$

and we've used that $\cosh x$ is an even function. For Ising spins we have the useful identity

$$e^{K\sigma} = \cosh K + \sigma \sinh K ,$$

(6.70)

and thus $\sum_{\sigma=-1}^{1} e^{K\sigma} = 2\cosh K$. The notational convention $\sum_{\{\sigma\}}$ saves writing—it indicates a sum over all 2^N spin configurations. The final equality in Eq. (6.69) is a *recursion relation* which is easily iterated to obtain an expression for the partition function:

$$Z_N(K) = 2^N \cosh^{N-1}(K) .$$

(6.71)

A good check on a formula like Eq. (6.71) is to set $K = 0$ ($T \to \infty$), corresponding to uncoupled spins, for which $Z_N(K = 0) = 2^N$.

From $Z_N(K)$, we can find the internal energy and the entropy using Eqs. (4.40) and (4.58):

$$U = -\frac{\partial}{\partial \beta} \ln Z_N = -J(N-1)\tanh K$$

$$\frac{S}{k} = \frac{\partial}{\partial T}(T \ln Z_N) = N\ln 2 - (N-1)K\tanh K + (N-1)\ln\cosh K .$$

(6.72)

In the thermodynamic limit,[38]

$$\lim_{N\to\infty}\left(\frac{S}{Nk}\right) = \ln(2\cosh K) - K\tanh K \to \begin{cases} \ln 2 & |K| \to 0 \quad (T \to \infty) \\ 0 & |K| \to \infty \quad (T \to 0) . \end{cases}$$

(6.73)

Figure 6.20 is a plot of S versus T. Entropy is *maximized* at $Nk\ln 2$ for high temperatures, $kT \gtrsim 10|J|$: there are 2^N configurations associated with uncoupled spins. The entropy vanishes at low temperature, $kT \ll |J|$ (third law of thermodynamics). Note how the entropy of Ising spins differs from that of the ideal gas. In the ideal gas, entropy is related to the kinetic energy of atoms (the only kind of energy atoms of an ideal gas can have); it diverges logarithmically at high temperatures and is unbounded. For Ising spins there is no contribution from kinetic energy; the energy of interaction is all potential. "Phase space" for Ising spins (configuration space) is finite;[39] phase space for the ideal gas (momentum space) is unbounded.

The heat capacity for the one-dimensional Ising model is, from either expression in Eq. (6.72),

$$C_V = \frac{\partial U}{\partial T} = T\left(\frac{\partial S}{\partial T}\right) = k(N-1)K^2\left(1 - \tanh^2 K\right) .$$

(6.74)

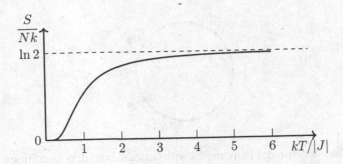

Figure 6.20: Entropy of a one-dimensional system of Ising spins versus temperature.

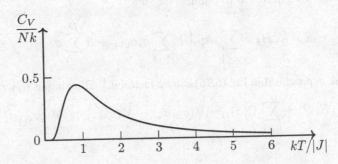

Figure 6.21: Heat capacity of a one-dimensional system of Ising spins versus temperature. Vertical scale the same as in Fig. 6.20.

This function is plotted in Fig. 6.21 for $N \gg 1$. Note that the maximum in C_V occurs at roughly the same temperature at which the slope of $S(T)$ in Fig. 6.20 is maximized. The configuration of the system associated with the maximum rate of change of entropy with temperature (heat capacity) is the configuration at which energy is most readily absorbed.

6.5.2 The transfer matrix method

The method of analysis leading to the recursion relation in Eq. (6.69) does not generalize to finite magnetic fields (try it!). We now present a more general technique for calculating the partition function of spin models, the *transfer matrix method*, which applies to systems satisfying periodic boundary conditions. Figure 6.22 shows a system of N Ising spins that wraps around on itself[40] with $\sigma_{N+1} \equiv \sigma_1$. The Hamiltonian

$$H = -J \sum_{i=1}^{N} \sigma_i \sigma_{i+1} - b \sum_{i=1}^{N} \sigma_i . \qquad (\sigma_{N+1} \equiv \sigma_1) \qquad (6.75)$$

The only difference between Eqs. (6.75) and (6.68) is the spin interaction $\sigma_N \sigma_1$.

---

[37]No factor of $N!$ in Eq. (6.69)? Spins attached to definite, identifiable lattice sites are distinguishable.

[38]The thermodynamic limit for this system is simply $N \to \infty$ (instead of $N, V \to \infty$, such that $N/V =$ fixed). For a crystalline lattice, the density of lattice points is fixed.

[39]Systems for which entropy saturates are candidates to demonstrate *negative absolute temperature*, which is a legitimate scientific concept.[3, Chapter 11]

[40]Periodic boundary conditions were introduced in Chapter 2 as a way to mathematically "fool" an electron so that it never encounters the potential energy environment associated with surfaces. Here periodic boundary conditions achieve a similar result: All lattice sites experience an equivalent energy environment (a spin to the right and the left—a symmetry broken with free boundary conditions), which builds in the *translational invariance* required of systems in the thermodynamic limit.

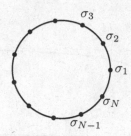

Figure 6.22: One-dimensional Ising model with periodic boundary conditions, $\sigma_{N+1} \equiv \sigma_1$.

The partition function requires us to evaluate the sum

$$Z_N(K, B) = \sum_{\{\sigma\}} \exp\left(K \sum_{i=1}^{N} \sigma_i\sigma_{i+1} + B \sum_{i=1}^{N} \sigma_i \right), \tag{6.76}$$

where $B = \beta b$. The exponential in Eq. (6.76) can be factored,[41] allowing us to write

$$Z_N(K, B) = \sum_{\{\sigma\}} V(\sigma_1, \sigma_2) V(\sigma_2, \sigma_3) \cdots V(\sigma_{N-1}, \sigma_N) V(\sigma_N, \sigma_1), \tag{6.77}$$

where

$$V(\sigma_i, \sigma_{i+1}) \equiv \exp\left(K\sigma_i\sigma_{i+1} + \frac{1}{2}B(\sigma_i + \sigma_{i+1}) \right) \tag{6.78}$$

is symmetric in its arguments,[42] $V(\sigma_i, \sigma_{i+1}) = V(\sigma_{i+1}, \sigma_i)$.

Equation (6.77) is in the form of a product of matrices. We can regard $V(\sigma, \sigma')$ as the elements of a 2×2 matrix, V, the *transfer matrix*,[43] which, in the "up-down" basis $\sigma_j = \pm 1$, has the form

$$V = \begin{matrix} (+) \\ (-) \end{matrix} \begin{pmatrix} e^{K+B} & e^{-K} \\ e^{-K} & e^{K-B} \end{pmatrix}. \tag{6.79}$$

Holding σ_1 fixed in Eq. (6.77) and summing over $\sigma_2, \cdots, \sigma_N$, Z is related to the trace of an N-fold matrix product,

$$Z_N(K, B) = \sum_{\sigma_1} V^N(\sigma_1, \sigma_1) = \operatorname{Tr} V^N. \tag{6.80}$$

Finding an expression for the Nth power of V, while it can be done (see Exercise 6.24), is unnecessary. The trace is independent of the basis in which a matrix is represented;[44] matrices are diagonal in a basis of their eigenfunctions, with their eigenvalues λ_i occurring as the diagonal elements—so choose that basis. The Nth power of a diagonal matrix D is itself diagonal with elements the Nth power of the elements of D. The transfer matrix Eq. (6.79) has two eigenvalues, $\lambda_\pm$. The trace operation in Eq. (6.80) is therefore easily evaluated,[45] with

$$Z_N(K, B) = \lambda_+^N + \lambda_-^N, \tag{6.81}$$

---

[41]We use the property of exponentials that $e^x e^y = e^z$ implies $z = x + y$. This familiar property does not hold for non-commuting operators. Lamentably, methods of classical physics may not be applicable in quantum mechanics.

[42]$V(\sigma, \sigma')$ could be multiplied by $e^{x(\sigma-\sigma')}$ (for any x) and not affect Eq. (6.77).

[43]The transfer matrix was introduced by Kramers and Wannier[71, 72] in 1941. As happens historically, useful ideas are often conceived nearly simultaneously by others. The transfer matrix approach was also developed by Montroll[73] in 1941.

[44]See any book on linear algebra; might I suggest [16, p36].

[45]Chalk one up for knowing some linear algebra.

where the eigenvalues of V are readily ascertained (show this),

$$\lambda_{\pm}(K, B) = e^K \cosh B \pm e^{-K} \sqrt{1 + e^{4K} \sinh^2 B}.$$ (6.82)

Because $\lambda_+ > \lambda_-$,

$$Z_N(K, B) = \lambda_+^N \left[1 + \left(\frac{\lambda_-}{\lambda_+} \right)^N \right] \overset{N \to \infty}{\sim} \lambda_+^N.$$ (6.83)

For large N, Z_N and hence all thermodynamic information is contained in the largest eigenvalue, λ_+. For $B = 0$, $\lambda_+(K, B = 0) = 2 \cosh K$, and $Z_N(K, 0)$ from Eq. (6.83) agrees with $Z_N(K)$ from Eq. (6.71) in the thermodynamic limit.[46]

The average magnetization $M \equiv \langle \sum_i \sigma_i \rangle = N \langle \sigma \rangle$ can be calculated from $Z_N(K, B)$:

$$\langle \sigma \rangle = \lim_{N \to \infty} \left(\frac{1}{N} \frac{\partial \ln Z_N}{\partial B} \right) = \frac{e^{2K} \sinh B}{\sqrt{1 + e^{4K} \sinh^2 B}}.$$ (6.84)

The system is paramagnetic (even with interactions between spins): As $B \to 0$, $\langle \sigma \rangle \to 0$. The zero-field susceptibility per spin, when calculated from Eq. (6.84), has the value

$$\chi = \frac{1}{N} \frac{\partial M}{\partial B} \Big|_{B=0} = e^{2K}.$$ (6.85)

The susceptibility can be calculated in another way. It's readily shown for $N \to \infty$ that

$$\chi = \frac{1}{N} \frac{\partial M}{\partial B} \Big|_{B=0} = \frac{1}{N} \sum_i \sum_j \langle \sigma_i \sigma_j \rangle_{B=0} = \frac{1 + \tanh K}{1 - \tanh K}.$$ (6.86)

The susceptibility is therefore the sum of all two-spin *correlation functions* $\langle \sigma_i \sigma_j \rangle$, a topic to which we now turn. It's straightforward to show that Eq. (6.86) is equivalent to Eq. (6.85).

6.5.3 Correlation functions

Correlation functions (such as $\langle \sigma_i \sigma_j \rangle$) play an important role in statistical mechanics and will increasingly occupy our attention in this book; they provide spatial, *structural* information that cannot be obtained from partition functions.[47,48] We can calculate correlation functions of Ising spins using the transfer matrix method, as we now show. Translational invariance (built into periodic boundary conditions, but attained in any event in the thermodynamic limit) implies that $\langle \sigma_i \sigma_j \rangle$ is a function of the separation between sites (i, j), $\langle \sigma_i \sigma_j \rangle = f(|i - j|)$. The quantity $\langle \sigma_i \sigma_j \rangle$ is in some sense a conditional probability: Given that the spin at site i has value σ_i, what is the probability that the spin at site j has value σ_j? That is, to what extent is the value of σ_j *correlated*[49] with the value of σ_i? We expect the closer spins are spatially, the more they are correlated. Correlation functions establish a length, the *correlation length*, ξ, a measure of the range over which correlations persist. We expect for separations far in excess of the correlation length, $|i - j| \gg \xi$, that $\langle \sigma_i \sigma_j \rangle \to \langle \sigma \rangle^2$.

---

[46]Which begs the question: Can the transfer matrix be used for free boundary conditions? See Exercise 6.24.

[47]If one were to introduce an inhomogeneous magnetic field that couples to each spin, $\sum_i b_i \sigma_i$, and be able to solve for the associated partition function, *then* one could generate correlation functions from derivatives of Z. One often hears that "everything" can be found from the partition function, true for thermodynamic quantities but not for structural information.

[48]We show in Section 6.6 that the two-particle correlation function can be inferred experimentally.

[49]We can't say that the spin at site i with value σ_i *causes* the spin at site j to have value σ_j; the best we can do is to say that σ_j is associated, or correlated, with σ_i. There is a competition between the inter-spin coupling energy, J, with the thermal energy kT, a measure of random energy exchanges between the spin system and its environment.

Because $V(\sigma_1, \sigma_2) \cdots V(\sigma_N, \sigma_1)/Z_N$ is the probability the system is in state $(\sigma_1, \cdots, \sigma_N)$ (for periodic boundary conditions), the average we wish to calculate is:

$$\langle \sigma_i \sigma_j \rangle = \frac{1}{Z_N} \sum_{\{\sigma\}} V(\sigma_1, \sigma_2) \cdots V(\sigma_{i-1}, \sigma_i) \sigma_i V(\sigma_i, \sigma_{i+1}) \cdots$$

$$\cdots V(\sigma_{j-1}, \sigma_j) \sigma_j V(\sigma_j, \sigma_{j+1}) \cdots V(\sigma_N, \sigma_1) . \quad (6.87)$$

We've written Eq. (6.87) using transfer matrix symbols, but it's not in the form of a matrix product (compare with Eq. (6.77)). We need a matrix representation of Ising spins. A matrix represents the action of a linear operator *in a given basis* of the vector space on which the operator acts. And of course bases are not unique—any set of linearly independent vectors that span the space will do. In Eq. (6.79) we used a basis of up and down spin states, which span a two dimensional space, $|+\rangle \equiv \left(\begin{smallmatrix}1\\0\end{smallmatrix}\right)$ and $|-\rangle \equiv \left(\begin{smallmatrix}0\\1\end{smallmatrix}\right)$. With a nod to quantum mechanics,[50] measuring σ at a given site is associated with an operator S that, in the "up-down" basis, is represented by a diagonal matrix with elements[51]

$$S(\sigma, \sigma') \equiv \begin{pmatrix} 1 & 0 \\ 0 & -1 \end{pmatrix} , \quad (6.88)$$

which is one of the Pauli spin matrices.[52]

Equation (6.87) can be written in a way that's independent of basis: For $0 \le j - i \le N$,

$$\langle \sigma_i \sigma_j \rangle = \frac{1}{Z_N} \operatorname{Tr} S V^{j-i} S V^{N-(j-i)} , \quad (6.89)$$

where we've used the cyclic invariance of the trace, $\operatorname{Tr} ABC = \operatorname{Tr} CAB = \operatorname{Tr} BCA$. While Eq. (6.89) is basis independent, it behooves us to choose the basis in which it's most easily evaluated:[53] Work in a basis of the eigenvectors of V (which we have yet to find)—in that way the N copies of V in Eq. (6.89) are diagonal. As is well known,[54] an $N \times N$ matrix V with N linearly independent eigenvectors can be diagonalized through a similarity transformation $P^{-1}VP = \Lambda$, where the columns of P are the eigenvectors of V, with Λ a diagonal matrix with the eigenvalues of V on the diagonal. Assume that we've found P; in that case Eq. (6.89) is equivalent to (show this)

$$\langle \sigma_i \sigma_j \rangle = \frac{1}{Z_N} \operatorname{Tr}(P^{-1}SP)\Lambda^{j-i}(P^{-1}SP)\Lambda^{N-(j-i)} . \quad (6.90)$$

We know the eigenvalues of V (Eq. (6.82)); let $\psi_\pm$ denote the eigenvectors corresponding to $\lambda_\pm$,

$$\begin{pmatrix} e^{K+B} & e^{-K} \\ e^{-K} & e^{K-B} \end{pmatrix} \psi_\pm = \lambda_\pm \psi_\pm . \quad (6.91)$$

It can be shown (after some algebra) that the normalized eigenvectors are

$$\psi_+ = \begin{pmatrix} \cos\phi \\ \sin\phi \end{pmatrix} \qquad \psi_- = \begin{pmatrix} \sin\phi \\ -\cos\phi \end{pmatrix} , \quad (6.92)$$

where ϕ is related to the parameters of the model through[55]

$$\cot 2\phi = e^{2K} \sinh B . \qquad (0 < \phi < \pi/2) \quad (6.93)$$

---

[50]We're not "doing" quantum mechanics, yet methods of reasoning developed in quantum mechanics are highly useful. The commonality is linear algebra, the *lingua franca* of quantum mechanics.

[51]One can readily show that $\left(\begin{smallmatrix}1&0\\0&-1\end{smallmatrix}\right)\left(\begin{smallmatrix}1\\0\end{smallmatrix}\right) = (+1)\left(\begin{smallmatrix}1\\0\end{smallmatrix}\right)$ and $\left(\begin{smallmatrix}1&0\\0&-1\end{smallmatrix}\right)\left(\begin{smallmatrix}0\\1\end{smallmatrix}\right) = (-1)\left(\begin{smallmatrix}0\\1\end{smallmatrix}\right)$.

[52]See any textbook on quantum mechanics for a derivation of the Pauli matrices.

[53]Reminiscent of the theory of relativity—even though certain theoretical expressions are independent of reference frame, for simplicity choose to work in the frame in which it's most easily evaluated.

[54]See for example [16, p41].

[55]For those wishing to verify Eq. (6.92), the key substitution is $\tan\phi = \sqrt{1+y^2} - y$, where $y \equiv e^{2K} \sinh B$.

For $B = 0$, $\phi = \pi/4$; for $B \to +\infty$, $\phi \to 0$, for $B \to -\infty$, $\phi \to \pi/2$. The transformation matrix P is therefore

$$P = \begin{pmatrix} \cos\phi & \sin\phi \\ \sin\phi & -\cos\phi \end{pmatrix}. \tag{6.94}$$

We note that P is its own inverse: $P^{-1} = P$. Thus,

$$P^{-1}SP = \begin{pmatrix} \cos\phi & \sin\phi \\ \sin\phi & -\cos\phi \end{pmatrix} \begin{pmatrix} 1 & 0 \\ 0 & -1 \end{pmatrix} \begin{pmatrix} \cos\phi & \sin\phi \\ \sin\phi & -\cos\phi \end{pmatrix} = \begin{pmatrix} \cos 2\phi & \sin 2\phi \\ \sin 2\phi & -\cos 2\phi \end{pmatrix} \equiv \widetilde{S}. \tag{6.95}$$

The quantity $\widetilde{S}$ is the Pauli matrix $\begin{pmatrix} 1 & 0 \\ 0 & -1 \end{pmatrix}$ expressed in a basis of the eigenvectors of the transfer matrix. The eigenvalues of the transfer matrix allow us to obtain the partition function, Eq. (6.81); its eigenvectors allow us to find correlation functions.

With $\widetilde{S}$ substituted into Eq. (6.90), we have

$$\langle \sigma_i \sigma_j \rangle = \frac{1}{Z_N} \text{Tr} \begin{pmatrix} \cos 2\phi & \sin 2\phi \\ \sin 2\phi & -\cos 2\phi \end{pmatrix} \begin{pmatrix} \lambda_+^{j-i} & 0 \\ 0 & \lambda_-^{j-i} \end{pmatrix} \begin{pmatrix} \cos 2\phi & \sin 2\phi \\ \sin 2\phi & -\cos 2\phi \end{pmatrix} \begin{pmatrix} \lambda_+^{N-(j-i)} & 0 \\ 0 & \lambda_-^{N-(j-i)} \end{pmatrix}$$

$$= \frac{1}{Z_N} \text{Tr} \begin{pmatrix} \lambda_+^N \cos^2 2\phi + \lambda_+^{N-(j-i)} \lambda_-^{j-i} \sin^2 2\phi & \lambda_-^{N-(j-i)} \left(\lambda_+^{j-i} - \lambda_-^{j-i} \right) \sin 2\phi \cos 2\phi \\ \lambda_+^{N-(j-i)} \left(\lambda_+^{j-i} - \lambda_-^{j-i} \right) \sin 2\phi \cos 2\phi & \lambda_-^N \cos^2 2\phi + \lambda_-^{N-(j-i)} \lambda_+^{j-i} \sin^2 2\phi \end{pmatrix}$$

$$= \cos^2 2\phi + \sin^2 2\phi \left(\frac{\lambda_+^{N-(j-i)} \lambda_-^{j-i} + \lambda_+^{j-i} \lambda_-^{N-(j-i)}}{\lambda_+^N + \lambda_-^N} \right),$$

where we've used Eq. (6.81) in the final equality. In the thermodynamic limit (keeping $j - i$ fixed),

$$\langle \sigma_i \sigma_j \rangle = \cos^2 2\phi + \sin^2 2\phi \left(\frac{\lambda_-}{\lambda_+} \right)^{j-i}. \qquad (j > i) \tag{6.96}$$

We can evaluate any spin average using the transfer matrix method; in particular the single spin average $\langle \sigma \rangle$—we don't have to use a derivative of $Z(K, B)$. Thus,

$$\langle \sigma \rangle = \frac{1}{Z_N} \text{Tr}\, S V^N = \frac{1}{Z_N} \text{Tr}\, \widetilde{S} \Lambda^N = \cos 2\phi \frac{\lambda_+^N - \lambda_-^N}{\lambda_+^N + \lambda_-^N} \overset{N \to \infty}{\to} \cos 2\phi. \tag{6.97}$$

It's straightforward to show that $\cos 2\phi$ is the same as the expression in Eq. (6.84). We therefore identify the role of $\cos^2 2\phi$ in Eq. (6.96):

$$\langle \sigma_i \sigma_j \rangle = \langle \sigma \rangle^2 + \sin^2 2\phi \left(\frac{\lambda_-}{\lambda_+} \right)^{j-i} = \langle \sigma \rangle^2 + \frac{1}{1 + e^{4K} \sinh^2 B} \left(\frac{\lambda_-}{\lambda_+} \right)^{j-i}. \qquad (j > i) \tag{6.98}$$

In the absence of an external magnetic field,

$$\langle \sigma_i \sigma_j \rangle_{B=0} = (\tanh K)^{|i-j|}. \tag{6.99}$$

The correlation of *fluctuations* plays an important role in Chapters 7 and 8. Let $\delta\sigma_i \equiv \sigma_i - \langle \sigma_i \rangle$ denote the local fluctuation at site i. From Eq. (6.98),

$$\langle \delta\sigma_i \delta\sigma_j \rangle = \sin^2 2\phi \left(\frac{\lambda_-}{\lambda_+} \right)^{|i-j|} \equiv \sin^2 2\phi \, e^{-|i-j|/\xi}, \tag{6.100}$$

where the correlation length,

$$\xi = -1/\ln(\lambda_-/\lambda_+) \overset{B=0}{=} -1/\ln(\tanh K). \tag{6.101}$$

If λ_+, λ_- are degenerate, we can't use Eq. (6.101). By *Perron's theorem* [74, p64], the largest eigenvalue of a finite positive matrix is real, positive, and non-degenerate for finite K. From Eq. (6.82), $\lambda_\pm$ are *asymptotically degenerate*, $\lambda_+ \sim \lambda_-$ for $B = 0$ and $K \to \infty$.

6.5.4 Beyond nearest-neighbor interactions

6.5.4.1 *Next-nearest-neighbor model*

The transfer matrix method allows us to treat models having interactions that extend beyond nearest-neighbors.[56] We show how to set up the transfer matrix for a model with nearest and next-nearest neighbor interactions,[57] with Hamiltonian

$$H(\sigma) = -J_1 \sum_{i=1}^{N} \sigma_i \sigma_{i+1} - J_2 \sum_{i=1}^{N} \sigma_i \sigma_{i+2} , \qquad (6.102)$$

where we adopt periodic boundary conditions, $\sigma_{N+1} \equiv \sigma_1$ and $\sigma_{N+2} \equiv \sigma_2$. Figure 6.23 shows the

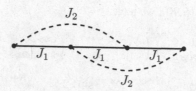

Figure 6.23: Nearest-neighbor (J_1, solid lines) and next-nearest neighbor (J_2, dashed) interactions.

two types of interactions and their connectivity in one dimension.

To set up the transfer matrix, we group spins into *cells* of two spins apiece, as shown in Fig. 6.24. We label the spins in the kth cell $(\sigma_{k,1}, \sigma_{k,2})$, $1 \le k \le N/2$. A key step is to associate

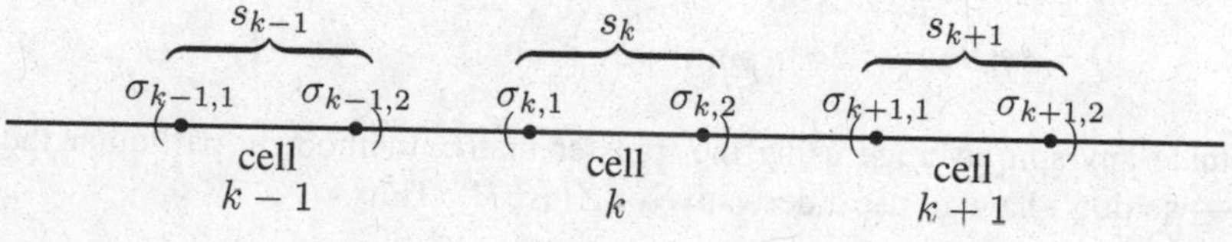

Figure 6.24: Grouping of spins into cells of two spins apiece.

the degrees of freedom of each cell with a new variable s_k representing the four configurations $(+,+), (+,-), (-,+), (-,-)$. This is a mapping from the 2^N degrees of freedom of Ising spins $\{\sigma_i\}$, $1 \le i \le N$, to an equivalent number $4^{N/2}$ degrees of freedom associated with the cell variables $\{s_k\}$, $1 \le k \le N/2$.

Besides grouping spins into cells, we also classify interactions as those associated with *intra*-cell couplings (see Fig. 6.25)

$$V_0(s_k) \equiv -J_1 \sigma_{k,1} \sigma_{k,2} , \qquad (6.103)$$

and *inter*-cell couplings,

$$V_1(s_k, s_{k+1}) \equiv -J_1 \sigma_{k,2} \sigma_{k+1,1} - J_2 \left(\sigma_{k,1} \sigma_{k+1,1} + \sigma_{k,2} \sigma_{k+1,2} \right) . \qquad (6.104)$$

With these definitions, the Hamiltonian can be written as the sum of two terms, one containing all intra-cell interactions and the other containing all inter-cell interactions,

$$H(s) = \sum_{k=1}^{N/2} \left(V_0(s_k) + V_1(s_k, s_{k+1}) \right) \equiv H_0(s) + H_1(s) , \qquad (6.105)$$

---

[56] In the Takahashi gas, the method of solution allows the treatment of systems with nearest-neighbor interactions only.

[57] While there may be applications where second-neighbor interactions are important, we're mainly interested in how the transfer matrix is set up because it allows us to see the generalization by which the transfer matrix is implemented for two-dimensional systems, and it lends insight to the method of the renormalization group, the subject of Chapter 8.

where $s_{(N/2)+1} \equiv s_1$. Comparing Eqs. (6.102) and (6.105), we see that all interactions of the model

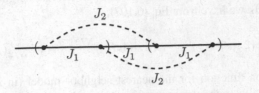

Figure 6.25: Intra- and inter-cell spin couplings.

are accounted for; we have simply rewritten the Hamiltonian by grouping the spins into cells.[58]
The partition function, expressed in terms of cells variables, is then

$$Z_N(K_1, K_2) = \sum_{\{s\}} \exp\left[-\beta\left(\sum_{k=1}^{N/2}(V_0(s_k) + V_1(s_k, s_{k+1}))\right)\right], \qquad (6.106)$$

where $K_i = \beta J_i$, $i = 1, 2$, and $\sum_{\{s\}} \equiv \sum_{s_1} \cdots \sum_{s_{N/2}}$ indicates a summation over cell degrees of freedom. In analogy with the step between Eqs. (6.76) and (6.77), we can write Eq. (6.106)

$$Z_N(K_1, K_2) = \sum_{\{s\}} T(s_1, s_2) T(s_2, s_3) \cdots T(s_N, s_1) = \operatorname{Tr} T^{N/2} = \sum_{i=1}^{4} \lambda_i^{N/2}, \qquad (6.107)$$

where we've introduced a transfer matrix T that couples adjacent cells, with matrix elements,

$$T(s, s') = e^{-\beta V_0(s)} e^{-\beta V_1(s,s')} = e^{K_1 \sigma_1 \sigma_2} e^{K_1 \sigma_2 \sigma_1'} + K_2(\sigma_1 \sigma_1' + \sigma_2 \sigma_2'), \qquad (6.108)$$

and where $\{\lambda_i\}_{i=1}^{4}$ are its eigenvalues. Clearly T is a 4×4 matrix because the sums in Eq. (6.107) are over the four degrees of freedom represented by the variable s_k. The explicit form of T is

$$T = \begin{array}{c} \\ (++) \\ (+-) \\ (--) \\ (-+) \end{array} \begin{array}{cccc} (++) & (+-) & (--) & (-+) \\ \left(\begin{array}{cccc} e^{2K_1+2K_2} & e^{2K_1} & e^{-2K_2} & 1 \\ e^{-2K_1} & e^{-2K_1+2K_2} & 1 & e^{-2K_2} \\ e^{-2K_2} & 1 & e^{2K_1+2K_2} & e^{2K_1} \\ 1 & e^{-2K_2} & e^{-2K_1} & e^{-2K_1+2K_2} \end{array}\right) \end{array}. \qquad (6.109)$$

The matrix in Eq. (6.109) is in block-symmetric form $\left(\begin{smallmatrix} A & B \\ B & A \end{smallmatrix}\right)$ because of the order with which we've written the basis elements in Eq. (6.109) with up-down symmetry. Explicit expressions for the eigenvalues of T are (after some algebra),

$$\lambda_1 = e^{2K_2} \cosh 2K_1 + e^{-2K_2} + 2\cosh K_1 \sqrt{e^{4K_2} \sinh^2 K_1 + 1}$$

$$\lambda_2 = e^{2K_2} \cosh 2K_1 + e^{-2K_2} - 2\cosh K_1 \sqrt{e^{4K_2} \sinh^2 K_1 + 1}$$

$$\lambda_3 = e^{2K_2} \cosh 2K_1 - e^{-2K_2} + 2\sinh K_1 \sqrt{e^{4K_2} \cosh^2 K_1 - 1}$$

$$\lambda_4 = e^{2K_2} \cosh 2K_1 - e^{-2K_2} - 2\sinh K_1 \sqrt{e^{4K_2} \cosh^2 K_1 - 1}. \qquad (6.110)$$

The partition function $Z_N(K_1, K_2)$ follows by combining Eq. (6.110) with Eq. (6.107).

---

[58]Compare our treatment here with that of Section 8.5. We're treating intra-cell and inter-cell couplings on equal footing; in the real-space renormalization method, intra and inter-cell couplings are not treated in an equivalent manner.

If we set $K_1 = 0$, the system separates into two inter-penetrating yet uncoupled sublattices, where, within each sublattice, the spins interact through nearest-neighbor couplings. Using the eigenvalues in Eq. (6.110), we have from Eq. (6.107),

$$Z_N(0, K_2) = \left[(2 \cosh K_2)^{N/2} + (2 \sinh K_2)^{N/2} \right]^2 = \left(Z_{N/2}^{\text{nn}}(K_2) \right)^2, \tag{6.111}$$

where Z^{nn} is the partition function for the nearest-neighbor model (in zero magnetic field), Eq. (6.80). Equation (6.111) illustrates the general result that the partition function of noninteracting subsystems is the product of the subsystem partition functions. If we set $K_2 = 0$, we're back to the N-spin nearest-neighbor Ising model in zero magnetic field. It's readily shown that[59]

$$Z_N(K_1, 0) = Z_N^{\text{nn}}(K_1). \tag{6.112}$$

6.5.4.2 Further-neighbor interactions

It's straightforward to generalize to a one-dimensional system with arbitrarily distant interactions.[60] Define an N-spin model with up to p^{th}-neighbor interactions, where p is arbitrary,

$$H(\sigma) = - \sum_{m=1}^{p} \sum_{k=1}^{N} J_m \sigma_k \sigma_{k+m}, \tag{6.113}$$

where we invoke periodic boundary conditions, $\sigma_{N+m} \equiv \sigma_m$, $1 \le m \le p$. Clearly we should have $N \gg p$ for such a model to be sensible.

We break the system into cells of p contiguous spins,[61] $(\sigma_{k,1}, \sigma_{k,2}, \cdots, \sigma_{k,p})$, $1 \le k \le N/p$. We associate the 2^p spin configurations of the kth cell with the symbol s_k. The transfer matrix will then be a $2^p \times 2^p$ matrix. We rewrite the Hamiltonian, making the distinction between intra- and inter-cell couplings,

$$V_0(s_k) = - \sum_{m=1}^{p-1} \sum_{j=1}^{p-m} J_m \sigma_{k,j} \sigma_{k,j+m} \tag{6.114}$$

$$V_1(s_k, s_{k+1}) = - \sum_{m=0}^{p-1} \sum_{j=1}^{p-m} J_{p-m} \sigma_{k,j+m} \sigma_{k+1,j}, \tag{6.115}$$

so that

$$H(s) = \sum_{k=1}^{N/p} (V_0(s_k) + V_1(s_k, s_{k+1})) \equiv H_0(s) + H_1(s). \tag{6.116}$$

Equations (6.103) and (6.104) are special cases of Eqs. (6.114) and (6.115) with $p = 2$. By counting terms in Eqs. (6.114) and (6.115), there are $p(p-1)/2$ intra-cell couplings and $p(p+1)/2$ inter-cell couplings for a total of p^2 couplings associated with each cell. The total number of spin interactions is thus preserved by the grouping of spins into cells. The Hamiltonian Eq. (6.113) represents a total of Np spin interactions, the same number represented by Eq. (6.116): $Np = (N/p)p^2$. The division into cells also preserves the number of spin configurations: $2^N = 2^{p(N/p)}$.

---

[59]In the limit $J_2 \to 0$, the two-spin cell for the transfer matrix is suddenly twice as large as it needs to be, at the same time two eigenvalues of the 4×4 transfer matrix vanish ($\lambda_2 = \lambda_4 = 0$ in Eq. (6.110)), while the other two reduce to the *square* of the eigenvalues of the 2×2 transfer matrix associated with the nearest-neighbor model. This behavior of the eigenvalues occurs whenever the size of the cell set up for the transfer matrix exceeds the range of the spin interactions.[75]

[60]We're not actually interested in the physics of a one-dimensional system with arbitrarily distant interactions. What we learn here will help us in Chapter 8.

[61]We assume for simplicity that N is an integer multiple of p.

To complete the discussion, the partition function for the pth-neighbor model is

$$Z_N(K_1, \cdots, K_p) = \sum_{\{s\}} \exp\left[-\beta \sum_{k=1}^{N/p} (V_0(s_k) + V_1(s_k, s_{k+1}))\right] = \sum_s T(s_1, s_2) \cdots T(s_{N/p}, s_1)$$

$$(6.117)$$

$$= \sum_{s_1} T^{N/p}(s_1, s_1) = \sum_{k=1}^{2^p} \lambda_k^{N/p},$$

where $(\lambda_1, \cdots, \lambda_{2^p})$ are the eigenvalues of the matrix $T(s, s') = \exp(-\beta V_0(s)) \exp(-\beta V_1(s, s'))$.

6.5.4.3 The Ising spin ladder

We now apply the transfer matrix to a more complicated one-dimensional system shown in Fig. 6.26, a "ladder" of $2N$ Ising spins satisfying periodic boundary conditions. To set up the transfer

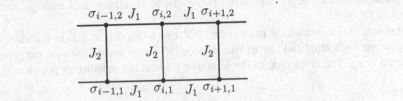

Figure 6.26: A $2 \times N$ Ising model with couplings J_1 and J_2.

matrix, which groups of spins can we treat as adjacent cells? With the spins labeled as in Fig. 6.26, the Hamiltonian can be written

$$H = -J_2 \sum_{i=1}^{N} \sigma_{i,1}\sigma_{i,2} - J_1 \sum_{i=1}^{N} (\sigma_{i,1}\sigma_{i+1,1} + \sigma_{i,2}\sigma_{i+1,2}) .$$

$$(6.118)$$

The transfer matrix is therefore

$$T(\sigma, \sigma') = e^{K_2\sigma_1\sigma_2} e^{K_1(\sigma_1\sigma_1' + \sigma_2\sigma_2')} .$$

$$(6.119)$$

We can write the elements of $T(\sigma, \sigma')$ as we did in Eq. (6.109),

$$\boldsymbol{T} = \begin{array}{c} (++) \\ (+-) \\ (--) \\ (-+) \end{array} \overset{\begin{array}{cccc} (++) & (+-) & (--) & (-+) \end{array}}{\begin{pmatrix} e^{K_2+2K_1} & e^{K_2} & e^{K_2-2K_1} & e^{K_2} \\ e^{-K_2} & e^{-K_2+2K_1} & e^{-K_2} & e^{-K_2-2K_1} \\ e^{K_2-2K_1} & e^{K_2} & e^{K_2+2K_1} & e^{K_2} \\ e^{-K_2} & e^{-K_2-2K_1} & e^{-K_2} & e^{-K_2+2K_1} \end{pmatrix}} .$$

$$(6.120)$$

The next step would be to find the eigenvalues of $\boldsymbol{T}$ in Eq. (6.120), but we stop here.

6.6 SCATTERING, FLUCTUATIONS, AND CORRELATIONS

Much of what we know about macroscopic systems comes from scattering experiments. In X-ray scattering, electromagnetic radiation scatters from charges in the system; in neutron scattering, neutrons scatter from magnetic moments in the system (see Appendix E). Figure 6.27 shows the geometry of a scattering experiment. A beam of monochromatic radiation of wave vector k_i and angular

Figure 6.27: Scattering geometry: Incoming and outgoing wave vectors k_i, k_f with $q = k_f - k_i$.

frequency ω is incident upon a sample and is scattered towards a detector in the direction of the outgoing wave vector k_f at angle θ relative to k_i. If the energy $\hbar\omega$ is much larger than the characteristic excitation energies of the molecules of the system, scattering occurs without change of frequency (*elastic scattering*, our concern here) and thus k_f has magnitude $|k_f| = |k_i|$. In elastic scattering, the *wave vector transfer*

$$q \equiv k_f - k_i \tag{6.121}$$

has magnitude $|q| = 2|k_i|\sin(\theta/2)$. A record of the scattering intensity as a function of θ provides one with the Fourier transform of the two-particle correlation function (as we'll show), the *static structure factor*.[62]

Assume, for a particle at position r_j (relative to an origin inside the sample) that an incident plane wave with amplitude proportional to $e^{ik_i \cdot r_j}$ is scattered into an outgoing spherical wave[63] centered at r_j. The amplitude of the scattered wave at the detector at position R is proportional to

$$\alpha e^{ik_i \cdot r_j} \frac{e^{ik_f|R-r_j|}}{|R-r_j|}, \tag{6.122}$$

where $k_f \equiv |k_f|$ and α is the scattering efficiency[64] of the particle at r_j. The detector is far removed from the sample with $|R| \equiv R \gg |r_j| \equiv r_j$ (for all j), implying that $|R - r_j| \approx R - \hat{R} \cdot r_j$, where $\hat{R} \equiv R/R$ (show this). In the denominator of (6.122) we can approximate $|R - r_j| \approx R$, but not in the phase factor. With $k_f = k_f \hat{R}$, we have for the amplitude at the detector:

$$e^{ik_i \cdot r_j} \frac{e^{ik_f|R-r_j|}}{|R-r_j|} \approx \frac{e^{ik_f R}}{R} e^{-iq \cdot r_j},$$

where q is defined in Eq. (6.121). The detector receives scattered waves from all particles of the sample, and thus the total amplitude A at the detector is

$$A = A_0 \sum_j e^{-iq \cdot r_j}, \tag{6.123}$$

where A_0 includes $e^{ik_f R}/R$, together with any other constants we've swept under the rug. The intensity at the detector is proportional to the square of the amplitude, $I \propto |A|^2$. Data is collected in scattering experiments over times large compared with microscopic time scales associated with fluctuations, and thus what we measure is the ensemble average,

$$I(q) = |A_0|^2 \left\langle \left| \sum_j e^{-iq \cdot r_j} \right|^2 \right\rangle = |A_0|^2 \left\langle \sum_j \sum_k e^{iq \cdot (r_k - r_j)} \right\rangle \equiv N|A_0|^2 S(q) \equiv I_0 S(q), \tag{6.124}$$

---

[62]The elastic scattering structure factor is known as the static structure factor; for inelastic scattering (not treated here), the structure factor is the Fourier transform of spatial and temporal variables known as the *dynamic structure factor*.

[63]A *spherical wave*, a solution of the Helmholtz equation in three dimensions using spherical coordinates, has the form e^{ikr}/r, not $e^{ik \cdot r}/r$, a source of mistakes.

[64]The parameter α can depend on the scattering angle θ if the beam is polarized in the direction of the scattering plane (see for example [76, p94]). We ignore such complications. Because α is the same for each scatterer (for a system of identical particles), it can be absorbed into overall proportionality constants; effectively we can set $\alpha = 1$.

where

$$S(q) = \frac{1}{N} \sum_j \sum_k \left\langle e^{iq \cdot (r_k - r_j)} \right\rangle \tag{6.125}$$

provisionally defines the static structure factor (see Eq. (6.130)). Note the separation that Eq. (6.124) achieves between I_0, which depends on details of the experimental setup, and $S(q)$, the intrinsic response of the system.

Let's momentarily put aside Eq. (6.125). Define the instantaneous *local* density of particles,

$$n(r) \equiv \sum_{j=1}^{N} \delta(r - r_j) . \tag{6.126}$$

The reader should understand how Eq. (6.126) works: The three-dimensional Dirac delta function has dimension V^{-1}, where V is volume. In summing over the positions r_j of all particles in Eq. (6.126), the delta function counts[65] the number of particles at r, and hence we have the local number density, $n(r)$. Its Fourier transform is[66]

$$n(q) \equiv \int e^{-iq \cdot r} n(r) \mathrm{d}^3 r = \sum_j e^{-iq \cdot r_j} , \tag{6.127}$$

where we've used Eq. (6.126). Note that $(n(q))^* = n(-q)$. By substituting Eq. (6.127) in Eq. (6.125), we have an equivalent expression for the structure factor,

$$S(q) = \frac{1}{N} \langle n(q) n(-q) \rangle = \frac{1}{N} \int \int \mathrm{d}^3 r \mathrm{d}^3 r' e^{iq \cdot (r' - r)} \langle n(r') n(r) \rangle . \tag{6.128}$$

The scattering intensity (into the direction associated with q) is therefore related to the Fourier transform (at wave vector q) of the correlation function, $\langle n(r') n(r) \rangle$. Given enough scattering measurements, the complete Fourier transform can be established, which can be inverted to find the correlation function. The point here is that the two-particle correlation function can be measured.

Equation (6.128) has a flaw that's easily fixed. Define *local* density fluctuations $\delta n(r) \equiv n(r) - \langle n(r) \rangle = n(r) - n$, where $n \equiv N/V$ is the average density (which because of translational invariance is independent of r).[67] Note that $\langle \delta n(r) \rangle = 0$. Substituting $n(r) = n + \delta n(r)$ in Eq. (6.128),

$$S(q) = \frac{1}{N} \int \int \mathrm{d}^3 r \mathrm{d}^3 r' e^{iq \cdot (r' - r)} \langle \delta n(r') \delta n(r) \rangle + 8\pi^3 n \delta(q) , \tag{6.129}$$

where we've used the integral representation of the delta function, $\int_{-\infty}^{\infty} e^{iqx} \mathrm{d}x = 2\pi \delta(q)$. The presence of $\delta(q)$ in Eq. (6.129) indicates a strong signal in the direction of $q = 0$, the *forward direction* defined by $k_f = k_i$. Because radiation scattered into the forward direction cannot be distinguished from no scattering at all, we cannot expect $S(q)$ as given by Eq. (6.129) to represent scattering data at $q = 0$. It's conventional to subtract this term, redefining $S(q)$,

$$S(q) \equiv \frac{1}{N} \int \int \mathrm{d}^3 r \mathrm{d}^3 r' e^{iq \cdot (r' - r)} \langle \delta n(r') \delta n(r) \rangle . \tag{6.130}$$

---

[65]The Dirac function (a *generalized function*) has meaning only "inside" an integral (see any book on mathematical methods, such as [16, p64]), so that $\int_{v(r)} n(r') \mathrm{d}^3 r' = N(r)$, the number of particles within a small volume $v(r)$. While Dirac delta functions have meaning only inside integrals, that doesn't stop people (like us) from writing formulas such as Eq. (6.126) that treat the delta function as an ordinary function, knowing that it will eventually show up inside an integral.

[66]We follow the customary (sloppy) habit of physicists of using the same symbol for two functions, $n(r)$, the density function, and $n(q)$, its Fourier transform. Such a practice might make mathematicians apoplectic, but you can handle it.

[67]Thus, we're distinguishing microscopic, local fluctuations, from macroscopic, thermodynamic fluctuations.

The out-of-beam scattering intensity at $q \neq 0$ is therefore related to the Fourier transform of the correlation function of fluctuations, $\langle \delta n(r') \delta n(r) \rangle$. The extent to which *spatially separated* local fluctuations are correlated determines the scattering strength.

The value of $S(q)$ as $q \to 0$ is, from Eq. (6.130) (after subtracting the delta function),

$$\lim_{q \to 0} S(q) = \frac{1}{N} \int \int d^3 r d^3 r' \langle \delta n(r) \delta n(r') \rangle = \frac{1}{N} \left\langle (N - \langle N \rangle)^2 \right\rangle = kTn\beta_T , \qquad (6.131)$$

where we've used Eq. (4.81) and $n = \langle N \rangle / V$. Thus, $S(q = 0)$ represents long-wavelength, *thermodynamic* fluctuations in the total particle number, $\langle (N - \langle N \rangle)^2 \rangle$, whereas $S(q \neq 0)$ represents correlations of spatially separated microscopic (local) fluctuations, $\langle \delta n(r) \delta n(r') \rangle$.

Example. Evaluate $S(q)$ for the one-dimensional, nearest-neighbor Ising model in zero magnetic field, using Eq. (6.99) for the correlation functions. From Eq. (6.130), using sums over lattice sites instead of integrals, we have (where a is the lattice constant)

$$S(q) = \frac{1}{N} \sum_n \sum_m \langle \sigma_n \sigma_m \rangle e^{iqa(n-m)} = \sum_{m=-\infty}^{\infty} \langle \sigma_0 \sigma_m \rangle e^{-iqma}$$

$$= 1 + \sum_{m=1}^{\infty} u^m e^{-iqma} + \sum_{m=-\infty}^{-1} u^{|m|} e^{-iqma} = \frac{1 - u^2}{1 - 2u \cos qa + u^2} , \qquad (6.132)$$

where we've let $N \to \infty$, we've used translational invariance, and we've introduced the abbreviation $u \equiv \tanh K$. $S(q)$ is peaked at $q = 0$ for $K > 0$, and at $q = \pm\pi/a$ for $K < 0$. It's straightforward to show that $S(q = 0) = \chi$, where the susceptibility χ is given in Eq. (6.86).

The structure factor can be written in yet another way by returning to Eq. (6.125) and recognizing that $e^{iq \cdot (r_k - r_j)} = \int d^3 r e^{iq \cdot r} \delta(r - (r_k - r_j))$. Thus we have the equivalent expression

$$S(q) = \frac{1}{N} \int d^3 r e^{iq \cdot r} \left\langle \sum_j \sum_k \delta(r - (r_k - r_j)) \right\rangle . \qquad (6.133)$$

Separate the terms in the double sum for which $k = j$:

$$\left\langle \sum_j \sum_k \delta(r - (r_k - r_j)) \right\rangle = N\delta(r) + \left\langle \sum_j \sum_{k \neq j} \delta(r - (r_k - r_j)) \right\rangle . \qquad (6.134)$$

The second term on the right of Eq. (6.134) defines the *radial distribution function*,

$$ng(r) \equiv \frac{1}{N} \left\langle \sum_j \sum_{k \neq j} \delta(r - (r_k - r_j)) \right\rangle , \qquad (6.135)$$

where n (density) is included in the definition to make $g(r)$ dimensionless. The delta functions in Eq. (6.135) count the number of pairs of particles that are separated by r. If the system is isotropic, $g(r)$ is a function only of r: $g = g(r)$. Because of translational invariance,[68] the definition in Eq. (6.135) is equivalent to[69] (taking r_j as the origin) $ng(r) = \sum_{k \neq 0} \langle \delta(r - r_k) \rangle$. As $r \to \infty$ the sum

---

[68]The terms isotropy (no preferred direction) and translational invariance (no unique location) are often bandied about together, yet they're logically distinct concepts. One could have isotropic systems that are not translationally invariant (a unique origin) and translationally invariant systems that are not isotropic (a preferred direction exists everywhere).

[69]Note the distinction between $n(r) = \sum_j \delta(r - r_j)$, Eq. (6.126), and $g(r) = n^{-1} \sum_{k \neq 0} \langle \delta(r - r_k) \rangle$. One is the instantaneous local density at r, the other involves an ensemble average of the number of particles at distance r from another particle at the origin. The radial distribution function $g(r)$ can be considered a conditional probability—given a particle at the origin, what's the probability of finding another particle at distance r.

captures all particles of the system and $g(r) \to 1$. Combining Eqs. (6.135) and (6.134) with Eq. (6.133),

$$S(q) = 1 + n \int d^3r e^{iq \cdot r} g(r) . \qquad (6.136)$$

Equation (6.136) suffers from the same malady as Eq. (6.128). So that the Fourier transform in Eq. (6.136) be well defined,[70] we add and subtract unity (the value of g as $r \to \infty$),

$$\int d^3r e^{iq \cdot r} g(r) = \int d^3r e^{iq \cdot r} [g(r) - 1 + 1] = \int d^3r e^{iq \cdot r} [g(r) - 1] + 8\pi^3 \delta(q) .$$

Just as in Eq. (6.129), we subtract the delta function. Thus, another definition of $S(q)$, equivalent to Eq. (6.130), is

$$S(q) = 1 + n \int d^3r e^{iq \cdot r} [g(r) - 1] . \qquad (6.137)$$

For an isotropic system,

$$S(q) = 1 + 4\pi n \int_0^\infty r^2 \left(\frac{\sin qr}{qr} \right) [g(r) - 1] \, dr . \qquad (6.138)$$

If the wavelength $\lambda = 2\pi/k_i$ is large compared with the range ξ over which $[g(r) - 1]$ is finite, i.e., $q\xi \ll 1$, one can replace $\sin qr/(qr)$ in Eq. (6.138) with unity, in which case the scattering is isotropic—*even for systems with correlated fluctuations*. The frequency must be chosen so that $\hbar\omega$ is large compared with excitation energies, *and* the wavelength is small,[71] $2\pi c/\omega \ll \xi$. The extent to which fluctuations are correlated can be probed experimentally only if $\lambda \ll \xi$. For noninteracting particles, $g(r) = 1$ (as one can show), and the scattered radiation is isotropic with $S(q) = 1$.

6.7   ORNSTEIN-ZERNIKE THEORY OF CRITICAL CORRELATIONS

To observe scattering from correlated fluctuations requires the wavelength to be smaller than the correlation length, $\lambda \ll \xi$, and for that reason X-rays are used to probe the distribution of molecules in fluids. Near critical points, however,[72] strong scattering of *visible light* occurs, where a normally transparent fluid appears cloudy or opalescent, a phenomenon known as *critical opalescence*. The wavelength of visible light is $\approx 10^4$ times as large as that for X-rays, implying that *fluctuations become correlated over macroscopic lengths at the critical point*. In 1914, L.S. Ornstein and F. Zernike made an important step in attempting to explain the development of long-range, critical correlations,[73] one that's relevant to our purposes and which we review here.

Ornstein and Zernike proposed a mechanism by which correlations can be established between particles of a fluid. They distinguished two *types* of correlation function: $c(r)$, the *direct correlation function*, a new function, and $h(r) \equiv g(r) - 1$, termed the *total correlation function* (with $g(r)$ the radial distribution function, Eq. (6.135)). The direct correlation function accounts for contributions to the correlation between points of a fluid that are not mediated by other particles, such as that caused by the potential energy of interaction, $v(r)$ (see Eq. (6.1)). Ornstein and Zernike posited a connection between the two types of correlation function (referring to Fig. 6.28):

$$h(r_2 - r_1) = c(r_2 - r_1) + n \int c(r_3 - r_1) h(r_2 - r_3) d^3 r_3 . \qquad (6.139)$$

Equation (6.139) is the *Ornstein-Zernike equation*. In addition to the direct correlation between

---

[70]Fourier integrals are defined for integrable functions, but issues related to integrability can be subtle. A good source for the fine print on Fourier integrals is Titchmarsh.[77, Chapter 1]

[71]X-rays meet these criteria, where a typical wavelength is 0.1 nm, corresponding to an energy per photon $\approx 10^4$ eV, which is large compared with thermal energies per particle, $\approx 10^{-1}$ eV.

[72]The critical point is defined in Chapter 7, which occurs at a well defined, material-specific temperature T_c.

[73]"Accidental deviations of density and opalescence at the critical point of a single substance," reprinted in Frisch and Lebowitz.[78, ppIII-3–III-25]. Today we would use the words *random fluctuations* instead of accidental deviations.

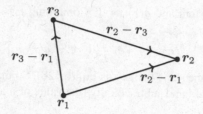

Figure 6.28: Geometry of the Ornstein-Zernike equation.

particles at r_1, r_2 (the first term of Eq. (6.139)), the integral sums the influence from all other particles of the fluid at positions r_3. The quantity $n d^3 r_3$ in Eq. (6.139) represents the number of particles in an infinitesimal volume at r_3, each "directly" correlated to the particle at r_1, which set up the full (total) correlation with the particle at r_2. Equation (6.139) is an integral equation[74,75] that defines $c(r)$ (given $h(r)$). The function $c(r)$ *can* be given an independent definition as a sum of a certain class of connected diagrams,[76, p99] a topic we lack sufficient space to develop.

By taking the Fourier transform of Eq. (6.139) and applying the convolution theorem[16, p111], we find, where $c(q) = \int d^3 r e^{i q \cdot r} c(r)$,

$$h(q) = c(q) + nc(q)h(q) \qquad \Longrightarrow \qquad c(q) = \frac{h(q)}{1 + nh(q)} . \qquad (6.140)$$

Equation (6.140) indicates that $c(q)$ does not show singular behavior at the critical point.[76] From Eq. (6.137), $S(q) = 1 + nh(q)$, and, because $S(q)$ diverges as $q \to 0$ at $T = T_c$ (see Section 7.6), $c(q = 0)$ remains finite at $T = T_c$. Using Eq. (6.131),

$$\frac{1}{kTn\beta_T} = \frac{1}{S(q = 0)} = \frac{1}{1 + nh(q = 0)} = 1 - nc(q = 0) = 1 - n \int d^3 r c(r) . \qquad (6.141)$$

Thus, *the direct correlation function is short ranged*, even at the critical point.[77] If we're interested in critical phenomena characterized by long-wavelength fluctuations (which we will be in coming chapters), approximations made on the short-ranged function $c(r)$ should prove rather innocuous[78] (at least that's the thinking[79]). Molecular dynamics simulations have confirmed the short-ranged nature of $c(r)$ [79]. An approximate form for $c(r)$ introduced by Percus and Yevick[80] gives good agreement with experiment and displays its short-ranged character:

$$c(r) \approx \left(1 - e^{\beta v(r)}\right) g(r) ,$$

so that $c(r)$ vanishes for distances outside the range of the pair potential.[80]

---

[74]Equation (6.139) is a Fredholm integral equation of the second kind [16, Chapter 10]. Liquid-state theory (which we won't pursue further in this book) is rife with integral equations.

[75]Integral equations occur in other areas of physics, such as quantum scattering theory. Equation (6.139) resembles the *Dyson equation* for the Green function, $G = G_0 + \int G_0 \Sigma G$, where G_0 is the zeroth-order propagator and Σ is the self-energy operator. The Ornstein-Zernike equation is like the Dyson equation without the self-energy term.

[76]We discuss in Chapter 7 the quantities that exhibit singular behavior at the critical point—critical phenomena; among others, heat capacity, susceptibility, compressibility, and $\lim_{q \to 0} S(q)$.

[77]At the critical point (where $\beta_T \to \infty$), $\int c(r) d^3 r \to (1/n)$; at high temperatures, if we approximate the system as an ideal gas, $\beta_T = (1/P)$, and $\int c(r) d^3 r \to 0$.

[78]What Ornstein and Zernike did in introducing the direct correlation function is excellent theoretical physics: introduce an unknown function, and then convince yourself that approximations to it cause little harm—sweep what you don't know under the rug, where it can't hurt you.

[79]This idea works in three dimensions, but requires modification in two dimensions as we discuss in Chapter 7.

[80]For $r \to 0$, $c(r)$ is negative and bounded; in this limit $g(r) \to 0$ from the short-ranged repulsion between atoms.

Because $c(q)$ is well behaved at T_c, it may be expanded in a Taylor series about $q = 0$. For an isotropic system, $c(q) = \sum_{k=0}^{\infty} c_k q^k$, where the coefficients c_k are functions of (n, T) and are related to the moments of the real-space correlation function $c(r)$:

$$c_k = \frac{1}{k!} \frac{\partial^k c(q)}{\partial q^k}\bigg|_{q=0} = \begin{cases} \dfrac{4\pi i^k}{(k+1)!} \int_0^{\infty} r^{k+2} c(r) \, dr & k = 0, 2, 4, \cdots \\ 0 & k = 1, 3, 5, \cdots . \end{cases} \tag{6.142}$$

The Ornstein-Zernike *approximation* consists of replacing $c(q)$ with the first two terms of its Taylor series:

$$c(q) = c_0 + c_2 q^2 + O(q^4) .$$

Higher-order terms are dropped because we're interested in the low-q, long-range behavior. Combining terms,

$$S(q) = \frac{1}{1 - nc(q)} \approx \frac{1}{1 - n(c_0 + c_2 q^2)} \equiv \frac{1}{R_0^2(q_0^2 + q^2)} , \tag{6.143}$$

where $R_0^2 \equiv -nc_2$ and $q_0^2 \equiv (1 - nc_0)/R_0^2$. This *approximate*, small-q form for $S(q)$ is based on the short-range nature of the direction correlation function $c(r)$, that $c(q)$ can be expanded in a power series, and that terms of $O(q^4)$ can be ignored. The inverse Fourier transform of Eq. (6.143) leads to an asymptotic form for the total correlation function, valid at large distances:

$$h(r) \sim \frac{1}{R_0^2} \frac{e^{-q_0 r}}{r} . \tag{6.144}$$

Thus, we identify $q_0 \equiv \xi^{-1}$ as the inverse correlation length. We return to Ornstein-Zernike theory in Chapter 7.

SUMMARY

We considered systems featuring inter-particle interactions: The classical gas, the Tonks-Takahashi gas, and the one-dimensional Ising model. We showed how the equation of state of real gases in the form of the virial expansion, Eq. (6.29), and the nature of the interactions giving rise to the virial coefficients, can be derived in the framework of the Mayer cluster expansion, the prototype of many-body perturbation theories. The Tonks gas features a hard-core repulsive potential (required for the existence of the thermodynamic limit), and the Takahashi gas allows nearest-neighbor attractive interactions in addition to a hard core (but only nearest-neighbor interactions). We introduced the Ising model of interacting degrees of freedom on a lattice, and the transfer matrix method of solution. We introduced the correlation length ξ, the characteristic length over which fluctuations are correlated, which plays an essential role in theories of critical phenomena. We discussed how scattering experiments probe the structure of systems of interacting particles, and the approximate Ornstein-Zernike theory of the static structure factor.

EXERCISES

6.1 We said in Section 6.1 that perturbation theory can't be used if the inter-particle potential $v(r)$ diverges as $r \to 0$. Why not use the Boltzmann factor $e^{-\beta v(r)}$ as a small parameter for $r \to 0$? What's wrong with that idea for applying perturbation theory to interacting gases?

6.2 Referring to Fig. 6.29, we have, in calculating the diagram, the integral

$$I \equiv \int d\mathbf{r}_1 d\mathbf{r}_2 d\mathbf{r}_3 f(|\mathbf{r}_1 - \mathbf{r}_2|) f(|\mathbf{r}_2 - \mathbf{r}_3|) f(|\mathbf{r}_3 - \mathbf{r}_1|) .$$

Change variables. Let $R \equiv \frac{1}{3}(r_1 + r_2 + r_3)$ be the center-of-mass-coordinate, and let $u \equiv r_1 - r_3$, $v \equiv r_2 - r_3$. Show that

$$ I = \int d\mathbf{R} du dv\, f(|u|) f(|v|) f(|u - v|) = V \int du dv\, f(|u|) f(|v|) f(|u - v|) \,, $$

where $V = \int d\mathbf{R}$. Hint: What is the Jacobian of the transformation?

Figure 6.29: Irreducible cluster of three particles.

6.3 Show that Eq. (6.25) reduces to Eq. (4.79) for the case of non-interacting particles.

6.4 Show for the interacting classical gas that $Z_G = e^{PV/kT}$. Hint: Eqs. (6.25) and (6.27).

6.5 Show that the chemical potential of a classical gas can be written in the form

$$ \mu = kT \left[\ln(n\lambda_T^3) + \ln(1 - 2b_2 n + \cdots) \right] \overset{|b_2 n| \ll 1}{\approx} kT \left[\ln(n\lambda_T^3) - 2b_2 n + O(n^2) \right] \,. $$

Hint: Eq. (6.28). The chemical potential is modified relative to the ideal gas, either positively or negatively, depending on the sign of b_2.

6.6 Derive an expression for the fourth virial coefficient, $B_4(T)$, i.e., work out the next contribution to the series in Eq. (6.29). A: $-20b_2^3 + 18b_2 b_3 - 3b_4$.

6.7 Derive Eq. (6.33) for the second virial coefficient associated with the van der Waals equation of state, B_2^{vdw}. What is the expression for B_3^{vdw}? A: b^2.

6.8 Fill in the steps from Eq. (6.34) to Eq. (6.35) using the Lennard-Jones potential.

6.9 Derive Eqs. (6.44), (6.45), and (6.49).

6.10 Verify the claim that $C_3 = 0$ for the diagrams in Fig. 6.14.

6.11 Show that $C_3 = 0$ for the diagram in the right-most part of Fig. 6.15. Hint: Make use of Eq. (6.45).

6.12 Derive Eq. (6.53). Hint: Start with $P = -(\partial F/\partial V)_{T,N}$ (see Table 1.2) and show that (with $n = N/V$)

$$ P = n^2 \left(\frac{\partial (F/N)}{\partial n} \right)_T . $$

Then use the relation $-\beta F = \ln Z$, Eq. (4.57). Without fanfare, we've been working with the canonical ensemble in Section 6.3.

6.13 Derive Eqs. (6.59)–(6.62). Then show the thermodynamic limits, Eq. (6.63)

6.14 Derive an expression for the internal energy of the Tonks gas. Does your result make sense? Hint: $F = U - TS$.

6.15 Verify the identity shown in Eq. (6.70).

6.16 Derive the expressions in Eq. (6.72). Hint: First show that $T(\partial/\partial T) = -K(\partial/\partial K)$.

6.17 Show that the heat capacity of one-dimensional Ising spins has the form at low temperature,

$$C_V \overset{T\to 0}{\sim} \frac{1}{T^2}e^{-2|J|/(kT)},$$

the same low-temperature form of the heat capacity of rotational and vibrational modes in molecules (see Chapter 5). Compare with the low-temperature heat capacity of free fermions, Eq. (5.106), free massive bosons, Eq. (5.166), and photons, Eq. (5.134).

6.18 Find the eigenvalues of the transfer matrix in Eq. (6.79) and show they agree with Eq. (6.82).

6.19 Show that the magnetization of the one-dimensional Ising model (see Eq. (6.84)) demonstrates saturation, that $\langle\sigma\rangle \to 1$ for $\sinh^2 B \gg 1$.

6.20 Suppose there were no near-neighbor interactions in the Ising model ($K = 0$), but we keep the coupling to the magnetic field. Show that in this case $\langle\sigma\rangle$ reduces to one of the Brillouin functions studied in Section 5.2. Which one is it, and does that make sense?

6.21 Find the eigenvalues of the transformed Pauli matrix $\tilde{S}$ in Eq. (6.95). Are you surprised by the result?

6.22 Consider the single-spin average in Eq. (6.97) for the finite system (before we take the thermodynamic limit). Suppose $N = 1$, then the resulting expression for $\langle\sigma\rangle$ should be independent of the near-neighbor coupling constant K. Show that the limiting form of Eq. (6.97) for $N = 1$ is the correct expression.

6.23 Show in the one-dimensional Ising model for $B = 0$ that ξ diverges for $T \to 0$ as,

$$\xi \overset{T\to 0}{\sim} \tfrac{1}{2}\exp(2K). \tag{P6.1}$$

6.24 The Ising partition function for periodic boundary conditions, Eq. (6.80), is the customary result derived using the transfer matrix, so much so one might conclude periodic boundary conditions are essential to the method. That's not the case, as this exercise shows, where we derive the partition function associated with free boundary conditions, Eq. (6.71), using the transfer matrix.

a. Start with Eq. (6.68) as the Hamiltonian for free boundary conditions (take $b = 0$ for simplicity). Set up the calculation of the partition function using the transfer matrix:

$$Z_N(K) = \sum_{\sigma_1}\cdots\sum_{\sigma_N} V(\sigma_1,\sigma_2)\cdots V(\sigma_{N-1},\sigma_N) = \sum_{\sigma_1}\sum_{\sigma_N} V^{N-1}(\sigma_1,\sigma_N).$$

Clearly, we can't "wrap around" to take the trace. To finish the calculation, we require a general expression for powers of the transfer matrix.

b. The transfer matrix V is diagonalizable—there exists a matrix P such that $P^{-1}VP = \Lambda$, where Λ is diagonal. That implies $V = P\Lambda P^{-1}$ (show this). Show that for integer n, $V^n = P\Lambda^n P^{-1}$. Using the transformation matrix in Eq. (6.94) for $\phi = \pi/4$ (zero field), show that

$$V^n = \frac{1}{2}\begin{pmatrix} 1 & 1 \\ 1 & -1 \end{pmatrix}\begin{pmatrix} \lambda_+^n & 0 \\ 0 & \lambda_-^n \end{pmatrix}\begin{pmatrix} 1 & 1 \\ 1 & -1 \end{pmatrix} = \frac{1}{2}\begin{pmatrix} \lambda_+^n + \lambda_-^n & \lambda_+^n - \lambda_-^n \\ \lambda_+^n - \lambda_-^n & \lambda_+^n + \lambda_-^n \end{pmatrix}.$$

c. Put it together to show that

$$Z_N(K) = 2\lambda_+^{N-1} = 2^N \cosh^{N-1}(K),$$ (P6.2)

the same as Eq. (6.71). Equation (P6.2) should be compared with the partition function for periodic boundary conditions, Eq. (6.81), $Z_N(K) = \lambda_+^N + \lambda_-^N$.

d. Show that thermodynamic quantities derived from the two partition functions agree in the thermodynamic limit.

6.25 Show the results in Eqs. (6.111) and (6.112).

6.26 For the Ising ladder problem, rewrite the Hamiltonian, Eq. (6.118), in the equivalent form

$$H = -\frac{1}{2}J_2\sum_{i=1}^{N}(\sigma_{i,1}\sigma_{i,2} + \sigma_{i+1,1}\sigma_{i+1,2}) - J_1\sum_{i=1}^{N}(\sigma_{i,1}\sigma_{i+1,1} + \sigma_{i,2}\sigma_{i+1,2}).$$

Show that the transfer matrix has the form

$$
\boldsymbol{T} = \begin{array}{c} (++) \\ (+-) \\ (--) \\ (-+) \end{array}
\begin{pmatrix}
\overset{(++)}{e^{K_2+2K_1}} & \overset{(+-)}{1} & \overset{(--)}{e^{K_2-2K_1}} & \overset{(-+)}{1} \\
1 & e^{-K_2+2K_1} & 1 & e^{-K_2-2K_1} \\
e^{K_2-2K_1} & 1 & e^{K_2+2K_1} & 1 \\
1 & e^{-K_2-2K_1} & 1 & e^{-K_2+2K_1}
\end{pmatrix}.
$$ (P6.3)

The eigenvalues of the matrix in Eq. (P6.3) (which is symmetric) are the same as those of the matrix in Eq. (6.120) (which is block symmetric)—they describe the same physical system. There must be a similarity transformation between the two matrices.

6.27 Show, using Eq. (6.125), that $S(q) = S^*(q)$, i.e., the structure factor is a real-valued function. Hint: j, k are dummy indices.

CHAPTER 7

Phase transitions and critical phenomena

P HASE, as the term is used in thermodynamics, refers to a spatially uniform equilibrium system.[1] A body of given chemical composition can exist in a number of phases. H_2O, for example, can exist in the familiar liquid and vapor phases, as well as several forms of ice having different crystal structures. Substances undergo *phase transitions*, changes in phase that occur upon variations of state variables. Phases can coexist in physical contact (such as an ice-water mixture). Figure 7.1 is

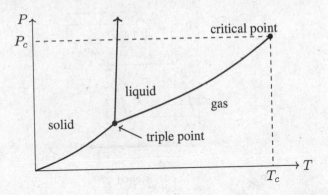

Figure 7.1: Phase diagram in the P-T plane. Thick lines are coexistence curves.

a generic *phase diagram*, the values of T, P for which phases exist and coexist along *coexistence curves*. Note the *triple point*, a unique combination of T, P at which three phases coexist.[2] At the *critical point*, the liquid-vapor coexistence curve *ends* at T_c, P_c (critical temperature and pressure), where the distinction between liquid and gas disappears.[3] For $T > T_c$, a gas cannot be liquefied regardless of pressure. In the vicinity of the critical point (the *critical region*), properties of materials undergo dramatic changes as the distinction between liquid and gas disappears. Lots of interesting physics occurs near the critical point—*critical phenomena*—basically the remainder of this book.[4]

---

[1] Phases exist over a range of values of state variables, T, P, V, etc., and are therefore supersets of equilibrium states. The existence of phases is implied by the existence of equilibrium.

[2] The triple point of H_2O is used in the definition of the Kelvin temperature scale[3, p31].

[3] Every substance has a critical temperature, above which only one *fluid* phase can exist at any volume and pressure. To quote D.L. Goodstein[81, p228]: "We shall use all three terms—fluid, liquid, and gas—hoping that in each instance either the meaning is clear or the distinction is not important."

[4] Our focus is on *static* critical phenomena, not involving time. *Dynamic* critical phenomena are not treated in this book.

7.1 PHASE COEXISTENCE, GIBBS PHASE RULE

7.1.1 Equilibrium conditions

We start by asking whether there is a limit to the number of phases that can coexist. An elegant answer is provided by the *Gibbs phase rule*, Eq. (7.13). The chemical potential of substances in coexisting phases has the same value in each of the phases in which coexistence occurs.[5] Consider two phases of a substance, I and II. Because matter and energy can be exchanged between phases in physical contact, equilibrium is achieved when T and P are the same in both phases, and when the chemical potentials are equal, $\mu^I = \mu^{II}$ (see Section 1.12). We know from the Gibbs-Duhem equation,[6] (P1.1), that $\mu = \mu(T, P)$, and thus chemical potential can be visualized as a surface $\mu = \mu(T, P)$ (see Fig. 7.2). Two phases of the same substance coexist when

$$\mu^I(T, P) = \mu^{II}(T, P) . \tag{7.1}$$

The intersection of the two surfaces defines the locus of points $P = P(T)$ for which Eq. (7.1) is satisfied—the coexistence curve (see Fig. 7.2). Three coexisting phases (I, II, III) would require the equality of three chemical potential functions,

$$\mu^I(T, P) = \mu^{II}(T, P) = \mu^{III}(T, P) . \tag{7.2}$$

Equation (7.2) implies two equations in two unknowns and thus three phases can coexist at a unique

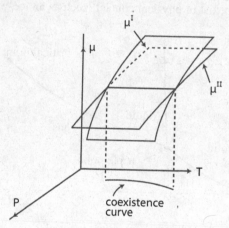

Figure 7.2: Coexistence curve defined by the intersection of chemical potential surfaces associated with phases I and II of a single substance.

combination of T and P, the triple point. By this reasoning, it would not be possible for four phases of a single substance to coexist (which would require three equations in two unknowns). Coexistence of four phases of the same substance is not known to occur.

Multicomponent phases have more than one chemical species. Let μ_j^γ denote the chemical potential of species j in the γ phase (we use Roman letters to label species and Greek letters to label phases). Assume k chemical species, $1 \le j \le k$, and π phases, $1 \le \gamma \le \pi$, where π is an integer.

---

[5]This point becomes obvious with a little thought—the chemical potential of two phases in equilibrium must be the same (Eq. (1.67)), and by extension therefore (using the zeroth law of thermodynamics) to any other coexisting phases. We demonstrate the point using the requirement that the Gibbs energy be a minimum in equilibrium; see Eq. (7.12).

[6]The Gibbs-Duhem equation doesn't generally find a lot of use in thermodynamics, until we come to where find ourselves right now, in considerations of phase coexistence. The fact that $\mu = \mu(T, P)$ for a single substance is crucial.

The first law for multiphase, multicomponent systems is the generalization of Eq. (1.21):

$$dU = TdS - PdV + \sum_{\gamma=1}^{\pi} \sum_{j=1}^{k} \mu_j^\gamma dN_j^\gamma . \tag{7.3}$$

Extensivity implies the scaling property $U(\lambda S, \lambda V, \lambda N_j^\gamma) = \lambda U(S, V, N_j^\gamma)$. By Euler's theorem,

$$U = S \left(\frac{\partial U}{\partial S}\right)_{V,N_j^\gamma} + V \left(\frac{\partial U}{\partial V}\right)_{S,N_j^\gamma} + \sum_{\gamma=1}^{\pi} \sum_{j=1}^{k} N_j^\gamma \left(\frac{\partial U}{\partial N_j^\gamma}\right)_{S,V} = TS - PV + \sum_{\gamma=1}^{\pi} \sum_{j=1}^{k} \mu_j^\gamma N_j^\gamma , \tag{7.4}$$

where the derivatives follow from Eq. (7.3) and $\overline{N_j^\gamma}$ indicates to hold fixed particle numbers except N_j^γ. Equation (7.4) implies

$$G = \sum_{\gamma=1}^{\pi} \sum_{j=1}^{k} \mu_j^\gamma N_j^\gamma , \tag{7.5}$$

the generalization of Eq. (1.54). Taking the differential of Eq. (7.4) and using Eq. (7.3), we have the multicomponent, multiphase generalization of the Gibbs-Duhem equation, (P1.1),

$$\sum_{\gamma=1}^{\pi} \sum_{j=1}^{k} N_j^\gamma d\mu_j^\gamma = -SdT + VdP . \tag{7.6}$$

By taking the differential of G in Eq. (7.5), and making use of Eq. (7.6),

$$dG = -SdT + VdP + \sum_{\gamma=1}^{\pi} \sum_{j=1}^{k} \mu_j^\gamma dN_j^\gamma , \tag{7.7}$$

and thus we have the alternate definition of chemical potential,[7] $\mu_j^\gamma = (\partial G/\partial N_j^\gamma)_{T,P,\overline{N_j^\gamma}}$, the energy to add a particle of type j in phase γ holding fixed T, P, and the other particle numbers. The Gibbs energy is a minimum in equilibrium.[8] From Eq. (7.7),

$$[dG]_{T,P} = \sum_{\gamma=1}^{\pi} \sum_{j=1}^{k} \mu_j^\gamma [dN_j^\gamma]_{T,P} = 0 . \tag{7.8}$$

If the particle numbers N_j^γ could be independently varied, one would conclude from Eq. (7.8) that $\mu_j^\gamma = 0$. But particle numbers are not independent. The number of particles of each species spread among the phases is a constant,[9] $\sum_\gamma N_j^\gamma = $ constant, and thus there are k equations of constraint

$$\sum_{\gamma=1}^{\pi} dN_j^\gamma = 0 . \qquad j = 1, \cdots, k \tag{7.9}$$

---

[7]See Section 1.6.

[8]A stability analysis of a multicomponent system (requiring entropy to be a maximum in equilibrium), as in Section 1.12, would show that $C_V > 0$ and $\beta_T > 0$ (thermal and mechanical stability), just as for single-component phases. The requirement $\partial\mu/\partial N > 0$ (compositional stability) generalizes so that the *matrix* of derivatives $\mu_{j,k} \equiv \partial\mu_j/\partial N_k$ is positive definite, with $\sum_{j,k} \mu_{j,k}\delta N_j\delta N_k > 0$ for any deviations δN_m. The matrix is symmetric, $\mu_{j,k} = \mu_{k,j}$, which is a set of Maxwell relations. From the stability condition on S follows that H, F, and G are a minimum in equilibrium.

[9]We're not allowing chemical reactions to take place.

Constraints are handled through the method of Lagrange multipliers[10] λ_j, which when multiplied by Eq. (7.9) and added to Eq. (7.8) leads to

$$\sum_{j=1}^{k}\sum_{\gamma=1}^{\pi}\left(\mu_j^\gamma + \lambda_j\right) dN_j^\gamma = 0. \tag{7.10}$$

We can now treat the particle numbers as unconstrained, so that Eq. (7.10) implies

$$\mu_j^\gamma = -\lambda_j. \tag{7.11}$$

The chemical potential of each species is *independent of phase*. Equation (7.11) is equivalent to

$$\mu_j^1 = \mu_j^2 = \cdots = \mu_j^\pi. \qquad j = 1, \cdots, k \tag{7.12}$$

There are $k(\pi - 1)$ equations of equilibrium for k chemical components in π phases.

7.1.2 The Gibbs phase rule

How many independent state variables can exist in a multicomponent, multiphase system? In each phase there are $N^\gamma \equiv \sum_{j=1}^{k} N_j^\gamma$ particles, and thus there are $k - 1$ independent *concentrations* $c_j^\gamma \equiv N_j^\gamma / N^\gamma$, where $\sum_{j=1}^{k} c_j^\gamma = 1$. Among π phases there are $\pi(k-1)$ independent concentrations. Including P and T, there are $2 + \pi(k - 1)$ independent intensive variables.

There are $k(\pi - 1)$ equations of equilibrium, Eq. (7.12). The *variance* of the system is the difference between the number of independent variables and the number of equations of equilibrium,

$$f \equiv 2 + \pi(k - 1) - k(\pi - 1) = 2 + k - \pi. \tag{7.13}$$

Equation (7.13) is the Gibbs phase rule.[11, p96] It specifies the number of intensive variables that can be independently varied without disturbing the number of coexisting phases ($f \geq 0$).

- $k = 1, \pi = 1 \implies f = 2$: a single substance in one phase. Two intensive variables can be independently varied; T and P in a gas.

- $k = 2, \pi = 1 \implies f = 3$: two substances in a single phase, as in a mixture of gases. We can independently vary T, P, and one mole fraction.

- $k = 1, \pi = 2 \implies f = 1$: a single substance in two phases; a single intensive variable such as the density can be varied without disrupting phase coexistence.

- $k = 1, \pi = 3 \implies f = 0$: a single substance in three phases; we cannot vary the conditions under which three phases coexist in equilibrium. Unique values of T and P define a triple point.

One should appreciate the *generality* of the phase rule, which doesn't depend on the *type* of chemical components, only that the Gibbs energy is a minimum in equilibrium.

7.1.3 The Clausius-Clapeyron equation

The latent heat, L, is the heat released or absorbed during a phase change, and is measured either as a molar quantity (per mole) or as a specific quantity (per mass). One has the latent heat of vaporization (boiling), fusion (melting), and sublimation.[11] Latent heats are also called the enthalpy of vaporization (or fusion or sublimation).[12] At a given T,

$$L(T) = h^v - h^l = T\left[s^v(T, P(T)) - s^l(T, P(T))\right], \tag{7.14}$$

---

[10]Lagrange multipliers in this context are reviewed in [3, p81].

[11]In sublimation the solid phase converts directly to the gaseous phase, without going through the liquid state.

[12]Measurements are made at fixed pressure, and enthalpy is the heat added at constant pressure; see Table 1.2.

where v and l refer to vapor and liquid, lower-case quantities such as $s \equiv S/n$ indicate molar values, and $P(T)$ describes the coexistence curve.[13] The difference in molar entropy between phases is denoted $\Delta s \equiv s^v - s^l$; likewise with Δh. Equation (7.14) is written compactly as $L = \Delta h = T\Delta s$.

As we move along a coexistence curve, T and P vary in such a way as to maintain the equality $\mu^I(T, P) = \mu^{II}(T, P)$. Variations in T and P induce changes $\delta\mu$ in the chemical potential, and thus along the coexistence curve, $\delta\mu^I = \delta\mu^{II}$. For a single substance, $d\mu = -sdT + vdP$ (Gibbs-Duhem equation). At a coexistence curve, $-s^I dT + v^I dP = -s^{II} dT + v^{II} dP$, implying

$$\left(\frac{dP}{dT}\right)_{\text{coexist}} = \frac{s^I - s^{II}}{v^I - v^{II}} = \frac{\Delta s}{\Delta v} = \frac{\Delta h}{T\Delta v} = \frac{L}{T\Delta v}. \tag{7.15}$$

Equation (7.15) is the *Clausius-Clapeyron equation*; it tells us the local slope of the coexistence curve. If one had enough data for L as a function of T, P and the volume change Δv, Eq. (7.15) could be integrated to obtain the coexistence curve in a P-T diagram.

7.2 THE CLASSIFICATION OF PHASE TRANSITIONS

Phase coexistence at a given value of (T, P) requires the equality $\mu^I(T, P) = \mu^{II}(T, P)$, but that says nothing about the continuity of *derivatives* of μ at coexistence curves. For a single substance, $\mu = G/N$, and thus from Eq. (1.14) (or the Gibbs-Duhem equation (P1.1)),

$$\frac{V}{N} = \left(\frac{\partial\mu}{\partial P}\right)_{T,N} \qquad \frac{S}{N} = -\left(\frac{\partial\mu}{\partial T}\right)_{P,N}. \tag{7.16}$$

For second derivatives, using Eqs. (1.33) and (P1.6), together with Eq. (7.16),[14]

$$\frac{\beta_T}{N} = -\frac{1}{V}\left(\frac{\partial(V/N)}{\partial P}\right)_{T,N} = -\frac{1}{V}\left(\frac{\partial^2\mu}{\partial P^2}\right)_{T,N}$$

$$\frac{C_P}{N} = T\left(\frac{\partial(S/N)}{\partial T}\right)_{P,N} = -T\left(\frac{\partial^2\mu}{\partial T^2}\right)_{P,N}. \tag{7.17}$$

The behavior of these derivatives at coexistence curves allows a way to classify phase transitions.

• If the derivatives in Eq. (7.16) are discontinuous at the coexistence curve (i.e., $\mu(T, P)$ has a kink at the coexistence curve), the transition is called *first order* and the specific volume and entropy are not the same between phases, $\Delta v \neq 0$, $\Delta s \neq 0$. The Clausius-Clapeyron equation is given in terms of the change in specific volume Δv and entropy, $\Delta s = L/T$. If a latent heat is involved, there are discontinuities in specific volume and entropy, and the transition is first order.

• If the derivatives in Eq. (7.16) are continuous at the coexistence curve, but higher-order derivatives are discontinuous, the transition is called *continuous*. For historical reasons, continuous transitions are also referred to as *second-order* phase transitions.[15] At continuous phase transitions, $\Delta v = 0$ and $\Delta s = 0$—there is no latent heat. Entropy is continuous, but its first derivative, the heat capacity, is discontinuous. Whether or not there is a latent heat seems to be the best way of distinguishing phase transitions.

---

[13] While Eq. (7.14) has been written in terms of the liquid-gas transition, it applies for any two coexisting phases.

[14] Because $\beta_T, C_P > 0$, the second derivatives in Eq. (7.17) are negative.

[15] The *Ehrenfest classification scheme*, introduced in 1933, proposed to classify phase transitions according to the lowest order of the differential coefficients of the Gibbs function that show a discontinuity. Pippard[82, p137] illustrates how C_P would behave at first, second, and third-order transitions. There's no reason phase transitions obeying the Ehrenfest scheme can't occur, yet the idea hasn't proved useful in practice, other than introducing the terminology of first and second-order phase transitions into the parlance of physics. By focusing only on discontinuities in thermodynamic derivatives, the scheme ignores the possibility of *divergences* in measurable quantities at phase transitions.

7.3 VAN DER WAALS LIQUID-GAS PHASE TRANSITION

The van der Waals equation of state, (6.32), is a cubic equation in the specific volume, V/N:

$$\left(\frac{V}{N}\right)^3 - \left(b + \frac{kT}{P}\right)\left(\frac{V}{N}\right)^2 + \frac{a}{P}\left(\frac{V}{N}\right) - \frac{ab}{P} = 0 . \qquad (7.18)$$

Cubic equations have three types of solutions: a real root together with a pair of complex conjugate roots, all roots real with at least two equal, or three real roots.[50, p17] The equation of state is therefore not necessarily *single-valued*;[16] it could predict several values of V/N for given values of P and T. For large V/N (how large?—see below), we can write Eq. (7.18) in the form

$$\left(\frac{V}{N}\right)^2\left[\frac{V}{N} - \left(b + \frac{kT}{P}\right) + \frac{aN^2}{PV^2}\left(\frac{V}{N} - b\right)\right] = 0 . \qquad (7.19)$$

Large V/N is the regime where the volume per particle is large compared with microscopic volumes, $V/N \gg (a/(kT), b)$ (the van der Waals parameter a has the dimension of energy-volume). In this limit, the nontrivial solution[17] of Eq. (7.19) is the ideal gas law, $PV = NkT$. For small V/N, we can guess a solution of Eq. (7.18) as $V/N \approx b$ (the smallest volume in the problem), valid at low temperature; see Exercise 7.3. (We can't have $V/N = b$, which implies infinite pressure.) For $V/N \approx b$, densities approach the close-packing density. We have no reason to expect the van der Waals model to be valid at such densities, yet it indicates a condensed phase, an incompressible form of matter with $V/N \approx$ constant, independent of P. At low temperature, $kT \ll Pb$, Eq. (7.18) has one real root, with a pair of complex conjugate roots (see Exercise 7.4).

7.3.1 The van der Waals critical point

Thus, we pass from the low temperature, high density system (low specific volume), in which Eq. (7.18) has one real root, to the high temperature, low density system (large specific volume) which has one nontrivial root (ideal gas law). We shouldn't be surprised if in between there is a regime for which Eq. (7.18) has multiple roots (which hopefully can be interpreted physically). Detailed studies show there is a unique set of parameters P_c, T_c, and $(V/N)_c$ (which as the notation suggests are the pressure, temperature, and specific volume of a critical point[18]), at which Eq. (7.18) has a triple root. For $(V/N)_c \equiv v_c$ a triple root of Eq. (7.18), the cubic equation

$$(v - v_c)^3 = v^3 - 3v_c v^2 + 3v_c^2 v - v_c^3 = 0 \qquad (7.20)$$

must be equivalent to Eq. (7.18) evaluated at P_c and T_c. Comparing coefficients of identical powers of v in Eqs. (7.18) and (7.20), we infer the correspondences

$$3v_c = b + \frac{kT_c}{P_c} \qquad 3v_c^2 = \frac{a}{P_c} \qquad v_c^3 = \frac{ab}{P_c} .$$

These equations imply the values of the critical parameters (show this):

$$v_c = 3b \qquad P_c = \frac{a}{27b^2} \qquad kT_c = \frac{8a}{27b} . \qquad (7.21)$$

The relations in Eq. (7.21) predict P_c and T_c reasonably well, but overestimate v_c (Exercise 7.5). Equation (7.21) also predicts $P_c v_c/(kT_c) = \frac{3}{8}$ (show this), a universal number, the same for all

---

[16]The ideal gas equation of state is a linear relation, $V/N = kT/P$. For nonlinear equations of state (such as in the van der Waals model), the possibility of multiple valued solutions is a way of life.

[17]For $a = 0$, the roots of Eq. (7.18) are $V/N = (0, 0, b + kT/P)$, and we're assuming $V/N \gg b$.

[18]Thus, there is a third variable associated with critical points—the density. What we see in a diagram like Fig. 7.1 is the projection of a three-dimensional P-T-v surface onto the P-T plane.

gases. In actuality, $P_c v_c/(kT_c) \approx 0.2\text{–}0.3$ for real gases.[19] Should we be discouraged by this disagreement? Not at all! The van der Waals model is the simplest model of interacting gases—if we want better agreement with experiment, we have to include more relevant physics in the model.

7.3.2 The law of corresponding states and the Maxwell equal area rule

Express $P, T,$ and v in units of the critical parameters. Let

$$\overline{P} \equiv P/P_c \qquad\qquad \overline{T} \equiv T/T_c \qquad\qquad \overline{v} \equiv v/v_c \,. \qquad (7.22)$$

The dimensionless quantities $\overline{P}, \overline{T}, \overline{v}$ are known as the *reduced* state variables of the system. With $P = P_c \overline{P}, T = T_c \overline{T},$ and $v = v_c \overline{v},$ the van der Waals equation of state can be written (show this)

$$\left(\overline{P} + \frac{3}{\overline{v}^2}\right)(3\overline{v} - 1) = 8\overline{T} \,. \qquad (7.23)$$

Equation (7.23) is a remarkable development: It's independent of material-specific parameters, indicating that when P, T, v are scaled in units of P_c, T_c, v_c (which are material specific), *the equation of state is the same.* Systems having their own values of (a, b), yet the same values of $\overline{P}, \overline{T}, \overline{v},$ are said to be in *corresponding states.* Equation (7.23) is the *law of corresponding states*: Systems in corresponding states behave the same. Thus, argon at $T = 300$ K, for which $T_c = 151$ K, will behave the same as ethane (C_2H_6) at $T = 600$ K, for which $T_c = 305$ K (both have $\overline{T} \approx 2$).[20] Figure 7.3 is a compelling illustration of the law of corresponding states. For ρ_l (ρ_g) the density of liquid (vapor) in equilibrium at $T < T_c$, by the law of corresponding states we expect ρ_l/ρ_c and ρ_g/ρ_c to be universal functions of T/T_c, where ρ_c denotes the critical density. Figure 7.3 shows data for eight substances (of varying complexity[21] from noble-gas atoms Ne, Ar, Kr, and Xe, to diatomic molecules N_2, O_2, and CO, to methane, CH_4), which, when plotted in terms of reduced variables, ostensibly fall on the same curve![83] The solid line is a fit to the data points of Ar (solid circles) assuming the critical exponent $\beta = \frac{1}{3}$ (see Section 7.4). Figure 7.3 shows the *coexistence region*—values of temperature and density for which phase coexistence occurs.[22]

Because the equation of state is predicted to be the same for all systems (when expressed in reduced variables), any thermodynamic properties derived from it are also predicted to be the same. *Universality* is the term used to indicate that the behavior of systems near critical points is largely independent of the details of the system.[23] Universality has emerged as a key aspect of modern theories of critical phenomena, as we'll see in Section 8.4.[24]

Figure 7.4 shows the isotherms calculated from Eq. (7.23). For $T \geq T_c$, there is a unique specific volume for every pressure—the equation of state is single valued.[25] For $T < T_c$, however, there are three possible volumes associated with a given pressure, such as the values of v/v_c shown at A, B, C for $T = 0.9T_c$. How do we interpret these multiple solutions?

---

[19] Values of T_c, P_c, v_c are listed in [57, pp6-37–6-51], from which one can calculate $P_c v_c/(kT_c)$.

[20] Note that comparisons of temperatures such as these are possible only when working with absolute temperature.

[21] Note that H and He are *not* included Fig. 7.3. The law of corresponding states holds for systems in which the distinction between Fermi-Dirac and Bose-Einstein statistics has a negligible effect. It also does not apply to highly polar molecules.

[22] The coexistence *curve* in Fig. 7.1 is the projection of the coexistence region (in the P-v plane) onto the P-T plane.

[23] Experimental evidence for universality is presented in the review article of Kadanoff *et al*[84].

[24] The occurrence of universality in the van der Waals model illustrates a general process: Discoveries applicable to a wide range of physical systems are often made through specific examples. Entropy was discovered through an analysis of the efficiency of heat engines, yet it's not restricted to heat engines—it's a universal feature of the physical world, applying to all forms of matter (solid, liquid, gas), but also to such disparate systems as information and black holes. Experimentally, the law of corresponding states applies to fluids that do not obey the van der Waals equation of state; the law of corresponding states is a more comprehensive theory. We note the distinction between the law of corresponding states, which applies anywhere on a phase diagram, and universality, which concerns systems near critical points.

[25] From the Gibbs phase rule, two intensive variables can be freely varied for a single-component gas—T and P; the density is uniquely determined by the equation of state.

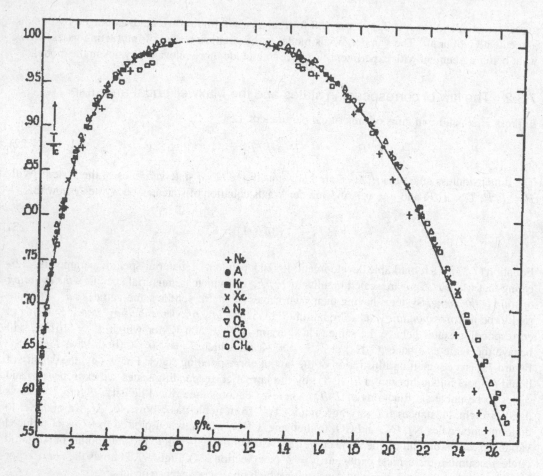

Figure 7.3: The law of corresponding states. Coexistence regions are the same when plotted in terms of reduced state variables. Reproduced from E.A. Guggenheim, J. Chem. Phys. **13**, p. 253 (1945), with the permission of AIP Publishing.

Figure 7.5 shows an expanded view of an isotherm for $T < T_c$. We note that $(\partial P/\partial v)_T$ is *positive* along the segment *bcd*, which is unphysical. An *increase* in pressure upon an isothermal expansion would imply a violation of the second law of thermodynamics. The root c in Fig. 7.5 (or B in Fig. 7.4) is unphysical and can be discarded.

What to make of the other two roots? Figure 7.6 indicates what we expect phenomenologically and hence what we want from a model of phase transitions. As a gas is compressed, pressure rises.[26] When (for $T < T_c$) the gas has been compressed to point B, *condensation* first occurs (coalescence of gas-phase atoms into liquid-phase clusters). At B, the specific volume v_g represents the maximum density a gas can have at $T < T_c$. Upon further isothermal compression, the pressure remains constant along the horizontal segment AB (at the *saturation pressure*, $P_{\text{sat}}(T)$) as more gas-phase atoms are absorbed into the liquid phase.[27] At point A, when all gas-phase atoms have condensed into the liquid, the specific volume $v = v_l$ is the minimum density a liquid can have at $T < T_c$. Further compression at this point results in a rapid rise in pressure.

---

[26] We know from experience that the pressure of compressed gases rises. More formally we know that a *requirement* for the stability of the equilibrium state is $(\partial P/\partial v)_T < 0$.

[27] The latent heat released upon the absorption of gas-phase atoms into the liquid is removed by the heat reservoir used to maintain the system at constant temperature.

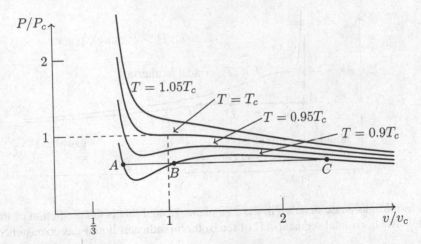

Figure 7.4: Isotherms of the van der Waals equation of state. Points A, B, and C are three possible volumes associated with the same pressure (and temperature).

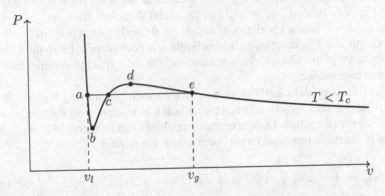

Figure 7.5: Subcritical isotherm of the van der Waals equation of state ($T = 0.85T_c$), with its unphysical segment bcd. The point c (root of the equation of state) is unphysical and can be discarded.

At a point in the coexistence region, such as point D in Fig. 7.6, there are N_l atoms in the liquid phase and N_g in the vapor phase, where $N = N_l + N_g$ is the total number of atoms. Let $x_l \equiv N_l/N$ and $x_g \equiv N_g/N$ denote the mole fractions of the amount of material in the liquid and gas phases. Clearly $x_l + x_g = 1$. The volume occupied by the gas (liquid) phase at this temperature is $V_g = N_g v_g$ ($V_l = N_l v_l$), implying $V = N_g v_g + N_l v_l$. The specific volume v_d associated with point D is:[28]

$$v_d = \frac{V}{N} = \frac{N_g v_g + N_l v_l}{N} = x_g v_g + x_l v_l = v_d (x_g + x_l) \, , \tag{7.24}$$

implying

$$x_l (v_d - v_l) = x_g (v_g - v_d) \, . \tag{7.25}$$

Equation (7.25) is the *lever rule*.[29] As $v_d \to v_g$, $x_l \to 0$, and vice versa. See Exercise 7.7.

---

[28] We're assuming the volumes occupied by liquid and gas-phase molecules, in a mixture of coexisting phases, are the same as their values in the pure phases at the same pressure and temperature (their values at the two sides of the coexistence region). Such a property doesn't hold in general—there is a *volume of mixing* effect, where the molar volume of substances in mixtures is different from that of the pure substances [3, p87]. An *ideal mixture* is one without any volume of mixing.

[29] Equation (7.25) is known as the lever rule in analogy with the *law of the lever* of basic mechanics in which each end of a balanced lever has the same value of the product of the force and the distance to the fulcrum.

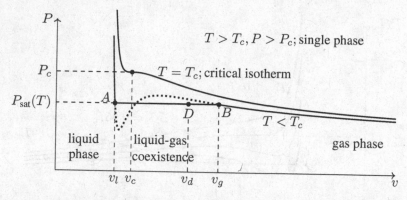

Figure 7.6: Liquid-gas phase diagram in the P-v plane. Dotted line is the prediction of the van der Waals model. The horizontal segment AB of the isotherm indicates liquid-gas coexistence.

Note that points A and B in Fig. 7.6, which occur at separate volumes in a P-v phase diagram, are, in a P-T phase diagram (such as Fig. 7.1), located *at the same point on a coexistence curve*. Along AB in Fig. 7.6, $\mu_l(T, P, v_l) = \mu_g(T, P, v_g)$, and therefore from $N d\mu = -S dT + V dP$ (Gibbs-Duhem), $d\mu = 0$ (phase coexistence) and $dT = 0$ (isotherm) implies $dP = 0$. A horizontal, subcritical isotherm in a P-v diagram indicates liquid-gas coexistence. From the Gibbs phase rule, for a single substance in two phases, one intensive variable (the specific volume) can be varied and not disrupt phase coexistence.

The van der Waals model clearly does not have isotherms featuring horizontal segments. In 1875, Maxwell suggested a way to reconcile undulating isotherms with the requirement of phase coexistence [85]. From the Gibbs-Duhem equation applied to an isotherm ($dT = 0$), $N d\mu = V dP$, and thus for an isotherm to represent phase coexistence, we require

$$0 = \mu_l - \mu_g = \int_l^g d\mu = \int_l^g v dP = \int_l^g [d(Pv) - Pdv] = (Pv)\Big|_l^g - \int_l^g P dv \,, \qquad (7.26)$$

implying

$$P_{\text{sat}}(T)(v_g - v_l) = \int_l^g P(v) dv \,. \qquad (7.27)$$

Equation (7.27) is the *Maxwell equal area rule*. It indicates that v_l, v_g, and $P_{\text{sat}}(T)$ should be chosen so that the area under the curve of an undulating isotherm is the same as the area under a horizontal isotherm.[30] Maxwell in essence modified the van der Waals model so that it describes phase coexistence:

$$P(v) = \begin{cases} \dfrac{kT}{v - b} - \dfrac{a}{v^2} & T \geq T_c, \text{all } v; T < T_c, v > v_g, v < v_l \\ P_{\text{sat}}(T) & T < T_c, v_l < v < v_g \,. \end{cases} \qquad (7.28)$$

The Maxwell rule is an ad hoc fix that allows one to locate the coexistence region in a P-v diagram. One could object that it uses an unphysical isotherm to draw physical conclusions. It's not a theory of phase transitions in the sense that condensation is shown to occur from the partition function.[31] In an important study, Kac, Uhlenbeck, and Hemmer showed rigorously that a one-dimensional model with a weak, long-range interaction (in addition to a hard-core, short-range potential), has the van der Waals equation of state *together* with the Maxwell construction.[88] This work was generalized by Lebowitz and Penrose to an arbitrary number of dimensions.[89] Models

---

[30]Details of the implementation of the Maxwell construction for the van der Waals model are given in Mazenko[86].
[31]The Percus-Yevick approximation mentioned in Section 6.7 *does* predict a liquid-vapor phase transition.[87]

featuring weak, long-range interactions are known as *mean field theories* (Section 7.9); the van der Waals-Maxwell model is within a *class* of theories, mean field theories. The energy per particle of the van der Waals gas is modified from its ideal-gas value ($\frac{3}{2}kT$) to include a contribution that's proportional to the density, $-an$, the hallmark of a mean-field theory; see Exercise 7.8.

7.3.3 Heat capacity of coexisting liquid-gas mixtures

At a point in the coexistence region (such as D in Fig. 7.6), the total internal energy of the system,

$$U = N_l u_l + N_g u_g , \qquad (7.29)$$

where u_l (u_g) is the *specific energy*, the energy per particle $u \equiv U/N$, of the liquid (gas) phase at the same pressure and temperature as the coexisting phases.[32] Dividing Eq. (7.29) by N,

$$u = x_l u_l + x_g u_g . \qquad (7.30)$$

Compare Eq. (7.30) with Eq. (7.24), which has the same form.

To calculate the heat capacity it suffices to calculate the specific heat, $c \equiv C/N$. Starting from $C_V = (\partial U/\partial T)_{V,N}$ (Eq. (1.38)), the specific heat $c_v \equiv C_V/N = (\partial u/\partial T)_v$, where $v = V/N$. One might think it would be a simple matter to find c_v by differentiating Eq. (7.30) with respect to T, presuming that the mole fractions x_l, x_g are temperature independent. The mole fractions, however, vary with temperature at fixed volume. As one increases the temperature along the line associated with fixed volume v_d in Fig. 7.6, x_l, x_g vary because the shape of the coexistence region changes with temperature; consider the lever rule, Eq. (7.25). Of course, $dx_l = -dx_g$ because of the constraint $x_l + x_g = 1$. We have, using Eq. (7.30),

$$c_v = \left(\frac{\partial u}{\partial T}\right)_v = x_l \left(\frac{\partial u_l}{\partial T}\right)_{\text{coex}} + x_g \left(\frac{\partial u_g}{\partial T}\right)_{\text{coex}} - (u_g - u_l)\left(\frac{\partial x_l}{\partial T}\right)_{\text{coex}} , \qquad (7.31)$$

where $(\partial/\partial T)_{\text{coex}}$ indicates a derivative taken at the sides of the coexistence region.[33] Evaluating these derivatives is relegated to Exercise 7.10. The final result is:

$$c_v = x_l\left[c_{v_l} - T\left(\frac{\partial P}{\partial v_l}\right)_T \left(\frac{\partial v_l}{\partial T}\right)^2_{\text{coex}}\right] + x_g\left[c_{v_g} - T\left(\frac{\partial P}{\partial v_g}\right)_T \left(\frac{\partial v_g}{\partial T}\right)^2_{\text{coex}}\right] . \qquad (7.32)$$

Equation (7.32) is a general result, not specific to the van der Waals model.

7.4 A BESTIARY OF CRITICAL EXPONENTS: $\alpha, \beta, \gamma, \delta$

As T approaches T_c from below, $T \to T_c^-$, the specific volumes v_g, v_l of coexisting gas and liquid phases tend to v_c, which we can express mathematically as $(v_g - v_c) \to 0$, $(v_c - v_l) \to 0$ as $(T_c - T) \to 0$. Can we *quantify* how $v_g, v_l \to v_c$ as $T \to T_c^-$? Because that's something that can be measured, and calculated from theory. The *critical exponent*[34] β characterizes how $v_g \to v_c$ as $T \to T_c^-$ through the relation $(v_g - v_c) \propto |T - T_c|^\beta$. Other quantities such as the heat capacity and the compressibility show singular behavior as $v_g, v_l \to v_c$ (and the distinction between liquid and gas disappears), *and each is associated with its own critical exponent*, as we'll see.

---

[32]Equation (7.29) ignores any "surface" interactions between liquid and gas phases; that was swept under the rug in our assumption of weakly interacting subsystems in Chapter 4, and is presumed for the existence of the thermodynamic limit.

[33]We're assuming the thermodynamic properties of liquid and vapor at a point in the coexistence region are the same as their values at the respective sides of the coexistence region. See Footnote 28.

[34]The notation for β as an exponent is standard; β here does not mean $(kT)^{-1}$ nor β_T, the isothermal compressibility.

Define deviations from the critical point (which are also referred to as reduced variables)

$$t \equiv \frac{T - T_c}{T_c} = \overline{T} - 1 \qquad \phi \equiv \frac{v - v_c}{v_c} = \overline{v} - 1 , \qquad (7.33)$$

so that $t \to 0^-$ as $T \to T_c^-$, and $\phi \to 0$, but can be of either sign as $T \to T_c^-$. Substituting $\overline{T} = 1 + t$ and $\overline{v} = 1 + \phi$ into the van der Waals equation of state (7.23), we find, to third order in small quantities,

$$\overline{P} = 1 + 4t - 6t\phi + 9t\phi^2 - \frac{3}{2}\phi^3 + O(t\phi^3, \phi^4) . \qquad (7.34)$$

We'll use Eq. (7.34) to show that $\phi \sim \sqrt{|t|}$ as $t \to 0$ (see Eq. (7.36)).

7.4.1 Shape of the critical coexistence region, the exponent β

The Maxwell rule can be used to infer an important property of the volumes v_l, v_g in the vicinity of the critical point. From its form in Eq. (7.26), $\int_l^g v \mathrm{d}P = 0$,

$$0 = \int_l^g \overline{v}\mathrm{d}\overline{P} = \int_l^g (\overline{v} - 1 + 1)\,\mathrm{d}\overline{P} = \int_l^g \phi \mathrm{d}\overline{P} + \overbrace{\int_l^g \mathrm{d}\overline{P}}^{0} = \int_l^g \phi \left(\frac{\partial \overline{P}}{\partial \phi}\right)_t \mathrm{d}\phi$$

$$= -\int_{\phi_l}^{\phi_g} \phi \left(6t + \frac{9}{2}\phi^2\right)\mathrm{d}\phi , \qquad (7.35)$$

where $\int_{P_l}^{P_g} \mathrm{d}P = 0$ for coexisting phases, we've used Eqs. (7.33) and (7.34), and we've retained terms up to first order in $|t|$ in $(\partial \overline{P}/\partial \phi)_t$ ($\phi \sim \sqrt{|t|}$). The integrand of the final integral in Eq. (7.35) is an odd function of ϕ under $\phi \to -\phi$, and the simplest way to ensure the vanishing of this integral for any $|t| \ll 1$ is to take $\phi_l = -\phi_g$, i.e., the limits of integration are symmetrically placed about $\phi = 0$. We can make this conclusion only sufficiently close to the critical point where higher-order terms in Eq. (7.34) can be neglected; close to the critical point the coexistence region is symmetric about v_c (which is not true in general—see Fig. 7.3).[35]

The symmetry $P(t, -\phi_g) = P(t, \phi_g)$ applied to Eq. (7.34) implies $3\phi_g \left(-4|t| + \phi_g^2\right) = 0$, or

$$\phi_g = 2\sqrt{|t|} \sim |T - T_c|^{1/2} . \qquad |t| \to 0 \qquad (7.36)$$

In the van der Waals model, therefore, $v_g \to v_c$ with the square root of $|T - T_c|$, and thus it predicts $\beta = \frac{1}{2}$. The measured value of β for liquid-gas critical points is $\beta \approx 0.32$–0.34 [90, 91]. We see from Fig. 7.3 that the coexistence region is well described using $\beta = \frac{1}{3}$.

7.4.2 The heat capacity exponent, α

How does the specific heat behave in the critical region? Let's approach the critical point along the *critical isochore*, a line of constant volume, $v = v_c$. As $T \to T_c^-$, $c_{v_g}, c_{v_l} \to c_{v_c}$, and $x_l, x_g \to \frac{1}{2}$ (use the lever rule and the symmetry of the critical region, $v_c - v_l = v_g - v_c$). From Eq. (7.32),

$$c_v(T_c^-) = c_v(T_c^+) - \frac{T_c}{2} \lim_{T \to T_c^-} \left[\left(\frac{\partial P}{\partial v_l}\right)_T \left(\frac{\partial v_l}{\partial T}\right)_{\text{coex}}^2 + \left(\frac{\partial P}{\partial v_g}\right)_T \left(\frac{\partial v_g}{\partial T}\right)_{\text{coex}}^2\right] , \qquad (7.37)$$

where we've set $c_v(T_c^+) \equiv \frac{1}{2}\left(c_{v_l} + c_{v_g}\right)$ as the limit $T \to T_c^+$, i.e., the terms in square brackets disappear for $T > T_c$. For the van der Waals gas, c_v has the ideal-gas value for $T > T_c$, $c_v^0 \equiv \frac{3}{2}k$ (see Exercise 7.8). Equation (7.37) is a general result, and is not specific to the van der Waals model.

---

[35] An approximate rule of thumb, the *law of rectilinear diameters*, is that the critical density is the average of the densities of the liquid and gas phases. It holds exactly if the coexistence region is symmetric about the critical point.

For the van der Waals model, we find using Eq. (7.36),

$$\left(\frac{\partial v}{\partial T}\right)_{\text{coex}} = \mp \frac{v_c}{T_c} \frac{1}{\sqrt{|t|}}, \tag{7.38}$$

where the upper (lower) sign is for gas (liquid). For the other derivative in Eq. (7.37), we can use Eq. (7.34),

$$\left(\frac{\partial P}{\partial v}\right)_T = \frac{P_c}{v_c}\left(\frac{\partial \overline{P}}{\partial \phi}\right)_t = \frac{P_c}{v_c}\left(6|t| - 18|t|\phi - \frac{9}{2}\phi^2\right) = -12\frac{P_c}{v_c}|t| + O(|t|^{3/2}), \tag{7.39}$$

where we've used Eq. (7.36). From Eq. (7.37), taking the limit,

$$c_v(T_c^-) - c_v^0 = -T_c\left(-12\frac{P_c}{v_c}|t|\right)\left(\frac{v_c}{T_c}\right)^2 \frac{1}{|t|} = 12\frac{P_c v_c}{T_c} = \frac{9}{2}k,$$

where we've used Eqs. (7.38) and (7.39), and that $P_c v_c / T_c = \frac{3}{8}k$ (see Eq. (7.21)). Thus,

$$c_v - c_v^0 = \begin{cases} \frac{9}{2}k & T \to T_c^- \\ 0 & T \to T_c^+ . \end{cases} \tag{7.40}$$

The specific heat is *discontinuous* at the van der Waals critical point—a second-order phase transition in the Ehrenfest classification scheme.

For many systems C_V does not show a discontinuity at $T = T_c$, it *diverges* as $T \to T_c$ (from above or below),[36] such as in liquid ^{4}He at the superfluid transition,[37] the so-called "λ-transition." To cover these cases, a heat capacity critical exponent α is introduced,[38]

$$C_V \sim |T - T_c|^{-\alpha} . \qquad T \to T_c \tag{7.41}$$

The value $\alpha = 0$ is assigned to systems showing a discontinuity in C_V at T_c, such as for the van der Waals model. The measured value of α for liquid-gas phase transitions is $\alpha \approx 0.1$. Extracting an exponent from measurements can be difficult. If a thermodynamic quantity $f(t)$ shows singular behavior as $t \to 0$ in the form $f(t) \sim At^x$, the exponent is obtained from the limit

$$x \equiv \lim_{t \to 0} \frac{\ln f(t)}{\ln t} . \tag{7.42}$$

If $x < 0$, $f(t)$ diverges at the critical point; if $x > 0$, $f(t) \to 0$ as $T \to T_c$.

From thermodynamics, $C_P > C_V$, Eq. (1.42). The heat capacity C_V for the van der Waals gas is the same as for the ideal gas (Exercise 7.8). C_P, however, has considerable structure for $T > T_c$; see Eq. (P7.4). In terms of reduced variables, Eq. (P7.4) is equivalent to (show this):

$$\frac{C_P - C_V}{Nk} = \frac{(1 + \phi)^3 (1 + t)}{\frac{3}{4}\phi^2 + \phi^3 + t(1 + \phi)^3} . \tag{7.43}$$

Because $\phi \to 0$ as $\sqrt{|t|}$, Eq. (7.36), we see that C_P diverges as $T \to T_c^+$, with

$$C_P \sim |T - T_c|^{-1} . \qquad (T \to T_c^+) \tag{7.44}$$

Equation (7.44) is valid for $T > T_c$; C_P isn't well defined in the coexistence region.[39]

---

[36] The existence of divergent thermodynamic quantities at the critical point obviates the Ehrenfest classification scheme.
[37] See London[92, p3], a great source of information about superfluid helium.
[38] One can introduce separate critical exponents to cover the cases $T \to T_c^+$ (exponent α) and $T \to T_c^-$ (exponent α').
[39] The latent heat $[dQ]_P = C_P dT$, but $dT = 0$ at constant P, implying C_P isn't defined in the coexistence region.

7.4.3 The compressibility exponent, γ

We've introduced two exponents that characterize critical behavior: α for the heat capacity and β for $v_g - v_l$. Is there a γ? One can show that at the van der Waals critical point,

$$\left.\frac{\partial P}{\partial v}\right|_{T=T_c,v=v_c} = \left.\frac{\partial^2 P}{\partial v^2}\right|_{T=T_c,v=v_c} = 0 . \tag{7.45}$$

The critical isotherm therefore has an inflection point at $v = v_c$, as we see in Figs. 7.4 or 7.6. The two conditions in Eq. (7.45) *define* critical points in fluid systems. One can show that

$$\left(\frac{\partial P}{\partial v}\right)_t = -\frac{2a}{27b^3}\left[\frac{\frac{3}{4}\phi^2 + \phi^3 + t(1+\phi)^3}{(1+\phi)^3(1+\frac{3}{2}\phi)^2}\right] . \tag{7.46}$$

Thus, $(\partial P/\partial v)_t \sim t$ as $t \to 0$. Its *inverse*, the compressibility $\beta_T = -(1/v)(\partial v/\partial P)_T$ therefore diverges at the critical point,[40] which is characterized by the exponent γ,

$$\beta_T \sim |T - T_c|^{-\gamma} . \tag{7.47}$$

The van der Waals model predicts $\gamma = 1$; experimentally its value is closer to $\gamma = 1.25$.

7.4.4 The critical isotherm exponent, δ

The equation of state on the critical isotherm is obtained by setting $t = 0$ in Eq. (7.34), $\overline{P} \approx 1 - \frac{3}{2}\phi^3$, implying

$$\overline{P} - 1 = \frac{P - P_c}{P_c} = -\frac{3}{2}\left(\frac{v - v_c}{v_c}\right)^3 \sim \left(\frac{v - v_c}{v_c}\right)^\delta , \tag{7.48}$$

where δ is the conventional symbol for the critical isotherm exponent. The van der Waals model predicts $\delta = 3$, whereas for real fluids [91], $\delta \approx 4.7$–5.

The definition of the critical exponents $\alpha, \beta, \gamma, \delta$ and their values are summarized in Table 7.1.

Table 7.1: Critical exponents $\alpha, \beta, \gamma, \delta$ for liquid-vapor phase transitions

Exponent	Definition	Van der Waals model	Measured
α	$C_V \sim |T - T_c|^{-\alpha}$	$\alpha = 0$	$\alpha \approx 0.1$
β	$v_g - v_l \sim |T - T_c|^\beta$	$\beta = \frac{1}{2}$	$\beta \approx 0.33$
γ	$\beta_T \sim |T - T_c|^{-\gamma}$	$\gamma = 1$	$\gamma \approx 1.25$
δ	$|P - P_c| \sim |V - V_c|^\delta$	$\delta = 3$	$\delta \approx 4.7 - 5$

7.5 WEISS MOLECULAR FIELD THEORY OF FERROMAGNETISM

We studied paramagnetism in Section 5.2, where independent magnetic moments interact with an external magnetic field. The hallmark of paramagnetism is that the magnetization M vanishes as the applied field is turned off, at any temperature. We saw in the one-dimensional Ising model that adding interactions between spins does not change the paramagnetic nature of that system, Eq. (6.84). Ferromagnets display *spontaneous magnetization*—a phase transition—in zero applied field, from a state of $M = 0$ for temperatures $T > T_c$, to one of $M \neq 0$ for $T < T_c$. The two-dimensional

---

[40]Because $(\partial v/\partial P)_T \sim \phi^{-2}$ as $\phi \to 0$ and $\phi \sim t^\beta$ as $t \to 0$, we would conclude that $\gamma = 2\beta$, which is true in the van der Waals model, but does not hold in general.

Ising model shows spontaneous magnetization (see Section 7.10), so that dimensionality is a relevant factor affecting phase transitions. Indeed, we'll show that one-dimensional systems cannot support phase transitions (Section 7.12). The two-dimensional Ising model has a deserved reputation for mathematical difficulty. Is there a simple model of spontaneous magnetization?

In 1907, P. Weiss introduced a model that bears his name, the *Weiss molecular field theory*. Figure 7.7 shows part of a lattice of magnetic moments. Weiss argued that if a system is magnetized,

Figure 7.7: Molecular field B_{mol} is produced by the surrounding magnetic dipoles of the system.

the aligned dipole moments of the system would produce an internal magnetic field, B_{mol}, the *molecular field*,[41] that's proportional to the magnetization, $B_{\text{mol}} = \lambda M$, where the proportionality constant λ will be inferred from the theory. The internal field B_{mol} would add to the external field, B. Reaching for Eq. (5.18) (the magnetization of independent magnetic moments),

$$M = N\langle \mu_z \rangle = N\mu L\left(\beta\mu(B + \lambda M)\right), \tag{7.49}$$

where $L(x)$ is the Langevin function. Equation (7.49) is a nonlinear equation of state, one that's typical of *self-consistent* fields where the system responds to the same field that it generates.[42]

Because we're interested in spontaneous magnetization, set $B = 0$ in Eq. (7.49):

$$M = N\mu L\left(\beta\mu\lambda M\right). \tag{7.50}$$

Does Eq. (7.50) possess solutions? It has the trivial solution $M = 0$ ($L(0) = 0$). If Eq. (7.50) is to represent a continuous phase transition, we expect that it has solutions for $M \neq 0$. Using the small-argument form of $L(x) \approx \frac{1}{3}x - \frac{1}{45}x^3$ (see Eq. (P5.1)), Eq. (7.50) has the approximate form for small M,

$$M \approx aM - bM^3 \quad \Longrightarrow \quad M\left(1 - a + bM^2\right) = 0, \tag{7.51}$$

where $a \equiv \beta N\mu^2\lambda/3$ and $b \equiv \beta^3\mu^4 N\lambda^3/45$ are positive if $\lambda > 0$. If $a \leq 1$, Eq. (7.51) has the trivial solution. If, however, $a > 1$, it has the nontrivial solution $M^2 \approx (a-1)/b$. Clearly $a = 1$ represents the critical temperature. Let[43]

$$kT_c \equiv \frac{1}{3}N\mu^2\lambda \quad \Longrightarrow \quad \lambda = \frac{3kT_c}{N\mu^2}. \tag{7.52}$$

Combining Eqs. (7.52) and (7.50), and letting $y \equiv M/(N\mu)$, we have the dimensionless equation:

$$y = L\left(3\frac{T_c}{T}y\right). \tag{7.53}$$

Figure 7.8 shows a graphical solution of Eq. (7.53). For $T \geq T_c$, there is only the trivial solution, whereas for $T < T_c$, there is a solution with $M \neq 0$ for $B = 0$. Spontaneous magnetization thus occurs in the Weiss model. Just as with liquid-vapor phase transitions, exponents characterize the critical behavior of magnetic systems. We now consider the critical exponents of the Weiss model.

---

[41] And, of course, if the system is not magnetized, randomly oriented dipoles produce no molecular field.

[42] Nonlinear theories are among the most difficult in physics. Finding the self-consistent charge distribution in semiconductor structures is a nonlinear problem—charge carriers respond to the same field they generate [93]. The general theory of relativity is nonlinear for the same reason—masses respond to the gravitational potential they generate.[13]

[43] $B_{\text{mol}} = \lambda M$ implies that λ must scale inversely with the size of the system (M extensive, B_{mol} intensive).

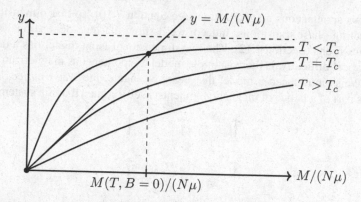

Figure 7.8: Graphical solution of Eq. (7.53). Spontaneous magnetization occurs for $T < T_c$.

7.5.1 The magnetization exponent, β

To find β is easy—it's implied by Eq. (7.51) for $T < T_c$. In the limit $T \to T_c^-$,

$$M^2 = \left(\frac{5}{3}N^2\mu^2\right)\left(\frac{T_c - T}{T_c}\right) \sim \left(\frac{T_c - T}{T_c}\right)^{2\beta}, \qquad (T \to T_c^-) \qquad (7.54)$$

implying $\beta = \frac{1}{2}$ for the Weiss model. Experimental values of β for magnetic systems are in the range 0.30–0.325 [94, 95].

7.5.2 The critical isotherm exponent, δ

At $T = T_c$, $M \to 0$ as $B \to 0$; the vanishing of M in characterized by the critical exponent δ, *defined* as $B \sim M^\delta$ at $T = T_c$. To extract the exponent, we need to "dig out" B from the equation of state, Eq. (7.49), for small M. Combining Eq. (7.52) with Eq. (7.49), where $y = M/(N\mu)$,

$$y = L\left(\beta\mu B + 3\frac{T_c}{T}y\right). \qquad (7.55)$$

We can isolate the field term B in Eq. (7.55) by invoking the *inverse Langevin function* $L^{-1}(y)$:

$$\beta\mu B = L^{-1}(y) - 3\frac{T_c}{T}y. \qquad (7.56)$$

A power series representation of L^{-1} can be derived from the *Lagrange inversion theorem* [50, p14], where for $-1 < y < 1$,

$$L^{-1}(y) = \sum_{k=1}^{\infty}\frac{y^k}{k!}\left[\frac{d^{k-1}}{dx^{k-1}}\left(\frac{x}{L(x)}\right)^k\right]_{x=0} \equiv \sum_{k=1}^{\infty}c_k y^k = 3y + \frac{9}{5}y^3 + \frac{297}{175}y^5 + O(y^7). \quad (7.57)$$

Additional coefficients c_k are tabulated in [96]. Combining the small-y form of L^{-1} from Eq. (7.57), with Eq. (7.56), we find

$$\beta\mu B = \frac{3t}{1+t}y + \frac{9}{5}y^3 + O(y^5), \qquad (7.58)$$

where t is defined in Eq. (7.33). On the critical isotherm $t = 0$, and thus from Eq. (7.58),

$$B = \left(\frac{9kT_c}{5\mu}\right)\left(\frac{M}{N\mu}\right)^3. \qquad (|M| \ll N\mu) \qquad (7.59)$$

Equation (7.59) should be compared with Eq. (7.48), the analogous result for fluids. The Weiss model predicts $\delta = 3$, whereas experimentally $\delta \approx 4$–5.

7.5.3 The susceptibility exponent, γ

The magnetic susceptibility $\chi = (\partial M/\partial B)_T$, see Eq. (1.50), is the magnetic analog of the compressibility of fluids, Eq. (1.33). For small M, we have by implicitly differentiating Eq. (7.58),

$$\beta\mu = \frac{t}{1+t}\frac{1}{N\mu}\chi + O\left(M^2\chi\right) .$$

We defined χ in Eq. (1.50) as the isothermal susceptibility, but the value of B was unrestricted. What's usually referred to as the magnetic susceptibility is the *zero-field* susceptibility. As $T \to T_c$,

$$\chi = \left(\frac{\partial M}{\partial B}\right)_{T,B=0} = \frac{N\mu^2}{3k}\frac{1}{T-T_c} \equiv \frac{C}{T-T_c} \sim |T-T_c|^{-\gamma} , \tag{7.60}$$

where $C = N\mu^2/(3k)$ is the Curie constant[44] for this system (see Section 5.2). Equation (7.60) is the *Curie-Weiss law*;[45] it generalizes the Curie law, Eq. (5.15), $M = CH/T$. The Weiss model predicts $\gamma = 1$.

7.5.4 The heat capacity exponent, α

From the Helmholtz energy $F = U - TS$ and the magnetic Gibbs energy $G_m = F - BM$ (see Section 1.9), we have the identities (using $dU = TdS + BdM$)

$$S = -\left(\frac{\partial F}{\partial T}\right)_M \qquad B = \left(\frac{\partial F}{\partial M}\right)_T$$

$$S = -\left(\frac{\partial G_m}{\partial T}\right)_B \qquad M = -\left(\frac{\partial G_m}{\partial B}\right)_T . \tag{7.61}$$

Combining Eqs. (P1.13) and (P1.14) with the results in Eq. (7.61),

$$C_M = T\left(\frac{\partial S}{\partial T}\right)_M = -T\left(\frac{\partial^2 F}{\partial T^2}\right)_M$$

$$C_B = T\left(\frac{\partial S}{\partial T}\right)_B = -T\left(\frac{\partial^2 G_m}{\partial T^2}\right)_B . \tag{7.62}$$

Which formula[46] should be used to calculate the heat capacity in the Weiss model?

Equation (7.58) is a power series for $B(M)$, which when combined with $B = (\partial F/\partial M)_T$ (Eq. (7.61)) provides an expression that can be integrated term by term:

$$B = \left(\frac{\partial F}{\partial M}\right)_T = \frac{1}{\beta\mu}\frac{3t}{1+t}y + \frac{9kT}{5\mu}y^3 + O(y^5) = \frac{3k}{N\mu^2}(T-T_c)M + \frac{9kT}{5\mu}\frac{1}{(N\mu)^3}M^3 + O(M^5),$$

implying,

$$F(T, M) = F_0(T) + \frac{3k}{2N\mu^2}(T-T_c)M^2 + \frac{9kT}{20\mu}\frac{1}{(N\mu)^3}M^4 + O(M^6) , \tag{7.63}$$

where $F_0(T)$ is the free energy of the *non-magnetic* contributions to the equilibrium state of the system "hosting" the magnetic moments (and which cannot be calculated without further information about the system).

---

[44] We defined the Curie constant in Eq. (5.16) as $C = N(\mu_B g)^2 J(J+1)/(3k)$, which agrees with the present result when $\mu = \mu_B\sqrt{J(J+1)}$ is recognized as the magnitude of the classical magnetic moment and $g = 1$ for classical spins.

[45] What we have been referring to as the critical temperature, T_c, is also known as the *Curie temperature*.

[46] Wouldn't it be easier to use $C_M = (\partial U/\partial T)_M$, Eq. (P1.11)? We've avoided finding the internal energy U in the Weiss model in favor of working from the equation of state, Eq. (7.49).

For $T \geq T_c$, $M = 0$, and for $T \lesssim T_c$, M^2 is given by Eq. (7.54), so that, from Eq. (7.63),

$$F(T) = \begin{cases} F_0(T) & T \geq T_c \\ F_0(T) - \frac{5}{2}NkT_c t^2 & T \to T_c^- \end{cases} \tag{7.64}$$

Using Eq. (7.62),

$$C_M - C^0 = \begin{cases} 0 & T \geq T_c \\ 5Nk & T \to T_c^- \end{cases} \tag{7.65}$$

where C^0 is the heat capacity of the non-magnetic degrees of freedom. Thus, there is a discontinuity in C_M for the Weiss model.

The critical exponent α for magnetic systems is, by definition, associated with C_B (not C_M) as $B \to 0$,

$$C_{B=0} \sim |T - T_c|^{-\alpha} . \tag{7.66}$$

One might wonder why C_B is used to define the exponent α, given the correspondence with fluid systems $V \leftrightarrow M$, $-P \leftrightarrow B$. For $T \geq T_c$, $B = 0$ implies $M = 0$, and for $T < T_c$ the "two phase" $M = 0$ heat capacity also corresponds to $B = 0$ because with ferromagnets there is up-down symmetry. For fluids, we considered the heat capacity of coexisting phases, each of which has a different heat capacity; see Eq. (7.32). For the Weiss model, we can use Eq. (1.51) to conclude that C_B also has a discontinuity at the critical point. Thus, $\alpha = 0$ for the Weiss model.

The critical exponents of the Weiss model have the same values as those in the van der Waals model. Magnets and fluids are different physical systems, yet they have the same critical behavior. Understanding why the critical exponents of different systems can be the same is the central issue of modern theories of critical phenomena; see Chapter 8. Before proceeding, we introduce two more exponents that are associated with correlations in the critical region.[47]

7.6 THE CRITICAL CORRELATION EXPONENTS: ν, η

The long-wavelength limit of the static structure factor for fluids is related to a thermodynamic quantity, β_T, Eq. (6.131) (there is a similar relation for magnets involving χ):

$$\lim_{q \to 0} S(q) = nkT\beta_T . \tag{7.67}$$

As $T \to T_c$, $\beta_T \to \infty$, Eq. (7.47), and thus $S(q = 0) = N^{-1} \int \int d^3r d^3r' \langle \delta n(r) \delta n(r') \rangle$ diverges[48] as $T \to T_c$, indicating that fluctuations become correlated over macroscopic distances,[49] *even though inter-particle interactions are short ranged*. The Ornstein-Zernike equation (6.139) is a model of how long-range correlations can develop through short-range interactions.[50]

The correlation function has the asymptotic form (Ornstein-Zernike theory, Eq. (6.144)),

$$g(r) \sim \frac{1}{R_0^2} \frac{e^{-r/\xi}}{r} , \tag{7.68}$$

where $R_0^2 = -nc_2$ and the correlation length $\xi = R_0/\sqrt{1 - nc_0}$, with c_0, c_2 moments of the direct correlation function $c(r)$, Eq. (6.142), $c_0 = \int c(r)d^3r$ and $c_2 = -\frac{1}{6}\int r^2 c(r)d^3r$. At the critical

---

[47]Critical phenomena—the behavior of physical systems near critical points—are characterized by six numbers: $\alpha, \beta, \gamma, \delta, \nu, \eta$. Six numbers isn't so bad—you learned the alphabet at some point in your life. If you want to do research in this field, you'll need to know these numbers for a handful of representative systems.

[48]The integrals in Eq. (6.131) are over an infinite volume (thermodynamic limit), yet they usually give finite results because the correlation function quickly decays to zero outside the correlation length ξ.

[49]And of course that's consistent with the interpretation of critical opalescence. See Section 6.7.

[50]Here's an urban analogy: Imagine a barking dog in your neighborhood. The audible range of a dog bark is finite, yet if the density of dogs in the town is sufficiently great, dog barking can become correlated over large distances.

point, $c_0 \to 1/n$, Eq. (6.141). Because R_0 is finite at the critical point,[51] we conclude that $\xi \to \infty$ as $T \to T_c$. The divergence of ξ as $T \to T_c$ is characterized by another critical exponent, ν, with

$$\xi \sim |T - T_c|^{-\nu} . \tag{7.69}$$

Like all critical exponents, ν has been introduced phenomenologically. Within the confines of Ornstein-Zernike theory, ν can be related to the exponent γ. Using Eq. (6.141), we find

$$\xi^2 = R_0^2 k T n \beta_T , \qquad \text{(Ornstein-Zernike)} \tag{7.70}$$

and thus, using Eqs. (7.47) and (7.69), $2\nu = \gamma$. This relation does not hold in general, however.

In Ornstein-Zernike theory, $S(q)$ has the form (from Eq. (6.142)),

$$S(q, T) = \frac{R_0^{-2}}{\xi^{-2} + q^2 + O(q^4)} . \tag{7.71}$$

At $T = T_c$, $S(q)$ diverges as $q \to 0$. This divergence is characterized by another (our final) critical exponent, η, such that

$$S(q, T = T_c) \sim q^{-2+\eta} . \tag{7.72}$$

Clearly in the Ornstein-Zernike model, $\eta = 0$. If one could always ignore the higher-order terms in Eq. (7.71) (the Ornstein-Zernike approximation), one would always have $\eta = 0$. It's found, however, that this is not justified in general. The necessity for introducing η is based on a more rigorous analysis of the correlation function that results from the inverse Fourier transform of the Ornstein-Zernike structure factor, particularly in regard to the role of the spatial dimension, d. There are two regimes:[97, p106]

(i) $r \gg \xi$. For fixed $T > T_c$ (and thus finite ξ), we have the asymptotic result for $r \to \infty$,

$$g(r) \sim \frac{e^{-r/\xi}}{r^{(d-1)/2}} \left[1 + O\left(\frac{d-3}{r/\xi}\right) \right] , \tag{7.73}$$

which clearly is of the form of Eq. (7.68) for $d = 3$.

(ii) $\xi \gg r$. For fixed r, as $T \to T_c^+$ (and hence as ξ gets large),

$$g(r) \sim \begin{cases} (\ln r) e^{-r/\xi} & d = 2 \\ \dfrac{e^{-r/\xi}}{r^{d-2}} & d \geq 3 . \end{cases} \tag{7.74}$$

The two limiting forms agree for $d = 3$, with $g(r) \sim e^{-r/\xi}/r$. They do not agree for $d > 3$, however. It would be natural to ask, who cares about dimensions higher than three? In the subject of critical phenomena it pays to understand the role of dimension, as we'll see. Something is clearly amiss for $d = 2$. For $T = T_c$ ($\xi \to \infty$), Eq. (7.74) indicates that correlations in a two-dimensional fluid would *increase* with distance, which is unphysical; Ornstein-Zernike theory breaks down in two dimensions. A qualitative conclusion from Eq. (7.74) (for $d \neq 2$) is that at $T = T_c$, correlations decay with distance inversely as a power of r: For $T \neq T_c$, correlations decay exponentially; for $T = T_c$, correlations decay algebraically.[52] In the two-dimensional Ising model, the two-spin correlation function evaluated at $T = T_c$ has the asymptotic form[98] $g(r) \sim r^{-1/4}$. To accommodate the actual behavior of the correlation function at $T = T_c$ (and not that predicted by Ornstein-Zernike theory), the exponent η is introduced in Eq. (7.72). The correlation function that results from inverse Fourier transformation of the structure factor shown in Eq. (7.72) has the form

$$g(r)|_{T=T_c} \sim r^{-(d-2+\eta)} . \tag{7.75}$$

For the two-dimensional Ising model, therefore, $\eta = \frac{1}{4}$.

---

[51] The quantity R_0 remains finite at the critical point because the direct correlation function $c(r)$ is short-ranged.

[52] If we set $d = 2$ in the form for $d \geq 3$ in Eq. (7.74), we have, formally, r^0 in the denominator. It's often the case that a logarithmic form corresponds to a power-law r^ψ for $\psi = 0$, as we see in Eq. (7.74) for $d = 2$.

7.7 LANDAU THEORY OF PHASE TRANSITIONS

The van der Waals treatment of the liquid-gas transition and the Weiss theory of ferromagnetism predict the same critical exponents, $\alpha = 0, \beta = \frac{1}{2}, \gamma = 1, \delta = 3$, the so-called *classical exponents*. Is that a coincidence? In 1937, L. Landau[53] proposed a model that purports to describe any phase transition (Landau and Lifshitz[99, Chapter 14]) and which predicts the classical exponents (as we'll show). Bearing in mind the classical exponents are not in agreement with experiment, the value of Landau theory could be questioned. Landau theory provides a unified language of phase transitions that's proven quite useful, and is a point of departure for more realistic theories.

An *order parameter*, let's call it ϕ, is a quantity that's identically zero for $T \geq T_c$, nonzero for $T < T_c$, and is a monotonically decreasing function of T as $T \to T_c^-$. In Bose-Einstein condensation (Section 5.9), the fraction of particles in the ground state is nonzero below the Bose temperature T_B, smoothly goes to zero as $T \to T_B^-$ (see Fig. 5.13), and is identically zero for $T \geq T_B$. For the liquid-gas transition, $\phi = v_g - v_l$ is the difference in specific volumes between coexisting phases; for the ferromagnet ϕ is the spontaneous (zero-field) magnetization. Other examples of order parameters and their associated phase transitions could be given.[54] Through the concept of order parameter, Landau theory subsumes the phase-transition behavior of different systems into one theory.

Systems become more ordered as the temperature is lowered, as random thermal motions become less effective in disrupting the contingent order implied by inter-particle interactions. It's a fact of experience that transitions occur abruptly at $T = T_c$ between a disordered high-temperature phase characterized by $\phi = 0$, and an ordered low-temperature phase with $\phi \neq 0$. For that reason, continuous phase transitions are referred to as *order-disorder* transitions. Another way of characterizing phase transitions is in terms of *symmetry breaking*. In ferromagnets, the spherical symmetry of the orientation of magnetic moments (for $T \geq T_c$ and zero field) is lost in the transition to a state of spontaneous magnetization for $T < T_c$ wherein moments single out a unique spatial direction. The concept of symmetry breaking is a unifying theme among different branches of physics—quantum field theory and cosmology, for example.[55]

7.7.1 Landau free energy

The central idea of Landau theory is that *the free energy is an analytic function*[56] of the order parameter ϕ, and as such possesses a Taylor expansion about $\phi = 0$. Denote with $F(T, \phi)$ the *Landau free energy*,[57] which under the assumption of analyticity can be written

$$F(T, \phi) = F_0(T) + a_2(T)\phi^2 + a_3(T)\phi^3 + a_4(T)\phi^4 + \cdots, \qquad (7.76)$$

where the expansion coefficients a_k are functions of T (and other thermodynamic variables) and $F_0(T)$ pertains to the system associated with $\phi = 0$. In equilibrium free energy is a minimum. The symbol ϕ in $F(T, \phi)$ is a placeholder for *conceivable* values of ϕ; the actual values of ϕ representing equilibrium states at temperature T are those that minimize $F(T, \phi)$. For this reason, $F(T, \phi)$ is re-

---

[53]Landau received the 1962 Nobel Prize in Physics for work in condensed matter theory, especially that of liquid helium.

[54]See the review article by Kadanoff *et al.*[84]

[55]The 2008 Nobel Prize in Physics was awarded to Y. Nambu, M. Kobayashi, and T. Maskawa for discoveries of spontaneous symmetry breaking in subatomic physics.

[56]There are two senses in which the word *analytic* is used. A formula is said to be analytic when we have a closed-form expression. Another sense in which the term is used, apropos of Landau theory, is that a function $f(x)$ is analytic at $x = x_0$ if it possesses a power-series representation for $|x - x_0| < r$, the interval of convergence of the series. An analytic function has derivatives of all orders, which are themselves analytic at $x = x_0$. See for example [16].

[57]The distinction between Gibbs and Helmholtz energies is dropped at the level of Landau theory; we simply refer to the Landau free energy. It's assumed that thermodynamic quantities can be found from the Landau free energy through differentiation *when evaluated in an equilibrium state*.

ferred to as the Landau free energy *functional*.[58] The free energy $F(T, \phi)$ has a minimum whenever two conditions are met:

$$\left(\frac{\partial F}{\partial \phi}\right)_T = 0 \qquad \left(\frac{\partial^2 F}{\partial \phi^2}\right)_T \geq 0 . \qquad (7.77)$$

There is no term linear in ϕ in Eq. (7.76), which would imply an equilibrium state with $\phi \neq 0$ for $T > T_c$. The types of terms appearing in Landau functionals reflect the symmetries of the system and the tensor character of the order parameter.[59] For magnetic systems, the cubic term can be ruled out. The order parameter (magnetization) is a vector that changes sign under time reversal,[60] yet the free energy must be time-reversal invariant—hence we eliminate the cubic term; $a_3 = 0$.

For $T > T_c$, equilibrium is represented by $\phi = 0$; $F(T, \phi)$ has a minimum at $\phi = 0$ if $a_2 > 0, a_4 > 0$. For $T < T_c$, the state of the system is represented by $\phi \neq 0$; $F(T, \phi)$ has a minimum for $\phi \neq 0$ if $a_2 < 0, a_4 > 0$. (We *require* $a_4 > 0$ for stability—as ϕ gets large, so does the free energy.) By continuity, therefore, at $T = T_c$, $a_2 = 0$. Figure 7.9 shows $F(T, \phi) - F_0(T)$ for $a_2 > 0$

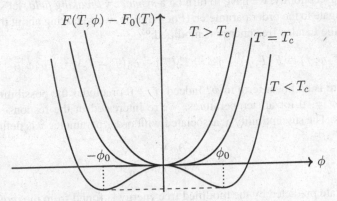

Figure 7.9: Landau functional $F(\phi)$ as a function of ϕ for $T > T_c$, $T = T_c$, and $T < T_c$. For $T \geq T_c$, equilibrium is associated with $\phi = 0$; for $T < T_c$, the equilibrium state is one of broken symmetry with $\phi = \pm\phi_0$. The dashed line indicates the convex envelope.

$(T \geq T_c)$, $a_2 = 0$ $(T = T_c)$, and $a_2 < 0$ $(T < T_c)$. We can "build in" the phase transition by taking[61]

$$a_2(T) = a_0(T)(T - T_c) , \qquad (7.78)$$

where $a_0(T)$ is a slowly varying, positive function of T. For magnetic systems,

$$F(T, \phi) = F_0(T) + a_0(T)(T - T_c)\phi^2 + a_4(T)\phi^4 + O(\phi^6) . \qquad (7.79)$$

Equation (7.79) should be compared with Eq. (7.63) (free energy of the Weiss ferromagnet); they have the same form *for small values of the order parameter*.[62] The free energy is a convex function of ϕ (see Section 4.3). The dashed line in Fig. 7.9 is the *convex envelope*, a line of points at which $\partial^2 F/\partial \phi^2 = 0$, connecting the two points at which $\partial F/\partial \phi = 0$.

---

[58] Just as the Lagrangian $L(q, \dot{q})$ is a function of all possible coordinates and velocities $(q, \dot{q})$, where the actual coordinates and velocities are selected through the Euler-Lagrange equation (C.18) to minimize the action, the Landau free energy $F(\phi)$ is like a Lagrangian in that it satisfies a variational principle to minimize the free energy; see Section 8.6.2 .

[59] The order parameter can be a scalar, vector, or tensor depending on the system (see for example de Gennes and Prost[100]), and there are ways of contracting tensors so as to produce invariants reflecting the symmetries of the system (the free energy is a scalar quantity).

[60] Currents would reverse upon a reversal of time, and therefore the magnetic field. Think Ampère's law.

[61] One should appreciate the phenomenological character of the Landau theory.

[62] Equation (7.63) comes from integrating the thermodynamic relation $B(M) = (\partial F/\partial M)_T$, where $B(M)$ is found from the equation of state in the form of a power series in M, Eq. (7.56), and as such, is known only for the first few terms associated with small values of the order parameter.

7.7.2 Landau critical exponents

7.7.2.1 Order parameter exponent, β

For $T < T_c$, we require $F(T, \phi)$ to have a minimum at $\phi = \phi_0 \neq 0$; from $(\partial F/\partial \phi)_{T,\phi=\phi_0} = 0$ we find that

$$\phi_0 = \pm\sqrt{\frac{a_0}{2a_4}(T_c - T)}. \qquad (T < T_c) \qquad (7.80)$$

Thus, $\beta = \frac{1}{2}$. Compare Eq. (7.80) (from Landau theory) with Eqs. (7.36) or (7.54), the analogous formulas in the van der Waals and Weiss theories. In Section 7.4.1, we referred to β as the "shape of the critical region" exponent; we didn't have the language of order parameter in place yet.

7.7.2.2 Susceptibility exponent, χ

To calculate the susceptibility, we have to turn on a *symmetry breaking field*, let's call it f, a generalized force conjugate to the order parameter. (For magnets, we're talking about the magnetic field.) For this purpose, the Landau free energy is modified:[63]

$$F(T, \phi; f) \equiv F(T, \phi) - f\phi = F_0(T) + a_0(T - T_c)\phi^2 + a_4\phi^4 - f\phi. \qquad (7.81)$$

Didn't we say there is no linear term in ϕ? Indeed, $f \neq 0$ precludes the possibility of a continuous phase transition: $\phi \neq 0$ for all temperatures. We're interested in the response of the system to infinitesimal fields. The susceptibility χ associated with order parameter ϕ is defined as

$$\chi \equiv \left(\frac{\partial \phi}{\partial f}\right)_{T, f=0}. \qquad (7.82)$$

The equilibrium state predicted by the modified free energy is found from the requirement

$$\left(\frac{\partial F}{\partial \phi}\right)_T = 0 = 2a_0(T - T_c)\phi + 4a_4\phi^3 - f. \qquad (7.83)$$

Equation (7.83) is the equation of state near the critical point. We can calculate χ by differentiating Eq. (7.83) implicitly with respect to f:

$$\left(\frac{\partial \phi}{\partial f}\right)_T \left[2a_0(T - T_c) + 12a_4\phi^2\right] - 1 = 0 \implies \chi = \left.\frac{1}{2a_0(T - T_c) + 12a_4\phi^2}\right|_{f=0}. \qquad (7.84)$$

Because we're to evaluate Eq. (7.84) for $f = 0$, we can use the known solutions for ϕ in that limit. We have

$$\chi = \begin{cases} [2a_0(T - T_c)]^{-1} & T > T_c \\ [4a_0(T_c - T)]^{-1} & T < T_c. \end{cases} \qquad (7.85)$$

Thus, $\gamma = 1$ in Landau theory.

7.7.2.3 Critical isotherm exponent, δ

At $T = T_c$, $\phi \to 0$ as $f \to 0$ such that $f \sim \phi^\delta$, implying from Eq. (7.83) with $T = T_c$,

$$f = 4a_4\phi^3. \qquad (7.86)$$

Thus, $\delta = 3$ in Landau theory.

---

[63]Equation (7.81) would appear to be the Gibbs free energy, a Legendre transformation of the Helmholtz free energy. In the context of Landau theory, the distinction between Helmholtz and Gibbs energies turns out not to be important.

7.7.2.4 Heat capacity exponent, α

By definition,

$$C_X \equiv -T \left(\frac{\partial^2 F}{\partial T^2} \right)_X . \tag{7.87}$$

See Eqs. (P1.5), (P1.6), and (7.62) for examples of C_X obtained from the second partial derivative of a thermodynamic potential with respect to T. Differentiate Eq. (7.81) twice (set $f = 0$) assuming that a_0, a_4 vary slowly with temperature and can effectively be treated as constants:

$$\frac{\partial^2 F}{\partial T^2} = \frac{\partial^2 F_0}{\partial T^2} + 2a_0(T - T_c) \left[\phi \frac{\partial^2 \phi}{\partial T^2} + \left(\frac{\partial \phi}{\partial T} \right)^2 \right] + 4a_0 \phi \frac{\partial \phi}{\partial T} + 4a_4 \phi^3 \frac{\partial^2 \phi}{\partial T^2} + 12a_4 \phi^2 \left(\frac{\partial \phi}{\partial T} \right)^2 .$$

Combining with Eq. (7.87), there is a discontinuity at $T = T_c$:

$$C_X(T_c^-) - C_X(T_c^+) = T_c \frac{a_0^2}{2a_4} , \tag{7.88}$$

where a_0, a_4 can be functions of the variables implied by X. Thus, we have a discontinuity and not a divergence; $\alpha = 0$ in Landau theory.

7.7.3 Analytic or nonanalytic, that is the question

The critical exponents predicted by Landau theory are not in good agreement with experiment.[64] Its key assumption, that the free energy function is analytic, *is not true at $T = T_c$*: partial derivatives of F do not generally exist at critical points. *Phase transitions are associated with points in thermodynamic state space where the free energy is singular.* The partition function, however,

$$Z(N, V, T) = \frac{1}{N! \lambda_T^{3N}} \int_V d\boldsymbol{r}_1 \cdots \int_V d\boldsymbol{r}_N \exp \left[-\beta \sum_{1 \leq i < j \leq N} v(|\boldsymbol{r}_i - \boldsymbol{r}_j|) \right]$$

is an analytic function of T for *finite N, V* (the integrand is an analytic function of β and the integral is over a finite domain). For statistical mechanics to have anything to say about phase transitions (at which $F = -kT \ln Z$ shows singular behavior), *we must consider infinite systems*, in particular the thermodynamic limit of $N, V \to \infty$ with N/V fixed. Up until the 1940s, it was not known whether the "algorithm" of statistical mechanics, the partition function, is sufficient to predict phase transitions. In 1944, L. Onsager[65] published an exact evaluation of the partition function for the $d = 2$ Ising model (see Section 7.10), showing that a continuous phase transition is indeed obtained in the thermodynamic limit.[102]

The free energy of the two-dimensional Ising model (after the thermodynamic limit has been taken) has the form near the critical point[103, p668]

$$F(T, M = 0) = F(T_c, 0) + a(T - T_c) + b(T - T_c)^2 \ln |T - T_c| + \cdots , \quad (d = 2 \text{ Ising}) \tag{7.89}$$

where a, b are constants. Because of the logarithmic term, $F(T, M = 0)$ does not possess a Taylor expansion about $T = T_c$. The heat capacity derived from Eq. (7.89) has a logarithmic singularity[66]

$$C \approx -2bT \ln |T - T_c| + \cdots . \quad (d = 2 \text{ Ising}) \tag{7.90}$$

The logarithmic singularity at $T = T_c$ (not a power law) implies $\alpha = 0$ (see Eq. (7.42)). A critical exponent of zero cannot distinguish a discontinuity as in Eq. (7.88) from a logarithmic singularity.

---

[64]What about the critical exponents ν, η associated with correlation functions? They can be calculated in an extension of Landau theory to include spatial inhomogeneities, *Ginzburg-Landau theory*, a topic we touch on in Section 8.6.3. It can be shown that $\nu = \frac{1}{2}$ and $\eta = 0$; see for example Goldenfeld[101, Section 5.7]. It should be noted that there *are* systems for which Landau theory provides a satisfactory description, such as superconductors and liquid crystals.

[65]Onsager received the 1968 Nobel Prize in Chemistry for work on the thermodynamics of irreversible processes.

[66]Recall that $\ln x \to -\infty$ as $x \to 0^+$. Heat capacity is always positive.

7.7.4 Internal energy in Landau theory

In the van der Waals and Weiss theories, we started with an equation of state (Eqs. (6.32) and (7.49)), and not a Hamiltonian (in essence a dividing line between thermodynamics and statistical mechanics). We found expressions for the free energy in these theories by integrating the equation of state; Eqs. (7.63) and (P7.1). Landau theory *starts* by positing the form of the free energy in the critical region. Can the internal energy be found if the free energy is known? Consider that

$$F = U - TS = U + T\left(\frac{\partial F}{\partial T}\right) \implies U = -T^2 \frac{\partial(F/T)}{\partial T} . \tag{7.91}$$

Equation (7.91) is the *Gibbs-Helmholtz equation* [3, p93]. If we know F, we can find U. Applying Eq. (7.91) to the Landau free energy, Eq. (7.79), we find, near the critical point,

$$U(T < T_c) = U(T > T_c) - a_0 T \phi_0^2 , \tag{7.92}$$

where ϕ_0 is the order parameter, Eq. (7.80). Equation (7.92) should be compared with Eq. (P7.3) (internal energy of the van der Waals gas)—in both cases the energy is reduced in the ordered phase by a term that uniformly couples to all particles of the system, the signature of a mean field theory.

7.8 MEAN-FIELD THEORY: UNCORRELATED FLUCTUATIONS

Landau theory posits a free energy function $F(T, \phi)$ for small values of the order parameter ϕ, such as in Eq. (7.79). Is there a type of system that, *starting from a Hamiltonian*,[67] is found to possess the Landau free energy in the critical region? As we show, Landau theory applies to systems in which the energy of interaction between fluctuations is negligible, in which fluctuations are independent. Consider a lattice model, which for simplicity we take to be the Ising model,

$$H = -\frac{1}{2}\sum_{i=1}^{N}\sum_{j=1}^{N} J_{ij}\sigma_i\sigma_j - b\sum_{i=1}^{N}\sigma_i , \tag{7.93}$$

where the factor of $\frac{1}{2}$ prevents overcounting. The Ising Hamiltonian in Eq. (7.93) is more general than that in Eq. (6.68). It's not restricted to one spatial dimension—it holds in any number of dimensions and it indicates that every spin interacts with every other spin. Usually one takes $J_{ij} = 0$ except when sites i, j are nearest neighbors. A model with interactions of arbitrary range, as in Eq. (7.93), allows us to discuss the molecular field.

Make the substitution $\sigma_i = \sigma_i - \langle\sigma\rangle + \langle\sigma\rangle \equiv \langle\sigma\rangle + \delta\sigma_i$ in the two-spin terms in Eq. (7.93):

$$\begin{aligned}
H &= -\frac{1}{2}\sum_{i=1}^{N}\sum_{j=1}^{N} J_{ij}\left[(\langle\sigma\rangle + \delta\sigma_i)(\langle\sigma\rangle + \delta\sigma_j)\right] - b\sum_{i=1}^{N}\sigma_i \\
&= -\frac{1}{2}\sum_{i=1}^{N}\sum_{j=1}^{N} J_{ij}\left[\langle\sigma\rangle^2 + \langle\sigma\rangle\delta\sigma_i + \langle\sigma\rangle\delta\sigma_j + \underbrace{\delta\sigma_i\delta\sigma_j}_{\text{ignore}}\right] - b\sum_{i=1}^{N}\sigma_i \\
&\approx -\frac{1}{2}\langle\sigma\rangle^2\sum_{i=1}^{N}\sum_{j=1}^{N} J_{ij} - \langle\sigma\rangle\sum_{i=1}^{N}\sum_{j=1}^{N} J_{ij}(\sigma_j - \langle\sigma\rangle) - b\sum_{i=1}^{N}\sigma_i \\
&= \frac{1}{2}\langle\sigma\rangle^2\sum_{i=1}^{N}\sum_{j=1}^{N} J_{ij} - \sum_{i=1}^{N}\left(b + \langle\sigma\rangle\sum_{j=1}^{N} J_{ij}\right)\sigma_i .
\end{aligned} \tag{7.94}$$

---

[67] If there's a Hamiltonian involved, we're doing statistical mechanics; otherwise it's thermodynamics.

The *mean field approximation* ignores couplings between fluctuations (as in Eq. (7.94)). Define $q \equiv \sum_{j=1}^{N} J_{ij}$, the sum of all coupling coefficients connected to site i, which because of translational invariance is independent of i. The value of q depends on the spatial dimension d of the space that the lattice is embedded in. For nearest-neighbor models, $q = 2J$ in one dimension, $q = 4J$ for the square lattice, and $q = 6J$ for the simple cubic lattice. The *mean field Hamiltonian* is[68]

$$H^{\mathrm{mf}} \equiv \frac{1}{2} N q \langle \sigma \rangle^2 - (b + q\langle \sigma \rangle) \sum_{i=1}^{N} \sigma_i \,. \tag{7.95}$$

Ignoring interactions between fluctuations therefore produces a model (like the Weiss model) in which each spin couples to the external field b and to an effective field $q\langle\sigma\rangle$ established by the other spins (the molecular field).[69] The partition function associated with Eq. (7.95) is easy to evaluate:

$$Z = e^{-\frac{1}{2}\beta N q \langle \sigma \rangle^2} \prod_{i=1}^{N} \left(\sum_{\sigma_i = -1}^{1} e^{\beta(b + q\langle\sigma\rangle)\sigma_i} \right) = e^{-\frac{1}{2}\beta N q \langle \sigma \rangle^2} \left(2 \cosh \left[\beta(b + q\langle\sigma\rangle) \right] \right)^N \,. \tag{7.96}$$

The free energy per spin is therefore ($F = -\beta \ln Z$)

$$\frac{1}{N} F = \frac{1}{2} q \langle \sigma \rangle^2 - kT \ln \left(2 \cosh \left[\beta(b + q\langle\sigma\rangle) \right] \right) \,. \tag{7.97}$$

Equation (7.97)—the mean-field free energy—is not manifestly in the Landau form.[70] To bring that out, we need to know the critical properties. From $\langle\sigma\rangle = -(1/N)(\partial F/\partial b)_T$, we find

$$\langle\sigma\rangle = \tanh \left[\beta(b + q\langle\sigma\rangle) \right] \,, \tag{7.98}$$

the same as Eq. (7.49) (Weiss theory), except that here we have a spin-$\frac{1}{2}$ system.[71] Equation (7.98) predicts[72] $T_c = q/k$ (see Exercise 7.19). The order parameter in this model is therefore:

$$\langle\sigma\rangle = \begin{cases} 0 & T \geq T_c \\ \sqrt{3} \dfrac{T}{T_c} \sqrt{\dfrac{T_c - T}{T_c}} & T < T_c \,. \end{cases} \tag{7.99}$$

For small $\langle\sigma\rangle$, Eq. (7.97) has the form, for $b = 0$ and expanding[73] $\ln(\cosh x)$,

$$\frac{1}{N} F = -kT \ln 2 + \frac{1}{2} k \left(\frac{T_c}{T} \right) (T - T_c) \langle\sigma\rangle^2 + \frac{1}{12} kT_c \left(\frac{T_c}{T} \right)^3 \langle\sigma\rangle^4 + O(\langle\sigma\rangle^6) \,, \tag{7.100}$$

which is in the Landau form, Eq. (7.79).

---

[68] Students sometimes say that the term $\frac{1}{2} N q \langle\sigma\rangle^2$ in Eq. (7.95) can be dropped because a constant added to a Hamiltonian has no effect on the dynamical behavior of the system. While true in classical mechanics, it's not the case here—see Exercise 7.17. The "constant" is a function of N, T. Such reasoning is a side effect of calling the Ising Hamiltonian a Hamiltonian, whereas it's an energy function of spin configurations. Ising spins are not canonical variables.

[69] It should not be construed that mean field theory is "just" a system of spins interacting with a magnetic field (which was covered in Section 5.2). The mean field is determined self-consistently through the equation of state, Eq. (7.98). It's often said that mean field theory "ignores fluctuations," which is true at the level of the Hamiltonian where the energy of interaction between fluctuations is ignored, but it's not the case that fluctuations are absent from mean field theory. The susceptibility in Landau theory diverges at the critical point, just not with the correct exponent.

[70] Mean field theory is more general than Landau theory; it reduces to the latter in the critical region.

[71] The Brillouin function $B_{\frac{1}{2}}(x) = \tanh x$, while the Langevin function $L(x) = \lim_{J \to \infty} B_J(x)$. See Exercise 5.9.

[72] Up to this point, the value of T_c has been taken as an experimentally determined quantity. The *existence* of a critical temperature is shown to occur within the models we have set up, with its *value* taken from experiment. In mean field theory, we have, for the first time, a *prediction* of T_c in terms of the microscopic energies of the system, $kT_c = q$. The accuracy of that prediction is another matter—see Table 7.2.

[73] For small x, $\ln(\cosh x) \approx \frac{1}{2}x^2 - \frac{1}{12}x^4 + O(x^6)$.

Mean-field theory predicts continuous phase transitions, even in one dimension, which is unphysical; see Section 7.12. For nearest-neighbor models, $q = zJ$, where z is the *coordination number* of the lattice, the number of nearest neighbors it has. For such models, the critical temperature predicted by mean field theory, $T_c = zJ/k$, would be the same for lattices with the same coordination number. Table 7.2 compares the mean-field critical temperature T_c^{mf} with the critical

Table 7.2: Mean field critical temperature T_c^{mf} and T_c for the nearest-neighbor Ising model on various lattices. $K_c \equiv J/(kT_c)$. Source: R.J. Creswick, H.A. Farach, and C.P. Poole, *Introduction to Renormalization Group Methods in Physics.*[104, p241]

Lattice	Dimension d	Coordination number z	$(T_c^{mf}/T_c) = zK_c$
Honeycomb	2	3	1.976
Square	2	4	1.763
Triangular	2	6	1.643
Diamond	3	4	1.479
Simple cubic	3	6	1.330
Body centered cubic	3	8	1.260
Face centered cubic	3	12	1.225

temperature T_c of the Ising model on various lattices.[74] The mean field approximation does not accurately account for the effects of dimensionality[75] for $d = 1, 2, 3$. Our question has been answered, however: Landau theory applies to systems in which the energy of interaction between spatially separated fluctuations is negligible—noninteracting fluctuations. The message is clear: If we want more accurate predictions of critical phenomena we have to include couplings between fluctuations.

7.9 WHEN IS MEAN FIELD THEORY EXACT?

Is there a type of system for which mean field theory is exact, wherein there's no need to invoke the mean field approximation? That question was answered in Section 7.3.2: The Lebowitz-Penrose theorem shows generally that mean field theory is obtained in models featuring infinite-ranged interactions, in any dimension. In this section, we illustrate how that comes about using the Ising model. We also show that mean field theory becomes exact in systems of spatial dimension $d > 4$.

7.9.1 Weak, long-range interactions

Consider an Ising model,

$$H = -2\frac{J}{N} \sum_{1 \leq i < j \leq N} \sigma_i \sigma_j , \qquad (7.101)$$

where the factor of N^{-1} keeps the energy extensive (H would scale as N^2 without it) and the factor of two is for convenience. In contrast to Eq. (7.93), the interaction strength is independent of inter-spin separation; all spins couple with the same strength regardless of separation (it's not a physical model, yet we learn something). The critical temperature of this model is finite, $kT_c = q = \sum_{i=1}^{N}(2J/N) = N(2J/N) = 2J$ (see Section 7.8). We note the identity for Ising spins:

$$\left(\sum_{i=1}^{N} \sigma_i\right)^2 = \sum_{i=1}^{N}\sum_{j=1}^{N} \sigma_i \sigma_j = \sum_{i=1}^{N}(\sigma_i)^2 + 2 \sum_{1 \leq i < j \leq N} \sigma_i \sigma_j = N + 2 \sum_{1 \leq i < j \leq N} \sigma_i \sigma_j .$$

---

[74] See any text on solid state physics for a discussion of crystal lattices.

[75] However, a trend is evident in Table 7.2: Mean field theory becomes more accurate, the greater the number of connections that spins have with the other spins of the lattice.

The Hamiltonian Eq. (7.101) can therefore be rewritten, exactly,

$$H = J - \frac{J}{N} \left(\sum_{i=1}^{N} \sigma_i \right)^2 . \tag{7.102}$$

The partition function associated with this Hamiltonian is:

$$Z = \prod_{i=1}^{N} \sum_{\sigma_i=-1}^{1} e^{-\beta H} = e^{-K} \prod_{i=1}^{N} \sum_{\sigma_i=-1}^{1} \exp \left(\frac{K}{N} \left(\sum_{i=1}^{N} \sigma_i \right)^2 \right) , \tag{7.103}$$

where $K \equiv \beta J$. Equation (7.103) appears to have presented us with a sum that can't be evaluated in closed form (in contrast to Eq. (7.96)), were it not for a well chosen identity,

$$\exp(a^2) = \frac{1}{\sqrt{2\pi}} \int_{-\infty}^{\infty} e^{-x^2/2} e^{\sqrt{2}ax} dx . \tag{7.104}$$

Equation (7.104) has a fancy name (even though it follows from completing the square in the argument of the integrand), the *Hubbard-Stratonovich transformation*,[76] an integral representation of the exponential of the square of a quantity a that involves the exponential of a term *linear* in a. The variable x in Eq. (7.104) has no direct physical meaning.

To make use of Eq. (7.104), let $a \equiv \sqrt{(K/N)} \sum_{i=1}^{N} \sigma_i$. Combining Eqs. (7.103) and (7.104),

$$Z = \frac{1}{\sqrt{2\pi}} e^{-K} \int_{-\infty}^{\infty} e^{-x^2/2} \prod_{i=1}^{N} \sum_{\sigma_i=-1}^{1} \exp \left[\left(\sqrt{\frac{2K}{N}} \sum_{i=1}^{N} \sigma_i \right) x \right] dx$$

$$= \frac{1}{\sqrt{2\pi}} e^{-K} \int_{-\infty}^{\infty} e^{-x^2/2} \left[2 \cosh \left(\sqrt{\frac{2K}{N}} x \right) \right]^N dx \tag{7.105}$$

$$= 2^N \sqrt{\frac{N}{2\pi}} e^{-K} \int_{-\infty}^{\infty} \left[e^{-y^2/2} \cosh \left(\sqrt{2K} y \right) \right]^N dy \equiv 2^N \sqrt{\frac{N}{2\pi}} e^{-K} \int_{-\infty}^{\infty} e^{Ng(y)} dy ,$$

where we've changed variables, $x = \sqrt{N} y$, and where

$$g(y) = \ln \left(e^{-y^2/2} \cosh \left(\sqrt{2K} y \right) \right) = -\frac{y^2}{2} + \ln \cosh \left(\sqrt{2K} y \right) . \tag{7.106}$$

The final step in Eq. (7.105) sets us up for the *method of steepest descent*, a method for approximately evaluating integrals of the form $\int_{-\infty}^{\infty} e^{Ng(y)} dy$, where N is large and $g(x)$ has a local maximum.[16, p233] The method requires us to find the maximum of $g(y)$. From Eq. (7.106),

$$g'(y) = -y + \sqrt{2K} \tanh(\sqrt{2K} y) = 0 \implies y_0 = \sqrt{2K} \tanh(\sqrt{2K} y_0) \tag{7.107}$$

such that $g'(y_0) = 0$. By a method of analysis that should be familiar by now, Eq. (7.107) has a nontrivial solution for $K > \frac{1}{2}$ and the trivial solution for $K < \frac{1}{2}$, implying $kT_c = 2J$. For $K \gtrsim \frac{1}{2}$, the solution of Eq. (7.107) has the form,

$$y_0 \sim \sqrt{3 \left(\frac{T}{T_c} \right) \left(\frac{T_c - T}{T_c} \right)} \sim \sqrt{-3t} , \qquad (T \to T_c^-, t \to 0^-) \tag{7.108}$$

---

[76] The Hubbard-Stratonovich transformation (one of the most useful tricks in the toolbox of theoretical physics) is used to convert problems involving interacting degrees of freedom to an integration or sum over non-interacting degrees of freedom. It's used in the *bosonization* of interacting fermions—the partition function for a system of interacting fermions is transformed to a sum over noninteracting bosons—a topic beyond the intended scope of this book.

virtually the same as Eq. (7.99). We must verify that $g(y)$ has a maximum at $y = y_0$. It's shown in Exercise 7.20 that $g''(y_0) < 0$ for $K > \frac{1}{2}$ and $g''(0) < 0$ for $K < \frac{1}{2}$. It's straightforward to show that $g''(y_0) = 2K - 1 - y_0^2$. In the vicinity of $y = y_0$, we have the first few terms of a Taylor series:

$$g(y) \approx g(y_0) + (y - y_0)g'(y_0)^{\,0} - \frac{1}{2}(y - y_0)^2 |g''(y_0)| \ .$$

Therefore,

$$\int_{-\infty}^{\infty} e^{Ng(y)} dy \approx e^{Ng(y_0)} \int_{-\infty}^{\infty} e^{-\frac{1}{2}N|g''(y_0)|(y-y_0)^2} dy = 2\sqrt{\frac{2\pi}{N|g''(y_0)|}} e^{Ng(y_0)} \ . \qquad (7.109)$$

where we've used a Gaussian integral (Appendix B). Combining Eq. (7.109) with Eq. (7.105),

$$Z = 2\frac{e^{-K}}{\sqrt{|g''(y_0)|}} 2^N e^{Ng(y_0)} \ . \qquad (7.110)$$

The free energy per spin, is, in the thermodynamic limit,

$$\lim_{N \to \infty} \left(\frac{F}{N} \right) = -kT \lim_{N \to \infty} \left(\frac{1}{N} \ln Z \right) = \frac{1}{2} kT y^2 - kT \ln \left(2\cosh(\sqrt{2K}y) \right) , \qquad (7.111)$$

which should be compared with Eq. (7.97) (for $b = 0$). The two formulas agree if we make the correspondence between the order parameter $\langle \sigma \rangle$ and that of the infinite range model, $y_0 \leftrightarrow \sqrt{2K}\langle \sigma \rangle$.

7.9.2 The Ginzburg criterion; upper critical dimension

Mean field theory results either from ignoring interactions between fluctuations (Section 7.8) or by having all spins coupled with the same interaction strength[77] (Section 7.9.1); each is equivalent in its predictions with that of Landau theory in the critical region, which, as we've noted, are not in good agreement with experiment. Away from the critical region, however, mean field theory does an adequate job in treating the thermodynamic properties of interacting systems.[78] Is there a physical argument why Landau theory fails in the critical region?

In 1961, V.L. Ginzburg offered a criterion,[79] the *Ginzburg criterion*, for the conditions under which the approximation of uncorrelated fluctuations is justified.[105] Consider the following ratio,

$$R \equiv \frac{\int_{\xi^d} g(r) d^d r}{\int_{\xi^d} \phi^2(r) d^d r} , \qquad (7.112)$$

where $g(r) \equiv \langle \delta\phi(0)\delta\phi(r) \rangle$ denotes the two-spin correlation function, and $\phi(r)$ is the order parameter.[80] The numerator in Eq. (7.112) is an average over a d-dimensional region whose linear dimension is of order ξ; it's important not to average over a region larger than ξ^d, otherwise we have uncorrelated fluctuations. The denominator is a measure of the square of the order parameter, ϕ^2, averaged over the same volume, ξ^d. The ratio R characterizes the strength of correlated fluctuations in a d-dimensional ball of radius ξ, $\langle (\delta\phi)^2 \rangle_\xi$, relative to the square of the order parameter, $\langle \phi^2 \rangle_\xi$ averaged over the same hypervolume. If $R \ll 1$ (Ginzburg criterion), Landau theory applies; if not, it fails in the critical region.[81]

---

[77] Mean field theory is not unique; there is more than one way to convert a model of interacting degrees of freedom to one in which the effects of interactions are replaced with an approximate average field.

[78] Given the relative simplicity of mean field theory, it's the first approach one tries in investigating phase transitions.

[79] Ginzburg received the 2003 Nobel Prize in Physics for work on superconductors and superfluids.

[80] In Landau theory the order parameter ϕ is independent of position; in Ginzburg-Landau theory the order parameter $\phi(r)$ is spatially varying; see Section 8.6.3.

[81] See the article by Amit for a deeper look at the Ginzburg criterion, [106].

In the critical region, the denominator in Eq. (7.112) can be approximated $\int_{\xi^d} \phi^2(r) d^d r \sim \xi^d t^{2\beta} \sim \xi^{d-(2\beta/\nu)}$, where we've used $\phi \sim t^\beta$ along with $\xi \sim t^{-\nu}$, where t is the reduced temperature. Similarly, for the numerator $\int_{\xi^d} g(r) d^d r \sim \chi \sim t^{-\gamma} \sim \xi^{\gamma/\nu}$. The ratio R in Eq. (7.112) therefore scales with ξ as

$$R \sim \xi^{-d+(\gamma+2\beta)/\nu} . \tag{7.113}$$

For $d > (\gamma + 2\beta)/\nu$, $R \to 0$ as $\xi \to \infty$, and Landau theory is valid. Using the classical exponents, $(\gamma + 2\beta)/\nu = 4$; *Landau theory gives a correct description of critical phenomena in systems for which $d > 4$.* For $d < 4$, the Ginzburg criterion is not satisfied and Landau theory does not apply in the critical region. The case of $d = 4$ is marginal and requires more analysis; Landau theory is not quite correct in this case. The special dimension $d = 4$ is referred to as the *upper critical dimension*.

7.10 THE TWO-DIMENSIONAL ISING MODEL

The two-dimensional Ising model occupies an esteemed place in statistical mechanics as one of the few exactly solvable, nontrivial models of phase transitions; no textbook on the subject can be complete without it. Onsager's 1944 derivation is a mathematical tour de force, of which the search for simplifications has become a field of research, but even the simplifications are rather complicated. Books have been written on the two-dimensional Ising model [107, 108]. We outline the method of solution and then examine the thermodynamic properties of this model.

7.10.1 Partition function, free energy, internal energy, and specific heat

Figure 7.10 shows an $N \times M$ array of Ising spins $\sigma_{n,m}$ in the nth row and m^{th} column, $1 \le n \le N$,

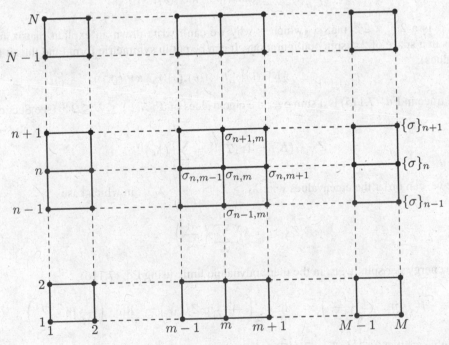

Figure 7.10: $N \times M$ array of Ising spins $\sigma_{n,m}$. All spins in row p are denoted $\{\sigma\}_p \equiv \{\sigma_{p,m}\}\big|_{m=1}^{M}$.

$1 \le m \le M$. The nearest-neighbor Hamiltonian can be written[82]

$$H = -J \underbrace{\sum_{n=1}^{N} \sum_{m=1}^{M-1} \sigma_{n,m}\sigma_{n,m+1}}_{\text{intra-row couplings}} -J \underbrace{\sum_{n=1}^{N-1} \sum_{m=1}^{M} \sigma_{n,m}\sigma_{n+1,m}}_{\text{inter-row couplings}} . \qquad (7.114)$$

Periodic boundary conditions are assumed (the lattice is on the surface of a torus), and thus the summation limits in Eq. (7.114), $N-1$, $M-1$ can be replaced with N, M. Give names to the terms containing intra and inter-row couplings,[83]

$$V_1\left(\{\sigma\}_n\right) \equiv \sum_{m=1}^{M} \sigma_{n,m}\sigma_{n,m+1} \qquad V_2\left(\{\sigma\}_n, \{\sigma\}_{n+1}\right) \equiv \sum_{m=1}^{M} \sigma_{n,m}\sigma_{n+1,m} ,$$

where $\{\sigma\}_n \equiv (\sigma_{n,1}, \cdots, \sigma_{p,M})$ denotes all spins in the nth row (see Fig. 7.10). We must evaluate the sum

$$Z_{N,M}(K) = \sum_{\{\sigma\}_1, \cdots, \{\sigma\}_N} \exp\left[K\left(\sum_{n=1}^{N} V_1\left(\{\sigma\}_n\right) + V_2\left(\{\sigma\}_n, \{\sigma\}_{n+1}\right)\right)\right] \qquad (7.115)$$

$$\equiv \sum_{\{\sigma\}_1, \cdots, \{\sigma\}_N} T_{\{\sigma\}_1,\{\sigma\}_2} T_{\{\sigma\}_2,\{\sigma\}_3} \cdots T_{\{\sigma\}_{N-1},\{\sigma\}_N} T_{\{\sigma\}_N,\{\sigma\}_1} = \text{Tr}\,(\boldsymbol{T})^N ,$$

where $K \equiv \beta J$ and we've introduced the transfer matrix $\boldsymbol{T}$ (see Section 6.5), with elements

$$T_{\{\sigma\},\{\sigma\}'} = e^{KV_1(\{\sigma\})} e^{KV_2(\{\sigma\},\{\sigma\}')} .$$

$T_{\{\sigma\},\{\sigma\}'}$ is a $2^M \times 2^M$ matrix (which is why we can't write down an explicit matrix form); it operates in a space of 2^M spin configurations. It can be put in symmetric form (and thus it has real eigenvalues):

$$T_{\{\sigma\},\{\sigma\}'} = e^{\frac{1}{2}KV_1(\{\sigma\})} e^{KV_2(\{\sigma\},\{\sigma\}')} e^{\frac{1}{2}KV_1(\{\sigma\}')} .$$

The trace in Eq. (7.115) is a sum over the eigenvalues of $\boldsymbol{T}$, λ_k, $1 \le k \le 2^M$ (see Section 6.5):

$$Z_{N,M}(K) = \text{Tr}\,(\boldsymbol{T})^N = \sum_{k=1}^{2^M} (\lambda_k)^N .$$

Assume we can order the eigenvalues with $\lambda_1 \ge \lambda_2 \ge \cdots \ge \lambda_{2^M}$, in which case

$$Z_{N,M} = \lambda_1^N \sum_{k=1}^{2^M} \left(\frac{\lambda_k}{\lambda_1}\right)^N . \qquad (7.116)$$

The free energy per spin, ψ, is, in the thermodynamic limit, using Eq. (7.116),

$$-\beta\psi \equiv -\beta \lim_{M,N\to\infty} \left(\frac{F}{MN}\right) = \lim_{M,N\to\infty} \left(\frac{1}{MN} \ln Z_{N,M}\right) = \lim_{M\to\infty} \left(\frac{1}{M} \ln \lambda_1^{(M)}\right) , \qquad (7.117)$$

where we've written $\lambda_1^{(M)}$ in Eq. (7.117) to indicate that it's the largest eigenvalue of the $2^M \times 2^M$ transfer matrix. "All" we have to do is find the largest eigenvalue $\lambda_1^{(M)}$ in the limit $M \to \infty$!

---

[82]One could define Eq. (7.114) with anisotropic couplings, J_x, J_y. We treat the simplest case of $J = J_x = J_y$.

[83]We're going to build a transfer matrix that "transfers" between rows of the lattice; it could just as well be constructed so that it transfers between columns.

Through a masterful piece of mathematical analysis, Onsager did just that. He showed, for large M, that

$$\lambda_1^{(M)}(K) = (2\sinh 2K)^{M/2} \, e^{\frac{1}{2}(\gamma_1(K)+\gamma_3(K)+\cdots+\gamma_{2M-1}(K))} \,, \qquad (7.118)$$

where

$$\cosh\gamma_k(K) \equiv \cosh 2K \coth 2K - \cos\left(\frac{\pi k}{M}\right) . \qquad (7.119)$$

A derivation of Eq. (7.118) is given in Thompson.[109, Appendix D] Combining Eqs. (7.118) and (7.117),

$$-\beta\psi = \frac{1}{2}\ln(2\sinh 2K) + \lim_{M\to\infty}\left(\frac{1}{2M}\sum_{k=0}^{M-1}\gamma_{2k+1}\right) . \qquad (7.120)$$

As $M\to\infty$, the sum in Eq. (7.120) approaches an integral (let $\pi k/M \approx \theta$):

$$-\beta\psi = \frac{1}{2}\ln(2\sinh 2K) + \frac{1}{2\pi}\int_0^\pi \cosh^{-1}[\cosh 2K \coth 2K - \cos\theta]\,\mathrm{d}\theta \,, \qquad (7.121)$$

where $\cosh^{-1}$ comes from Eq. (7.119) for γ_k. The integral in Eq. (7.121) would appear hopeless, were it not for a well chosen identity.[84] An integral representation of inverse hyperbolic cosine is:[85]

$$\cosh^{-1}|z| = \frac{1}{\pi}\int_0^\pi \ln[2(z - \cos\phi)]\mathrm{d}\phi \,. \qquad (7.122)$$

Combine Eq. (7.122) with Eq. (7.121), and we have an expression for the free energy per spin,

$$-\beta\psi = \frac{1}{2}\ln(2\sinh 2K) + \frac{1}{2\pi^2}\int_0^\pi\int_0^\pi \ln[2(\cosh 2K \coth 2K - \cos\theta - \cos\phi)]\,\mathrm{d}\theta\mathrm{d}\phi \,. \qquad (7.123)$$

It's a simple step to put all terms in Eq. (7.123) associated with K "inside" the integral:

$$-\beta\psi = \ln 2 + \frac{1}{2\pi^2}\int_0^\pi\int_0^\pi \ln\left[\cosh^2 2K - \sinh 2K\,(\cos\theta + \cos\phi)\right]\,\mathrm{d}\theta\mathrm{d}\phi \,. \qquad (7.124)$$

Examining Eq. (7.124) (which holds for any T), we find that the integrand has a logarithmic singularity at the origin ($\theta, \phi \to 0$) when the temperature has a special value T_c such that $\cosh^2 2K_c = 2\sinh 2K_c$ ($K_c \equiv J/(kT_c)$), *which we identify as the critical temperature*. There is a unique solution of $\cosh^2 2K_c = 2\sinh 2K_c$,

$$\sinh 2K_c = 1 \,, \qquad (7.125)$$

implying $K_c \approx 0.4407$. The largest eigenvalue of a *finite* transfer matrix is an analytic function of K (see Section 6.5.3). Given that the free energy is related to the largest eigenvalue, and that phase transitions are associated with singularities in the free energy (Section 7.7.3), we conclude there is no phase transition in one dimension for $T > 0$ (see Section 7.12). For the two-dimensional Ising model, the largest eigenvalue of the transfer matrix becomes singular in the thermodynamic limit[86] for $K = K_c$.

---

[84]See the same remark near Eq. (7.104).

[85]Equation (7.122) is not a standard identity. The inverse hyperbolic cosine has a logarithmic representation, $\cosh^{-1}|z| = \ln[z + \sqrt{z^2 - 1}]$,[50, p87] the same as the value of the integral in Eq. (7.122). Because that integral shows up again in the Onsager solution, we note that $\int_0^\pi \ln(a + b\cos x)\,\mathrm{d}x = \pi\ln\left(\frac{1}{2}\left(a + \sqrt{a^2 - b^2}\right)\right)$, $a > |b| > 0$. The integrand is periodic with period 2π, so the limit of integration can be extended to 2π, doubling the value of the integral.

[86]The finite sum in Eq. (7.118) specifies an analytic function $\lambda_1^{(M)}(K)$ for all K and finite M; the resulting integral in Eq. (7.124) for $M \to \infty$ has a singularity when $\cosh^2 2K = 2\sinh 2K$. In taking limits, mathematical properties can be lost. Consider the Taylor series for e^x. Set $x = 1$. For any finite number of terms in the series, the result specifies a rational number. Only in the limit of an infinite number of terms does one obtain the irrational number, e. I thank Professor Chris Frenzen for that observation.

Equation (7.124) can be simplified through the trigonometric identity $\cos\theta + \cos\phi = 2\cos\left(\frac{1}{2}(\theta+\phi)\right)\cos\left(\frac{1}{2}(\theta-\phi)\right)$, which suggests a change of variables. Let

$$\theta_1 \equiv \frac{1}{2}(\theta+\phi) \qquad\qquad \theta_2 \equiv \theta - \phi. \qquad\qquad (7.126)$$

No factor of $\frac{1}{2}$ is included in θ_2 so that the Jacobian of the transformation is unity (see Section C.9):

$$d\theta_1 d\theta_2 = \frac{\partial(\theta_1,\theta_2)}{\partial(\phi,\theta)}d\phi d\theta = d\phi d\theta.$$

The geometry of the coordinate transformation Eq. (7.126) is shown in Fig. 7.11. You should verify

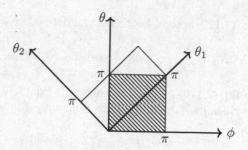

Figure 7.11: Geometry of the coordinate transformation in Eq. (7.126).

that $\theta_1 = 0, \theta_2 = \pi$ corresponds to $\theta = \pi/2, \phi = -\pi/2$, and $\theta_1 = \pi, \theta_2 = 0$ corresponds to $\phi, \theta = \pi$. The transformation is not a rigid rotation, although it is area preserving (Exercise 7.25). Making the substitutions, we have from Eq. (7.124),

$$-\beta\psi = \ln 2 + \frac{1}{2\pi^2}\int_0^\pi d\theta_1 \int_0^\pi d\theta_2 \ln\left(\cosh^2 2K - 2\sinh 2K \cos\theta_1 \cos(\theta_2/2)\right)$$

$$= \ln(2\cosh 2K) + \frac{1}{\pi^2}\int_0^\pi d\theta_1 \int_0^{\pi/2} d\omega \ln(1 - \kappa(K)\cos\theta_1 \cos\omega), \qquad (7.127)$$

where $\omega \equiv \theta_2/2$ and

$$\kappa(K) \equiv \frac{2\sinh 2K}{\cosh^2 2K}. \qquad\qquad (7.128)$$

Figure 7.12 shows a plot of $\kappa(K)$; it equals unity at $K = K_c$. It's worth stating that $\kappa(K)$ is an

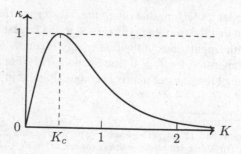

Figure 7.12: Function $\kappa(K)$ versus K.

analytic function. Because $|\kappa\cos\omega| \le 1$ (Exercise 7.21), we can integrate over θ_1 in Eq. (7.127):

$$-\beta\psi = \ln(2\cosh 2K) + \frac{1}{\pi}\int_0^{\pi/2}\ln\left(\frac{1}{2}\left(1 + \sqrt{1 - \kappa^2\cos^2\omega}\right)\right)d\omega. \qquad (7.129)$$

The value of the integral in Eq. (7.129) is unchanged[87] if $\cos^2 \omega$ is replaced with $\sin^2 \omega$, which is more convenient. Thus, we have our final expression for the free energy of the $d = 2$ Ising model,

$$-\beta\psi = \ln(2\cosh 2K) + \frac{1}{\pi}\int_0^{\pi/2} \ln\left(\frac{1}{2}\left(1 + \sqrt{1 - \kappa^2 \sin^2 \omega}\right)\right) d\omega . \qquad (7.130)$$

Combining Eq. (7.130) with Eq. (7.91) (Gibbs-Helmholtz equation), we have the internal energy per spin (specific energy), u, which, after a few steps has the form[88]

$$u = -J\frac{\partial}{\partial K}(-\beta\psi) = -J\coth 2K\left[1 + \frac{2}{\pi}\left(2\tanh^2 2K - 1\right)F(\kappa)\right], \qquad (7.131)$$

where F is the complete elliptic integral function of the first kind,[50, p590]

$$F(y) \equiv \int_0^{\pi/2} \frac{d\omega}{\sqrt{1 - y^2 \sin^2 \omega}}, \qquad (0 \le y < 1) \qquad (7.132)$$

the values of which are tabulated. We'll also require the complete elliptic integral function of the second kind,

$$E(y) \equiv \int_0^{\pi/2} \sqrt{1 - y^2 \sin^2 \omega}\, d\omega , \qquad (0 \le y \le 1) \qquad (7.133)$$

which shows up through the derivative of $F(y)$ (see Exercise 7.23). $E(y)$ has no singularities.

The specific heat c is obtained from one more derivative:

$$c = \frac{\partial u}{\partial T} = -\frac{K}{T}\frac{\partial}{\partial K}u \qquad (7.134)$$

$$= \frac{4}{\pi}k\left(K\coth 2K\right)^2\left[F(\kappa) - E(\kappa) - \left(1 - \tanh^2 2K\right)\left[\frac{\pi}{2} + \left(2\tanh^2 2K - 1\right)F(\kappa)\right]\right] .$$

The elliptic integral $F(y)$ diverges logarithmically as $y \to 1^-$, which we see from the asymptotic result[110, p905]

$$F(y) \overset{y \to 1^-}{\sim} \ln\left(\frac{4}{\sqrt{1 - y^2}}\right) . \qquad (7.135)$$

The heat capacity therefore has a logarithmic singularity at $T = T_c$, as we see in Eq. (7.90), implying $\alpha = 0$ for the two-dimensional Ising model.

7.10.2 Spontaneous magnetization and correlation length

The Onsager solution is a landmark achievement in theoretical physics, an exact evaluation of the partition function of interacting degrees of freedom in a two-dimensional system, from which we obtain the free energy, internal energy, and specific heat. We identified T_c as the temperature at which the free energy becomes singular ($\kappa = 1$). A more physical way of demonstrating the existence of a phase transition would be to calculate the order parameter. The first published derivation of the spontaneous magnetization appears to be that of C.N. Yang,[89] who showed[111]

$$\langle\sigma\rangle = \begin{cases} 0 & T > T_c \\ \left(1 - \frac{1}{\sinh^4 2K}\right)^{1/8} & T \le T_c . \end{cases} \qquad (7.136)$$

---

[87]Consider, for a function f: $\int_0^\pi f(\cos^2 \omega)d\omega = \int_0^{\pi/2} f(\cos^2 \omega)d\omega + \int_0^{\pi/2} f(\sin^2 \omega)d\omega$ (let $\omega \to \omega + \pi/2$ in the second integral). But, $\int_0^\pi f(\cos^2 \omega)d\omega = \int_{-\pi/2}^{\pi/2} f(\sin^2 \omega)d\omega = 2\int_0^{\pi/2} f(\sin^2 \omega)d\omega$ (let $\omega \to \omega - \pi/2$ in the first integral, the second integral is even about the origin). Thus, $\int_0^{\pi/2} f(\cos^2 \omega)d\omega = \int_0^{\pi/2} f(\sin^2 \omega)d\omega$.

[88]Note that the internal energy per spin in the one-dimensional Ising model is far simpler: $u = -J\tanh K$.

[89]Yang received the 1957 Nobel Prize in Physics for work on parity nonconservation in weak interactions.

Clearly the temperature at which $\langle\sigma\rangle \to 0$ is $\sinh 2K_c = 1$, the same as Eq. (7.125). We infer from Eq. (7.136) that the order parameter critical exponent $\beta = \frac{1}{8}$, our first non-classical exponent. The small ($\ll 1$) value of β implies a rapid rise in magnetization for $T \lesssim T_c$, as we see in Fig. 7.13. Yang's derivation was later simplified,[112] yet even the simplification is rather complicated.

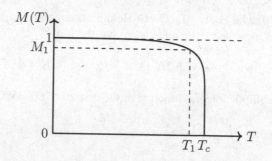

Figure 7.13: Spontaneous magnetization in the two-dimensional Ising model; $T_1 = 0.9T_c$.

In the one-dimensional Ising model, we found that the correlation length ξ is determined by the largest and second-largest eigenvalues of the transfer matrix (see Eq. (6.101)), with the result $\xi(K) = -a/\ln(\tanh K)$, where a is the lattice constant. The same holds in the two-dimensional Ising model, with the result[108, Chapter 7]

$$\xi(K) = \begin{cases} \dfrac{-2a}{\ln\sinh^2 2K} & T > T_c \ (K < K_c) \\[2ex] \dfrac{a}{\ln\sinh^2 2K} & T < T_c \ (K > K_c). \end{cases} \qquad (7.137)$$

For $K < K_c$ ($> K_c$), $\sinh^2 2K < 1$ (> 1), which is useful in understanding $\xi(K)$ as $\sinh^2 2K$ passes through the critical value $\sinh^2 2K_c = 1$. In either case, $\xi \to \infty$ for $K \to K_c$, such that

$$\xi \sim |T - T_c|^{-1}, \qquad (7.138)$$

and thus $\nu = 1$ in the two-dimensional Ising model. Note that $\xi(K) \overset{K \to 0}{\to} 0$ and $\xi(K) \overset{K \to \infty}{\to} 0$.

We now have four of the six critical exponents, α, β, ν, η, where $\eta = \frac{1}{4}$ (Section 7.6). Each is distinct from the predictions of Landau theory. The heat capacity exponent $\alpha = 0$ is seemingly common to both, but Landau theory predicts a discontinuity in specific heat rather than a logarithmic singularity. The remaining two exponents γ, δ will follow with a little more theoretical development (Section 7.11), but we can just state their values here (for the $d = 2$ Ising model) $\gamma = \frac{7}{4}$ and $\delta = 15$.

7.11 CRITICAL EXPONENT INEQUALITIES

Critical exponents are not independent, which is not surprising given all the thermodynamic interrelations among state variables. We saw in Section 7.4 that $\gamma = 2\beta$ in the van der Waals model and $\gamma = 2\nu$ in Ornstein-Zernike theory (Section 7.6), relations that hold in those particular theories, but which do not apply in general. Finding general relations among critical exponents should start with thermodynamics. As we now show, thermodynamics provides *inequalities*, but not equalities, among critical exponents that might otherwise reduce the number of independent exponents. Before getting started, we note an inequality among functions that implies an inequality among exponents. Suppose two functions $f(x), g(x)$ are such that $f(x) \le g(x)$ for $x \ge 0$, and that $f(x) \sim x^\psi$ and $g(x) \sim x^\phi$ as $x \to 0^+$. Then, $\psi \ge \phi$. See Exercise 7.26.

Equation (1.51) implies an inequality on the heat capacity of magnetic systems,

$$C_{B=0} \geq \frac{T}{\chi} \left(\frac{\partial M}{\partial T}\right)^2_{B=0}. \tag{7.139}$$

Because $C_{B=0} \sim |T - T_c|^{-\alpha}$ (Eq. (7.66)), $\chi \sim |T - T_c|^{-\gamma}$ (Eq. (7.60)), and $M \sim |T - T_c|^\beta$ (Eq. (7.54)), inequality (7.139) immediately implies *Rushbrooke's inequality*,[113]

$$\alpha + 2\beta + \gamma \geq 2. \tag{7.140}$$

Because it follows from thermodynamics, it applies to all systems.[90] For the $d = 2$ Ising model it implies $\gamma \geq \frac{7}{4}$. It's been shown rigorously that $\gamma = \frac{7}{4}$ for the two-dimensional Ising model [114], and thus Rushbrooke's inequality is satisfied as an equality among the exponents of the $d = 2$ Ising model. We also have the same equality among classical exponents: $\alpha + 2\beta + \gamma = 2$.

Numerous inequalities among critical exponents have been derived.[91] We prove one more, *Griffiths inequality*,[115]

$$\alpha + \beta(1 + \delta) \geq 2. \tag{7.141}$$

To show this, we note for magnetic systems[92] that M, T are independent variables of the free energy, $F(T, M)$. Equation (7.61), $B = (\partial F/\partial M)_T$, as applied to the coexistence region (which is associated with $B = 0$), implies

$$\left(\frac{\partial F}{\partial M}\right)_T = 0, \qquad (T \leq T_c, 0 \leq M \leq M_0(T))$$

where we use $M_0(T)$ to denote the spontaneous magnetization at temperature T (such as shown in Fig. 7.13). The free energy in the coexistence region is therefore independent of M:

$$F(T, M) = F(T, 0). \qquad (T \leq T_c, 0 \leq M \leq M_0(T)) \tag{7.142}$$

It may help to refer to Fig. 7.9, where F is independent of the order parameter along the convex envelope. Entropy is also independent of M in the coexistence region. From the Maxwell relation $(\partial S/\partial M)_T = -(\partial B/\partial T)_M$ (Exercise 1.15), we conclude, because $B = 0$, that

$$S(T, M) = S(T, 0). \qquad (T \leq T_c, 0 \leq M \leq M_0(T)) \tag{7.143}$$

$F(T, M)$ is a concave function[93] of T for fixed M. The tangent to a concave function always lies above the curve (see Fig. 4.2), implying the following inequalities derived from the slope F' evaluated at a lower temperature $T_2 < T_1 \leq T_c$, and F' evaluated at a higher temperature, $T_1 > T_2$:

$$F(T_1, M) \leq F(T_2, M) + (T_1 - T_2)F'(T_2, M) = F(T_2, M) - (T_1 - T_2)S(T_2, M)$$
$$F(T_2, M) \leq F(T_1, M) - (T_1 - T_2)F'(T_1, M) = F(T_1, M) + (T_1 - T_2)S(T_1, M), \tag{7.144}$$

where $F' \equiv (\partial F/\partial T)_M = -S(T, M)$, Eq. (7.61). With $T_1 = T_c$ and $T_2 \equiv T$, and using Eqs. (7.142) and (7.143), the inequalities in (7.144) imply the inequality

$$F(T_c, M) - F(T_c, 0) \leq (T_c - T) \left[S(T_c, 0) - S(T, 0)\right]. \tag{7.145}$$

---

[90] Inequality (7.139) pertains to magnetic systems, implying Rushbrooke's inequality holds only for magnets. It holds for fluid systems as well; see Exercise 7.28. One has to develop the analog of inequality (7.139) for fluids, inequality (P7.13).

[91] A list is given in Stanley[97, p61].

[92] Griffiths's inequality is easier to show for magnets; deriving it for fluids is more difficult, but the result is the same.[116]

[93] Section 4.3 covers the convexity properties of the free energy. We can see directly that $F(T, M)$ is a concave function of T. Equation (7.62), $C_M = -T \left(\partial^2 F/\partial T^2\right)_M$ implies, because $C_M \geq 0$, that $\left(\partial^2 F/\partial T^2\right)_M < 0$.

Note that for $M = 0$ and $T = T_c$, inequality (7.145) reduces to $0 \leq 0$, and students sometimes cry foul—it's a tautology. We're interested in *how* the two sides of the inequality vanish in the limit that the critical point is approached. As $M \to 0$ and $T \to T_c^-$, inequality (7.145) has the structure

$$M \left(\frac{\partial F}{\partial M} \right) \sim M \frac{F(T_c, M) - F(T_c, 0)}{M} \leq (T_c - T)^2 \frac{S(T_c, 0) - S(T, 0)}{T_c - T} \sim (T_c - T)^2 \left(\frac{\partial S}{\partial T} \right) .$$

Using Eqs. (7.61) and (7.62), and that $B \sim M^\delta \sim |T - T_c|^{\beta\delta}$ and $C \sim |T - T_c|^{-\alpha}$, inequality (7.145) implies $MB \sim M^{\delta+1} \sim |T - T_c|^{\beta(\delta+1)} \lesssim |T - T_c|^{2-\alpha}$. We must have therefore that $\beta(\delta + 1) \geq 2 - \alpha$, which is Griffiths inequality, (7.141). For the $d = 2$ Ising model it implies $\delta \geq 15$. It's found numerically that $\delta = 15.00 \pm 0.08$,[103, p694] from which we conclude that $\delta = 15$. For the classical exponents, Griffiths inequality is an equality, $\alpha + \beta(\delta + 1) = 2$.

Other inequalities among critical exponents have been derived, but none are as general as the Rushbrooke and Griffiths inequalities which follow from thermodynamics.[94] Under plausible assumptions, additional inequalities have been derived (and this list is not exhaustive):

$$\gamma \geq \beta(\delta - 1) \qquad \text{(Griffiths)}$$
$$(2 - \eta)\nu \geq \gamma \qquad \text{(Fisher)} \qquad\qquad (7.146)$$
$$d\frac{\delta - 1}{\delta + 1} \geq 2 - \eta . \qquad \text{(Buckingham-Gunton)}$$

These inequalities are satisfied as equalities for the critical exponents of the $d = 2$ Ising model; the last inequality fails for classical exponents if $d < 4$. Table 7.3 shows the values of critical exponents, experimental and theoretical. Exponents for the $d = 3$ Ising model are obtained from numerical investigations and are known only approximately.

Table 7.3: Critical exponents. Source: L.E. Reichl, *A Modern Course in Statistical Physics*[117].

Exponent	Experimental value	$d = 2$ Ising	$d = 3$ Ising	Mean field theory
α	0–0.2	0 (logarithmic)	0.110	0 (discontinuity)
β	0.3–0.4	1/8	0.326	1/2
γ	1.2–1.4	7/4	1.237	1
δ	4–5	15	4.790	3
ν	0.6–0.7	1	0.623	1/2
η	0.1	1/4	0.036	0

7.12 THE IMPOSSIBILITY OF PHASES IN ONE DIMENSION

We noted in Section 7.10 that phase transitions do not occur in one-dimensional systems. We based that conclusion on an analysis of the largest eigenvalue of the transfer matrix for the $d = 1$ Ising model, which by Perron's theorem is an analytic function of K (and phase transitions are associated with singularities in the free energy function). Perron's theorem does not apply to infinite matrices, as in the two-dimensional Ising model, where the largest eigenvalue is not an *entire function*,[95] and we have a phase transition in two dimensions. One could object that the Ising model is too specialized to draw conclusions about phase transitions in all one-dimensional systems. The

---

[94]Inequalities occur in thermodynamics generally as a consequence of the stability condition, $d^2S < 0$ for fluctuations, which leads to restrictions on the sign of response functions, such as $C_V > 0$ and $\beta_T > 0$; see [3, Section 3.10]. Inequalities also occur through the convexity properties of the free energy, such as discussed in Section 4.3.

[95]An entire function is analytic throughout the finite complex plane. See [16, p202].

Mermin-Wagner-Hohenberg theorem [118, 119] states that spontaneous ordering by means of *continuous symmetry breaking*[96] is precluded in models with short-range interactions in dimensions $d \leq 2$. The requirement of continuous symmetry breaking is essential—the two-dimensional Ising model in zero magnetic field (discrete symmetry) possesses a spontaneous magnetization. In either case (continuous or discrete symmetries), phase transitions in one dimension are precluded.

It would be nice if a physical argument could be given illustrating the point. Landau and Lifshitz, on the final page of *Statistical Physics*[99], present such an argument that we adapt for Ising spins.[97] Figure 7.14 shows a one-dimensional array of Ising spins consisting of *domains* of aligned spins

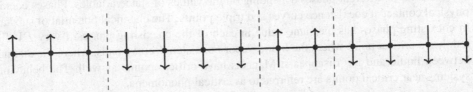

Figure 7.14: One-dimensional domains of Ising spins, separated by domain walls (dashed lines).

separated by interfaces, "domain walls" (dashed lines).[98] We can consider domains phases, call them A and B or *up* and *down*. The question is, is there a finite temperature for which the system consists entirely of one phase, i.e., can there be a phase with long-range order?

Assume N Ising spins in one dimension, as in Fig. 7.14, coupled through near-neighbor interactions, and let there be m domain walls. The energy of the system is (show this)

$$U = -J(N - 1) + 2mJ. \tag{7.147}$$

It requires energy $2J$ to create a domain wall. There are $\binom{N-1}{m}$ ways of arranging m domain walls within the system of N spins, and thus the entropy of the collection of domain walls is, for N large,

$$\frac{1}{k}S = \ln\binom{N-1}{m} \overset{N \to \infty}{\sim} \ln\left(\frac{N^m}{m!}\right) = m\ln\left(\frac{Ne}{m}\right), \tag{7.148}$$

where we've used Eq. (3.44) and Stirling's approximation. The free energy is therefore

$$F = -J(N - 1) + 2mJ + kTm\ln\left(\frac{m}{Ne}\right). \tag{7.149}$$

Free energy is a minimum in equilibrium. From Eq. (7.149),

$$\frac{\partial F}{\partial m} = 2J + kT\ln\left(\frac{m}{N}\right). \tag{7.150}$$

Setting $\partial F/\partial m = 0$, the average number of domain walls in thermal equilibrium is

$$\overline{m} = N\exp(-2J/kT). \tag{7.151}$$

There is no temperature for which $\overline{m} = 0$; entropy wins—*an ordered phase in one dimension cannot exist*. For the one-dimensional Ising model, $\xi(K) \sim \frac{1}{2}e^{2K}$ for $K \gg 1$ (Eq. (P6.1)), and hence,

$$\overline{m} \overset{T \to 0}{\sim} \frac{N}{2\xi}. \tag{7.152}$$

---

[96] A continuous symmetry is a symmetry associated with continuous, smooth changes in some parameter, such as the orientation of magnetic moments in the high-temperature phase. Continuous symmetries are distinct from *discrete symmetries*, where the system is symmetric under discrete (non-continuous) changes in parameters, such as $\sigma_i \to -\sigma_i$ in the Ising model, or time-reversal, $t \to -t$.

[97] Pay attention to this argument, a favorite question on PhD oral qualifying exams.

[98] Magnetic domains are regions within a magnet in which the magnetization has a uniform direction. The magnetization within each domain points in a certain direction, but the magnetizations of different domains are not aligned. See for example Ashcroft and Mermin[18, p718].

As $T \to 0$, $\xi \to \infty$ implying that $\overline{m} \to 0$ in the limit. In one dimension, $T = 0$ is a critical temperature, but there is no "low temperature" phase.[99]

SUMMARY

This chapter introduced phase transitions and critical phenomena, setting the stage for Chapter 8.

- Phases are spatially homogeneous, macroscopic systems in thermodynamic equilibrium. Substances exist in different phases depending on the values of state variables. Phases coexist in physical contact at coexistence curves and triple points. The chemical potential of substances in coexisting phases has the same value in each of the coexisting phases. Every PVT system has a critical point where the liquid-vapor coexistence curve ends, where the distinction between liquid and gas disappears. Magnets have critical points as well. The behaviors of systems near critical points are referred to as critical phenomena.

- The Gibbs phase rule, Eq. (7.13), specifies the number of intensive variables that can be independently varied without disturbing the number of coexisting phases.

- The latent heat L is the energy released or absorbed in phase transitions. Phase transitions are classified according to whether $L \neq 0$—first-order phase transitions—and those for which $L = 0$, continuous phase transitions. For historical reasons, continuous phase transitions are referred to as second-order phase transitions. At first-order phase transitions, there is a change in specific entropy, $\Delta s = L/T$. At continuous transitions, $\Delta s = 0$.

- We discussed the liquid-gas phase transition using the van der Waals equation of state, from which we discovered the law of corresponding states, that systems in corresponding states behave the same. Corresponding states are specified by reduced state variables—state variables expressed in units of their critical values, Eq. (7.22). Substances obey the law of corresponding states even if they're not well described by the van der Waals model; see Fig. 7.3.

- The Maxwell construction is a rule by which the coexistence region can be located in a P-v diagram, that the area under the curve of sub-critical isotherms (obtained from the equation of state) be the same as the area under the flat isotherms of coexisting liquid and gas phases. The van der Waals equation of state (motivated phenomenologically, see Section 6.2) and the Maxwell construction occur rigorously in theories featuring weak, long-range interactions (Lebowitz-Penrose theorem), a class of theories known as mean field theory. Besides long-range interactions, mean field theories result when the energy of interaction between fluctuations is ignored—fluctuations are independent and uncorrelated in mean field theories.

- The prototype mean field theory is the Weiss molecular field theory of ferromagnetism. It asserts that the aligned dipole moments in the magnetized state contribute to an internal magnetic field, the molecular field. By postulating a proportionality between the molecular field and the magnetization, one arrives at a theory with a nonlinear equation of state, Eq. (7.49), a self-consistent theory where spins respond to the same field they generate. The molecular field is a proxy for the ordering of spins that's otherwise achieved through inter-spin interactions. Statistical mechanics is particularly simple for mean field theories, while the statistical mechanics of interacting degrees of freedom is not so simple, as we've seen.

- The order parameter ϕ is a quantity that's zero for $T \geq T_c$, nonzero for $T < T_c$, and vanishes smoothly as $T \to T_c^-$. The vanishing of ϕ as $T \to T_c$ is characterized in terms of an

---

[99]There are exceptional systems characterized by negative absolute temperature, but negative absolute temperature does not imply a temperature colder than zero. Negative temperatures are hotter than all positive temperatures.[3, p163]

exponent, β, such that $\phi \sim |T - T_c|^\beta$, one of several exponents by which critical phenomena are described. The critical isotherm exponent δ characterizes how the field f conjugate to ϕ behaves in the critical region, with $f \sim \phi^\delta$ as $\phi \to 0$ and $T = T_c$. Other quantities diverge as $T \to T_c$, with each divergence characterized by its own exponent. As $T \to T_c$, the heat capacity $C \sim |T - T_c|^{-\alpha}$; for the compressibility (or susceptibility χ) with $f = 0$, $\beta_T \sim |T - T_c|^{-\gamma}$; for the correlation length, $\xi \sim |T - T_c|^{-\nu}$; for the static structure factor $S(q, T)$, which for $T = T_c$ diverges as $S(q, T = T_c) \sim q^{-(2-\eta)}$ as $q \to 0$. Critical exponents can be measured and calculated from theory. In standard statistical mechanics, there are six critical exponents: $\alpha, \beta, \gamma, \delta, \nu, \eta$.

- Landau theory describes phase transitions in terms of a model free energy function, $F(\phi, T)$, Eq. (7.79). The graph of $F(\phi, T)$, shown in Fig. 7.9, is one of the more widely known diagrams in physics. One sees that for $T \geq T_c$, the minimum (equilibrium state) occurs at $\phi = 0$; for $T < T_c$, the equilibrium state is one of broken symmetry with $\phi = \pm \phi_0$. Landau theory is a mean field theory. The critical exponents predicted by mean field theories are all the same, the classical exponents: $\alpha = 0, \beta = \frac{1}{2}, \gamma = 1, \delta = 3, \nu = \frac{1}{2}, \eta = 0$.

- The classical exponents are not in good agreement with experiment; see Table 7.3. The key assumption of Landau theory is that the free energy function is analytic, which is not correct—partial derivatives of F generally do not exist at the critical point. Phase transitions are associated with points in thermodynamic state space where the free energy function is singular. For finite N and V, the partition function is an analytic function of its arguments; to describe phase transitions, the thermodynamic limit must be taken. Landau theory does an adequate job in treating the thermodynamics of interacting systems away from the critical region. From the Ginzburg criterion, Landau theory applies in the critical region of systems of dimension $d > 4$. We presented the Landau-Lifshitz argument on the impossibility of the existence of long-range order for $d = 1$.

- We outlined the Onsager solution of the two-dimensional Ising model in zero magnetic field, which requires that one find the largest eigenvalue of the $2^M \times 2^M$ transfer matrix in the limit $M \to \infty$. Starting from Onsager's formula for the largest eigenvalue, Eq. (7.118), one can see how the free energy develops a singularity in the thermodynamic limit. Exact expressions can be found for the partition function, free energy, internal energy, and heat capacity, which displays a logarithmic divergence at $T = T_c$, $\alpha = 0$. The theory predicts the critical temperature as that at which the free energy is singular, $K_c = 0.4407$. From Eq. (7.136) for the spontaneous magnetization we infer the non-classical critical exponent $\beta = \frac{1}{8}$.

- Critical exponents are not independent of each other. Thermodynamics provides relations among critical exponents in the form of inequalities, such as the Rushbrooke and Griffiths inequalities. Explaining why these inequalities are satisfied as equalities is the starting point of the next chapter.

EXERCISES

7.1 An approximate solution of the Clausius-Clapeyron equation can be derived if we assume that: 1) the latent heat L is a constant; 2) $v_{gas} \gg v_{liquid}$, so that $\Delta v \approx v_{gas}$; 3) the gas is ideal. With these assumptions, show that the coexistence curve has the form $P = P_0 e^{-L/(kT)}$.

7.2 Referring to Fig. 5.14, what is the order in the Ehrenfest classification scheme of the Bose-Einstein condensation?

7.3 We argued that $V/N \approx b$ is a solution of Eq. (7.18) for small V/N. In what regime would such a guess be correct? Write $V/N = b(1 + \delta)$, where δ is dimensionless and presumed small, $\delta \ll 1$. Show at lowest order that $\delta = kT/(Pb + a/b)$. Our guess is correct for temperatures such that $kT \ll (Pb, a/b)$.

7.4 Show for low temperatures, $kT \ll Pb$, there is only one real root of Eq. (7.18), together with a pair of unphysical, complex conjugate roots. Hint: Look up the formula for the roots of cubic equations (such as in [50, p17]).

7.5 The van der Waals model predicts the critical parameters v_c, P_c, T_c reasonably well. A gas for which the constants (a, b) are known, as well as the critical parameters is argon (there are many others). For argon, $a = 1.355$ bar L^2/mol^2 and $b = 0.0320$ L/mol.[57, p6-33] To convert to SI units, 1 bar $L^2/mol^2 = 0.1$ Pa m^6/mol^2 and 1 L/mol$= 10^{-3}$ m^3/mol. Use Eq. (7.21) to calculate v_c, P_c, T_c and compare with the experimental results $T_c = 150.87$ K, $P_c = 4.898$ MPa, and $v_c = 75$ cm^3 mol^{-1} (see [57, p6-37]). Note that you have to convert from moles to particle number (Avogadro's number) to find T_c.

7.6 Derive the law of corresponding states, Eq. (7.23).

7.7 Show that for liquid-gas phase coexistence, the mole fractions in the gas and liquid phases are given by the expressions

$$x_g = \frac{v_d - v_l}{v_g - v_l} \qquad x_l = \frac{v_g - v_d}{v_g - v_l} .$$

Hint: Use the lever rule, Eq. (7.25).

7.8 Thermodynamic properties of the van der Waals gas.

a. Derive the Helmholtz energy, which can be found by integrating the equation of state. Show, from Eq. (4.58), $P = -(\partial F/\partial V)_{T,N}$, that, using Eq. (6.32),

$$F(T, V, N) = -\int P dV + \phi(T, N) = -NkT \ln(V - Nb) - a\frac{N^2}{V} + \phi(T, N) ,$$

where $\phi(T, N)$, the integration "constant," is a function of T, N. We can determine $\phi(T, N)$ by requiring that for $a = b = 0$, we recover the Helmholtz energy of the ideal gas. Show, from $-\beta F = \ln Z$, Eq. (4.57), that using Eq. (5.1) for Z and the Stirling approximation, we have for the ideal gas $F = -NkT \left[1 + \ln\left(V/(N\lambda_T^3)\right)\right]$, implying $\phi(T, N) = NkT\left(\ln(N\lambda_T^3) - 1\right)$, and therefore

$$F(T, V, N) = -NkT\left[\ln\left(\frac{V - Nb}{N\lambda_T^3}\right) + 1\right] - a\frac{N^2}{V} . \qquad (P7.1)$$

b. Use the other two expressions in Eq. (4.58) to find the chemical potential and entropy:

$$\mu = -kT \ln\left(\frac{V - Nb}{N\lambda_T^3}\right) + \frac{NbkT}{V - Nb} - 2a\frac{N}{V}$$

$$S = Nk\left[\frac{5}{2} + \ln\left(\frac{V - Nb}{N\lambda_T^3}\right)\right] . \qquad (P7.2)$$

The entropy of the van der Waals gas is the same as the ideal gas, except for the excluded volume. Note that the excluded volume Nb (which can be considered a repulsive interaction) makes a positive contribution to the chemical potential, whereas the attractive part, characterized by the parameter a, works to decrease μ. See the remarks on the sign of the chemical potential in Section 1.6.

c. Find the internal energy. Hint: $U = F + TS$.

$$U = \frac{3}{2}NkT - a\frac{N^2}{V} . \tag{P7.3}$$

The energy per particle U/N is modified from its value for the ideal gas (all kinetic energy) to include a contribution from the potential energy of interactions, simply proportional to the average density, $-an$.

d. Show that the heat capacity C_V of the van der Waals gas is the same as for the ideal gas. Hint: The parameter a is independent of temperature. You could make use of Eq. (P1.7) to show that C_V for the van der Waals model has no volume dependence.

e. From the thermodynamic identity Eq. (1.42),

$$C_P = C_V - T\left(\frac{\partial P}{\partial V}\right)_T \left(\frac{\partial V}{\partial T}\right)_P^2 ,$$

show for the van der Waals gas that,

$$\frac{C_P - C_V}{Nk} = \frac{1}{1 - \dfrac{2a(v-b)^2}{v^3 kT}} . \tag{P7.4}$$

7.9 Show that the Helmholtz energy of the van der Waals gas, an expression for which is given in Eq. (P7.1), can be written in terms of reduced variables

$$-\frac{F}{NkT} = \ln\left((3\bar{v} - 1)\overline{T}^{3/2}\right) + 1 + \ln\left(\frac{b}{\lambda_{T_c}^3}\right) + \frac{9}{8\overline{T}\bar{v}} . \tag{P7.5}$$

Show from Eq. (P7.5) that $Pv/(kT)$ is a universal function. Hint: $P = kT\left(\partial(-F/NkT)/\partial v\right)_{T,N}$

7.10 Guided exercise to derive Eq. (7.32).

a. The specific energy u is a function of specific volume and temperature, $u = u(v, T)$. To see this, assume that $U = U(T, V, N)$. You should convince yourself of this—U is "naturally" a function of (S, V, N) (see Eq. (1.21)), but we can take S as a function of (T, V) (show that $dS = (C_V/T)dT + (\alpha/\beta_T)dV$), and thus, invoking the thermodynamic limit, $U = Nf(v, T)$ and hence $u = u(v, T)$.

b. Referring to Eq. (7.31), for $u = u(v, T)$, show from Eq. (1.32) that

$$\left(\frac{\partial u}{\partial T}\right)_{\text{coex}} = \left(\frac{\partial u}{\partial T}\right)_v + \left(\frac{\partial u}{\partial v}\right)_T \left(\frac{\partial v}{\partial T}\right)_{\text{coex}} = c_v + \left(\frac{\partial u}{\partial v}\right)_T \left(\frac{\partial v}{\partial T}\right)_{\text{coex}} . \tag{P7.6}$$

c. Use Eq. (1.41) to show that

$$\left(\frac{\partial u}{\partial v}\right)_{T,N} = T\left(\frac{\partial P}{\partial T}\right)_{V,N} - P . \tag{P7.7}$$

d. We require $\Delta u \equiv u_g - u_l$. Use Eq. (7.15) to show that the Clausius-Clapeyron equation can be written

$$\left(\frac{dP}{dT}\right)_{\text{coex}} = \frac{\Delta h}{T\Delta v} = \frac{1}{T\Delta v}(\Delta u + P\Delta v) = \frac{\Delta u}{T\Delta v} + \frac{P}{T} ,$$

implying that

$$\Delta u = \left[T\left(\frac{dP}{dT}\right)_{\text{coex}} - P\right]\Delta v . \tag{P7.8}$$

e. Differentiate the lever rule, Eq. (7.25), to show that

$$(v_g - v_l)\left(\frac{\partial x_l}{\partial T}\right)_{\text{coex}} = x_g\left(\frac{\partial v_g}{\partial T}\right)_{\text{coex}} + x_l\left(\frac{\partial v_l}{\partial T}\right)_{\text{coex}}. \qquad \text{(P7.9)}$$

Show that Eq. (P7.9) is symmetric under the interchange of labels, $g \leftrightarrow l$.

f. Show from Eq. (1.32) that for $P_{\text{coex}} = P_{\text{coex}}(T, v)$,

$$\left(\frac{dP}{dT}\right)_{\text{coex}} = \left(\frac{\partial P}{\partial T}\right)_v + \left(\frac{\partial P}{\partial v}\right)_T \left(\frac{\partial v}{\partial T}\right)_{\text{coex}}. \qquad \text{(P7.10)}$$

The coexistence curve P_{coex} is a clearly a function of T (see Fig. 7.1), but it's also a function of v: The coexistence curve is the projection onto the P-T plane of the coexistence region in the P-v plane.

g. Show that by combining Eqs. (P7.6) through (P7.10), one arrives at Eq. (7.32) starting from Eq. (7.31).

7.11 Derive Eq. (7.34), the van der Waals equation of state in the vicinity of the critical point, by substituting Eq. (7.33) into Eq. (7.23) and keeping track of small quantities. Hint: For $x \ll 1$, $(1+x)^{-1} \approx 1 - x + x^2 - x^3 + \cdots$.

7.12 Derive Eq. (7.43), the form of Eq. (P7.4) when expressed in terms of reduced variables.

7.13 From Eq. (7.44), C_P for the van der Waals gas diverges as $T \to T_c^+$. From Exercise 1.7,

$$C_P = T\left(\frac{\partial S}{\partial T}\right)_P.$$

Referring to Eq. (P7.2), the expression for S for the van der Waals gas doesn't appear to have any terms that would contribute to a singularity in C_P. By differentiating S in Eq. (P7.2), find out what physical quantity contributes to the divergence of C_P as $T \to T_c^+$. Hint: Find $(\partial V/\partial T)_P$ using the van der Waals equation of state.

7.14 The two equations in (7.45) are a way to define the critical point in fluids. Show using the van der Waals equation of state that these relations lead to the same prediction of critical parameters that we found in Eq. (7.21).

7.15 Derive Eq. (7.85). Hint: $\phi = 0$ for $T \geq T_c$, and for $T < T_c$, use Eq. (7.80).

7.16 Derive Eq. (7.88) for the heat capacity discontinuity in Landau theory. Be mindful of minus signs.

7.17 Consider the effect of adding a function of thermodynamic variables to the Hamiltonian, such as we have in the mean-field Hamiltonian, Eq. (7.95). Under the transformation $H \to \tilde{H} = H + \alpha(T)$, show that $Z \to \tilde{Z} = e^{-\beta\alpha}Z$, $U \to \tilde{U} = U + \alpha - T\partial\alpha/\partial T$, $S \to \tilde{S} - \partial\alpha/\partial T$, and $F \to \tilde{F} = F + \alpha$.

7.18 Derive Eq. (7.97) starting from Eq. (7.96).

7.19 Show, starting from Eq. (7.98), that the mean-field Ising model predicts a spontaneous magnetization for $T < T_c = q/k$. Set $b = 0$ and use $\tanh x \approx x - \frac{1}{3}x^3$ for $x \ll 1$.

7.20 Show that $g(y)$ defined in Eq. (7.106) is concave for small values of y_0 for $T < T_c$, and is concave at $y = 0$ for $T > T_c$. Hint: Use the identity $(\cosh x)^{-2} = 1 - \tanh^2 x$.

7.21 a. Show that Eq. (7.125) defining the critical coupling constant, $\sinh 2K_c = 1$, is the unique solution of $\cosh^2 2K - 2\sinh 2K = 0$. Hint: $\cosh^2 x - \sinh^2 x = 1$.

b. Show that Eq. (7.125) can be written in the form

$$\tanh K_c = \sqrt{2} - 1 . \tag{P7.11}$$

c. From Eq. (7.128), $\kappa \equiv 2\sinh 2K / \cosh^2 2K$. Show that $|\kappa| \le 1$.

7.22 Consider the high-temperature limit of Eq. (7.130), the free energy per spin in the two-dimensional Ising model. Set $K = 0$, corresponding to uncoupled spins. Interpret the resulting expression, $\psi = -kT \ln 2$. Hint: What is the entropy of a system of uncoupled Ising spins? What is the $K \to 0$ limit of Eq. (7.131) for the internal energy per spin? Set $\kappa = 0$ in the integral first.

7.23 The complete elliptic integral of the second kind, $E(y)$, Eq. (7.133), occurs through the derivative of $F(y)$, the complete elliptic integral of the first kind, Eq. (7.132):[110, p907]

$$E(y) = (1 - y^2) \left(y \frac{dF}{dy} + F(y) \right) . \tag{P7.12}$$

This identity can be vexing to verify, if you haven't seen the strategy. Change variables; let $x = \sin^2 \omega$. Show using the definitions Eqs. (7.132) and (7.133) that

$$y \frac{dF}{dy} + F(y) - \frac{E(y)}{1 - y^2} = -\frac{y^2}{1 - y^2} \int_0^1 \left(\frac{d}{dx} \sqrt{\frac{x(1 - x)}{1 - y^2 x}} \right) dx .$$

You should therefore find that you've established Eq. (P7.12). Why?

7.24 Derive Eq. (7.138) starting from Eq. (7.137). Hint: Show for $K \approx K_c$ that $\sinh 2K \approx 1 + \sqrt{2} \sinh 2(K - K_c)$. Then use an approximate result every student of advanced physics should know, $\ln(1 + x) \approx x$ for $x \ll 1$.

7.25 Does Eq. (7.126) imply a rigid rotation? Write the equation in matrix form. Is it an orthogonal matrix, i.e., is its transpose the same as its inverse? Show that the transformation in Eq. (7.126) can be written as a rigid rotation through $45°$, followed by a scale transformation,

$$\begin{pmatrix} \frac{1}{2} & \frac{1}{2} \\ -1 & 1 \end{pmatrix} = \begin{pmatrix} \frac{1}{2} & 0 \\ 0 & 1 \end{pmatrix} \begin{pmatrix} 1 & 1 \\ -1 & 1 \end{pmatrix} .$$

7.26 Prove the statement made in Section 7.11 that if $f(x) \le g(x)$ and $f(x) \sim x^\psi$ and $g(x) \sim x^\phi$ as $x \to 0^+$, then $\psi \ge \phi$. Hint: $\ln(x) \le 0$ for $x \le 1$.

7.27 Derive Rushbrooke's inequality, (7.140), starting from the inequality (7.139).

7.28 Derive Rushbrooke's inequality for fluid systems, which takes a few steps.

a. We derived in Section 7.3.3 an expression for the specific heat of coexisting liquid and gas phases, Eq. (7.32). Show that Eq. (7.32) can be written

$$c_v = x_g c_{v_g} + x_l c_{v_l} + \frac{x_g T}{\beta_T^g V} \left(\frac{\partial v_g}{\partial T} \right)^2_{\text{coex}} + \frac{x_l T}{\beta_T^l V} \left(\frac{\partial v_l}{\partial T} \right)^2_{\text{coex}} ,$$

where β_T^l, β_T^g are the isothermal compressibilities of the liquid and gas phases.

b. Conclude that

$$c_v \geq \frac{x_g T}{V} \frac{1}{\beta_T^g} \left(\frac{\partial v_g}{\partial T} \right)_{\text{coex}}^2, \tag{P7.13}$$

which is the fluid analog of the inequality for magnets, (7.139). Hint: You have an expression for c_v as a sum of four positive terms, $c_v \equiv A + B + C + D$. Show that the inequality (P7.13) follows.

c. As $T \to T_c^-$, $x_l, x_g \to \frac{1}{2}$, and the order parameter $\phi = v_g - v_l \to 0$. Show that Rushbrooke's inequality (7.140) follows from the inequality (P7.13).

7.29 Derive the inequalities in (7.144). As is true in many problems: Draw a picture. Then derive the inequality (7.145).

7.30 Thermodynamics is rather heavy on relations involving derivatives (you may have noticed). Consider the Helmholtz and Gibbs energies for fluids, $F = F(T, V)$ and $G = G(T, P)$, the second-order derivatives of which are related to response functions. From Eq. (P1.5) and Exercise 4.35, $(\partial^2 F / \partial T^2)_V = -C_V / T$ and $(\partial^2 F / \partial V^2)_T = 1/(\beta_T V)$. From Exercise 4.36, $(\partial^2 G / \partial T^2)_P = -C_P / T$ and $(\partial^2 G / \partial P^2)_T = -\beta_T V$. For magnetic systems, with $F = F(T, M)$ and $G = G(T, B)$, we have from Eq. (7.62), $(\partial^2 F / \partial T^2)_M = -C_M / T$ and $(\partial^2 G / \partial T^2)_B = -C_B / T$. Complete this catalog of second derivatives. Show that

$$\left(\frac{\partial^2 F}{\partial M^2} \right)_T = \frac{1}{\chi} \qquad \left(\frac{\partial^2 G}{\partial B^2} \right)_T = -\chi. \tag{P7.14}$$

Note the correspondence $\chi \leftrightarrow \beta_T V$ (under the substitutions $M \leftrightarrow V$, $B \leftrightarrow -P$), between the derivatives in Eq. (P7.14) for magnets with those for fluids, $(\partial^2 F / \partial V^2)_T = 1/(\beta_T V)$ and $(\partial^2 G / \partial P^2)_T = -\beta_T V$.

CHAPTER 8

Scaling theories and the renormalization group

T HE primary goal of modern theories of critical phenomena is the evaluation of critical exponents from first principles, implying the need to get beyond mean field theories which predict
the same exponents independent of the nature of the system. Yet what recourse do we have, short
of exact evaluations of partition functions? A new approach emerged in the 1960s based on two
ideas—*scaling* and *renormalization*—the subject of this chapter. Mean field theory is valid in its
treatment of critical phenomena in spatial dimensions $d > 4$ (Section 7.9.2). Ordered phases do
not exist for $d = 1$ (Section 7.12), although some critical exponents can be defined at the critical
temperature $T_c = 0$[120] (see Exercise 8.1). Modern theories must account for the dependence of
exponents on dimensionality for $d < 4$. Furthermore, relations among exponents in the form of
inequalities (Section 7.11) are satisfied as equalities; see Table 8.1—another task for theory.

Table 8.1: Inequalities among critical exponents are satisfied as equalities (within numerical uncertainties for the $d = 3$ Ising model, using exponent values from Table 7.3).

	$\alpha + 2\beta + \gamma - 2$ (Rushbrooke, ≥ 0)	$\alpha + \beta(1 + \delta) - 2$ (Griffiths, ≥ 0)	$(2 - \eta)\nu - \gamma$ (Fisher, ≥ 0)
Mean field theory	0	0	0
$d = 2$ Ising model	0	0	0
$d = 3$ Ising model	-0.001	-0.002	0.000

8.1 THE WIDOM SCALING HYPOTHESIS

The Rushbrooke and Griffiths inequalities, (7.140) and (7.141), involve the critical exponents derived from the free energy, $\alpha, \beta, \gamma, \delta$. Can the model-independent equalities that we see in Table 8.1,
$\alpha + 2\beta + \gamma = 2$ and $\alpha + \beta(1 + \delta) = 2$, be explained? In 1965, B. Widom proposed a mechanism
to account for these relations.[121] Widom's hypothesis assumes the free energy function can be
decomposed into a *singular part*, which we'll denote[1] G_s, and a regular part, G_r, $G = G_r + G_s$.
The regular part G_r varies slowly in the critical region; critical phenomena are associated with G_s.

---

[1] We use the magnetic Gibbs energy, $G = G(T, B)$, for convenience (see Eq. (1.46)); it's not necessary, however—any
other thermodynamic potential could be used.

The *Widom scaling hypothesis* is that G_s is a *generalized homogeneous function*, such that

$$\lambda^{-1} G_s(\lambda^a t, \lambda^b B) = G_s(t, B) , \tag{8.1}$$

where $\lambda > 0$ is the *scaling parameter*, t denotes the reduced temperature, Eq. (7.33), and a, b are constants, the *scaling exponents*.[2,3] Equation (8.1) asserts a geometric property of $G_s(t, B)$ that, when t is "stretched" by λ^a, $t \to \lambda^a t$, and at the same time B is stretched by λ^b, $B \to \lambda^b B$, $G_s(\lambda^a t, \lambda^b B)$ has the value $G_s(t, B)$ when scaled by λ^{-1}. Let's see how this idea helps us.

Differentiate Eq. (8.1) with respect to B, which we can write

$$\lambda^b \frac{\partial}{\partial (\lambda^b B)} G_s(\lambda^a t, \lambda^b B) = \lambda \frac{\partial}{\partial B} G_s(t, B) . \tag{8.2}$$

Equation (8.2) implies (using Eq. (7.61)) that the critical magnetization satisfies a scaling relation,

$$M(t, B) = \lambda^{b-1} M(\lambda^a t, \lambda^b B) . \tag{8.3}$$

Equation (8.3) indicates, as a consequence of Eq. (8.1), that in the critical region, under $t \to \lambda^a t$ and $B \to \lambda^b B$, $M(\lambda^a t, \lambda^b B)$ has the value $M(t, B)$ when scaled by λ^{b-1}. The scaling hypothesis asserts the existence of the exponents a, b, but does not specify their values. They are chosen to be consistent with critical exponents, as we now show.

Two exponents are associated with the equation of state in the critical region: β as $t \to 0$ for $B = 0$, and δ as $B \to 0$ for $t = 0$. The strategy in working with scaling relations such as Eq. (8.3) is to recognize that if it holds for all values of λ, it holds for particular values as well. Set $B = 0$ in Eq. (8.3): $\lambda^b M(\lambda^a t, 0) = \lambda M(t, 0)$. Now let $\lambda = t^{-1/a}$, in which case (show this)

$$M(t, 0) = t^{(1-b)/a} M(1, 0) . \tag{8.4}$$

Comparing with the definition of β ($M(t, 0) \overset{t \to 0}{\sim} t^\beta$), we identify

$$\frac{1-b}{a} = \beta . \tag{8.5}$$

Now play the game again. Set $t = 0$ in Eq. (8.3) and let $\lambda = B^{-1/b}$: $M(0, B) = B^{(1/b)-1} M(0, 1)$. Comparing with the definition of δ ($M(0, B) \overset{B \to 0}{\sim} B^{1/\delta}$), we identify

$$\frac{1}{b} - 1 = \frac{1}{\delta} . \tag{8.6}$$

We therefore have two equations in two unknowns (Eqs. (8.5) and (8.6)), which are readily solved:

$$a = \frac{1}{\beta(\delta + 1)} \qquad b = \frac{\delta}{\delta + 1} . \tag{8.7}$$

If we know β, δ, we know a, b.

What about α, γ? Equations (8.5), (8.6) follow from the equation of state, obtained from the first derivative of G, $M = -(\partial G/\partial B)_T$, Eq. (7.61).[4] The critical exponents α, γ are associated

---

[2]Homogeneous functions F (defined in Section 1.10) have the property $F(\lambda x_1, \cdots, \lambda x_k) = \lambda^p F(x_1, \cdots, x_k)$. *Generalized* homogeneous functions are defined such that $F(\lambda^{a_1} x_1, \cdots, \lambda^{a_k} x_k) = \lambda F(x_1, \cdots, x_k)$. Why not define them so that $F(\lambda^{a_1} x_1, \cdots, \lambda^{a_k} x_k) = \lambda^p F(x_1, \cdots, x_k)$? Let $\lambda \to \lambda^{1/p}$, in which case $F(\lambda^{a_1} x_1, \cdots, \lambda^{a_k} x_k) = \lambda^p F(x_1, \cdots, x_k)$ implies $F(\lambda^{a_1'} x_1, \cdots, \lambda^{a_k'} x_k) = \lambda F(x_1, \cdots, x_k)$, where $a_i' \equiv a_i/p$; we're back to the beginning.

[3]The Legendre transformation of a generalized homogeneous function is another generalized homogeneous function; Exercise 8.3. The scaling hypothesis therefore applies to any of the thermodynamic potentials (see Section 1.4).

[4]What about the other partial derivative of G? From Eq. (7.61), $(\partial G/\partial T)_B = -S$, and there is no critical exponent associated with entropy. At continuous phase transitions, $\Delta s = 0$.

with *second* derivatives of G, $(\partial^2 G/\partial T^2)_B = -C_B/T$ and $(\partial^2 G/\partial B^2)_T = -\chi$, Eqs. (7.62) and (P7.14). Differentiating Eq. (8.1) twice with respect to B,

$$\lambda^{2b}\frac{\partial^2}{\partial(\lambda^b B)^2}G_s(\lambda^a t, \lambda^b B) = \lambda\frac{\partial^2}{\partial B^2}G_s(t, B) \implies \chi(t, B) = \lambda^{2b-1}\chi(\lambda^a t, \lambda^b B) . \tag{8.8}$$

Equation (8.8) is a scaling relation for χ in the critical region. Set $B = 0$ in Eq. (8.8) and let $\lambda = t^{-1/a}$, $\chi(t, 0) = t^{(1-2b)/a}\chi(1, 0)$. Comparing with the definition of γ, $\chi \sim t^{-\gamma}$, we identify

$$\frac{2b - 1}{a} = \gamma . \tag{8.9}$$

Differentiating Eq. (8.1) twice with respect to t, we find $C_B(t, B) = \lambda^{2a-1}C_B(\lambda^a t, \lambda^b B)$. Set $B = 0$ and $\lambda = t^{-1/a}$, implying $C_B(t, 0) = t^{(1/a)-2}C_B(1, 0)$, and thus for $C_B(t, 0) \sim t^{-\alpha}$,

$$2 - \frac{1}{a} = \alpha . \tag{8.10}$$

Equations (8.9) and (8.10) are two equations in two unknowns, implying

$$a = \frac{1}{2 - \alpha} \qquad b = \frac{1}{2}\left(1 + \frac{\gamma}{2 - \alpha}\right) . \tag{8.11}$$

If we know α, γ, we know a, b. Note that we also know a, b if we know β, δ; Eq. (8.7).

Thus, through application of thermodynamics to the proposed scaling form of the free energy, Eq. (8.1), we've found four equations in the two unknowns a, b (Eqs. (8.5), (8.6), (8.9), (8.10)), implying the existence of relationships among $\alpha, \beta, \gamma, \delta$. We've just shown that if we know β, δ, we know α, γ, and conversely (see Exercise 8.7). Using Eq. (8.7) for a, b in Eq. (8.9), we find

$$\gamma = \beta(\delta - 1) . \tag{8.12}$$

We noted in (7.146) the Griffiths inequality $\gamma \geq \beta(\delta-1)$, which is satisfied as an equality among the classical exponents and those for the $d = 2$ and $d = 3$ Ising models, within numerical uncertainties. The scaling hypothesis therefore accounts for the equality in (8.12) that otherwise we would have known only empirically. Equating the result for a from Eq. (8.7) with that in Eq. (8.11), we find

$$\alpha + \beta(\delta + 1) = 2 , \tag{8.13}$$

a relation that satisfies Griffiths inequality (7.141) as an equality. By eliminating δ between Eqs. (8.12) and (8.13), we find $\alpha + 2\beta + \gamma = 2$, Rushbrooke's inequality (7.140), satisfied as an equality.

The scaling hypothesis is phenomenological: It's designed to account for the relations among critical exponents that we've found to be true empirically. We have (at this point) no microscopic justification for scaling behavior of singular thermodynamic functions. Because of its simplicity, however, and because of its successful predictions, it stimulated research into a fundamental understanding of scaling (the subject of coming sections). If the scaling hypothesis is lacking theoretical support (at this point), what about experimental? In Eq. (8.3) let $\lambda = |t|^{-1/a}$,

$$M(t, B) = |t|^{-(b-1)/a}M\left(\frac{t}{|t|}, \frac{B}{|t|^{b/a}}\right) = |t|^\beta M\left(\pm 1, \frac{B}{|t|^{\beta+\gamma}}\right) , \tag{8.14}$$

where we've used Eq. (8.5) and $b/a = \beta + \gamma$ (Exercise 8.8). Equation (8.14) can be inverted,

$$\frac{B}{|t|^{\beta+\gamma}} = f_\pm\left(|t|^{-\beta}M(t, B)\right) , \tag{8.15}$$

i.e., $B/(|t|^{\beta+\gamma})$ is predicted to be a function f (with two branches $f_{\pm}$, corresponding to $t > 0$ and $t < 0$) of the single variable $|t|^{-\beta}M(t, B)$. Equation (8.15) is a definite prediction of the scaling hypothesis, subject to validation. Figure 8.1 is a plot of the scaled magnetic field versus the scaled magnetization for the ferromagnet $CrBr_3$ [122]. The critical exponents β, γ were independently measured in zero field, the values of which were used to calculate the scaled field and magnetization. Measurements of M, B were made along isotherms for $T < T_c$ and $T > T_c$, and the data beautifully fall on two curves. Figure 8.1 offers compelling evidence for the scaling hypothesis.

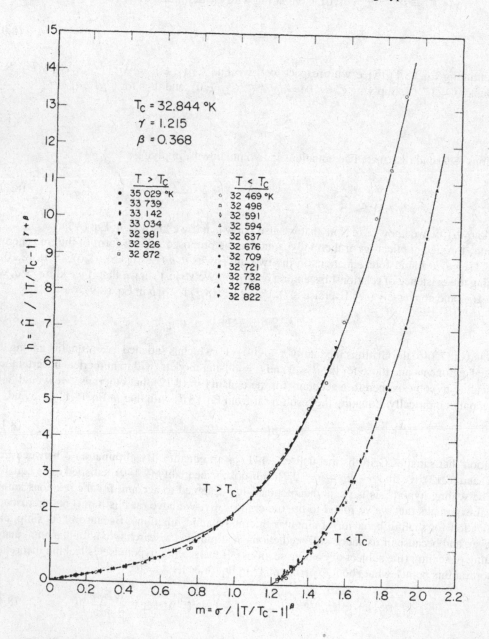

Figure 8.1: Scaled magnetic field $B/|t|^{\beta+\gamma}$ versus scaled magnetization $M/|t|^{\beta}$ for the ferromagnet $CrBr_3$. Reprinted figure with permission from J.T. Ho and J.D. Litster, Phys. Rev. Lett., **22**, p. 603, (1969). Copyright (2020) by the American Physical Society.

8.2   KADANOFF SCALING THEORY: SCALING EXPONENTS

The scaling hypothesis accounts for the relations we find among critical exponents and it has experimental support. It's incumbent upon us therefore to find a microscopic understanding of scaling. We present the physical picture developed by L.P. Kadanoff[123] that provides a conceptual basis for scaling and which has become part of the standard language of critical phenomena.

Consider a square lattice of lattice constant a, such as in the left part of Fig. 8.2, on which we

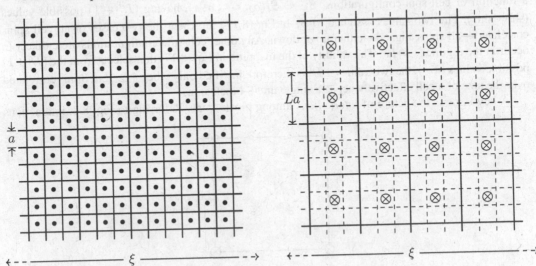

Figure 8.2: Two ways of looking at the same problem. Left: Square lattice of lattice constant a; circles (•) represent Ising spins. In the critical region, $\xi \gg a$. Right: Square lattice with scaled lattice constant La, such that $L \ll \xi/a$; crosses (⊗) represent block spins. $L = 3$ in this example.

have Ising spins (•) coupled through nearest-neighbor interactions, with Hamiltonian

$$H = -J \sum_{\langle ij \rangle} \sigma_i \sigma_j - b \sum_i \sigma_i \,, \tag{8.16}$$

where $\langle ij \rangle$ indicates a sum over nearest neighbors, on any lattice.[5] In the critical region the correlation length ξ becomes macroscopic (critical opalescence, Section 6.7), implying the obvious inequality $\xi \gg a$, yet this simple observation lies at the core of Kadanoff's argument. So far, the lattice constant has been irrelevant in our treatment of phase transitions (often set to unity for convenience and forgotten thereafter). In many areas of physics, the length scale determining the relevant physics is fixed by fundamental constants.[6] The correlation length ξ is an *emergent* length that develops macroscopic proportions as $T \to T_c$, and is not fixed once and for all. A divergent correlation length in essence describes a new state of matter of highly correlated fluctuations.

Following Kadanoff, redefine the lattice constant, $a \to a' \equiv La$, $L \ll \xi/a$, an operation specifying a new lattice with the same symmetries as the original; the case for $L = 3$ on the square lattice is shown in the right part of Fig. 8.2. On the original lattice there is one spin per unit cell. The unit cell of the scaled lattice contains Ld Ising spins, where we allow for an arbitrary dimension d, not just $d = 2$ as in Fig. 8.2. That is, spins are still in their places; we've just redefined the unit

---

[5]The sum is restricted so that nearest-neighbor pairs are counted only once.

[6]The de Broglie wavelength is determined by Planck's constant $\lambda = h/p$, the Schwarzschild radius $r_S = 2Gm/c^2$ is associated with mass m, and the Compton wavelength is associated with m, $\lambda_C = h/(mc)$. The de Broglie wavelength separates classical and (nonrelativistic) quantum mechanics, the Compton wavelength separates nonrelativistic from relativistic quantum mechanics, and the Schwarzschild radius determines whether a general-relativistic treatment is required.

of length (a *scale transformation*, or a *dilatation*). Label unit cells on the scaled lattice with capital Roman letters. Define a variable $\widetilde{S}_I$ representing the degrees of freedom in the Ith cell,

$$\widetilde{S}_I \equiv \sum_{i \in I} \sigma_i \overset{\xi \gg La}{\approx} \pm L^d \equiv S_I L^d , \qquad (8.17)$$

where $S_I = \pm 1$ is a new Ising spin, the *block spin* (denoted $\otimes$ in Fig 8.2). Technically $\widetilde{S}_I$ is a function of cell-spin configurations, $\widetilde{S}_I = \widetilde{S}_I(\sigma_1, \cdots, \sigma_{L^d})$, having $(L^d + 1)$ possible values (show this). The two values assigned to $\widetilde{S}_I$ in Eq. (8.17) represent the two cases of all cell spins correlated—all spins aligned—either up or down. Any other cell-spin configurations are unlikely to occur in the regime $\xi \gg La$. The validity of the mapping in Eq. (8.17) that selects from the $(L^d + 1)$ possible values of $\widetilde{S}_I$, the two values $\pm L^d$ is therefore justified only when $\xi \gg La$. Spins correlated over the size of a cell effectively act as a single unit—block spins.

A key issue is the nature of *interactions* among block spins. Referring to Fig. 8.3, interactions

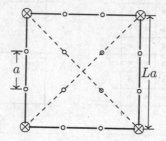

Figure 8.3: Block spins ($\otimes$) interact with nearest neighbors (solid lines) through the nearest-neighbor couplings of the spins of the original lattice (open circles); next nearest neighbor interactions among block spins (dashed lines) can be mediated through the correlated spins of the cell.

between near-neighbor block spins (solid lines) are induced through the interactions of the spins on the original lattice (shown as open circles). New types of interactions, however, not in the original Hamiltonian are possible among block spins: next-near neighbor (dashed lines) or four-spin interactions. That is a complication we'll need to address. For now, assume that block spins interact only with nearest neighbors and the magnetic field. Let's write, therefore, a Hamiltonian for block spins:

$$H_L \equiv -J_L \sum_{\langle IJ \rangle} S_I S_J - b_L \sum_I S_I , \qquad (8.18)$$

where J_L is the near-neighbor interaction strength between block spins associated with lattice constant La; likewise b_L is the coupling of S_I to the magnetic field. Clearly $J_1 \equiv J$ and $b_1 \equiv b$. How we might calculate J_L, b_L will concern us in upcoming sections; for now we take them as given.

We're considering, from a theoretical perspective, a change in the unit of length for systems in their critical regions, $a \to La$, where $\xi \gg La > a$. The correlation length is a function of $K \equiv \beta J$, $B \equiv \beta b$, $\xi = \xi(K, B)$. On the scaled lattice, $\xi = \xi(K_L, B_L)$. Because it's the *same length*, however, measured in two ways,[7]

$$\xi(K_L, B_L) = L^{-1} \xi(K, B) . \qquad (8.19)$$

For N spins on the original lattice, there are $N' \equiv N/L^d$ block spins on the scaled lattice. Because it's the same system described in two ways, the free energy is invariant. Let $g(t, B) \equiv G(t, B)/N$ denote the Gibbs energy per spin. Thus,

$$Ng(t, B) = N'g(t_L, B_L) \implies g(t, B) = L^{-d} g(t_L, B_L) , \qquad (8.20)$$

---

[7] A distance measured in kilometers is numerically a thousand times smaller than the same distance measured in meters.

where $t = (K_c/K) - 1$. Equations (8.19) and (8.20) constrain the form of K_L, B_L; such relations hold only in the critical region, $\xi \gg La$, and thus they apply for T near T_c and for small B.

Because $\xi(K_L, B_L) < \xi(K, B)$ (Eq. (8.19)), the transformed system is *further* from the critical point (where $\xi \to \infty$). Let's *assume*, in order to satisfy Eqs. (8.19), (8.20), that t_L and B_L are in the form

$$t_L = L^x t \qquad\qquad B_L = L^y B, \qquad\qquad (8.21)$$

where x, y are new scaling exponents (independent of L). We require $x, y > 0$ so that we move away from the critical point under $a \to La$ for $L > 1$. Combining Eq. (8.21) with Eq. (8.20),

$$g(t, B) = L^{-d} g(L^x t, L^y B). \qquad\qquad (8.22)$$

Under the assumptions of Kadanoff's analysis, the free energy per spin is a generalized homogeneous function. Whereas Eq. (8.1) is posited to hold for an arbitrary *mathematical* parameter λ, the parameter L in Eq. (8.22) has physical meaning (and is not entirely arbitrary: $L \ll \xi/a$). If we let $L \to L^{1/d}$, Eq. (8.22) has the form of Eq. (8.1):

$$g(t, B) = L^{-1} g(L^{x/d} t, L^{y/d} B). \qquad\qquad (8.23)$$

The exponents a, b in Eq. (8.1) correspond to Kadanoff's exponents, $a \leftrightarrow x/d$ and $b \leftrightarrow y/d$.

Kadanoff has shown therefore, *for systems in their critical regions*, how, *if* under a change in length scale $a \to La$ the couplings t_L, B_L transform as in Eq. (8.21), then the free energy exhibits Widom scaling. Of the two assumptions on which the theory rests, that block spins interact with nearest neighbors and t_L, B_L scale as in Eq. (8.21), the latter is most in need of justification, and we'll come back to it (Sections 8.4, 8.6). The larger point, however, is that *the block-spin picture implies a new paradigm*, that changes in length scale (a spatial quantity) *induce* changes in couplings, thereby providing a physical mechanism for scaling.

We've noted previously (Section 6.5.3) the distinction between thermodynamic and structural quantities (correlation functions); the former can be derived from the free energy, the latter cannot. Kadanoff's scaling picture, *which links thermodynamic quantities with spatial considerations*, can be used to develop a scaling theory of correlation functions, thereby relating the critical exponents ν, η to the others, $\alpha, \beta, \gamma, \delta$. Define the two-point correlation function for block spins,

$$C(r_L, t_L, B_L) \equiv \langle S_I S_J \rangle - \langle S_I \rangle \langle S_J \rangle, \qquad\qquad (8.24)$$

where r_L is the distance between the cells associated with block spins S_I, S_J (measured in units of La). Let r denote the same distance measured in units of a; thus, $r_L = L^{-1} r$ (see Eq. 8.19). In order for block-spin correlations to be well defined, we require the inter-block separation be much greater than the size of cells, $r_L \gg La \implies r \gg L^2 a > a$. A scaling theory of correlation functions is possible only for long-range correlations, $r \gg a$.

In formulating the block spin idea, we noted in Eq. (8.17) the approximate correspondence $S_I \approx L^{-d} \sum_{i \in I} \sigma_i$. How accurate is that association, and does it hold the same for all dimensions d? Consider the interaction of the spin system with a magnetic field, supposing the inter-spin couplings have been turned off. Because we're looking at the same system in two ways,

$$b_L \sum_I S_I \overset{\text{what we want}}{=} b \sum_i \sigma_i \overset{\text{by definition}}{=} b \sum_I \sum_{i \in I} \sigma_i \overset{\text{Eq. (8.17)}}{\approx} b L^d \sum_I S_I. \qquad (8.25)$$

Equation (8.25) invites us to infer $b_L = b L^d$, which, comparing with Eq. (8.21), implies the scaling exponent $y = d$. If that were the case, it would imply the Widom scaling exponent $b = y/d = 1$, and thus $\delta \to \infty$ (Eq. (8.6)). To achieve consistency, we take, instead of Eq. (8.17),

$$S_I = L^{-y} \sum_{i \in I} \sigma_i, \qquad\qquad (8.26)$$

where we expect $y \lesssim d$. Nowhere in our analysis did we make explicit use of Eq. (8.17).

Substituting Eq. (8.26) in Eq. (8.24),

$$C(r_L, t_L, B_L) = \langle S_I S_J \rangle - \langle S_I \rangle \langle S_J \rangle = L^{-2y} \sum_{i \in I} \sum_{j \in J} [\langle \sigma_i \sigma_j \rangle - \langle \sigma_i \rangle \langle \sigma_j \rangle]$$

$$= L^{-2y} L^{2d} [\langle \sigma_i \sigma_j \rangle - \langle \sigma_i \rangle \langle \sigma_j \rangle] \equiv L^{2(d-y)} C(r, t, B) . \tag{8.27}$$

Equation (8.27) implies, using Eq. (8.21) and $r_L = r/L$, the scaling form for correlation functions:

$$C(r, t, B) = L^{2(y-d)} C(r/L, L^x t, L^y B) . \tag{8.28}$$

Let $L = |t|^{-1/x}$, and thus, from Eq. (8.28),

$$C(r, t, B) = |t|^{2(d-y)/x} f_{\pm} \left(|t|^{1/x} r, \frac{B}{|t|^{y/x}} \right) = \frac{1}{r^{2(d-y)}} \left(|t|^{1/x} r \right)^{2(d-y)} f_{\pm} \left(|t|^{1/x} r, \frac{B}{|t|^{y/x}} \right) \tag{8.29}$$

$$\equiv \frac{1}{r^{2(d-y)}} g_{\pm} \left(|t|^{1/x} r, \frac{B}{|t|^{y/x}} \right) ,$$

where $f_{\pm}, g_{\pm}$ are scaling functions.

Set $B = 0$ in Eq. (8.29) and then let $|t| = 0$. Correlation functions decay algebraically with distance at $T = T_c$ (Section 7.6), and thus we identify, using Eq. (7.75),

$$2(d - y) = d - 2 + \eta \implies y = \frac{1}{2} (d + 2 - \eta) . \tag{8.30}$$

Again set $B = 0$; we know as $|t| \to 0$, long-range correlation functions are associated with the correlation length ξ. We require that $r/\xi \sim r|t|^{\nu} = r|t|^{1/x}$, implying

$$x = \frac{1}{\nu} . \tag{8.31}$$

With these identifications of x, y, combined with the Widom scaling exponents $x = ad$, $y = bd$, and using the expressions for a, b in Eqs. (8.7), (8.11), there are numerous interrelations involving ν, η with the other critical exponents. For example, one can show

$$2 - \eta = d \left(\frac{\delta - 1}{\delta + 1} \right) \qquad d\nu = 2 - \alpha \qquad \gamma = \nu(2 - \eta) . \tag{8.32}$$

8.3   RENORMALIZATION: A FIRST LOOK

The Widom scaling hypothesis Eq. (8.1) accounts for relations among the exponents $\alpha, \beta, \gamma, \delta$. The block-spin picture (Section 8.2) motivates Eq. (8.1) and shows the exponents η, ν are related to $\alpha, \beta, \gamma, \delta$ and d. The upshot is, that of the six critical exponents $\alpha, \beta, \gamma, \delta, \eta, \nu$, *only two are independent*. That statement hinges on the validity of Eq. (8.21), Kadanoff's scaling form for block-spin couplings. It's time we got down to the business of calculating block-spin couplings, a process known as *renormalization*. We'll see (Sections 8.4, 8.6) how understanding the physics behind Eq. (8.21) leads to a more general theory known as the *renormalization group*.[8]

---

[8]This subject provides a nice example of theory development. Kadanoff, in his original scaling article[123], says of the Widom hypothesis, "...his idea about the homogeneity of the singular part of the free energy is not given any very strong justification beyond the fact that it appears to work." Yet Kadanoff does not offer a particularly strong justification for his scaling form Eq. (8.21), beyond the fact that it motivates Widom scaling. In 1971, K.G. Wilson showed how Kadanoff scaling occurs quite generally in a more comprehensive theory (the renormalization group).[124] Wilson received the 1982 Nobel Prize in Physics for work on phase transitions and critical phenomena.

8.3.1 Decimation method for one-dimensional systems

We start with the $d = 1$ Ising model with near-neighbor couplings where we double the lattice constant, a system simple enough that we can carry out all steps exactly. Figure 8.4 shows a one-

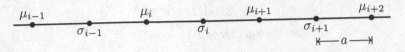

Figure 8.4: One-dimensional lattice of lattice constant a; block spins are denoted μ.

dimensional lattice with lattice constant a, with Ising spins σ on lattice sites, except that on every other site we've renamed the spins μ, which shall be the block spins. One way to define block spins (but not the only way) is simply to rename a subset of the original Ising spins, μ. One finds the interactions between μ-spins by summing over the degrees of freedom of the σ-spins in a *partial evaluation* of the partition function, a technique known as *decimation*. To show that, it's convenient to work with a dimensionless Hamiltonian, $\mathcal{H} \equiv -\beta H$. Thus, for the $d = 1$ Ising model, $\mathcal{H} = K \sum_i \sigma_i \sigma_{i+1} + B \sum_i \sigma_i$. Referring to Fig. 8.4, we can rewrite the Hamiltonian (exactly) assuming that even-numbered spins σ_{2i} are named μ_i,

$$
\mathcal{H} = \sum_{i=1}^{N} \left(K \sigma_i \sigma_{i+1} + B \sigma_i \right) \stackrel{\sigma_{2i} \to \mu_i}{\Rightarrow} \sum_{i=1}^{N/2} \left(\sigma_{2i+1} \left[K \left(\mu_i + \mu_{i+1} \right) + B \right] + \frac{1}{2} B \left(\mu_i + \mu_{i+1} \right) \right),
$$

(8.33)

where we've written the coupling of the μ-spins to the B-field in the form $\frac{1}{2}B(\mu_i + \mu_{i+1})$ to "share" the μ-spins surrounding each odd-numbered σ-spin.[9] For the partition function,

$$
Z_N(K, B) = \sum_{\{\sigma\}} e^{\mathcal{H}} \equiv \prod_{i=1}^{N} \sum_{\sigma_i = -1}^{1} \exp \left(\sum_{j=1}^{N} \left(K \sigma_j \sigma_{j+1} + B \sigma_j \right) \right)
$$

(8.34)

$$
\stackrel{\sigma_{2i} \to \mu_i}{\Rightarrow} \prod_{i=1}^{N/2} \sum_{\mu_i = -1}^{1} \sum_{\sigma_{2i+1} = -1}^{1} \exp \left(\sum_{n=1}^{N/2} \left[K \sigma_{2n+1} \left(\mu_n + \mu_{n+1} \right) + B \sigma_{2n+1} + \frac{1}{2} B \left(\mu_n + \mu_{n+1} \right) \right] \right).
$$

The strategy is to evaluate Eq. (8.34) in two steps: Sum first over the σ-degrees of freedom—a partial evaluation of the partition function—and then those associated with the μ-spins. The sum over σ_{2i+1} in Eq. (8.34) is straightforward:

$$
\sum_{\sigma_{2n+1} = -1}^{1} e^{\sigma_{2n+1} [K(\mu_n + \mu_{n+1}) + B]} = 2 \cosh \left[K \left(\mu_n + \mu_{n+1} \right) + B \right].
$$

(8.35)

Our goal is to *exponentiate* the terms we find after summing over σ_{2n+1}, i.e., fit the right side of Eq. (8.35) to an exponential form. We want the following relation to hold:

$$
2 \exp \left(\frac{1}{2} B \left(\mu_n + \mu_{n+1} \right) \right) \cosh \left[K \left(\mu_n + \mu_{n+1} \right) + B \right]
$$

$$
\equiv \exp \left(K_0 + K' \mu_n \mu_{n+1} + \frac{1}{2} B' \left(\mu_n + \mu_{n+1} \right) \right).
$$

(8.36)

---

[9] We used the same step in setting up the transfer matrix, Eq. (6.78).

Three parameters, K_0, K', B', are required to match the three independent configurations of μ_n, μ_{n+1}: both up, ↑↑; both down, ↓↓; and anti-aligned, ↑↓ or ↓↑. We require

$$
\begin{array}{lll}
\text{↑↑:} & 2e^B \cosh(2K + B) = e^{K_0+K'+B'} & \text{①} \\[2mm]
\text{↑↓:} & 2\cosh B = e^{K_0-K'} & \text{②} \\[2mm]
\text{↓↓:} & 2e^{-B} \cosh(2K - B) = e^{K_0+K'-B'}. & \text{③}
\end{array}
\tag{8.37}
$$

We can isolate K_0, K', B' through combinations of the equations in (8.37),

$$
\begin{aligned}
\text{①} \times \text{②}^2 \times \text{③} &\implies e^{4K_0} = 16\cosh^2 B \cosh(2K + B)\cosh(2K - B) \\[2mm]
\frac{\text{①} \times \text{③}}{\text{②}^2} &\implies e^{4K'} = \frac{\cosh(2K + B)\cosh(2K - B)}{\cosh^2 B} \\[2mm]
\frac{\text{①}}{\text{③}} &\implies e^{2B'} = e^{2B}\frac{\cosh(2K + B)}{\cosh(2K - B)}.
\end{aligned}
\tag{8.38}
$$

Thus, we have in Eq. (8.38) explicit expressions for K_0, K', B'; they are known quantities. Combining Eq. (8.36) with Eq. (8.34), we find

$$
Z_N(K, B) = \sum_{\{\sigma\}} e^{\mathcal{H}} = e^{NK_0/2}\sum_{\{\mu\}} e^{\mathcal{H}'} = e^{NK_0/2}Z_{N/2}(K', B'),
\tag{8.39}
$$

where $\sum_{\{\mu\}} \equiv \prod_{i=1}^{N/2}\sum_{\mu_i=-1}^{1}$ and $\mathcal{H}' \equiv \sum_{i=1}^{N/2}(K'\mu_i\mu_{i+1} + B'\mu_i)$. The quantities K', B' are known as *renormalized* couplings, with $\mathcal{H}'$ the renormalized Hamiltonian.[10] In stretching the lattice constant $a \to 2a$, we have at the same time "thinned," or *coarse grained*, the number of spins $N \to N/2$, which interact through effective couplings K', B'. Equation (8.39) indicates that while the number of states available[11] to the renormalized system is less than the original ($Z_{N/2} < Z_N$), the equality $Z_N(K, B) = e^{NK_0/2}Z_N(K', B')$ is maintained[12] through the factor of $e^{NK_0/2}$, where we note that $K_0 \neq 0$, even if $K = 0$ or $B = 0$ (see (8.38)).

We've achieved the first part of the Kadanoff construction. Starting with a near-neighbor model with couplings K, B, we have, upon $a \to 2a$, another near-neighbor model with couplings K', B'. By the scaling hypothesis, K', B' should occur further from the critical point at $T = 0, B = 0$ with $K' < K$ and $B' > B$. To examine the *recursion relations* $K' = K'(K, B)$, $B' = B'(K, B)$, it's easier in this case (because $T_c = 0$) to work with $x \equiv e^{-4K}$ and $y \equiv e^{-2B}$, in terms of which the critical point occurs at $x = 0, y = 1$. With $x' \equiv e^{-4K'}$, $y' \equiv e^{-2B'}$, we find from (8.38) an equivalent form of the recursion relations

$$
x' = \frac{x(1 + y)^2}{x(1 + y^2) + y(1 + x^2)} \qquad\qquad y' = y\left(\frac{y + x}{1 + xy}\right),
\tag{8.40}
$$

from which it's readily shown that $y' < y$ if $y < 1$ and $x' > x$ if $x < 1$. Figure 8.5 shows the *flows* that occur under iteration of the recursion relations in (8.40). Starting from a given point in the x-y plane, the renormalized parameters do indeed occur further from the critical point.

We see from Fig. 8.5 that systems with coupling constants $e^{-4K} \ll 1$ (for any B-field strength) are transformed under successive renormalizations into equivalent systems characterized by $K = 0$.

---

[10]We're borrowing a word—renormalization—from quantum field theory, a technique for eliminating divergences by replacing physical parameters with their observed values. Physics thrives on analogies, but we don't have to insist on this one. In the present context, the renormalized coupling constant K' replaces K, the coupling constant associated with one length scale, with that at a larger length scale.

[11]The partition function is the total number of states available to a system at temperature T (Chapter 4).

[12]The block-spin transformation looks at the same system in two ways; there must be invariants such as in Eq. (8.39).

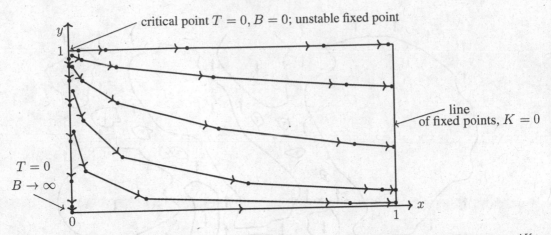

Figure 8.5: Flow of renormalized couplings for the $d = 1$ Ising model, $y \equiv e^{-2B}$, $x \equiv e^{-4K}$. Arrows are placed 90% of the way to the next point.

What may have been a difficult problem for $K \gg 0$ is, by this technique, transformed into an equivalent problem associated with $K = 0$, which is trivially solved.[13] What started as a way to provide a physical underpinning to scaling has turned into a method of solving problems in statistical mechanics. But, back to scaling. Let's check if Eq. (8.21) is satisfied.

We start by examining the recursion relations (8.40) for the occurrence of *fixed points*, values x^*, y^* invariant under the block-spin transformation:

$$x^* = \frac{x^*(1+y^*)^2}{x^*(1+y^{*2}) + y^*(1+x^{*2})} \qquad\qquad y^* = y^* \left(\frac{y^* + x^*}{1 + x^*y^*} \right). \qquad (8.41)$$

Analysis of (8.41) shows fixed points for $x^* = 0, y^* = 1$—the critical point—and a *line* of "high temperature" fixed points $x^* = 1$ for all y. From Fig. 8.5, for K, B near the critical point, the renormalization flows are *away* from the fixed point, an *unstable* fixed point, and toward the high-temperature fixed points. We therefore have another way of characterizing critical points as *unstable fixed points of the recursion relations for couplings*.[14] If for a fixed set of couplings (such as at a fixed point) there is no change in couplings under changes in length scale, *the correlation length is the same for all length scales*, implying *fluctuations of all sizes*, schematically illustrated in Fig. 8.6. The only way ξ can have the same value for any finite length scale is if $\xi \to \infty$. *Unstable fixed points imply infinite correlation lengths*. At the critical point (and only at the critical point) the system appears the same no matter what length scale you use to look at it. Such systems are said to be *self-similar* or *scale invariant*.[15] The critical point is a special state of matter, indeed!

Scale invariance is not typical. In most areas of physics, an understanding of the phenomena requires knowledge of the "reigning physics" at a single scale (length, time, mass, etc.). For example, modeling sound waves in a gas of uranium atoms does not involve subatomic physics. One cannot start with a model of nuclear degrees of freedom, and arrive at the equations of hydrodynamics by continuously varying the length scale. Most theories apply at a definite scale, such as the mean free path between collisions. There *are* systems besides critical phenomena exhibiting scale

---

[13] We already know how to solve the $d = 1$ Ising model; this method allows us to solve more general problems.

[14] Previously we identified critical points as points where the free energy is singular (Section 7.7.3). That will have to be reconciled with the idea of critical points as unstable fixed points. As we argue in the text, fixed points imply $\xi \to \infty$.

[15] Imagine looking at something under a microscope. A scale-invariant system would appear the same no matter the magnification of the microscope.

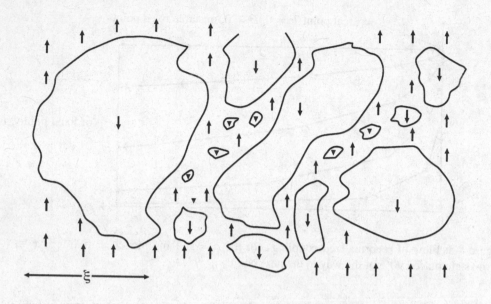

Figure 8.6: At critical points ($\xi \to \infty$) fluctuations of all sizes occur.

invariance—fully developed turbulence in fluids[125][126], in elementary particle physics,[16] and in *fractal* systems.[17] Scale invariance is also observed in systems where the framework of equilibrium statistical mechanics does not apply, such as the *jamming* and *yielding* transitions in granular media [128, 129]

The recursion relations in (8.40) are nonlinear; let's *linearize* them in the vicinity of a fixed point. Taylor expand[18] $x' = x'(x, y)$, $y' = y'(x, y)$ about x^*, y^*,

$$x' = x^* + (x - x^*) \left(\frac{\partial x'}{\partial x}\right)\Big|_{x^*, y^*} + (y - y^*) \left(\frac{\partial x'}{\partial y}\right)\Big|_{x^*, y^*} + \cdots$$

$$y' = y^* + (x - x^*) \left(\frac{\partial y'}{\partial x}\right)\Big|_{x^*, y^*} + (y - y^*) \left(\frac{\partial y'}{\partial y}\right)\Big|_{x^*, y^*} + \cdots . \qquad (8.42)$$

Defining $\delta x \equiv x - x^*$, $\delta y \equiv y - y^*$, and keeping terms to first order,

$$\begin{pmatrix} \delta x' \\ \delta y' \end{pmatrix} = \begin{pmatrix} (\partial x'/\partial x)^* & (\partial x'/\partial y)^* \\ (\partial y'/\partial x)^* & (\partial y'/\partial y)^* \end{pmatrix} \begin{pmatrix} \delta x \\ \delta y \end{pmatrix} = \begin{pmatrix} 4 & 0 \\ 0 & 2 \end{pmatrix} \begin{pmatrix} \delta x \\ \delta y \end{pmatrix}, \qquad (8.43)$$

where we've evaluated at $x^* = 0$, $y^* = 1$ the partial derivatives of x', y' obtained from (8.40). The fixed point associated with the critical point is indeed unstable: Small deviations δx, δy from the critical point are, upon $a \to 2a$, mapped into larger deviations $\delta x'$, $\delta y'$. Are the relations in Eq. (8.43) in the scaling form posited by Eq. (8.21)? Because $L = 2$ in this case, we can write (from Eq. (8.43)) $\delta x' = L^2 \delta x$ and $\delta y' = L \delta y$—nominally the scaling form we seek. The fact, however, that $T_c = 0$ for $d = 1$ complicates the analysis. For small B (near the critical point), $\delta y \equiv y - 1 \approx -2B$;

---

[16]The 2004 Nobel Prize in Physics was awarded to D. Gross, F. Wilczek, and D. Politzer for discoveries of scale invariance (pertaining to energy scales) in particle physics.

[17]Fractals are self-similar geometries, on which one can define physical processes. The standard reference is Mandelbrot[127], but the field has grown considerably since the 1980s.

[18]By differentiating the recursion relations, we're assuming $x' = x'(x, y)$, $y' = y'(x, y)$ are analytic functions. Coupling constants at length L are found by grouping together L^d spins. It's difficult to see how adding up interactions over a finite region can result in singular expressions for the renormalized couplings.

thus $\delta y' = L\delta y$ implies $B' = LB \equiv L^{y_s}B$, where y_s denotes the Kadanoff scaling exponent (to distinguish it from the variable y in use here). Thus, we find $y_s = 1$ for the $d = 1$ Ising model. Comparing with Eq. (8.30), we see that $y_s = 1$ is precisely what we expect because $\eta = 1$ exists for $d = 1$ (see Exercise 8.1). For the coupling K, $e^{-4K'} = 4e^{-4K}$ (Eq. (8.43)) implies $K' = K - \frac{1}{2}\ln 2$ (for large K). Thus, $(\partial K'/\partial K)^* = 1 = L^0$, implying the scaling exponent $x_s = 0$, which in turn implies $\nu \to \infty$ from Eq. (8.31). The critical exponent ν can't be defined for $d = 1$ because ξ doesn't diverge as a power law, but rather exponentially; see Eq. (P6.1). Kadanoff scaling holds in one dimension, but we need to find the "right" scaling variables, here $\delta x, \delta y$.

The scaling form $t_L = L^x t$ (Eq. (8.21)) is written in terms of $t \equiv (T - T_c)/T_c$, where $|t| \to 0$ as $T \to T_c$. We can, equivalently, develop a scaling variable involving $K = J/(kT)$, with

$$K_L = K_c - (K_c - K)L^x .\qquad (8.44)$$

Equation (8.44) indicates that $(K_c - K)$ is the quantity that gets small near the critical point;[19] see Exercise 8.20. We've been writing K_L, B_L in this section as K', B'. Using Eq. (8.44) and $B_L = L^y B$ from Eq. (8.21), we can infer the critical exponents η, ν by connecting these scaling forms with Eqs. (8.30) and (8.31):

$$\frac{1}{x} = \nu = \frac{\ln L}{\ln(\partial K'/\partial K)^*} \qquad y = \frac{1}{2}(d + 2 - \eta) = \frac{\ln(\partial B'/\partial B)^*}{\ln L} .\qquad (8.45)$$

The exponents η, ν can be calculated from the recursion relations at unstable fixed points. Once they're known, the other exponents follow (only two are independent). The renormalization method predicts K_c from the fixed point of the transformation, and, by connecting critical exponents with fixed-point behavior, *scaling emerges from the linearized recursion relations at the fixed point.* The Kadanoff construction, devised to support the scaling hypothesis, turns out to represent a more comprehensive theory (see Section 8.4).

Let's see what else we can do with recursion relations besides finding critical exponents. Define a dimensionless free energy per spin, $\mathcal{F} \equiv \lim_{N\to\infty}(-\beta F/N) = \lim_{N\to\infty}\left(\frac{1}{N}\ln Z_N\right)$. Combining $\mathcal{F}$ with Eq. (8.39),

$$\mathcal{F} = \frac{1}{2}K_0 + \frac{1}{2}\mathcal{F}' ,\qquad (8.46)$$

where $\mathcal{F}' \equiv \mathcal{F}(K', B')$. Make sure you understand the factor of $\frac{1}{2}$ multiplying $\mathcal{F}'$ in Eq. (8.46). Let's iterate Eq. (8.46) twice:

$$\mathcal{F} = \frac{1}{2}K_0 + \frac{1}{2}\mathcal{F}' = \frac{1}{2}K_0 + \frac{1}{2}\left(\frac{1}{2}K_0' + \frac{1}{2}\mathcal{F}''\right) = \frac{1}{2}K_0 + \frac{1}{4}K_0' + \frac{1}{8}K_0'' + \frac{1}{8}\mathcal{F}''' ,$$

where $K_0' \equiv K_0(K', B')$. Generalize to N iterations:

$$\mathcal{F} = \sum_{n=0}^{N} \frac{1}{2^{n+1}}K_0^{(n)} + \frac{1}{2^{N+1}}\mathcal{F}^{(N+1)} ,\qquad (8.47)$$

where $K_0^{(m)} \equiv K_0(K^{(m)}, B^{(m)})$ denotes the mth iterate of K_0, with $K_0^{(0)} \equiv K_0(K, B)$. Under successive iterations, K, B are mapped into fixed points at $K = 0$ and some value $B = B^*$. From (8.38), $K_0(K = 0, B) = \ln(2\cosh B)$. For the Ising model $\mathcal{F} = \ln\lambda_+$, where λ_+ is the largest eigenvalue of the transfer matrix. At the high-temperature fixed point, $\lambda_+ = 2\cosh B^*$ (Eq. (6.82)). Thus, $\mathcal{F}$ is mapped into $\mathcal{F} = \ln(2\cosh B^*)$, and hence the sum in Eq. (8.47) converges as $N \to \infty$,

$$\mathcal{F}(K, B) = \sum_{n=0}^{\infty} \frac{1}{2^{n+1}}K_0\left(K^{(n)}, B^{(n)}\right) .\qquad (8.48)$$

---

[19] We've written Eq. (8.44) so that $K_L < K$ for $T > T_c$.

Renormalization provides a way to calculate the free energy not involving a direct evaluation of the partition function—it's a new paradigm in statistical mechanics. The function K_0, the "constant" in the renormalized Hamiltonian, is essential for this purpose, as are the recursion relations $K' = K'(K, B)$, $B' = B'(K, B)$. Non-analyticities arise in the limit of an infinite number of iterations of recursion relations, in which all degrees of freedom in the thermodynamic limit have been summed over, as if we had exactly evaluated the partition function.

Recursion relations can be developed for quantities derivable from the free energy by taking derivatives. For the magnetization per spin $m \equiv (\partial \mathcal{F}/\partial B)_K$ (see Eq. (6.84)), we have by differentiating Eq. (8.46):

$$m = \frac{1}{2}\left(\frac{\partial K_0}{\partial B}\right) + \frac{1}{2}\left(\frac{\partial B'}{\partial B}\right) m' . \tag{8.49}$$

For the zero-field susceptibility per spin $\chi = (\partial m/\partial B)_{B=0}$,

$$\chi = \frac{1}{2}\left(\frac{\partial^2 K_0}{\partial B^2}\right)_{B=0} + \frac{1}{2}\left(\frac{\partial B'}{\partial B}\right)^2_{B=0} \chi' . \tag{8.50}$$

It's straightforward to write computer programs to iterate recursion relations such as these.

8.3.2 Decimation of the square-lattice Ising model

Let's try decimation on the square lattice for an Ising model having nearest-neighbor interactions in the absence of a magnetic field. From Fig. 8.7, a square lattice of lattice constant a can be de-

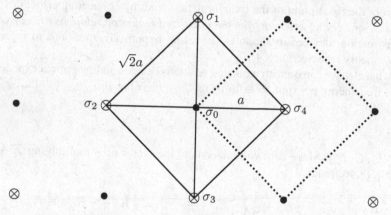

Figure 8.7: Square lattice of lattice constant a decomposed into interpenetrating square sublattices of circles (•) and crosses (⊗), each having lattice constant $\sqrt{2}a$.

composed into interpenetrating *sublattices*,[20] square lattices of lattice constant $\sqrt{2}a$ that have been rotated $45°$ relative to the original.[21,22] Let spins on one sublattice be the block spins. Interactions between them can be inferred by summing out the degrees of freedom associated with the other sublattice. Referring to Fig. 8.7, we sum over σ_0:

$$\sum_{\sigma_0=-1}^{1} \exp\left(K\sigma_0\left(\sigma_1 + \sigma_2 + \sigma_3 + \sigma_4\right)\right) = 2\cosh\left(K\left(\sigma_1 + \sigma_2 + \sigma_3 + \sigma_4\right)\right) . \tag{8.51}$$

---

[20]The simple cubic lattice can be viewed as two interpenetrating face-centered cubic lattices; a body centered cubic lattice can be viewed as two interpenetrating simple cubic lattices. The zincblende structure consists of two interpenetrating face-centered cubic lattices. The face-centered cubic lattice, however, cannot be decomposed into interpenetrating sublattices.

[21]The scaling factor L needn't be an integer.

[22]You can see from Fig. 8.4 that the one-dimensional lattice of lattice constant a has been decomposed into interpenetrating one-dimensional sublattices, of lattice constant $2a$.

Just as with Eq. (8.35), our job is to fit the right side of Eq. (8.51) to an exponential form:

$$2\cosh\left[K\left(\sigma_1 + \sigma_2 + \sigma_3 + \sigma_4\right)\right] \tag{8.52}$$

$$= \exp\left(K_0 + \frac{1}{2}K_1\left(\sigma_1\sigma_2 + \sigma_2\sigma_3 + \sigma_3\sigma_4 + \sigma_4\sigma_1\right) + K_2\left(\sigma_1\sigma_3 + \sigma_2\sigma_4\right) + K_4\sigma_1\sigma_2\sigma_3\sigma_4\right),$$

where, in addition to K_0, we've allowed for nearest-neighbor couplings K_1 (the factor of $\frac{1}{2}$ is because these interactions are "shared" with neighboring cells), next-nearest neighbor couplings K_2, and a four-spin interaction, K_4. Four parameters are required to match the four independent energy configurations of $\sigma_1, \sigma_2, \sigma_3, \sigma_4$: ↑↑↑↑, ↑↑↑↓, ↑↑↓↓, ↑↓↑↓. Thus, from Eq. (8.52),

$$\begin{aligned}
\text{↑↑↑↑:} && 2\cosh 4K &= e^{K_0 + 2K_1 + 2K_2 + K_4} \\
\text{↑↑↑↓:} && 2\cosh 2K &= e^{K_0 - K_4} \\
\text{↑↑↓↓:} && 2 &= e^{K_0 - 2K_2 + K_4} \\
\text{↑↓↑↓:} && 2 &= e^{K_0 - 2K_1 + 2K_2 + K_4} .
\end{aligned} \tag{8.53}$$

The latter two relations imply $K_2 = \frac{1}{2}K_1$. With $K_2 = \frac{1}{2}K_1$, we find

$$e^{8K_0} = 256\cosh 4K\cosh^4 2K \qquad e^{4K_1} = \cosh 4K \qquad e^{8K_4} = \frac{\cosh 4K}{\cosh^4 2K}. \tag{8.54}$$

Thus, starting with a near-neighbor model on the square lattice, we've found, using decimation, a model on the scaled lattice having near-neighbor interactions in addition to next-nearest-neighbor and four-spin couplings. To be consistent, we should start over with a model having first, second neighbor and four-spin interactions.[23] If one does that, however, progressively more types of interactions are generated under successive transformations. K.G. Wilson found[130] that more than 200 types of couplings are required to have a consistent set of recursion relations for the square-lattice Ising model.[24] For that reason, decimation isn't viable in two or more dimensions.[25] We'd still like a way, however, to illustrate renormalization in two dimensions as there are new features to be learned. We develop in Section 8.3.4 an *approximate* set of recursion relations for the square lattice. Before doing that, we touch on a traditional approach to critical phenomena, the method of *high-temperature series expansions*.

8.3.3 High-temperature series expansions—once over lightly

Suppose the temperature is such that $kT \gg H[\sigma_1, \cdots, \sigma_N]$ for all spin configurations. In that case one could try expanding the Boltzmann factor in a Taylor series,

$$Z_N = \sum_{\{\sigma\}} e^{-\beta H} = \sum_{\{\sigma\}}\left(1 - \beta H + \frac{1}{2}\beta^2 H^2 + \cdots\right) = 2^N - \beta\sum_{\{\sigma\}} H + \frac{1}{2}\beta^2\sum_{\{\sigma\}} H^2 + \cdots \tag{8.55}$$

$$\equiv 2^N\left(1 - \beta\langle H\rangle_0 + \frac{1}{2}\beta^2\langle H^2\rangle_0 + \cdots\right),$$

---

[23] If a given type of spin interaction is generated in a renormalization transformation, it appears in the starting Hamiltonian for the next. To be consistent, we must generalize the model to include all generated interactions.

[24] Kadanoff in his original paper[123] refers to the renormalized near-neighbor coupling as a "more subtle beast." He was aware that, if by stretching the lattice constant, *long-range* couplings are induced, it would invalidate the approach. Long-range interactions lead to mean field theory (Section 7.9), which we're trying to get away from. A more precise statement that the Kadanoff theory rests on the assumption that near-neighbor models are mapped into near-neighbor models, is that long-range interactions are not developed in the process. Whether nearest or next-nearest neighbor, what's important is that the renormalized interactions remain finite-ranged.

[25] It can, however, be applied to spin systems defined on fractal geometries.[131]

where for any spin function, $\langle f(\sigma) \rangle_0 \equiv (1/2^N) \sum_{\{\sigma\}} f(\sigma)$ denotes an average with respect to the uniform probability distribution, 2^{-N}. In terms of the average symbol $\langle \rangle_0$,

$$Z_N = 2^N \langle e^{-\beta H} \rangle_0 . \tag{8.56}$$

Example. As an example, consider a system of three Ising spins arranged on the vertices of an equilateral triangle, with Hamiltonian $H = -J(\sigma_1\sigma_2 + \sigma_2\sigma_3 + \sigma_3\sigma_1)$. Then,

$$\langle H \rangle_0 = -J \frac{1}{8} \sum_{\sigma_1=-1}^{1} \sum_{\sigma_2=-1}^{1} \sum_{\sigma_3=-1}^{1} (\sigma_1\sigma_2 + \sigma_2\sigma_3 + \sigma_3\sigma_1) = 0 .$$

As one can show, $H^2 = J^2(3 + 2\sigma_1\sigma_2 + 2\sigma_2\sigma_3 + 2\sigma_3\sigma_1)$, and thus

$$\langle H^2 \rangle_0 = \frac{1}{8} J^2 \sum_{\sigma_1=-1}^{1} \sum_{\sigma_2=-1}^{1} \sum_{\sigma_3=-1}^{1} (3 + 2\sigma_1\sigma_2 + 2\sigma_2\sigma_3 + 2\sigma_3\sigma_1) = 3J^2 .$$

Through second order in a high-temperature series, $Z_3 = 8\left(1 + \frac{3}{2}K^2 + \cdots\right)$, where $K \equiv \beta J$.

We obtain the partition function in statistical mechanics to calculate averages, a step facilitated by working with the free energy, $-\beta F = \ln Z_N$. From Eq. (8.56),

$$\ln Z_N = N \ln 2 + \ln \langle e^{-\beta H} \rangle_0 = N \ln 2 + \sum_{n=1}^{\infty} \frac{(-\beta)^n}{n!} C_n(\langle H \rangle_0) , \tag{8.57}$$

where the quantities C_n are cumulants (see Sections 3.7 and 6.3). Equation (8.57) is identical to Eq. (6.38) when we identify $-\beta F_{ideal}$ with $N \ln 2$. Explicit expressions for cumulants in terms of moments are listed in Eq. (3.62), e.g., $C_1 = \langle H \rangle_0$ and $C_2 = \langle H^2 \rangle_0 - \langle H \rangle_0^2$. The values of the cumulants C_n depend on the range of the interactions and the geometry of the lattice.[26] There is an extensive literature on high-temperature series expansions. To give just one example, 54 terms are known for the high-temperature series of the susceptibility of the square-lattice Ising model [132]. With high-temperature series, we're expanding about the state of uncoupled spins, $K = 0$. Can one find information this way about thermodynamic functions exhibiting *singularities* at $K = K_c$? Indeed, that's the art of this approach[27]—extracting information about critical phenomena at $K = K_c$ from a finite number of terms in a Taylor series about $K = 0$.

Series expansions can be developed for correlation functions, and to do that we introduce another approach to the partition function. For a nearest-neighbor Ising model ($H = -J \sum_{\langle ij \rangle} \sigma_i \sigma_j$),

$$Z_N = \sum_{\{\sigma\}} e^{-\beta H} = \sum_{\{\sigma\}} e^{K \sum_{\langle ij \rangle} \sigma_i \sigma_j} = \sum_{\{\sigma\}} \prod_{\langle ij \rangle} e^{K \sigma_i \sigma_j} . \tag{8.58}$$

We use the following identity for Ising spins, not necessarily near-neighbor pairs,

$$e^{K \sigma_k \sigma_l} = \cosh K + \sigma_k \sigma_l \sinh K = \cosh K (1 + u \sigma_k \sigma_l) , \tag{8.59}$$

---

[26]Cumulants are extensive quantities—they scale with N. See Section 6.3.

[27]Analytic functions are represented by power series within their interval of convergence. Can one use a power series (about $K = 0$) to represent a function that's non-analytic at $K = K_c$? The art of this approach consists of developing *Padé approximants*, *rational functions* the power series of which agree with the high-temperature series through a given order (a rational function $f(x) = P(x)/Q(x)$ is a ratio of functions where P, Q are polynomials[16, p228]). By this method, information can be obtained about functions represented by power series *outside their interval of convergence*. See [133].

where $u \equiv \tanh K$. Combining Eq. (8.59) with Eq. (8.58),

$$Z_N = \cosh^P K \sum_{\{\sigma\}} \prod_{\langle ij \rangle} (1 + u\sigma_i\sigma_j) , \tag{8.60}$$

where P is the number of distinct near-neighbor pairs on the lattice.[28] For systems satisfying periodic boundary conditions, $P = Nz/2$ where z is the coordination number of the lattice. There are P factors of $(1 + u\sigma_i\sigma_j)$ in the product in Eq. (8.60).

Let's illustrate Eq. (8.60) with some simple cases. Figure 8.8 shows two, three-spin Ising

Figure 8.8: Three-spin Ising systems with free boundary conditions (left) and periodic boundary conditions (right). Each near-neighbor bond is given weight $u = \tanh K$.

systems, one with free boundary conditions and the other with periodic boundary conditions. For free boundary conditions, there are two near-neighbor pairs ($P = 2$), and thus expanding $(1 + u\sigma_1\sigma_2)(1 + u\sigma_2\sigma_3)$ in Eq. (8.60),

$$Z_3^{\text{free b.c.}} = \cosh^2 K \sum_{\sigma_1} \sum_{\sigma_2} \sum_{\sigma_3} (1 + u\sigma_1\sigma_2 + u\sigma_2\sigma_3 + u^2\sigma_1\sigma_3) = 8\cosh^2 K . \tag{8.61}$$

It helps to draw a line connecting the spins in each near-neighbor pair, as in Fig. 8.8. Referring to these lines as "bonds," we associate a factor of $u = \tanh K$ with each bond. Before summing over spins in Eq. (8.61), we see a connection between spins σ_1, σ_3 of strength u^2; terms like this occur when, in expanding the product in Eq. (8.60), we have "overlap" products such as $(u\sigma_1\sigma_2)(u\sigma_2\sigma_3) = u^2\sigma_1\sigma_3$. In the left part of Fig. 8.8 there are two bonds between σ_1 and σ_3—one between σ_1, σ_2 and the other between σ_2, σ_3. We multiply the bond strengths for each link in any lattice-path connecting given spins consisting entirely of nearest-neighbor pairs.[29]

For periodic boundary conditions in Fig. 8.8, $P = 3$ and thus from Eq. (8.60),

$$Z_3^{\text{periodic b.c.}} = \cosh^3 K \sum_{\sigma_1} \sum_{\sigma_2} \sum_{\sigma_3} [1 + u\sigma_1\sigma_2 + u\sigma_2\sigma_3 + u\sigma_3\sigma_1$$
$$+ u^2\sigma_1\sigma_2 + u^2\sigma_2\sigma_3 + u^2\sigma_3\sigma_1 + u^3]$$
$$= 8\cosh^3 K \left(1 + u^3\right) = 8\left(\cosh^3 K + \sinh^3 K\right) . \tag{8.62}$$

Before summing over spins in Eq. (8.62) and referring to the right part of Fig. 8.8, we see a direct near-neighbor connection between spins σ_1, σ_2 of strength u, and another connection between σ_1, σ_2 of strength u^2 mediated by the path through σ_3. We also see a new feature in Eq. (8.62): a coupling of strength u^3 associated with a *closed path* of near-neighbor links from the overlap product $(u\sigma_1\sigma_2)(u\sigma_2\sigma_3)(u\sigma_3\sigma_1) = u^3$. Note that Eq. (8.61) agrees with Eq. (6.71) for $N = 3$, while Eq. (8.62) agrees with Eq. (6.81) for $N = 3$.

---

[28]Note that Eq. (8.60) is similar in mathematical form to Eq. (6.4); in the latter the notation $\prod_{i>j}$ generalizes $\prod_{\langle ij \rangle}$. In the Mayer expansion $f_{ij} \to 0$ in the noninteracting limit; in Eq. (8.60) $u \to 0$ in the noninteracting limit.

[29]One can surmise that the diagrammatic methods discussed in Chapter 6 play a significant role in high-temperature series expansions, a topic we forgo.

We can now see how expansions for correlation functions are developed. Consider, for any spins σ_l, σ_m, that for a nearest-neighbor Ising model,

$$\langle \sigma_l \sigma_m \rangle = \frac{1}{Z_N} \sum_{\{\sigma\}} \sigma_l \sigma_m e^{-\beta H} = \frac{\cosh^P K}{Z_N} \sum_{\{\sigma\}} \sigma_l \sigma_m \prod_{\langle ij \rangle} (1 + u \sigma_i \sigma_j) . \quad (8.63)$$

For $N = 3$, for example, we have for periodic boundary conditions (pbc)

$$\langle \sigma_1 \sigma_3 \rangle_{N=3}^{\text{pbc}} = \frac{\cosh^3 K}{Z_3^{\text{pbc}}} \sum_{\sigma_1} \sum_{\sigma_2} \sum_{\sigma_3} \sigma_1 \sigma_3 (1 + u \sigma_1 \sigma_2)(1 + u \sigma_2 \sigma_3)(1 + u \sigma_3 \sigma_1)$$

$$= \frac{\cosh^3 K}{Z_3^{\text{pbc}}} \sum_{\sigma_1} \sum_{\sigma_2} \sum_{\sigma_3} \sigma_1 \sigma_3 \left[1 + \cdots + u \sigma_1 \sigma_3 + u^2 \sigma_1 \sigma_3 + \cdots \right]$$

$$= \frac{8 \cosh^3 K (u + u^2)}{Z_3^{\text{pbc}}} = \frac{u + u^2}{1 + u^3} = u + u^2 + O(u^4) , \quad (8.64)$$

where we've used Eq. (8.62) for Z_3^{pbc}. In the second equality in Eq. (8.64), we've listed in square brackets only the terms proportional to the spin product $\sigma_1 \sigma_3$. We have in Eq. (8.64) the exact result for $\langle \sigma_1 \sigma_3 \rangle$ (simple enough in this case to obtain through direct calculation), and, in the final equality, the first two terms in a high-temperature series expansion, valid for $u \ll 1$. The general process is clear: Somewhere in the expansion of the product in Eq. (8.63), there are terms involving the product $\sigma_l \sigma_m$. *Only those terms survive the summation over spins* in Eq. (8.63). One must find all lattice-paths connecting σ_l, σ_m consisting entirely of near-neighbor links.

8.3.4 Approximate recursion relations for the square-lattice Ising model

Let's return to the square-lattice Ising model. In Section 8.3.2, we found that a near-neighbor model leads, under decimation, to a model with next-nearest neighbor and four-spin interactions. In this section we analyze a simplified, approximate set of recursion relations valid at high temperature for a model with nearest and next-nearest-neighbor interactions.[30]

The left part of Fig. 8.9 shows part of a square lattice (lattice constant a) having near-neighbor

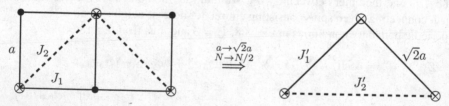

Figure 8.9: Left: Square lattice with nearest and next-nearest-neighbor interactions J_1, J_2 (solid and dashed lines). Right: Scaled lattice with nearest and next-nearest-neighbor interactions J_1', J_2'.

interactions J_1 (solid lines) and next-near-neighbor interactions J_2 (dashed lines). Under the block-spin transformation ($a \to \sqrt{2}a$ and the number of spins $N \to N/2$), we seek near and next-near-neighbor interactions J_1', J_2' (solid and dashed lines, right part of Fig. 8.9). There is "already" a direct interaction J_2 between the spins we're calling block spins ($\otimes$-spins, left part of Fig. 8.9). On the scaled lattice, that interaction (J_2 on the original lattice) is a partial contribution to the coupling J_1' between near-neighbor $\otimes$-spins. There is another contribution to J_1' from the near-neighbor couplings on the original lattice, $2J_1^2$ at high temperature—there are two near-neighbor

---

links separating near-neighbor $\otimes$-spins, and there are two such paths connecting them (see Section 8.3.3 for this reasoning). For the second-neighbor interaction J_2', there is a contribution J_1^2 (at high temperature) from the two nearest-neighbor links between $\otimes$-spins (left part of Fig. 8.9), in addition to an indirect contribution J_2^2. Working with dimensionless parameters, we simply write down the recursion relations

$$K_1' = 2K_1^2 + K_2$$
$$K_2' = K_1^2 , \tag{8.65}$$

where K_2^2 has been dropped (from K_2') because it's higher order in K. We have in (8.65) a consistent set of recursion relations at lowest order in K for the square lattice. They can be considered model recursion relations having enough complexity to illustrate more general recursion relations.

The first thing to do is find the fixed points. As is readily shown, the nontrivial fixed point is $K_1^* = \frac{1}{3}$, $K_2^* = \frac{1}{9}$. Next, we study the flows in the vicinity of that fixed point. Referring to Eq. (8.43), we have using Eq. (8.65) the linearized transformation

$$\begin{pmatrix} \delta K_1' \\ \delta K_2' \end{pmatrix} = \begin{pmatrix} (\partial K_1'/\partial K_1)^* & (\partial K_1'/\partial K_2)^* \\ (\partial K_2'/\partial K_1)^* & (\partial K_2'/\partial K_2)^* \end{pmatrix} \begin{pmatrix} \delta K_1 \\ \delta K_2 \end{pmatrix} = \begin{pmatrix} 4/3 & 1 \\ 2/3 & 0 \end{pmatrix} \begin{pmatrix} \delta K_1 \\ \delta K_2 \end{pmatrix} . \tag{8.66}$$

With Eq. (8.66) we encounter a new feature—the matrix representing the linearized transformation is not diagonal, as in Eq. (8.43). *To find the appropriate scaling variables, we must work in a basis in which the transformation matrix is diagonal.*

The eigenvalues of the matrix in Eq. (8.66) (easily found) are:

$$\lambda_1 = \frac{1}{3}\left(2 + \sqrt{10}\right) \approx 1.721 \qquad \lambda_2 = \frac{1}{3}\left(2 - \sqrt{10}\right) \approx -0.387 , \tag{8.67}$$

where it's customary to list the eigenvalues in descending order by magnitude, $|\lambda_1| > |\lambda_2|$. The associated eigenvectors (in the K_1, K_2 basis) are:

$$\psi_1 = \begin{pmatrix} 1 \\ 0.387 \end{pmatrix} \qquad \psi_2 = \begin{pmatrix} 1 \\ -1.721 \end{pmatrix} . \tag{8.68}$$

Figure 8.10 shows the fixed points of (8.65) in the K_1, K_2 plane, and the directions of the eigenvectors ψ_1, ψ_2 at the nontrivial fixed point. Because $\lambda_1 > 1$, the fixed point is unstable for deviations $\delta K_1, \delta K_2$ lying in the direction associated with ψ_1: small deviations in the direction of ψ_1 (positive

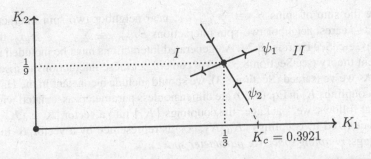

Figure 8.10: Fixed points ($\bullet$) of the recursion relations (8.65). Shown are the directions of the eigenvectors ψ_1 (positive eigenvalue) and ψ_2 (negative eigenvalue) of the linearized transformation (see (8.68)). Initial points in region I flow under iteration to the high-temperature fixed point; those in region II flow to larger values of K_1, K_2 (where the model isn't valid). $K_c \approx 0.3921$ is the numerically obtained extension to $K_2 = 0$ of the line separating regions I, II (dashed line).

or negative) are mapped into larger deviations. For deviations $\delta K_1, \delta K_2$ lying in the direction of ψ_2, however, the fixed point is *stable*—deviations are mapped closer to the fixed point ($|\lambda_2| < 1$). Region I in Fig. 8.10 represents couplings that are mapped under iteration to the high-temperature fixed point. In Region II, couplings are mapped to "infinity." Shown as the dashed line in Fig. 8.10 is the extension to $K_2 = 0$ of the line separating regions I, II, which occurs at $K_1 \equiv K_c \approx 0.3921$. Initial couplings $K_2 = 0$, $K_1 < K_c$ are mapped into the high-temperature fixed point; initial couplings $K_2 = 0$, $K_1 > K_c$ flow to infinity (see Exercise 8.28). If one started with couplings $K_1 = K_c$, $K_2 = 0$ (i.e., a near-neighbor model), one would have in a few iterations of (8.65) an equivalent model with the fixed-point couplings K_1^*, K_2^* (and it's the fixed point that controls critical phenomena). Thus, this method predicts $K_c \approx 0.3921$ for the near-neighbor square-lattice Ising model, within 11% of the exact value $K_c = \frac{1}{2}\sinh^{-1}(1) \approx 0.4407$, Eq. (7.125), and distinct from the predictions of mean field theory on the square lattice, $K_c = 0.25$ (see Section 7.8). That's not bad considering the approximations in setting up Eq. (8.65), and is therefore encouraging: It compels us to find better ways of doing block-spin transformations (see Section 8.5). Prior to the renormalization method, there was no recourse to mean field theory other than exact evaluations of partition functions, nothing in the middle; now we have a method, a path, by which we can seek better approximations in terms of how we do block-spin transformations.[31]

We find critical exponents from linearized recursion relations at fixed points, as in Eq. (8.45). Equation (8.45), however, utilized the diagonal transformation matrix Eq. (8.43) and must be modified for non-diagonal matrices as in Eq. (8.66). If we work in a basis of eigenvectors, the transformation matrix is diagonal with eigenvalues along the diagonal. Thus, instead of Eq. (8.45),

$$\nu = \frac{1}{x} = \frac{\ln L}{\ln \lambda_1}. \tag{8.69}$$

With $L = \sqrt{2}$ and $\lambda_1 = 1.721$ (Eq. (8.67)), we find $\nu \approx 0.638$. Again, not the exact value $\nu = 1$, but also not the mean field theory exponent, $\nu = \frac{1}{2}$ (see Table 7.3). Given the approximations, $\nu = 0.638$ is an encouraging first result, and will improve as we get better at block-spin transformations.

8.4 THE RENORMALIZATION GROUP

We now look at renormalization from a more general perspective. We start by enlarging the class of Hamiltonians to include all the types of spin interactions $S_i(\sigma)$ generated by renormalization transformations,[32]

$$-\beta H = \mathcal{H} = \sum_i K_i S_i(\sigma). \tag{8.70}$$

We could have the sum of spins $S_1 \equiv \sum_{k=1}^N \sigma_k$, near-neighbor two-spin interactions $S_{2,nn} \equiv \sum_{\langle ij \rangle} \sigma_i \sigma_j$, next-nearest neighbor two-spin interactions $S_{2,nnn} \equiv \sum_{ij \in nnn} \sigma_i \sigma_j$, three, and four-spin interactions, etc. See Exercise 8.29. All generated interactions must be included in the model to have a consistent theory (see Section 8.3.2). We can collect these interactions in a set $\{S_i\}$ labeled by an index i. As we've learned (Section 8.3), we should include a constant in the Hamiltonian; call it $S_0 \equiv 1$. The couplings K_i in Eq. (8.70) are dimensionless parameters associated with spin interactions S_i. In what follows, we "package" the couplings $\{K_i\}$ into a vector, $\boldsymbol{K} \equiv (K_0, K_1, K_2, \cdots)$. A system characterized by couplings $(K_0, K_1, \cdots)$ is represented by a vector $\boldsymbol{K}$ in a space of all possible couplings, *coupling space* (or *parameter space*).

---

[31]Theoretical physics is the art of approximation, something that can get lost in the blaze of mathematics. A scenario I pose to students is the following: Walk into a classroom with equations on the board. How would you infer whether it's a physics or a mathematics classroom? While there is no right answer, progress in theoretical physics occurs through physically motivated approximations, the validity of which are checked against the results of experiment; nature is the arbiter of truth, math is true by definition. As Einstein observed, "As far as the laws of mathematics refer to reality, they are not certain; as far as they are certain, they do not refer to reality."

[32]The range of the index i in Eq. (8.70), the number of required spin interactions, is left unspecified.

Renormalization transformations have two parts: 1) stretch the lattice constant $a \to La$ and thin the number of spins $N \to N/L^d$; and 2) find the effective couplings K' among the remaining degrees of freedom (such that Eqs. (8.19) and (8.20) are satisfied). Starting with a model characterized by couplings K, find the renormalized couplings K', an operation that we symbolize

$$K' = \mathcal{R}_L(K) \cdot \qquad (L > 1) \qquad (8.71)$$

That is, $\mathcal{R}_L$ is an operator that maps the coupling vector K to that of the transformed system K' associated with scale change L. Representing renormalizations with an operator is standard,[33] but it's abstract. An equivalent but more concrete way of writing Eq. (8.71) is

$$K_i' = f_i^{(L)}(K) , \qquad (8.72)$$

where i runs over the range of the index in Eq. (8.70). For each renormalized coupling K_i' associated with scale change L, there is a function $f_i^{(L)}$ of all couplings K (see Eqs. (8.38), (8.40), or (8.65)). $\mathcal{R}_L$ is thus a collection of functions that act on the components of K, $\mathcal{R}_L \leftrightarrow \{f_i^{(L)}\}$. We'll use both ways of writing the transformation.

A renormalization followed by a renormalization is itself a renormalization; let's show that. Starting from a model characterized by couplings K, do the transformation with $L = L_1$, followed by another with $L = L_2$. Then, with $K' = \mathcal{R}_{L_1}(K)$ and $K'' = \mathcal{R}_{L_2}(K')$,

$$K'' = \mathcal{R}_{L_2}(K') = \mathcal{R}_{L_2} \cdot \mathcal{R}_{L_1}(K) \equiv \mathcal{R}_{L_2 \times L_1}(K) \cdot \qquad (8.73)$$

The compound effect of two successive transformations (of scale changes L_1, L_2) is equivalent to a single transformation with the product $L_1 L_2$ as the scale change factor. Equation (8.73) implies the operator statement,

$$\mathcal{R}_{L_2} \cdot \mathcal{R}_{L_1} = \mathcal{R}_{L_2 L_1} \cdot \qquad (8.74)$$

Equation (8.74) indicates that the *order* in which we perform renormalizations is immaterial, i.e., renormalizations commute: $\mathcal{R}_{L_1 L_2} = \mathcal{R}_{L_2 L_1} \implies \mathcal{R}_{L_1} \cdot \mathcal{R}_{L_2} = \mathcal{R}_{L_2} \cdot \mathcal{R}_{L_1}$. This must be the case—we're rescaling the same lattice structure (having the same symmetries), the order in which we do that is immaterial. Equation (8.74) is the *group property*, a word borrowed from mathematics. A *group* is a set of mathematical objects (elements) equipped with a binary operation such that the composition of any two elements produces another member of the set, such as we see in Eq. (8.74) for the set of all renormalization transformations $\{\mathcal{R}_{L \geq 1}\}$. For that reason, the theory of renormalization is referred to as *renormalization group* theory, *even though renormalizations do not constitute a group*! To be a group, the set must satisfy additional criteria,[34] the most important of which for our purposes is that for each element of the set, there exists an inverse element in the set.[35] Renormalizations possess the group composition property, Eq. (8.74), but they do not possess inverses. The operator $\mathcal{R}_L$ is defined for $L > 1$ (Eq. (8.71)) and not for $0 < L < 1$; a definite "one-sided" progression in renormalized systems is implied by the requirement $L > 1$. Once we've coarse-grained the number of spins $N \to N/L^d$, there's no going back—we can't undo the transformation. Sets possessing the attributes of a group except for inverses are known as *semigroups*. The term renormalization group is a misnomer, but that doesn't stop people (like us) from using it.

---

[33] There are many excellent reviews of renormalization in statistical physics. See for example Ma[134], Wilson and Kogut[135], Fisher[136], and Wallace and Zia[137].

[34] See any text on algebra.

[35] The set of integers, for example, comprises a group under ordinary addition as the composition rule. The inverse element corresponding to any integer n is $-n$, so that $n + (-n) = 0$, where 0 is the identity element of the group, $n + 0 = n$.

Critical phenomena are associated with the behavior of recursion relations in the vicinity of fixed points (Section 8.3). Fixed points are solutions of the set of equations (from Eq. (8.72))[36]

$$K_i^* = f_i^{(L)}(\mathbf{K}^*) . \tag{8.75}$$

Solutions of Eq. (8.75) may be isolated points in $\mathbf{K}$-space, or solution families parameterized by continuous variables, or it may be that Eq. (8.75) has no solution. In what follows, we assume that a solution $\mathbf{K}^*$ is known to exist at an isolated point.

In the vicinity of the fixed point, we have, generalizing the relations in (8.42),

$$\delta K_i' = \sum_j \left(\frac{\partial f_i^{(L)}}{\partial K_j}\right)^* \delta K_j \equiv \sum_j R_{ij}^{(L)} \delta K_j , \tag{8.76}$$

where $R_{ij}^{(L)}$ are the elements of the linearized transformation matrix $R^{(L)}$. The matrix $R^{(L)}$ is in general not symmetric and may not be diagonalizable; even if it is diagonalizable, it may not have real eigenvalues. We assume the simplest case that $R^{(L)}$ is diagonalizable with real eigenvalues.

The eigenvalue problem associated with $R^{(L)}$ can be written

$$\sum_j R_{ij}^{(L)} \psi_j^{(k)} = \lambda_k^{(L)} \psi_i^{(k)} , \tag{8.77}$$

where k labels the eigenvalue $\lambda_k^{(L)}$ of $\mathcal{R}_L$ associated with eigenvector $\psi^{(k)}$ (having $\psi_j^{(k)}$ as its jth component). We don't label the eigenvectors in Eq. (8.77) with L because renormalizations commute. Equation (8.74), which holds for any renormalizations, holds for the linearized version involving matrices,

$$R^{(L_1)} R^{(L_2)} = R^{(L_1 L_2)} , \tag{8.78}$$

and thus, because commuting operators have a common set of eigenvectors, but not eigenvalues,[37] we can't label eigenvectors with L. Equation (8.78) implies for eigenvalues (see Exercise 8.30)

$$\lambda_k^{(L_1)} \lambda_k^{(L_2)} = \lambda_k^{(L_1 L_2)} . \tag{8.79}$$

Equation (8.79) is a functional equation for eigenvalues; it has the solution

$$\lambda_k^{(L)} = L^{y_k} , \tag{8.80}$$

where the exponent y_k is independent of L. Equation (8.80) is a *strong* prediction of the theory: The eigenvalues occur in the scaling form posited in Eq. (8.21). By setting $L_2 = 1$ in Eq. (8.79), we find $\lambda_k^{(1)} = 1$ (the $L = 1$ transformation is the identity transformation), and thus we have the explicit scaling relation $\lambda_k^{(L)} = L^{y_k} \lambda_k^{(1)}$. Scaling in the form assumed by Kadanoff occurs quite generally, relying only on the existence of fixed points and the group property, Eq. (8.74).

Scaling fields q_k are vectors in $\mathbf{K}$-space (coupling space) obtained from linear combinations of the deviations δK_i, weighted by eigenvector components,[38]

$$q_k \equiv \sum_l \psi_l^{(k)} \delta K_l . \tag{8.81}$$

---

[36]A question to ask of Eq. (8.75) is if the fixed point $\mathbf{K}^*$ is independent of L. It is in the limit $L \to \infty$, as the theory becomes independent of the microscopic details of the system, accounting for the universality of critical phenomena.

[37]Commuting linear operators have a common set of eigenfunctions (see any text on quantum mechanics) provided that each operator has a complete set of eigenfunctions, which is the case here because we've assumed $R^{(L)}$ is diagonalizable.

[38]Such reasoning should be familiar from the theory of *normal modes* in mechanics. Normal modes are patterns of vibration in which all parts of a system vibrate with the same frequency. Scaling fields are weighted sets of deviations from the critical point all associated with the same eigenvalue of $R^{(L)}$.

In terms of these variables (scaling fields), the linearized recursion relations *decouple* (instead of their form in Eq. (8.76)),

$$q'_k = \lambda_k^{(L)} q_k = L^{y_k} q_k \,, \tag{8.82}$$

where we've used Eq. (8.80) (see Exercise 8.31). Scaling fields are classified as follows:

1. Scaling fields q_k with $|\lambda_k| > 1$ ($y_k > 0$) are termed *relevant*. Under iteration, relevant scaling fields grow in magnitude, $q'_k > q_k$, with the new couplings K' further removed from the critical point at K^* than the original K;

2. Scaling fields q_k with $|\lambda_k| < 1$ ($y_k < 0$) are termed *irrelevant*. Under iteration, $q'_k < q_k$; the new couplings K' are closer to K^* than the original K;

3. Scaling fields q_k with $|\lambda_k| = 1$ ($y_k = 0$) are termed *marginal*. In the linear approximation, such variables neither grow nor decrease in magnitude under iteration. Further analysis is required to ascertain the fate of such variables under many iterations.

After many iterations, only the components of δK along the direction of relevant scaling fields will matter; components of δK along irrelevant or marginal scaling fields will either shrink or stay fixed at a finite value. The manifold spanned by irrelevant scaling fields is called the *critical manifold*, the existence of which accounts for universality, that different physical systems can exhibit the same critical phenomena (see Section 7.3.2). Irrelevant variables are simply that—irrelevant for the purposes of describing critical phenomena. Couplings K not lying in the critical manifold flow under successive iterations away from it. Measurable critical phenomena are associated with relevant variables. It's found that *most* interactions are irrelevant; critical phenomena are determined by a few relevant variables which can be grouped into *universality classes* determined by the dimensionality and the symmetries of the system.[39]

Renormalization group theory systematizes the framework of block-spin transformations and shows that the Kadanoff scaling form, Eq. (8.21), is exhibited by the eigenvalues of the linearized transformation matrix, Eq. (8.80). It should be kept in mind that the functions $f_i(K)$ on which the method is based (see Eq. (8.72)) can no more be found in general than we can evaluate partition functions. A strength of the renormalization approach is that the functions $f_i(K)$ are expected to be analytic, and hence amenable to approximation. Singularities associated with phase transitions occur only in the thermodynamic limit (Section 7.10), implying the need (in traditional statistical mechanics) to find the exact free energy as a function of N. In the renormalization method, singularities are built up iteratively. Two complementary approaches to the functions $f_i(K)$ have been developed: lattice-based methods (real space or position space renormalization), and field-theoretic methods (momentum space or k-space renormalization) that work with continuous spin fields. We treat the former in the next section and touch on the latter in Section 8.6.

8.5 REAL-SPACE RENORMALIZATION

Consider a set $\{\sigma\}$ of N interacting Ising spins on a lattice of lattice constant a, and suppose we wish to find a new set $\{\mu\}$ of N' Ising spins *and their interactions* on a scaled lattice with lattice constant La, where $N' = N/L^d$. For that purpose, we define the *real space mapping function* $T[\mu|\sigma]$, such that

$$e^{\mathcal{H}'(\mu)} = \sum_{\{\sigma\}} T[\mu|\sigma] e^{\mathcal{H}(\sigma)} \,. \tag{8.83}$$

---

[39] See for example Zinn-Justin[138].

A requirement on the function $T[\mu|\sigma]$ is that

$$\sum_{\{\mu\}} T[\mu|\sigma] \equiv \prod_{i=1}^{N'} \sum_{\mu_i=-1}^{1} T[\mu|\sigma] = 1 .$$

(8.84)

Equation (8.84) implies, from Eq. (8.83), that

$$Z_{N'}(\boldsymbol{K'}) = Z_N(\boldsymbol{K}) .$$

(8.85)

Equation (8.85) appears to differ from Eq. (8.39), but there is no discrepancy: In (8.39) we separated the constant term $N'K_0$ from the renormalized Hamiltonian $\mathcal{H}'$; in (8.85) the generated constant is part of $\mathcal{H}'$. By combining Eq. (8.85) with Eq. (8.83), we have for the block-spin probability distribution $P(\mu)$,

$$P(\mu) = \langle T[\mu|\sigma] \rangle .$$

(8.86)

The quantity $T[\mu|\sigma]$ therefore plays the role of a conditional probability (Section 3.2) that the μ-spins have their values, given that the configuration of σ-spins is known.

In principle there is a renormalization transformation for every function $T[\mu|\sigma]$ satisfying Eq. (8.84); in practice there are a limited number of forms for $T[\mu|\sigma]$. Defining transformations as in Eq. (8.83) (with a mapping function) formalizes and generalizes the decimation method, to which it reduces if $T[\mu|\sigma]$ is a product of delta functions. The transformation studied in Section 8.3.1 is effected by Eq. (8.83) if $T[\mu|\sigma] = \prod_{i=1}^{N/2} \delta_{\mu_i,\sigma_{2i}}$. We now illustrate the application of Eq. (8.83) to the triangular-lattice Ising model.

Figure 8.11 shows part of a triangular lattice with Ising spins ($\bullet$) on its vertices (solid lines).

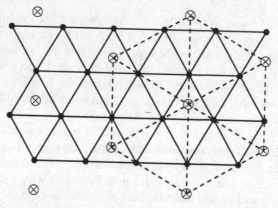

Figure 8.11: Part of 1) a triangular lattice (solid lines) with Ising spins ($\bullet$), and 2) another triangular lattice (dashed lines) for block spins ($\otimes$) scaled by $L = \sqrt{3}$ and rotated 30°.

A new triangular lattice for block spins ($\otimes$) can be formed by dilating the original with $L = \sqrt{3}$ and rotating it 30° (dashed lines). Figure 8.12 shows groups (in crosshatching) of three σ-spins that form the coarse-graining cells in which their degrees of freedom are mapped into those of the block spin. Every σ-spin is uniquely associated with one of these cells. As explained below, we distinguish *intracell* couplings (solid lines) and *intercell* couplings between σ-spins in near-neighbor cells (dashed lines). By writing $T[\mu|\sigma]$ as a product of *cell* mapping functions $T_i[\mu|\sigma]$,

$$T[\mu|\sigma] \equiv \prod_{i=1}^{N'} T_i[\mu|\sigma] ,$$

(8.87)

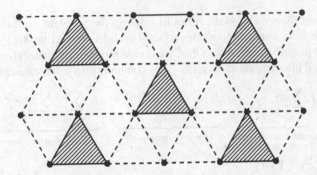

Figure 8.12: Three-spin coarse-graining cells on the triangular lattice (crosshatching). Intracell couplings indicated with solid lines; intercell couplings with dashed lines.

Eq. (8.84) is satisfied with

$$T_i[\mu|\sigma] \equiv \frac{1}{2}\left(1 + \mu_i\psi(\{\sigma\}_i)\right),\tag{8.88}$$

where ψ is a function of the cell spins $\{\sigma\}$. We will assume Eqs. (8.87) and (8.88) in what follows.

There is flexibility in the choice of ψ. Niemeijer and van Leeuwen[139] chose for the triangular lattice the so-called *majority rule* function,

$$\psi(\sigma_1, \sigma_2, \sigma_3) \equiv \frac{1}{2}\left(\sigma_1 + \sigma_2 + \sigma_3 - \sigma_1\sigma_2\sigma_3\right),\tag{8.89}$$

where $\sigma_1, \sigma_2, \sigma_3$ are the three σ-spins in each cell; see Fig. 8.13. There are four independent configurations of $\sigma_1, \sigma_2, \sigma_3$: $\uparrow\uparrow\uparrow, \uparrow\uparrow\downarrow, \uparrow\downarrow\downarrow, \downarrow\downarrow\downarrow$, for which ψ in Eq. (8.89) has the values $\psi(\uparrow\uparrow\uparrow) = 1$, $\psi(\uparrow\uparrow\downarrow) = 1$, $\psi(\uparrow\downarrow\downarrow) = -1$, and $\psi(\downarrow\downarrow\downarrow) = -1$. We note that the Kronecker delta function has the representation in Ising variables, $\delta_{\sigma,\sigma'} = \frac{1}{2}(1 + \sigma\sigma')$. Comparing with $T_i[\mu|\sigma]$ in Eq. (8.88), we see that the majority-rule function with $\psi_i = \pm1$ generalizes the decimation method with $T[\mu|\sigma] = \prod_{i=1}^{N'}\delta_{\mu_i,\psi_i}$. Another approach, that of Mazenko, Nolan, and Valls[140], is to map onto the block spin the slowest *dynamical* mode of cell spins.[40]

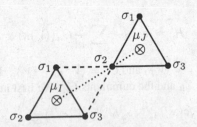

Figure 8.13: Interaction between nearest-neighbor block spins μ_I, μ_J (dotted line).

Real-space renormalization based on mapping functions $T[\mu|\sigma]$ is more in keeping with the intent of the block-spin transformation (a mapping of cell degrees of freedom onto block spins) than is decimation (a renaming of a subset of the original spins as block spins). If decimation generates an apparently unending set of new interactions on two-dimensional lattices (Section 8.3.2), does real-space renormalization (a generalization of decimation) sidestep that problem? As we show,

---

[40]Dynamical phenomena are outside the intended scope of this book. Ising spins do not possess a natural dynamics generated by the Hamiltonian, but, given the success of the Ising model in treating equilibrium phenomena, models for the stochastic dynamics of Ising spins have been developed, *kinetic Ising models*, starting with the work of R.J. Glauber[141]. Glauber received the 2005 Nobel Prize in Physics for work on optical coherence.

the use of perturbation theory allows us, if not to sidestep the problem, to forestall it and make it more systematic. We treat intracell couplings exactly (solid lines in Fig. 8.12), and intercell couplings (dashed lines) perturbatively. To do that, we divide the Hamiltonian $\mathcal{H}(\sigma)$ into a term $\mathcal{H}_0(\sigma)$ containing all intracell interactions and a term $V(\sigma)$ containing all intercell interactions:

$$\mathcal{H}(\sigma) = \underbrace{K \cdot \sum_{\langle ij \rangle} \sigma_i \sigma_j}_{\substack{\text{near-neighbor interactions} \\ \text{on the original lattice}}} = \underbrace{K \sum_I \sum_{i,j \in I} \sigma_i^{(I)} \sigma_j^{(I)}}_{\text{intracell interactions}} + \underbrace{K \sum_{\langle IJ \rangle} \sum_{i \in I, j \in J} \sigma_i^{(I)} \sigma_j^{(J)}}_{\text{intercell interactions}}$$

$$\equiv \mathcal{H}_0(\sigma) + V(\sigma) . \tag{8.90}$$

By substituting $\mathcal{H} = \mathcal{H}_0 + V$ from Eq. (8.90) in Eq. (8.83),

$$e^{\mathcal{H}'(\mu)} = \sum_{\{\sigma\}} T[\mu|\sigma] e^{\mathcal{H}(\sigma)} = \sum_{\{\sigma\}} e^{\mathcal{H}_0(\sigma)} T[\mu|\sigma] e^{V(\sigma)} = Z_0 \sum_{\{\sigma\}} \frac{e^{\mathcal{H}_0(\sigma)}}{Z_0} T[\mu|\sigma] e^{V(\sigma)} , \tag{8.91}$$

where

$$Z_0 \equiv \sum_{\{\sigma\}} T[\mu|\sigma] e^{\mathcal{H}_0(\sigma)} . \tag{8.92}$$

Equation (8.91) can be written

$$e^{\mathcal{H}'} = Z_0 \langle e^V \rangle_0 , \tag{8.93}$$

where for any spin function

$$\langle A \rangle_0 \equiv \frac{1}{Z_0} \sum_{\{\sigma\}} e^{\mathcal{H}_0} T[\mu|\sigma] A(\sigma) . \tag{8.94}$$

That is, $\langle \rangle_0$ denotes an average with respect to the probability distribution $e^{\mathcal{H}_0(\sigma)} T[\mu|\sigma]/Z_0$, which is normalized with respect to σ-variables. From Eq. (8.93), we have the renormalized Hamiltonian

$$\mathcal{H}' = \ln Z_0 + \ln \langle e^V \rangle_0 . \tag{8.95}$$

Hopefully you recognize the next step: $\ln \langle e^V \rangle$ is the cumulant generating function (see Eqs. (3.60), (6.39), or (8.57)). Thus,

$$\mathcal{H}' = \ln Z_0 + \sum_{n=1}^{\infty} \frac{1}{n!} C_n(\langle V \rangle_0) , \tag{8.96}$$

where the C_n are cumulants ($C_1 = \langle V \rangle_0$ and $C_2 = \langle V^2 \rangle_0 - \langle V \rangle_0^2$; Eq. (3.62)).

We must therefore evaluate Z_0 and the cumulants C_n. We first note that for each three-spin cell,

$$\sum_{\{\sigma\}_I} e^{\mathcal{H}_0^{(I)}} \equiv \sum_{\sigma_1^{(I)}} \sum_{\sigma_2^{(I)}} \sum_{\sigma_3^{(I)}} e^{K\left(\sigma_1^{(I)}\sigma_2^{(I)} + \sigma_2^{(I)}\sigma_3^{(I)} + \sigma_3^{(I)}\sigma_1^{(I)}\right)} = 2(e^{3K} + 3e^{-K}) \equiv 2z_0 , \tag{8.97}$$

where the intracell Hamiltonian is separable, $\mathcal{H}_0 \equiv \sum_I \mathcal{H}_0^{(I)}$. Then, from Eq. (8.92),

$$Z_0 = \prod_{I=1}^{N'} \left[\sum_{\{\sigma\}_I} \left(\frac{1}{2}\left[1 + \mu_I \psi(\{\sigma\}_I)\right] e^{\mathcal{H}_0^{(I)}} \right) \right] = (z_0)^{N'} = \left(e^{3K} + 3e^{-K}\right)^{N'} . \tag{8.98}$$

There is no μ-dependence to Z_0 (in this case) because of up-down symmetry in the absence of a magnetic field. Using Eq. (8.90), we have the first cumulant

$$C_1 = \langle V \rangle_0 = K \sum_{\langle IJ \rangle} \sum_{i \in I} \sum_{j \in J} \langle \sigma_i^{(I)} \sigma_j^{(J)} \rangle_0 . \tag{8.99}$$

Referring to Fig. 8.13, let's work out the generic term:

$$\langle \sigma_1^{(I)} \sigma_2^{(J)} \rangle_0 = \frac{1}{Z_0} \sum_{\{\sigma\}} T[\mu|\sigma] e^{\mathcal{H}_0(\sigma)} \sigma_1^{(I)} \sigma_2^{(J)}$$

$$= \frac{1}{Z_0} \left(\prod_{L \neq I, J} \sum_{\{\sigma\}_L} e^{\mathcal{H}_0^{(L)}} T_L[\mu|\sigma] \right) \left(\sum_{\{\sigma\}_I} T_I[\mu|\sigma] e^{\mathcal{H}_0^{(I)}} \sigma_1^{(I)} \right) \left(\sum_{\{\sigma\}_J} T_J[\mu|\sigma] e^{\mathcal{H}_0^{(J)}} \sigma_2^{(J)} \right)$$

$$= \left(\frac{1}{z_0} \sum_{\{\sigma\}_I} T_I[\mu|\sigma] e^{\mathcal{H}_0^{(I)}} \sigma_1^{(I)} \right) \left(\frac{1}{z_0} \sum_{\{\sigma\}_J} T_J[\mu|\sigma] e^{\mathcal{H}_0^{(J)}} \sigma_2^{(J)} \right). \tag{8.100}$$

Because the probability distribution $e^{\mathcal{H}_0(\sigma)} T[\mu|\sigma]/Z_0$ factorizes over cells, the average we seek is a product of cell averages, $\langle \sigma_1^{(I)} \sigma_2^{(J)} \rangle_0 = \langle \sigma_1^{(I)} \rangle_0 \langle \sigma_2^{(J)} \rangle_0$. We must evaluate

$$\overbrace{\sum_{\{\sigma\}_I} e^{\mathcal{H}_0^{(I)}} \sigma_1^{(I)} T_I[\mu|\sigma]}^{\psi(\{\sigma\}_I)} = \frac{1}{2} \sum_{\{\sigma\}_I} e^{\mathcal{H}_0^{(I)}} \sigma_1 \left(1 + \mu_I \frac{1}{2} [\sigma_1 + \sigma_2 + \sigma_3 - \sigma_1\sigma_2\sigma_3] \right)$$

$$= \frac{1}{2} \sum_{\{\sigma\}_I} e^{\mathcal{H}_0^{(I)}} \left(\sigma_1 + \frac{1}{2} \mu_I [1 + \sigma_1\sigma_2 + \sigma_1\sigma_3 - \sigma_2\sigma_3] \right)$$

$$= \frac{1}{4} \mu_I \sum_{\{\sigma\}_I} e^{\mathcal{H}_0^{(I)}} [1 + \sigma_1\sigma_2 + \sigma_1\sigma_3 - \sigma_2\sigma_3]$$

$$= \frac{1}{4} \mu_I \left[2z_0 + 2\left(e^{3K} - e^{-K} \right) \right] = \mu_I \left(e^{3K} + e^{-K} \right), \tag{8.101}$$

where we've used Eqs. (8.88) and (8.97) and the result of Exercise 8.36.

By combining Eq. (8.101) with Eq. (8.100), and then with Eq. (8.99), we have from Eqs. (8.96) and (8.98) the renormalized Hamiltonian to first order in the cumulant expansion:

$$\mathcal{H}' = N' \ln \left(e^{3K} + 3e^{-K} \right) + K' \sum_{\langle IJ \rangle} \mu_I \mu_J, \tag{8.102}$$

where

$$K' = 2K \left(\frac{e^{3K} + e^{-K}}{e^{3K} + 3e^{-K}} \right)^2. \tag{8.103}$$

The factor of two in Eq. (8.103) comes from the two dashed lines connecting near-neighbor cells in Fig. 8.13. Now that we have the recursion relation $K' = K'(K)$, what do we do with it? Hopefully you're saying: Find the fixed point and then linearize. First, we observe from Eq. (8.103) that $K' \approx \frac{1}{2}K$ for $K \to 0$, i.e., K' gets smaller for small K (high temperature), and $K' \approx 2K$ for large K, i.e., K' gets larger at low temperature. There must be a value K^* such that $K^{*\prime} = K^*$, the nontrivial solution of

$$K^* = 2K^* \left(\frac{e^{3K^*} + e^{-K^*}}{e^{3K^*} + 3e^{-K^*}} \right)^2. \tag{8.104}$$

The fixed point $K^* \approx 0.336$ should be compared with the exact critical coupling of the triangular-lattice Ising model, $K_c^{\text{exact}} = \frac{1}{4} \ln 3 \approx 0.275[142]$,[41] and the prediction of mean field theory,

---

[41] K_c for the triangular-lattice Ising model differs from that for the square lattice, Eq. (7.125). Is that a violation of universality? *No.* Universality asserts the same critical exponents for systems differing in their microscopic details; T_c is a nonuniversal property sensitive to details such as lattice structure.

$K_c^{\text{mean field}} = \frac{1}{6} \approx 0.167$ (Section 7.9). It's straightforward to evaluate from Eq. (8.103)

$$\left(\frac{\partial K'}{\partial K}\right)^* \approx 1.624 .$$ (8.105)

We then have, using Eq. (8.45), the critical exponent ν:

$$\nu = \frac{\ln L}{\ln\left(\partial K'/\partial K\right)^*} = \frac{\ln\sqrt{3}}{\ln 1.624} \approx 1.133 ,$$ (8.106)

a result not too far from the exact value $\nu = 1$ and distinct from the prediction of mean field theory, $\nu = \frac{1}{2}$ (see Table 7.3). Note that to have $\nu = 1$, we "want" $(\partial K'/\partial K)^* = \sqrt{3} \approx 1.732$.

The next step is to evaluate the cumulant C_2 to see whether the expansion in Eq. (8.96) leads to better results. These are laborious calculations; we omit the details and defer to [139]. In calculating C_2, next-nearest and third-neighbor interactions are generated, call them K_2, K_3, in addition to near-neighbor couplings, which we now call K_1. Figure 8.14 shows these interactions on the triangular lattice. One must start with a model having couplings K_1, K_2, K_3 on the original lat-

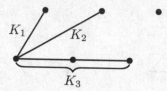

Figure 8.14: Near, next-nearest, and third-neighbor couplings K_1, K_2, K_3 on the triangular lattice.

tice to calculate K_1', K_2', K_3' on the scaled lattice. We omit expressions for the recursion relations $K_i' = f_i(\mathbf{K})$, $i = 1, 2, 3$ (included in f_1 are the results obtained at first order in the cumulant expansion, Eq. (8.103)). As per Niemeijer and van Leeuwen[139], the nontrivial fixed point is at $\mathbf{K}^* \equiv (K_1^* = 0.2789, K_2^* = -0.0143, K_3^* = -0.0152)$, with the transformation matrix

$$\left(\frac{\partial K_i'}{\partial K_j}\right)^* = \begin{pmatrix} 1.8966 & 1.3446 & 0.8964 \\ -0.0403 & 0 & 0.4482 \\ -0.0782 & 0 & 0 \end{pmatrix}$$ (8.107)

having eigenvalues $\lambda_1 = 1.7835$, $\lambda_2 = 0.2286$, $\lambda_3 = -0.1156$. There is one relevant and two irrelevant scaling fields. We see that λ_1 is closer to $\lambda_1^{\text{exact}} = \sqrt{3}$ than for $C_1, \approx 1.624$. As in Section 8.3.4, the line separating flows to the high and low-temperature fixed points can be extended to $K_2 = K_3 = 0$ at $K_c = 0.2514$, closer to $K_c^{\text{exact}} = 0.275$ than for $C_1, \approx 0.336$. From Eq. (8.69), $\nu = \ln\sqrt{3}/\ln \lambda_1 \approx 0.949$, closer to the exact result $\nu = 1$ than in Eq. (8.106).

Working to second order in the cumulant expansion therefore provides improved results for the triangular lattice. A third-order cumulant calculation on the triangular lattice found improvements over second-order results[143], and thus "so far, so good," it might appear that the cumulant expansion is a general method of real-space renormalization. The convergence of the expansion, however, is not understood—the series in Eq. (8.96) converges for sufficiently small K, but is K^* "sufficiently small"? We have no way of knowing until we evaluate the cumulants. A conceptual flaw in the cumulant expansion approach to real-space renormalization is that it treats intercell couplings perturbatively (dashed lines in Fig. 8.12), yet such couplings are not small "corrections" on uncoupled cells; its use of perturbation theory is formal in nature.[42] Other methods of real-space renormalization have been developed and are reviewed in Burkhardt and van Leeuwen[144]. Mazenko and

---

[42]The term *uncontrolled approximation* is applied to theories lacking a natural small parameter. The first cumulant $C_1 = \langle V \rangle_0$ is the average of the intercell energy of interaction. For N spins on the triangular lattice, there are $zN/2 = 3N$ distinct near-neighbor pairs (z is the coordination number). In each cell there are three near-neighbor pairs, and there are $N/(L^d)$ cells, implying $3(N/3) = N$ near-neighbor intracell pairs. That leaves $2N$ intercell near-neighbor pairs.

coworkers developed a perturbative scheme for real-space renormalization not based on the cumulant expansion that treats the mapping function in Eq. (8.88) as the zeroth-order mapping function for uncoupled cells, to which corrections can be calculated at first and second order, etc [145, 146].

Before leaving this topic, we note that by calculating the renormalized Hamiltonian, recursion relations for the magnetization and susceptibility can be obtained (Section 8.3.1), but what about our old friends, correlation functions? A method of constructing recursion relations for correlation functions was developed by Mazenko and coworkers [147, 148]. For any spin function $A(\sigma)$ on the original lattice, its counterpart on the scaled lattice $A(\mu)$ is defined such that

$$A(\mu)P(\mu) \equiv \sum_{\{\sigma\}} P(\sigma)T[\mu|\sigma]A(\sigma) = \langle T[\mu|\sigma]A(\sigma)\rangle . \qquad (8.108)$$

In that way, using Eq. (8.84), $\langle A(\mu)\rangle' = \langle A(\sigma)\rangle$. If one first finds $P(\mu)$ using Eq. (8.86), and then combines it with Eq. (8.108) to find $A(\mu)$, recursion relations are obtained connecting correlation functions (or any spin function) on the original and scaled lattices.

8.6 WILSON RENORMALIZATION GROUP IN A NUTSHELL

In 1971 K.G. Wilson published two articles[124][149] marking the introduction of the renormalization group to the study of critical phenomena.[43] In the first he reformulated Kadanoff's theory in an original way, showing that scaling and universality can be derived from assumptions not involving block spins. In the second he showed how renormalization can be done in k-space. The latter is sufficiently complicated that we can present it only in broad brushstrokes (see Section 8.6.3).

8.6.1 Renormalization group equations, Kadanoff scaling

In his first paper Wilson established *differential equations* for K_L, B_L considered as functions of L, and showed that their solutions in the critical region are the Kadanoff scaling forms, Eq. (8.21). To do that, he had to treat the scale factor L as a continuous quantity. In the block-spin picture, L is either an integer or is associated with one.[44] Consider a small change in length scale, $L \to L(1+\delta)$, $\delta \ll 1$. Assuming K_L, B_L differentiable functions of L, for infinitesimal δ,

$$K_{L(1+\delta)} - K_L \approx \delta L \left(\frac{dK_L}{dL}\right) \equiv \delta u$$

$$B_{L(1+\delta)} - B_L \approx \delta L \left(\frac{dB_L}{dL}\right) \equiv \delta v ,$$

with $u \equiv L(dK_L/dL)$, $v \equiv L(dB_L/dL)$. Wilson's key insight is that *the functions u, v depend on L only implicitly*, through the L-dependence of K_L, B_L, $u = u(K_L, B_L^2)$, $v = v(K_L, B_L^2)$.[45] In the Kadanoff construction, by assembling 2^d blocks to make a new block, couplings K_L, B_L are mapped to K_{2L}, B_{2L}, *a mapping independent of the absolute length of the initial blocks*. In Wilson's words, "... the Hamiltonian does not know what the size L of the old block was." By differentiating Eqs. (8.19) and (8.20) with respect to L, he showed that, indeed, u, v do not depend explicitly on L. Thus, we have differential equations known as the *renormalization group equations*,

$$\frac{dK_L}{dL} = \frac{1}{L}u\left(K_L, B_L^2\right)$$

$$\frac{dB_L}{dL} = \frac{1}{L}B_L v\left(K_L, B_L^2\right) , \qquad (8.109)$$

---

[43]A version of the renormalization group was developed in quantum electrodynamics in the 1950s [150, 151].

[44]L isn't restricted to be an integer in the block-spin picture, as we saw with $L = \sqrt{3}$ in our treatment of the triangular lattice, but the coarse-graining cells have an integer number of spins such that $L^d =$ integer.

[45]The dependence on B_L^2 is because the problem is symmetric under $B_L \to -B_L$.

where the factor of B_L in the second part of (8.109) is because v is an even function of B_L.[46]

Wilson identified the critical point as the point where no change in couplings occurs upon a change in length scale.[47] Thus, K_c is the solution of

$$u(K_c, 0) = 0 . \tag{8.110}$$

Note that $B_L = 0$ is automatically a stationary solution of Eq. (8.109). Having identified the critical point, linearize[48] about $u = 0$ and $B_L = 0$:

$$\frac{dK_L}{dL} = \frac{1}{L}(K_L - K_c)\frac{\partial u}{\partial K}\Big|_{(K_c, 0)} \equiv \frac{1}{L}(K_L - K_c)x$$

$$\frac{dB_L}{dL} = \frac{1}{L}B_L v(K_c, 0) \equiv \frac{1}{L}B_L y . \tag{8.111}$$

The numbers x, y are finite because u, v are analytic. These relations imply Eq. (8.21),

$$\delta K_L = L^x \delta K \qquad\qquad B_L = L^y B . \tag{8.112}$$

Kadanoff scaling therefore emerges quite generally from just a few assumptions:

1. The existence of a class of models having couplings that depend on a continuous scale parameter L and which have Ising-type partition functions,

$$Z = \sum_{\{\sigma\}} \exp\left(K_L \sum_{\langle ij \rangle} \sigma_i \sigma_j + B_L \sum_i \sigma_i\right) .$$

In today's parlance we would say this is the assumption of *renormalizability*.[49]

2. The validity of Eqs. (8.19) and (8.20) generalized to continuous L.

3. The existence of the differential equations for K_L, B_L, (8.109). Scaling emerges from the *form* of these equations, and not from the details of the Hamiltonian.[50]

Wilson's most surprising finding is that scaling is only loosely connected to the type of Hamiltonian; *scaling occurs as a consequence of the form of the renormalization group equations*, (8.109), which are not associated with the properties of a given Hamiltonian—they tell us how different Hamiltonians in the same class are related.

In his second paper Wilson set out an ambitious research program. Can L-scaling transformations be devised that are related to the structure of the partition function (rather than the Kadanoff construction), and is there a class of Hamiltonians invariant under such transformations? Can the renormalization group equations be derived (and not postulated)? Wilson worked with a spin model where variables are not restricted to the values ± 1. Before proceeding (see Section 8.6.3), we must make the acquaintance of more general spin models.

---

[46]We have in Eq. (8.109) nonlinear first-order differential equations (u, v are complicated functions of K_L, B_L), differential generalizations of the recursion relations K', B' that we found in previous sections. We don't know the actual functions u, v. Wilson established the *form* of the renormalization group equations, which is sufficient to derive scaling.

[47]This is the argument by which we infer that $\xi \to \infty$ at a fixed point; see Section 8.3.1.

[48]We're therefore assuming that u, v are analytic functions, even at the critical point.

[49]This is not a tautology ("renormalized models are renormalizable"). In quantum field theory there are criteria for theories to be renormalizable; see Peskin and Schroeder[152]. When renormalization was developed in the 1940s showing that the divergences of quantum electrodynamics could be eliminated by a change in parameterization (see the collection of articles in Schwinger[153]), many saw it as a "bandaid" solution covering up a deeper problem (Dyson, Dirac, Feynman). Compare that with today's thinking that the ability of theories to be renormalized (renormalizability) is a *requirement* on fundamental theories; see Weinberg, *Dreams of a Final Theory*[154].

[50]In classical mechanics, it's the form of Hamilton's equations of motion that gives rise to important properties such as canonical transformations, rather than the details of the Hamiltonian.

8.6.2 Ginzburg-Landau theory: Spatial inhomogeneities

Wilson sought a model appropriate for the *long-range* phenomena we have with critical phenomena involving many lattice sites. To do so, he approximated a lattice of spins as a continuous distribution of spins throughout space. He replaced a discrete system of interacting spins $\{\sigma_i\}$ with a continuous spin-density *field*[51] $S(r)$, the spin density at the point located by position vector r. *A field description is itself a coarse graining*, a view of a system from sufficiently large distances that it appears continuous. Consider (again) a block of spins of linear dimension L centered on the point located by r. Define a local magnetization density (σ_i here is not necessarily an Ising spin),

$$S_L(r) \equiv \frac{1}{N_L} \sum_{i \in \text{block at } r} \langle \sigma_i \rangle , \qquad (8.113)$$

where $N_L = (L/a)^d$ is the number of spins per block, with a the lattice constant. We parameterize $S_L(r)$ with the block size because L is not uniquely specified. One wants L large enough that $S_L(r)$ does not fluctuate wildly as a function of r, but yet small enough that we can use the methods of calculus; finding "physically" infinitesimal lengths is a generic problem in constructing macroscopic theories. How is the "blocking" in Eq. (8.113) different from Kadanoff block spins, Eq. (8.17)? $S_L(r)$ is defined in terms of averages $\langle \sigma_i \rangle$, no assertion is made that all spins are aligned, or that scaling is implied. In what follows we drop the coarse-graining block-size L as a parameter—we'll soon work in k-space where we consider only small wave numbers $k < L^{-1} \ll a^{-1}$.

We may think of the field $S(r)$ as an *inhomogeneous order parameter* in the language of Landau theory. The order parameter (Section 7.7) is a thermodynamic quantity such as magnetization that represents the average behavior of the system as a whole. With $S(r)$, we have a *local* magnetization density that allows for spatial inhomogeneities in equilibrium systems.[52] *Ginzburg-Landau theory is a generalization of Landau theory so that it has a more microscopic character*. The Landau free energy F (Eq. (7.79)) is an extensive thermodynamic quantity. It can always be written in terms of a density $\mathcal{F} \equiv F/V$, which for bulk systems is a number having no spatial dependence, but for which there is no harm in writing $F = \int d^d r \mathcal{F}$. If the magnetization density varies spatially, however, and sufficiently slowly, we can infer that its local value $S(r)$ (in equilibrium) represents a minimum of the free energy density *at that point*, implying that $\mathcal{F}$ varies spatially. The Ginzburg-Landau model finds the total free energy through an integration over spatial quantities,

$$F[S] = \int d^d r \mathcal{F}(S(r)) . \qquad (8.114)$$

Equation (8.114) presents us with a variational problem, the reason we referred to the Landau free energy as a *functional*[53] in Section 7.7.1: For given $\mathcal{F}$, what is the spatial configuration $S(r)$ that minimizes F? The calculus of variations answers such questions. We want the functional derivative to vanish (Euler-Lagrange equation, (C.18))

$$\frac{\delta F}{\delta S} = \frac{\partial \mathcal{F}}{\partial S} - \frac{d}{dr}\left(\frac{\partial \mathcal{F}}{\partial S'}\right) = 0 , \qquad (8.115)$$

where $S' \equiv dS/dr$.

---

[51] See Section C.12 for an example of how a displacement field emerges from a discrete system of interacting degrees of freedom. We use "displacement" loosely, the amplitude of vibration of an elastic medium or the spin density in space.

[52] Of course, we have to account for the *source* of such inhomogeneities; it must be energetically favorable for a system in thermal equilibrium to develop inhomogeneities, which could come from impurities, or boundary conditions, or from fluctuations.

[53] $F[S]$ in Eq. (8.114) is a functional: It maps a function $S(r)$ to a number by doing the integral (see Section C.2). The free energy density $\mathcal{F}$ is loosely referred to as a functional, even though it plays the role of a Lagrangian.

If the spatial variations of $S(r)$ are not too rapid, it's natural to assume that the Landau *form*, Eq. (7.81), written as a density, holds at every point in space with[54]

$$\mathcal{F}(S) = a(T - T_c)S^2 + bS^4 - BS \,, \tag{8.116}$$

where $a, b > 0$ are positive parameters. Combining Eq. (8.116) (which is independent of S') with Eq. (8.115),

$$\frac{\partial \mathcal{F}}{\partial S} = 0 = 2a(T - T_c)S + 4bS^3 - B \,.$$

For $B = 0$ and $T < T_c$, we have the mean field order parameter (Eq. (7.80))

$$S_{\mathrm{mf}}(r) = \pm\sqrt{\frac{a}{2b}|T - T_c|} \,.$$

Thus, taking $\mathcal{F}$ to be the Landau form leads to a spatially homogeneous system with the mean-field order parameter. Mean field theory ignores interactions between fluctuations (Section 7.8). To get beyond mean field theory, we must include an energy associated with fluctuations,[55] a piece of physics not contained in Eq. (8.116). Consider a term that quantifies the coupling of $S(r)$ to its near neighbors:

$$\sum_n \left(\frac{S(r) - S(r+n)}{L}\right)^2 \approx |\nabla S(r)|^2 \,,$$

where n is a vector of magnitude L pointing toward near-neighbor blocks. Ginzburg-Landau theory modifies the free energy functional to include inhomogeneities,

$$F[S] = \int \mathrm{d}^d r \left[a(T - T_c)S^2(r) + bS^4(r) - BS(r) + g\nabla S(r) \cdot \nabla S(r)\right] \,, \tag{8.117}$$

where g is a positive parameter characterizing the energy associated with gradients.[56] Our intent in establishing Eq. (8.117) is to prepare for Wilson's analysis; we don't consider any of its applications per se, of which there are many in the theory of superconductivity. See for example Schrieffer[155].

The partition function for Ginzburg-Landau theory generalizes what we have in bulk systems $Z = e^{-\beta F}$ to a summation over all possible field configurations $S(r)$:

$$Z = \int \mathcal{D}S e^{-F[S]} = \int \mathcal{D}S \exp\left(-\int \mathrm{d}^d r \mathcal{F}(S(r))\right) \,, \tag{8.118}$$

where $\int \mathcal{D}S$ indicates a *functional integral*[57] which means conceptually to sum over all smooth functions $S(r)$. The coarse-grained free energy $F[S]$ is called the *effective Hamiltonian*.

8.6.3 Renormalization in k-space, dimensionality as a continuous parameter

It's well known that the Fourier components of a function at wave vector k are associated with a range of spatial variations of order $2\pi/|k|$, a fact we can use to characterize the degrees of freedom associated with various length scales by working in k-space. Define the Fourier transform of $S(r)$

$$S_k \equiv \int \mathrm{d}^d r e^{ik \cdot r} S(r) \,, \tag{8.119}$$

---

[54]A constant term $F_0(T)$ has been omitted.

[55]The treatment of fluctuations is a key difference between statistical mechanics and thermodynamics; see Section 2.4. In statistical mechanics, we allow for *local* fluctuations that are correlated in space and time. The time dependence is treated in nonequilibrium statistical mechanics.

[56]The $|\nabla S|^2$ term in Eq. (8.117) is sometimes called the "kinetic energy" in analogy with quantum mechanics, where the Schrödinger equation kinetic energy is associated with curvature of the wave function, with the other terms (Eq. (8.116)) the "potential energy." Formal analogies such as these can be useful conceptually but should not be taken literally.

[57]See for example Feynman and Hibbs[156] or Popov[157].

with inverse

$$S(r) = \frac{1}{(2\pi)^d} \int d^d k e^{-ik \cdot r} S_k \, . \tag{8.120}$$

To show these equations imply each other, use $\int d^d k e^{ik \cdot (r-r')} = (2\pi)^d \delta(r - r')$. Note that $S_k^* = S_{-k}$. As one can show:

$$\int d^d r S(r) = \int d^d k S_k \delta(k) = S_{k=0} \, ;$$

$$\int d^d r S^2(r) = \frac{1}{(2\pi)^d} \int d^d k \int d^d k' S_k S_{k'} \delta(k + k') = \frac{1}{(2\pi)^d} \int d^d k |S_k|^2 \, ; \tag{8.121}$$

$$\int d^d r \nabla S(r) \cdot \nabla S(r) = -\frac{1}{(2\pi)^d} \int d^d k \int d^d k' (k \cdot k') S_k S_{k'} \delta(k + k') = \frac{1}{(2\pi)^d} \int d^d k k^2 |S_k|^2 \, ;$$

$$\int d^d r S^4(r) = \frac{1}{(2\pi)^{3d}} \int d^d k \int d^d k' \int d^d k'' \int d^d k''' S_k S_{k'} S_{k''} S_{k'''} \delta(k + k' + k'' + k''') \, .$$

It's traditional at this point to re-parameterize the Ginzburg-Landau functional so that $g = 1$, with new names given to the other parameters. We also set $B = 0$. Thus, we rewrite Eq. (8.117),

$$F[S] = \int d^d r \left(\widetilde{r} S^2(r) + u S^4(r) + |\nabla S(r)|^2 \right) , \tag{8.122}$$

where $\widetilde{r}$ can be either sign and $u \geq 0$. Combining the results in (8.121) with Eq. (8.122), we have F as a functional of the Fourier components S_k,

$$F[S_k] = \frac{1}{(2\pi)^d} \int d^d k \left(\widetilde{r} + k^2 \right) |S_k|^2 \tag{8.123}$$

$$+ \frac{u}{(2\pi)^{3d}} \int d^d k \int d^d k' \int d^d k'' \int d^d k''' S_k S_{k'} S_{k''} S_{k'''} \delta(k + k' + k'' + k''') \, .$$

Equation (8.123) is Wilson's starting point. At this point the trail, if followed, gets steep and rocky, with a few sections requiring hand-over-foot climbing. As with any good trail guide, let's give a synopsis of the journey. Under repeated renormalizations (which have yet to be specified), the free energy density retains the form it has in k-space (an essential requirement for a renormalization transformation). For $d > 4$ it's found that the quartic parameter u is associated with an irrelevant scaling field, with the fixed point at $\widetilde{r}^* = 0, u^* = 0$, the *Gaussian fixed point*. For $d < 4$, u drives the fixed point to the *Wilson-Fisher fixed point* at $\widetilde{r}^* \neq 0, u^* \neq 0$. The quartic term is handled (for $d < 4$) using a perturbation theory of a special kind, where the dimensionality d is *infinitesimally less than four*, i.e., *d is treated as a continuous parameter*. One defines a variable $\epsilon \equiv 4 - d$ and seeks perturbation expansions in ϵ for the critical exponents, the *epsilon expansion*. The Wilson-Fisher fixed point merges with the Gaussian fixed point as $\epsilon \to 0$. The ϵ-expansion might seem fanciful, but if one had enough terms in the expansion, one could attempt extrapolations to $\epsilon = 1$, i.e., $d = 3$. That, however, is beyond the intended scope of this book.

To show how renormalization in k-space works, we treat the case of $u = 0$ without the intricacies of the ϵ-expansion. The part of Eq. (8.123) with $u = 0$ is termed the *Gaussian model*; it describes systems with $d > 4$ and $T > T_c$. (If there is no quartic term, there is no ordered phase for $T < T_c$.)

We start by restricting the wave vectors k in Fourier representations of $S(r)$ to $|k| < \Lambda$, where Λ is the *cutoff parameter* (or simply *cutoff*). A natural choice is $\Lambda = L^{-1}$ (inverse coarse-graining length). A field defined as in Eq. (8.113) is smooth only for spatial ranges greater than L, which translates to $k < L^{-1}$ in the Fourier domain.[58] Thus, with $u = 0$ in Eq. (8.123) and with the cutoff

---

[58]Even if we're not defining fields through coarse graining, the range of wave vectors that describe lattice degrees of freedom such as phonons is restricted to a finite region of k-space, the *Brillouin zone*. See any text on solid-state physics.

displayed, we have the Gaussian model

$$F[S_k; \widetilde{r}, \Lambda] = \frac{1}{(2\pi)^d} \int_{|k|<\Lambda} \mathrm{d}^d k \, (\widetilde{r} + k^2) \, |S_k|^2 \, . \tag{8.124}$$

The notation indicates that F is a functional of the Fourier components S_k, but is a function of $\widetilde{r}, \Lambda$. The Gaussian model is a function of the one coupling parameter $\widetilde{r}$ ($u = 0$ by assumption) but also the cutoff parameter Λ. The model must include a specification of Λ, otherwise it's not well defined. Wave numbers must be prevented from becoming arbitrarily large, where the model is unphysical.

Renormalization transformations in real space are implemented in two steps (Sections 8.3, 8.5): sum out selected degrees of freedom and rescale lengths so that the Hamiltonian has its original form but with renormalized couplings. We do the same in k-space. As the first step, we integrate over wave vectors k such that, for $l > 1$ (l is not restricted to integer values),

$$\frac{1}{l}\Lambda < |k| < \Lambda \, . \tag{8.125}$$

The strategy is to sum out short-wavelength degrees of freedom in favor of long.

The dichotomy between long and short wavelengths can be treated mathematically by representing $S(r)$ as a sum of the real-space functions that are synthesized from the high and low-k components of its Fourier transform (see Eq. (8.120)):

$$S(r) = \frac{1}{(2\pi)^d} \int_0^{\Lambda/l} \mathrm{d}^d k \, e^{-i k \cdot r} S_k + \frac{1}{(2\pi)^d} \int_{\Lambda/l}^{\Lambda} \mathrm{d}^d k \, e^{-i k \cdot r} S_k$$
$$\equiv S'_l(r) + \sigma_l(r) \, . \tag{8.126}$$

The prime on $S'_l(r)$ is deliberate: The decomposition in Eq. (8.126) "filters" $S(r)$ into a term $S'_l(r)$ made up of long-wavelength components, which can be thought of as corresponding to the "block spin" of real-space renormalization, and $\sigma_l(r)$, composed of short-wavelength components, which corresponds to the "intracell" degrees of freedom. For convenience we represent S_k in terms of its short and long-wavelength parts, which we give new names to:

$$S_k \equiv \begin{cases} \widetilde{S}'_l(k) & 0 < |k| < \Lambda/l \\ \widetilde{\sigma}_l(k) & \Lambda/l < |k| < \Lambda \, . \end{cases} \tag{8.127}$$

Combining (8.127) with Eq. (8.124), $F[S_k]$ can be split into short and long wavelength parts,

$$F[S_k] = \frac{1}{(2\pi)^d} \int_0^{\Lambda/l} \mathrm{d}^d k \, (\widetilde{r} + k^2) \left| \widetilde{S}'_l(k) \right|^2 + \frac{1}{(2\pi)^d} \int_{\Lambda/l}^{\Lambda} \mathrm{d}^d k \, (\widetilde{r} + k^2) \, |\widetilde{\sigma}'_l(k)|^2$$
$$\equiv F_S[\widetilde{S}'_l(k)] + F_\sigma[\widetilde{\sigma}'_l(k)] \, . \tag{8.128}$$

The partition function Eq. (8.118) is, in the Fourier domain, a functional integral over S_k (instead of $S(r)$), where, using Eq. (8.128),

$$Z = \int \mathcal{D}S_k e^{-\{F_S[\widetilde{S}'_l(k)] + F_\sigma[\widetilde{\sigma}_l(k)]\}} = \left(\int \mathcal{D}\widetilde{S}'_l(k) e^{-F_S[\widetilde{S}'_l(k)]} \right) \left(\int \mathcal{D}\widetilde{\sigma}_l(k) e^{-F_\sigma[\widetilde{\sigma}_l(k)]} \right) \equiv Z_S Z_\sigma \, .$$

We leave Z_σ (partition function of short-wavelength components) as an unevaluated term. It can be treated as a constant—it doesn't affect the calculation of critical exponents (it impacts calculations of the free energy, however); Z_σ corresponds to the cell partition function Z_0 in real space renormalization, Eq. (8.92).

Restricting our attention to Z_S,

$$Z_S(\widetilde{r}) = \int \mathcal{D}\widetilde{S}'_l(\mathbf{k}) \exp\left(-\frac{1}{(2\pi)^d} \int_0^{\Lambda/l} d^d\mathbf{k}(\widetilde{r} + k^2)\left|\widetilde{S}'_l(\mathbf{k})\right|^2\right). \tag{8.129}$$

The integral in Eq. (8.129) is almost what would be obtained from Eq. (8.124): The cutoff has been modified, $\Lambda \to \Lambda/l$. The second step of the renormalization transformation is a length rescaling (with the first the integration over wave numbers in Eq. (8.125)). In Eq. (8.129), make the substitutions

$$k = \frac{1}{l}k' \qquad \widetilde{S}'_l(\mathbf{k}) = z\widetilde{S}_l(\mathbf{k}'), \tag{8.130}$$

where the scaling factor z is at this point unspecified, but will depend on l. (The latter scaling is referred to as *wave function renormalization* in quantum field theory.) Making the substitutions in Eq. (8.129),

$$Z_S(\widetilde{r}) = \int \mathcal{D}\widetilde{S}_l(\mathbf{k}') \exp\left(-\frac{1}{(2\pi)^d} \int_0^{\Lambda} d^d\mathbf{k}'(z^2\widetilde{r}/l^d + z^2 k'^2/l^{d+2})\left|\widetilde{S}_l(\mathbf{k}')\right|^2\right). \tag{8.131}$$

Choose z in Eq. (8.131) so that $z^2 = l^{d+2}$ (we keep the strength of the $|\nabla S(\mathbf{r})|^2$ term in Eq. (8.122) at unity). In that way (erasing the prime on $\mathbf{k}'$),

$$Z_S(\widetilde{r}) = \int \mathcal{D}\widetilde{S}_l(\mathbf{k}) \exp\left(-\frac{1}{(2\pi)^d} \int_0^{\Lambda} d^d\mathbf{k}\,(\widetilde{r}_l + k^2)\left|\widetilde{S}_l(\mathbf{k})\right|^2\right), \tag{8.132}$$

where

$$\widetilde{r}_l = l^2\widetilde{r}. \tag{8.133}$$

Equation (8.133) is the recursion relation we seek. We've found by integrating out a range of shorter wavelength degrees of freedom (Eq. (8.125)), and then rescaling lengths, Eq. (8.130), that we have an equivalent model characterized by a larger value of $\widetilde{r}$, i.e., the system is further from the critical point (the fixed point of Eq. (8.133) is clearly $\widetilde{r}^* = 0$).

From Eq. (8.133) we have the Kadanoff scaling exponent $x = 2$, implying from Eq. (8.31) the critical exponent $\nu = \frac{1}{2}$. One can include a magnetic field in the Gaussian model (a piece of analysis we forgo), where one finds the other scaling exponent $y = 1 + d/2$, implying $\eta = 0$ from Eq. (8.30). What about the others, $\alpha, \beta, \gamma, \delta$? Hopefully you're saying, use the relations we found among critical exponents from scaling theories. If one does that, however, one finds incorrect values of β, δ. What's gone wrong? The Gaussian model has no ordered phase for $T < T_c$ (just like $d = 1$), and β, δ refer to thermodynamic properties at or below T_c. With some more analysis it can be shown $\beta = \frac{1}{2}$ and $\delta = 3$ for the Gaussian model; see Goldenfeld[101, p359].

SUMMARY

This chapter provided an introduction to scaling and renormalization, two significant developments in theoretical condensed matter physics from the latter part of the 20th century.

- We started with the Widom scaling hypothesis, that in the critical region the free energy is a generalized homogeneous function, an idea which accounts for the many relations found among the critical exponents derivable from the free energy, $\alpha, \beta, \gamma, \delta$. Thermodynamics provides relations among critical exponents in the form of inequalities (Section 7.11); scaling provides an explanation for why these inequalities are satisfied as equalities. Widom scaling is supported experimentally (see Fig. 8.1). The Kadanoff block-spin picture lends theoretical support to the Widom hypothesis, and provides a scaling theory of correlation functions showing that the exponents ν, η are related to $\alpha, \beta, \gamma, \delta$ and the dimension d. Of the six critical exponents $\alpha, \beta, \gamma, \delta, \nu, \eta$, only two are independent.

- Kadanoff scaling theory boils down to the relations in Eq. (8.21), $t_L = L^x t$ and $B_L = L^y B$, which introduce the scaling exponents x, y. These relations link thermodynamic quantities with spatial considerations—scaling—how increasing the length scale for systems in their critical regions induces changes in the couplings such that the scaled system is further from the critical point. Wilson showed that the Kadanoff scaling forms emerge quite generally from assumptions not involving block spins (Section 8.6.1). The scaling exponents x, y are related to the critical exponents ν, η, Eqs. (8.30) and (8.31). Once we have found ν, η, the other critical exponents follow from the relations established by scaling theories. Finding the scaling exponents is the name of the game; that's how critical exponents are evaluated from first principles.

- Renormalization is the process by which new couplings K', B' are found under a change of length scale. There are two steps in a renormalization transformation: A partial evaluation of the partition function by summing over selected degrees of freedom, followed by a length rescaling bringing the Hamiltonian back to its original form, but with new couplings K', B'. The recursion relations $K' = K'(K, B)$, $B' = B'(K, B)$ are in general nonlinear functions of the original couplings K, B. One seeks the fixed points of these recursion relations, K^*, B^*. This serves two purposes. First, at the fixed point there are no changes in couplings with changes in length scale, implying the existence of a divergent correlation length, $\xi \to \infty$. Thus, we have a new way to characterize critical points, as fixed points of the recursion relations for coupling parameters. Second, scaling exponents are found from linearized versions of the recursion relations in the vicinity of the fixed point. Because a renormalization followed by a renormalization is itself a renormalization, the theory of renormalization is referred to as the renormalization group.

- Two complementary approaches have been developed for implementing the renormalization group: lattice-based methods (real-space or position-space renormalization), and field-theoretic methods (momentum-space or k-space renormalization). We gave examples of real-space renormalization in Sections 8.3 and 8.5. Real-space renormalization can be done in any number of dimensions, and is a general method for finding thermodynamic and structural quantities, i.e., it's not restricted to critical phenomena. We touched on k-space renormalization in Section 8.6. This method is restricted to dimensions "near" four dimensions; it treats dimensionality as a continuous parameter. Both methods (real and k-space renormalization) have a common theoretical strategy: Treat short-range degrees of freedom perturbatively and long-range degrees of freedom iteratively.

EXERCISES

8.1 For the $d = 1$ Ising model, the critical exponent $\eta = 1$. Let's see how that comes about.

  a. Equation (6.132) is an expression for the static structure factor $S(q, T)$ associated with the $d = 1$ Ising model. Show that it can be written (where a is the lattice constant)

$$S(q, T) = \frac{1 - u^2}{(1 - u)^2 + 4u \sin^2(qa/2)}, \tag{P8.1}$$

  where $u \equiv \tanh K$. Hint: Add and subtract one.

  b. Using Eq. (P8.1), show that $S(q, T)$ exhibits the limiting behaviors:

$$\lim_{\substack{T \to 0 \\ q \neq 0}} S(q, T) = 0 \qquad \lim_{\substack{q \to 0 \\ T \neq 0}} S(q, T) = \chi(T),$$

where $\chi(T)$ is given in Eq. (6.101). For the first limit, it helps to show that $1 - u = \xi^{-1} + O(\xi^{-2})$ for large K, where ξ is the correlation length for the one-dimensional Ising model; see Eq. (P6.1). Thus, $S(q,T)$ is not analytic in the neighborhood of $q = 0$, $T = 0$: The order of the limits matters, $\lim_{q \to 0} \lim_{T \to 0} \neq \lim_{T \to 0} \lim_{q \to 0}$. How should we investigate critical phenomena involving these two limits? The same issue occurs at any critical point, when T is replaced with the reduced temperature t.

c. Consider how the experiments are done. Correlated fluctuations can be probed in scattering experiments only if the wavelength λ of the radiation is much less than the correlation length (Section 6.6), $\lambda \ll \xi \implies q \gg \xi^{-1}$. Figure 8.15 schematically shows combina-

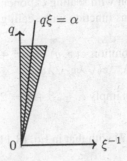

Figure 8.15: Critical region specified by $q \gg \xi^{-1}$ (shaded region). $\alpha \gg 1$ is a constant.

tions of q, ξ such that $q \gg \xi^{-1}$ (shaded region). In analyzing critical phenomena, we can approach $q = 0, \xi^{-1} = 0$ along the ray $q\xi = \alpha$ ($\alpha \gg 1$ is a constant) and remain in the region accessible to scattering experiments. Make sure you understand this argument.

d. Show from Eq. (P8.1), for small q and large ξ, that

$$S(q,T) \approx \frac{2}{(1/\xi) + (q\xi)a^2 q}$$

The structure factor is not in the Ornstein-Zernike form, Eq. (7.71). Comparing with the definition of η, Eq. (7.72), we can let $\xi \to \infty$ such that $q\xi = $ constant,

$$S(q, T = T_c) \overset{q \to 0}{\sim} q^{-1} \equiv q^{-(2-\eta)} ,$$

implying $\eta = 1$.

8.2 Consider a homogeneous function of order p, $f(x,y)$, such that

$$f(\lambda x, \lambda y) = \lambda^p f(x,y) . \tag{P8.2}$$

a. Show that a function $f(x,y)$ satisfying Eq. (P8.2) may be written in form

$$f(x,y) = y^p F(x/y) , \tag{P8.3}$$

where F is a function of a single variable. Hint: Set $\lambda = 1/y$. Equation (P8.2) holds for any value of λ; it's therefore true for a particular value. Repeat the argument to show that $f(x,y)$ can be written $f(x,y) = x^p G(y/x)$, where G is another function of a single variable.

b. Show that any function in the form of Eq. (P8.3) satisfies Eq. (P8.2). Hint: Consider $f(\lambda x, \lambda y)$ in Eq. (P8.3), and then use Eq. (P8.3) again to arrive at Eq. (P8.2).

8.3 Show that the Legendre transformation of a generalized homogeneous function is a generalized homogeneous function, but with a different scaling exponent. Let $f(x, y)$ be such that $f(\lambda^a x, \lambda^b y) = \lambda f(x, y)$. Define the Legendre transformation $g(u, y) \equiv f(x, y) - xu(x, y)$, where $u(x, y) \equiv (\partial f / \partial x)_y$. The Legendre transformation shifts emphasis from one variable to another (see Section C.1): If $u(x, y) = (\partial f / \partial x)_y$, then $dg = -xdu + (\partial f / \partial y)_x dy$ (show this), implying $g = g(u, y)$.

a. Show that $u(\lambda^a x, \lambda^b y) = \lambda^{1-a} u(x, y)$. Hint: Differentiate $f(\lambda^a x, \lambda^b y) = \lambda f(x, y)$.

b. Show that $g\left[u(\lambda^a x, \lambda^b y), \lambda^b y\right] = g(\lambda^{1-a} u, \lambda^b y) = \lambda g(u, y)$. Thus, if $f(x, y)$ is a generalized homogeneous function with scaling exponent a in the first argument, then $g(u, y)$ is a generalized homogeneous function with scaling exponent $1 - a$ in the first argument.

8.4 Verify that a paraboloid of revolution $z(x, y) = x^2 + y^2$ exemplifies the scaling behavior posited in Eq. (8.1), $z(x, y) = \lambda^{-1} z(\sqrt{\lambda} x, \sqrt{\lambda} y)$.

8.5 Show that Eqs. (8.12) and (8.13) imply $\alpha + 2\beta + \gamma = 2$, Rushbrooke's inequality satisfied as an equality.

8.6 Equate the result for b in Eq. (8.7) with that in Eq. (8.11), and show that

$$\frac{\gamma}{2 - \alpha} = \frac{\delta - 1}{\delta + 1}. \tag{P8.4}$$

Show that Eq. (P8.4) combined with Eq. (8.12) leads to $\alpha + \beta(\delta + 1) = 2$, Griffiths inequality (7.141) satisfied as an equality.

8.7 a. Show, assuming the validity of the scaling hypothesis, that if we know the critical exponents α, γ, we know β, δ:

$$\beta = \frac{1}{2}(2 - \alpha - \gamma) \qquad \delta = \frac{2 - \alpha + \gamma}{2 - \alpha - \gamma}. \tag{P8.5}$$

b. Show that if we know β, δ, we know γ, α:

$$\gamma = \beta(\delta - 1) \qquad \alpha = 2 - \beta(\delta + 1). \tag{P8.6}$$

8.8 Starting from either Eq. (8.7) or Eq. (8.11), show that the ratio of scaling exponents $b/a = \beta + \gamma$. Make liberal use of any identities we've derived for critical exponents.

8.9 a. Show that Eq. (8.23) implies the singular part of the free energy has the form

$$g(t, B) = |t|^{2-\alpha} f_\pm \left(\frac{B}{|t|^{\beta+\gamma}}\right). \tag{P8.7}$$

b. Show that the magnetization has the form

$$m(t, B) = |t|^\beta f'_\pm \left(\frac{B}{|t|^{\beta+\gamma}}\right). \tag{P8.8}$$

8.10 Derive the relations among critical exponents shown in Eq. (8.32).

8.11 The scaling theory implies that with knowledge of two critical exponents and the dimensionality d, the other critical exponents can be found. Show, assuming α, β and d are known, that one can find expressions for $\gamma, \delta, \eta, \nu$.

8.12 Derive the recursion relations in (8.40) starting from Eq. (8.38). Show that $x' > x$ and $y' < y$.

8.13 Referring to Eq. (8.38) for the renormalized magnetic coupling B', show that $B'(K, -B) = -B'(K, B)$. Hint: You may find it easier to first take the logarithm of Eq. (8.38). What are the symmetry properties of K_0, K' under $B \to -B$?

8.14 Examine the form of the of renormalized coupling B' for small B. Show from Eq. (8.38) that

$$B' = (1 + \tanh 2K) B + O(B^2).$$

As $K \to K_c$, $\tanh 2K \to 1$, and we have the scaling form $B' \approx 2B$.

8.15 Show that the number of degrees of freedom is preserved under a change of lattice constant for systems with $K = 0$ and $B = 0$. Using Eq. (8.39), show that $2^N = e^{NK_0/2} 2^{N/2}$ is an identity when $K_0 = \ln 2$; see Eq. (8.38).

8.16 Equation (8.39) is a relation between the partition function of the original system, $Z_N(K, B)$, and that of the renormalized system after a doubling of the lattice constant,

$$Z_N(K, B) = e^{NK_0/2} Z_{N/2}(K', B').$$

The partition function for the $d = 1$ Ising model is, for large N, $Z_N(K, B) = \lambda_+^N(K, B)$, Eq. (6.83), where $\lambda_+(K, B)$ is the largest eigenvalue of the transfer matrix, Eq. (6.82). Show that Eq. (8.39) implies

$$\lambda'_+ = e^{-K_0} \lambda_+^2,$$

where $\lambda'_+ \equiv \lambda_+(K', B')$. Verify this identity when $B = 0$. Note that $\lambda'_+ > 0$, which it must be by Perron's theorem applied to the transfer matrix of the renormalized system.

8.17 Find all the fixed points of the relations in Eq. (8.41).

8.18 Give a quick argument that Eq. (8.19) implies for fixed points either $\xi = 0$ or $\xi = \infty$.

8.19 Evaluate the four partial derivatives of the recursion relations in Eq. (8.40) (indicated in Eq. (8.43)), and verify that their values at $x^* = 0, y^* = 1$ are those shown in Eq. (8.43).

8.20 We have in Eq. (8.21) the scaling relation $t_L = L^x t$, where $t \equiv (T - T_c)/T_c$. The coupling constant $K \equiv J/(kT)$ can also be used as the scaling variable, as shown in Eq. (8.44). Equation (8.44) does *not* follow as an exact result of the correspondence $t \leftrightarrow (K_c/K) - 1$. Starting from $t' = L^x t$, first find the exact expression for K' that follows upon the change of variables $t = (K_c/K) - 1$. Then show that Eq. (8.44) is the lowest-order approximation in the quantity $(K_c - K)$. Scaling holds only in the critical region.

8.21 Derive the results in Eq. (8.45).

8.22 Derive Eq. (8.46). Get the factor of $\frac{1}{2}$ right.

8.23 Show by iterating Eq. (8.49), the recursion relation for the magnetization, that for $B = 0$, $m = 0$. This is as we expect, but it doesn't hurt to verify. Show explicitly that $(\partial B'/\partial B)_{B=0} \neq 0$. Is that what you would expect based on the symmetry of B' under $B \to -B$ (Exercise 8.13)? Also, don't forget scaling for small B—we can't have $(\partial B'/\partial B)_{B=0} = 0$.

8.24 Derive the expressions in (8.54) starting from (8.53) together with $K_2 = \frac{1}{2}K_1$.

8.25 Work out the high-temperature forms of the renormalized couplings K_1, K_4 in Eq. (8.54). Show to lowest order in small K that $K_1 \approx 2K^2$ and $K_4 \approx -2K^4$.

8.26 a. Referring to the example on page 280, find the exact partition function for the three-spin Ising system. A: $Z_3(K) = 2e^{3K} + 6e^{-K}$. Show that to lowest order in K one finds the same high-temperature expansion.

b. Show that an equivalent expression is $Z_3(K) = 8\left(\cosh^3 K + \sinh^3 K\right)$.

8.27 Figure 8.16 shows a four-spin Ising system. Assuming nearest neighbor interactions:

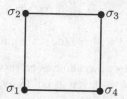

Figure 8.16: Four-spin Ising system.

a. Show using the high-temperature series method developed in Section 8.3.3, that

$$\langle \sigma_1\sigma_2 \rangle = \frac{u + u^3}{1 + u^4} \qquad \langle \sigma_2\sigma_4 \rangle = \frac{2u^2}{1 + u^4}.$$

Interpret the numerator of these expressions in terms of the paths between spins involving only nearest-neighbor links.

b. Using the high-temperature series method (adapt Eq. (8.63)), show that

$$\langle \sigma_1\sigma_2\sigma_3\sigma_4 \rangle = \frac{2u^2}{1 + u^4}.$$

Interpret the numerator of this expression in terms of the number of ways that two near-neighbor bonds cover the spins $\sigma_1, \sigma_2, \sigma_3, \sigma_4$ (for a given labelling of these spins).

c. Show that $\langle \sigma_1 \rangle = 0$. This one requires no explicit calculation.

8.28 Referring to Fig. 8.10, one might (erroneously) get the impression that starting with couplings $K_1 < K_c, K_2 = 0$, the flows towards the high-temperature fixed point would stay on the line $K_2 = 0$. Write a computer program to iterate the recursion relations in (8.65) starting from 1) $K_1 = 0.39, K_2 = 0$, and 2) $K_1 = 0.4, K_2 = 0$. Make a plot of the iterates in the K_1, K_2 plane. The flows do not approximate smooth trajectories.

8.29 Of the four Ising spins in Fig. 8.16, consider the following functions of these spins:

$$S_0 \equiv 1$$
$$S_1 \equiv \sigma_1 + \sigma_2 + \sigma_3 + \sigma_4$$
$$S_{nn} \equiv \sigma_1\sigma_2 + \sigma_2\sigma_3 + \sigma_3\sigma_4 + \sigma_4\sigma_1$$
$$S_{nnn} \equiv \sigma_1\sigma_3 + \sigma_2\sigma_4$$
$$S_3 \equiv \sigma_1\sigma_2\sigma_3 + \sigma_2\sigma_3\sigma_4 + \sigma_3\sigma_4\sigma_1 + \sigma_4\sigma_1\sigma_2$$
$$S_4 \equiv \sigma_1\sigma_2\sigma_3\sigma_4.$$

Refer to these functions (which are invariant under cyclic permutations, $1 \to 2 \to 3 \to 4 \to 1$) as S_i (i.e., S_i represents any of the functions $S_0, S_1, S_{nn}, S_{nnn}, S_3, S_4$). Show that the S_i comprise a *closed set* under multiplication. Fill out the following table:

Table 8.2: Product table of spin functions

	S_1	S_{nn}	S_{nnn}	S_3	S_4
S_1					
S_{nn}					
S_{nnn}					
S_3					
S_4					

8.30 Show that Eq. (8.79) follows from Eq. (8.78) using the fact that both sides of the equation have common eigenvectors. Verify that Eq. (8.79) has the solution given by Eq. (8.80). Is it possible that Eq. (8.80) should only be a proportionality, $\lambda_k^{(L)} \propto L^{y_k}$? Try $\lambda_k^{(L)} = aL^{y_k}$ and show it must be the case that $a = 1$.

8.31 Derive Eq. (8.82). Assume that the linearized transformation matrix $R^{(L)}$ is symmetric. Equation (8.82) holds for nonsymmetric matrices, but the analysis is more involved. See Goldenfeld [101, p255].

8.32 Show that Eq. (8.85) follows from Eqs. (8.83) and (8.84).

8.33 Referring to Fig. 8.11, confirm that the block spins sit on the sites of a new triangular lattice rotated from the original by $30°$ and stretched by the factor $L = \sqrt{3}$. Hint: Just *where* is the center of an equilateral triangle?

8.34 Verify that the "majority rule" function defined in Eq. (8.89) has the values ± 1 for all $2^3 = 8$ configurations of cell spins.

8.35 Fill in the details in deriving Eq. (8.98) from Eq. (8.92).

8.36 Show for Ising spins that $\sum_{\sigma_1} \sum_{\sigma_2} \sum_{\sigma_3} \sigma_1 \sigma_2 e^{K(\sigma_1 \sigma_2 + \sigma_2 \sigma_3 + \sigma_3 \sigma_1)} = 2\left(e^{3K} - e^{-K}\right)$.

8.37 Work through the steps in Eq. (8.101).

8.38 Find the fixed points of the recursion relation in Eq. (8.104).

8.39 Derive the formulas in Eq. (8.121).

8.40 One could model the effects of an inhomogeneous magnetic field $B(r)$ by generalizing the Landau free energy Eq. (8.117) to include $\int d^d r B(r) S(r)$. Show that

$$\int d^d r B(r) S(r) = \frac{1}{(2\pi)^d} \int d^d k \int d^d k' B_k S_{k'} \delta(k + k') = \frac{1}{(2\pi)^d} \int d^d k B_k S_{-k},$$

where one must introduce a Fourier transform of $B(r)$ as in Eq. (8.120).

Physical constants

Quantity	Symbol	Value (SI units)
Electron charge	e	1.602×10^{-19} C
Electron volt (eV)		1.602×10^{-19} J (eV)$^{-1}$
Speed of light	c	2.998×10^8 m s^{-1}
Planck's constant	h	6.626×10^{-34} J s 4.136×10^{-15} eV s
	hc	1.986×10^{-25} J m 1240 eV nm
Planck's constant	$\hbar$	1.055×10^{-34} J s 6.582×10^{-16} eV s
	$\hbar c$	3.162×10^{-26} J m 197.3 eV nm
Boltzmann's constant	k	1.381×10^{-23} J K^{-1} 8.617×10^{-5} eV K^{-1}
Gas constant	R	8.314 J K^{-1} mol^{-1}
Avogadro's number	N_A	6.022×10^{23}
Gravitational constant	G	6.674×10^{-11} m^3 kg^{-1} s^{-2}
Electron mass	m	9.109×10^{-31} kg 0.511 Mev/c^2
Proton mass	m_p	1.6726×10^{-27} kg 938.272 MeV/c^2

Physical constants, continued

Quantity	Symbol	Value (SI units)
Neutron mass	m_n	1.6749×10^{-27} kg 939.565 MeV$/c^2$
Bohr magneton $(\mu_B = e\hbar/(2m))$	μ_B	9.274×10^{-24} J T^{-1} 5.788×10^{-5} eV T^{-1}
Nuclear magneton $(\mu_N = e\hbar/(2m_p))$	μ_N	5.051×10^{-27} J T^{-1} 3.152×10^{-8} eV T^{-1}
Radiation constant $(a = \pi^2 k^4/(15\hbar^3 c^3))$	a	7.566×10^{-16} J m^{-3} K^{-4}
Stefan constant $(\sigma = ac/4)$	σ	5.670×10^{-8} W m^{-2} K^{-4}
Wien constant	b	2.898×10^{-3} m K
$\dfrac{\hbar^2}{2m}$		6.11×10^{-39} J m^2 3.81×10^{-20} eV m^2 0.0381 eV nm^2
Temperature equivalent of 1 eV		$11,605$ K
$\dfrac{\hbar c}{G m_p^2}$		1.693×10^{38}
$\dfrac{\hbar^2}{G m_n^3}$		3.55×10^{22} m

Useful integrals

W E review certain integrals and functions that occur in statistical mechanics.

B.1 THE GAMMA FUNCTION

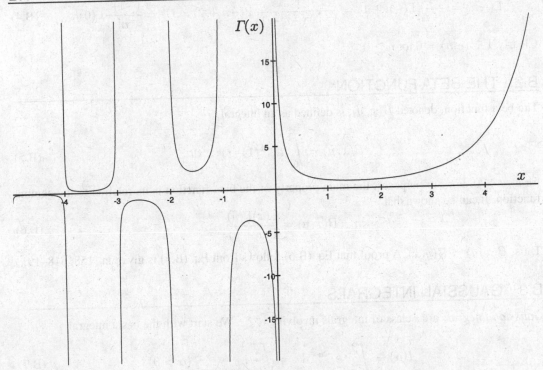

Figure B.1: $\Gamma(x)$ vs. x.

The gamma function, $\Gamma(x)$, the graph of which is shown in Fig. B.1, originated in the quest to find a generalization of the factorial to real numbers. Can one find a function $\Gamma(x)$ such that $\Gamma(x+1) = x\Gamma(x)$? The answer is in the affirmative; it's defined by the *Euler integral*

$$\Gamma(x) \equiv \int_0^\infty t^{x-1} e^{-t} \, dt . \tag{B.1}$$

From direct integration of Eq. (B.1) we have the special values $\Gamma(1) = 1$ and $\Gamma(\frac{1}{2}) = \sqrt{\pi}$. For $x = n$ a positive integer, Eq. (B.1) can be evaluated by repeated integration by parts, with the result

$$\Gamma(n) = (n-1)! . \qquad (n = 1, 2, \cdots) \qquad \text{(B.2)}$$

$\Gamma(x)$ is analytic except at $x = 0, -1, -2, \cdots$, where it has simple poles, implying that at these points $1/\Gamma(x) = 0$. There are no points x for which $\Gamma(x) = 0$.

One of the most useful properties of $\Gamma(x)$ is the recursion relation it satisfies. Integrate Eq. (B.1) by parts: $\Gamma(x) = -t^{x-1}e^{-t}\big|_0^\infty + (x-1)\int_0^\infty t^{x-2}e^{-t}dt = (x-1)\Gamma(x-1)$. This result is usually written

$$\Gamma(x+1) = x\Gamma(x) . \qquad \text{(B.3)}$$

Thus, $\Gamma(x)$ *is the generalization of the factorial to real numbers* x for $x \geq 1$.

Equation (B.3) is invaluable for numerical purposes. It also enables $\Gamma(x)$ to be analytically continued to negative values of x. We see from Eq. (B.3) that $\lim_{x\to 0} x\Gamma(x) = \Gamma(1) = 1$, and thus, as $x \to 0$, $\Gamma(x) \approx x^{-1}$, i.e., it has a simple pole at $x = 0$. Moreover it has simple poles at all negative integers. Let $x = -n$ in Eq. (B.3) and iterate:

$$\Gamma(-n) = \frac{1}{(-n)}\Gamma(-n+1) = \frac{1}{(-n)}\frac{1}{(-n+1)}\Gamma(-n+2) = \cdots = \frac{(-1)^n}{n!}\Gamma(0) . \qquad \text{(B.4)}$$

Clearly, $\Gamma^{-1}(-n) = 0$ for $n \geq 0$.

B.2   THE BETA FUNCTION

The beta function, denoted $B(x, y)$, is defined as an integral,

$$B(x, y) = \int_0^1 t^{x-1}(1-t)^{y-1}dt \qquad \text{(B.5)}$$

for $x, y > 0$. For our purposes, the main property of the beta function is its relation to the gamma function. It can be shown that

$$B(x, y) = \frac{\Gamma(x)\Gamma(y)}{\Gamma(x+y)} . \qquad \text{(B.6)}$$

Thus, $B(x, y) = B(y, x)$. A proof that Eq. (B.6) follows from Eq. (B.5) is given in [158, p18–19].

B.3   GAUSSIAN INTEGRALS

Gaussian integrals are a class of integrals involving e^{-x^2}. We start with the basic integral

$$I(\alpha) \equiv \int_{-\infty}^\infty e^{-\alpha x^2}dx = \sqrt{\frac{\pi}{\alpha}} . \qquad (\alpha > 0) \qquad \text{(B.7)}$$

There's a trick to evaluating this integral that every student should know. Let $I \equiv \int_{-\infty}^\infty e^{-x^2}dx$. Then $I^2 = \int_{-\infty}^\infty \int_{-\infty}^\infty e^{-(x^2+y^2)}dxdy$. Change variables to polar coordinates with $r^2 = x^2 + y^2$ and change the area element $dxdy \to rdrd\theta$, so that $I^2 = \int_0^{2\pi} d\theta \int_0^\infty e^{-r^2}rdr$. Change variables again with $u = r^2$, and you should find that $I^2 = \pi$.

Forms of this integral containing powers of x are straightforward to determine. The integrand in $I(\alpha)$ is an even function, and thus integrals involving odd powers of x vanish:

$$\int_{-\infty}^\infty x^{2n+1}e^{-\alpha x^2}dx = 0 . \qquad (n \geq 0) \qquad \text{(B.8)}$$

Integrals involving even powers of x can be generated by differentiating $I(\alpha)$ in Eq. (B.7) with respect to α:

$$\int_{-\infty}^{\infty} x^{2n} e^{-\alpha x^2} \, dx = (-1)^n \frac{\partial^n}{\partial \alpha^n} I(\alpha) = \sqrt{\frac{\pi}{\alpha}} \frac{1 \cdot 3 \cdot 5 \cdots (2n-1)}{(2\alpha)^n} \qquad (n \geq 1) \qquad \text{(B.9)}$$

Gaussian integrals are often specified over the range $[0, \infty)$. We state the following result and then discuss it,

$$\int_0^{\infty} x^n e^{-\alpha x^2} \, dx = \frac{1}{2} \Gamma\left(\frac{n+1}{2}\right) \frac{1}{\alpha^{(n+1)/2}} \cdot \qquad (n \geq 0) \qquad \text{(B.10)}$$

Equation (B.10) follows under the substitution $y = \alpha x^2$ and then noting that the integral so obtained is none other than $\Gamma(x)$, Eq. (B.1).

Sometimes we require the multidimensional generalization of Gaussian integrals,

$$I \equiv \int_{-\infty}^{\infty} \exp\left(-\sum_{i,j=1}^{n} A_{ij} x_i x_j\right) dx_1 \cdots dx_n, \qquad \text{(B.11)}$$

where A_{ij} are elements of an $n \times n$ matrix A that do not depend on the variables x_i. The strategy in evaluating this integral is to make a change of variables that orthonormalizes the quadratic form $\sum_{ij} A_{ij} x_i x_j$. First note that $\sum_{ij} A_{ij} x_i x_y$ is generated by the terms $x^T A x$, where x is a column vector of the variables $x_1, \cdots, x_n$ and T denotes transpose. Make a linear transformation $x_i = \sum_j L_{ij} y_j$, or, in matrix notation, $x = Ly$. We want to choose L so that $x^T A x = y^T y$. Thus we require $y^T L^T A L y = y^T y$, or that $L^T A L = I$, where I is the $n \times n$ identity matrix. Using the properties of determinants, $(\det L)^2 \det A = 1$, implying that $\det L = 1/\sqrt{\det A}$. The volume element in Eq. (B.11) transforms as $dx_1 \cdots dx_n = \det L \, dy_1 \cdots dy_n$ ($\det L$ is the Jacobian of the transformation). Thus, starting from Eq. (B.11), under the substitution $x = Ly$ where $L^T A L = I$,

$$\int_{-\infty}^{\infty} e^{\left(-\sum_{ij=1}^n A_{ij} x_i x_j\right)} dx_1 \cdots dx_n = \det L \int_{-\infty}^{\infty} e^{\left(-\sum_{i=1}^n y_i^2\right)} dy_1 \cdots dy_n = \frac{\pi^{n/2}}{\sqrt{\det A}}. \qquad \text{(B.12)}$$

B.4 THE ERROR FUNCTION

The *error function*, $\text{erf}(z)$, is the cumulative probability distribution of a normally distributed random variable, $P(x \leq z)$:

$$\text{erf}(z) \equiv \frac{2}{\sqrt{\pi}} \int_0^z e^{-t^2} \, dt. \qquad \text{(B.13)}$$

It's an odd function: $\text{erf}(-z) = -\text{erf}(z)$. It has the value $\text{erf}(z) \to 1$ as $z \to \infty$.

B.5 VOLUME OF A HYPERSPHERE

A *hypersphere* is a set of points at a constant distance from a given point (its center) in any number of dimensions. As we now show, the volume $V_n(R)$ enclosed by a hypersphere of radius R in n-dimensional Euclidean space $(\sum_{k=1}^n x_k^2 = R^2)$ is given by

$$V_n(R) = \frac{\pi^{n/2}}{\Gamma(\frac{n}{2} + 1)} R^n. \qquad \text{(B.14)}$$

As an example of Eq. (B.14), $V_3(R) = (\pi^{3/2}/\Gamma(1+\frac{3}{2})) R^3$. Using the recursion relation, $\Gamma(1+\frac{3}{2}) = \frac{3}{2}\Gamma(\frac{3}{2}) = \frac{3}{2}\Gamma(1+\frac{1}{2}) = \frac{3}{4}\Gamma(\frac{1}{2}) = 3\sqrt{\pi}/4$. Thus, $V_3(R) = 4\pi R^3/3$. For $n = 2$ and $n = 1$, Eq. (B.14) gives $V_2(R) = \pi R^2$ and $V_1(R) = 2R$.

To derive Eq. (B.14), we first define a function of n variables, $f(x_1, \cdots, x_n) \equiv \exp\left(-\frac{1}{2}\sum_{i=1}^{n} x_i^2\right)$. In Cartesian coordinates, the integral of f over all of $\mathbb{R}^n$ is

$$\int_{\mathbb{R}^n} f(x_1, \cdots, x_n) dV = \prod_{i=1}^{n} \left(\int_{-\infty}^{\infty} dx_i \exp(-\tfrac{1}{2}x_i^2)\right) = (2\pi)^{n/2}, \qquad (B.15)$$

where dV is the n-dimensional volume element, and we've used $\int_{-\infty}^{\infty} e^{-ax^2} dx = \sqrt{\pi/a}$. Now recognize that $\sum_{i=1}^{n} x_i^2 \equiv r^2$ defines the radial coordinate in n dimensions. Let $dV = A_n(r)dr$, where $A_n(r)$ is the surface area of an n-dimensional sphere. The surface area of an n-sphere can be written $A_n(r) = A_n(1)r^{n-1}$. The same integral in Eq. (B.15) can then be expressed

$$\int_{\mathbb{R}^n} f dV = \int_0^\infty A_n(r) \exp(-\tfrac{1}{2}r^2)dr = A_n(1)\int_0^\infty r^{n-1} \exp(-\tfrac{1}{2}r^2)dr \qquad (B.16)$$

$$= 2^{(n/2)-1} A_n(1)\int_0^\infty t^{(n/2)-1}e^{-t}dt = 2^{(n/2)-1} A_n(1)\Gamma\left(\frac{n}{2}\right),$$

where in the second line we have changed variables $t = \frac{1}{2}r^2$. Comparing Eqs. (B.16) and (B.15), we conclude that $A_n(1) = 2\pi^{n/2}/\Gamma(\frac{n}{2})$. This gives us the surface area of an n-sphere of radius R:

$$A_n(R) = \frac{2\pi^{n/2}}{\Gamma(\frac{n}{2})} R^{n-1}. \qquad (B.17)$$

From Eq. (B.17), $A_3(R) = 4\pi R^2$, $A_2(R) = 2\pi R$, and $A_1(R) = 2$. By integrating $dV = A_n(r)dr$ from 0 to R, we obtain Eq. (B.14). Note that $A_n(R) = (n/R)V_n(R)$.

B.6 BOSE-EINSTEIN INTEGRALS

In the treatment of identical bosons we encounter the *Bose-Einstein integral functions* of order n,

$$G_n(z) \equiv \frac{1}{\Gamma(n)} \int_0^\infty \frac{x^{n-1}}{z^{-1}e^x - 1} dx, \qquad (n \geq 1, 0 \leq z \leq 1) \qquad (B.18)$$

where the fugacity $z = e^{\beta\mu}$. For bosons μ is nonpositive, $\mu \leq 0$ (Section 5.5), implying $0 \leq z \leq 1$. The factor of $\Gamma(n)$ is included so that as $z \to 0$, $G_n(z) \to z$ for all n (show this). In applications n is usually an integer or a half-integer, but the integral in Eq. (B.18) is defined for arbitrary $n \geq 1$. For $n = 1$, Eq. (B.18) can be evaluated exactly, with

$$G_1(z) = -\ln(1 - z). \qquad (B.19)$$

Note that $G_1(z)$ diverges logarithmically as $z \to 1^-$. For $n < 1$, it can be shown that $G_n(z)$ diverges like $(1 - z)^{n-1}$ as $z \to 1^-$. Values of G_n are tabulated in London[92, Appendix].

The functions $G_n(z)$ satisfy a recursion relation of sorts. Starting from Eq. (B.18),

$$\frac{\partial G_n(z)}{\partial z} = \frac{1}{\Gamma(n)} \int_0^\infty x^{n-1} \frac{e^x}{(e^x - z)^2} dx = -\frac{1}{\Gamma(n)} \int_0^\infty x^{n-1} \frac{d}{dx}\left(\frac{1}{e^x - z}\right) dx.$$

Integrate by parts,

$$\frac{\partial G_n(z)}{\partial z} = -\frac{1}{\Gamma(n)} \left[\frac{x^{n-1}}{e^x - z}\Bigg|_{x=0}^{x=\infty} - (n-1)\int_0^\infty \frac{x^{n-2}}{e^x - z} dx\right].$$

For the integrated part to vanish (in particular at the lower limit), we *require* $n > 1$. Thus, for $n > 1$,

$$\frac{\partial}{\partial z}G_n(z) = \frac{1}{z}G_{n-1}(z). \qquad (n > 1) \qquad (B.20)$$

Equation (B.18) can be rewritten,

$$\Gamma(n)G_n(z) = z \int_0^\infty \frac{e^{-x}x^{n-1}}{1 - ze^{-x}}dx = \sum_{k=1}^\infty \int_0^\infty x^{n-1}\left(ze^{-x}\right)^k dx = \sum_{k=1}^\infty z^k \int_0^\infty x^{n-1}e^{-xk}dx$$

$$= \sum_{k=1}^\infty \frac{z^k}{k^n} \int_0^\infty u^{n-1}e^{-u}du = \Gamma(n)\sum_{k=1}^\infty \frac{z^k}{k^n},$$

where we've expanded the denominator in the integrand as a geometric series; this works because $ze^{-x} < 1$ for $0 < x < \infty$ (for $z \le 1$). The Bose-Einstein functions therefore have power-series representations (*fugacity expansions*):

$$G_n(z) = \sum_{k=1}^\infty \frac{z^k}{k^n} = z + \frac{z^2}{2^n} + \frac{z^3}{3^n} + \cdots. \qquad (0 \le z \le 1) \qquad (B.21)$$

For $n = 1$, Eq. (B.21) is the Taylor series of $-\ln(1 - z)$ about $z = 0$. The functions $G_n(z)$ are monotonically increasing in z for all n. Their largest value[1] occurs for $z = 1$, in which case, from Eq. (B.21),

$$G_n(1) = \sum_{k=1}^\infty \frac{1}{k^n} \equiv \zeta(n), \qquad (n > 1) \qquad (B.22)$$

where $\zeta(n)$ is the *Riemann zeta function*, which is convergent for $n > 1$ (but which diverges for $n = 1$). Some values of the zeta function are listed in Table B.1.

Table B.1: Values of the Riemann zeta function, $\zeta(n)$

$\zeta(3/2) \approx 2.612$	$\zeta(4) = \dfrac{\pi^4}{90} \approx 1.082$
$\zeta(2) = \dfrac{\pi^2}{6} \approx 1.645$	$\zeta(5) \approx 1.037$
$\zeta(5/2) \approx 1.341$	$\zeta(6) = \dfrac{\pi^6}{945} \approx 1.017$
$\zeta(3) \approx 1.202$	$\zeta(7) \approx 1.008$
$\zeta(7/2) \approx 1.127$	$\zeta(8) = \dfrac{\pi^8}{9450} \approx 1.004$

The power series in Eq. (B.21) converge too slowly to be of practical use in the limit $z \to 1^-$, which is the low-temperature limit where interesting physics occurs (Bose-Einstein condensation). In that case, it's better to rely on expansions that have been developed in the variable $\alpha = -\ln z = -\beta\mu$, which is a small positive number. As tabulated in London[92, Appendix],

$$G_{1/2}(\alpha) = \sqrt{\pi}\alpha^{-1/2} - 1.460 + 0.208\alpha - 0.0128\alpha^2 - 0.00142\alpha^3 + \cdots$$
$$G_{3/2}(\alpha) = \zeta(3/2) - 2\sqrt{\pi\alpha} + 1.460\alpha - 0.104\alpha^2 + 0.00425\alpha^3 - \cdots$$
$$G_{5/2}(\alpha) = \zeta(5/2) - \zeta(3/2)\alpha + \frac{4\sqrt{\pi}}{3}\alpha^{3/2} - 0.730\alpha^2 + 0.0347\alpha^3 - \cdots \qquad (B.23)$$

---

[1]One could mumble here that because $G_n(z)$ is monotonically increasing for $z \le 1$, and because the series in Eq. (B.21) is convergent for $z \le 1$, $G_n(z)$ is bounded by the value $G_n(1)$ (monotone convergence theorem).

These expansions provide better than 1% accuracy for $0 < \alpha \le 1$. Note that we've included $G_{1/2}$ in Eq. (B.23). This function is implicitly defined through Eq. (B.20) for $n = \frac{3}{2}$. Consider the generalization of Eq. (B.18) for $n = \frac{1}{2}$:

$$
G_{1/2}(z) \equiv \frac{1}{\Gamma(1/2)} \int_0^\infty \frac{1}{\sqrt{x}} \frac{1}{z^{-1}e^x - 1} dx . \qquad \text{(not defined for } z \to 1^-) \qquad \text{(B.24)}
$$

The denominator in the integrand of Eq. (B.24) vanishes as $x \to 0$, a square-root singularity that can be integrated over. It would also seemingly vanish for x such that $e^x = z$; such an equation, however, has no solution if $z < 1$. In the limit $z \to 1^-$, however, the denominator vanishes as $x^{3/2}$ for small x, which cannot be integrated. The limit $z \to 1^-$ for $G_{1/2}(z)$ therefore does not exist, as we see in Eq. (B.23) as $\alpha \to 0$.

It's shown in Chapter 5 that the density N/V of an ideal Bose gas is related to $G_{3/2}$, Eq. (5.69), with its pressure P governed by $G_{5/2}$, Eq. (5.67), and the internal energy per particle U/N controlled by the ratio $G_{5/2}/G_{3/2}$, Eq. (5.76), which gives U/N as a function of μ and T. Often we want to eliminate reference to μ and have a representation of U/N as a function of T and V. In that case, it proves useful to have a representation of $G_{5/2}$ as a function of $G_{3/2}$. To save writing, let's make the abbreviations $y \equiv G_{5/2}$ and $x \equiv G_{3/2}$. We seek a representation

$$
y = \sum_{k=1}^\infty a_k x^k , \qquad (B.25)
$$

where the coefficients are, from Taylor's theorem,

$$
a_k = \frac{1}{k!} \left(\frac{d^k y}{dx^k} \right)_{x=0} .
$$

We're expanding $G_{5/2}$ "around" the function $G_{3/2}$; $x = 0$ corresponds to $G_{3/2} = 0$, which occurs at $z = G_{3/2}^{-1}(0) = 0$. Because $G_{5/2}(0) = 0$, the expansion in Eq. (B.25) starts at $k = 1$ and not at $k = 0$. The hard work here is in obtaining the expansion coefficients a_k. Using the chain rule,

$$
\frac{d^k y}{dx^k} = \frac{d}{dx} \left(\frac{d^{k-1}y}{dx^{k-1}} \right) = \frac{d\alpha}{dx} \frac{d}{d\alpha} \left(\frac{d^{k-1}y}{dx^{k-1}} \right) = \frac{1}{(dx/d\alpha)} \frac{d}{d\alpha} \left(\frac{d^{k-1}y}{dx^{k-1}} \right) .
$$

One can show, through a change of variables, that Eq. (B.20) is equivalent to

$$
\frac{\partial}{\partial \alpha} G_n(\alpha) = -G_{n-1}(\alpha) , \qquad (n > 1) \qquad (B.26)
$$

implying

$$
\frac{dx}{d\alpha} = \frac{d}{d\alpha} G_{3/2}(\alpha) = -G_{1/2}(\alpha) .
$$

The coefficients a_k are therefore

$$
a_k = -\frac{1}{k!} \left[\frac{1}{G_{1/2}(\alpha)} \frac{d}{d\alpha} \left(\frac{d^{k-1}y}{dx^{k-1}} \right) \right] \Bigg|_{z=0} . \qquad (B.27)
$$

The coefficient a_1 is (the "zeroth" derivative of a function is the function)

$$
a_1 = - \left(\frac{1}{G_{1/2}(\alpha)} \frac{d}{d\alpha} G_{5/2}(\alpha) \right) \Bigg|_{z=0} = \frac{G_{3/2}}{G_{1/2}} \Bigg|_{z=0} = 1 .
$$

The next coefficient, a_2, is more difficult, and we show how it's derived. From Eq. (B.27), we must evaluate the limit

$$
-2!a_2 = \left[\frac{1}{G_{1/2}(\alpha)} \frac{d}{d\alpha} \left(\frac{dy}{dx} \right) \right]_{z=0} .
$$

The derivative

$$\frac{dy}{dx} = \frac{dy}{d\alpha} \cdot \frac{1}{dx/d\alpha} = \frac{dG_{5/2}/d\alpha}{dG_{3/2}/d\alpha} = \frac{G_{3/2}(\alpha)}{G_{1/2}(\alpha)},$$

where we've used Eq. (B.26). The next derivative

$$\frac{d}{d\alpha}\left(\frac{G_{3/2}}{G_{1/2}}\right) = \frac{1}{G_{1/2}^2}\left(G_{1/2}G'_{3/2} - G_{3/2}G'_{1/2}\right) = -\left(1 + \frac{G_{3/2}G'_{1/2}}{G_{1/2}^2}\right),$$

where we've used Eq. (B.26) for $G'_{3/2}$. We can't use Eq. (B.26) for $G'_{1/2}$, however. To evaluate that derivative, change variables again:

$$\frac{d}{d\alpha}G_{1/2} = \frac{dG_{1/2}}{dz} \cdot \frac{dz}{d\alpha} = -z\frac{dG_{1/2}}{dz},$$

where $z = e^{-\alpha}$. We're interested in the limit $z \to 0$, and thus we can use Eq. (B.21) generalized to $n = \frac{1}{2}$ to represent $G_{1/2}$ for small z:

$$\frac{dG_{1/2}}{dz} = \frac{d}{dz}\left(z + \frac{1}{\sqrt{2}}z^2 + \frac{1}{\sqrt{3}}z^3 + \cdots\right) = 1 + \sqrt{2}z + \sqrt{3}z^2 + \cdots. \qquad (z \ll 1)$$

Putting it all together, we have, for small z:

$$1 + \frac{G_{3/2}G'_{1/2}}{G_{1/2}^2} \overset{z \to 0}{\Longrightarrow} -\frac{z}{2^{3/2}} + O(z^2),$$

implying $a_2 = 2^{-5/2}$. The desired expansion therefore is,[92, p204] for $y = G_{5/2}$ and $x = G_{3/2}$,

$$y = x - \frac{1}{2^{5/2}}x^2 - \left(\frac{2\sqrt{3}}{27} - \frac{1}{8}\right)x^3 - \left(\frac{3}{32} + \frac{5\sqrt{2}}{64} - \frac{\sqrt{6}}{12}\right)x^4 - \cdots$$

$$= x - 0.17678x^2 - 0.00330x^3 - 0.00011x^4 - \cdots. \qquad (B.28)$$

We'll use Eq. (B.28) in Chapter 5.

B.7 FERMI-DIRAC INTEGRALS

In the treatment of identical fermions, we encounter the *Fermi-Dirac integral functions* of order n,

$$F_n(z) \equiv \frac{1}{\Gamma(n)} \int_0^\infty \frac{x^{n-1}}{z^{-1}e^x + 1}dx, \qquad (n > 0, 0 \le z < \infty) \qquad (B.29)$$

where (as with the Bose-Einstein integrals) $\Gamma(n)$ is included so that as $z \to 0$, $F_n(z) \to z$ for all $n > 0$. There is no restriction on the chemical potential of fermions, $0 \le z < \infty$. We'll treat the cases $z \le 1$ and $z > 1$ separately. The integral can be evaluated exactly for the special case of $n = 1$:

$$F_1(z) = \ln(1 + z). \qquad (B.30)$$

Following the same reasoning as with the Bose-Einstein integrals (see steps leading to Eq. (B.20)), it's straightforward to show, starting from Eq. (B.29), that

$$\frac{\partial}{\partial z}F_n(z) = \frac{1}{z}F_{n-1}(z). \qquad (n > 1) \qquad (B.31)$$

It can also be shown, by the same method of analysis leading to Eq. (B.21), that for $z \leq 1$,

$$F_n(z) = \sum_{k=1}^{\infty} (-1)^{k+1} \frac{z^k}{k^n} = z - \frac{z^2}{2^n} + \frac{z^3}{3^n} - \cdots . \qquad (z \leq 1) \qquad (B.32)$$

Note that, compared with Eq. (B.21), Eq. (B.32) is an alternating series. Values of $F_n(z)$ are tabulated in Blakemore[159, Appendix B] and in [160]. It's a property of the zeta function that[50, p807] for $n > 1$,

$$\sum_{k=1}^{\infty} (-1)^{k+1} k^{-n} = \left(1 - 2^{1-n}\right) \zeta(n) . \qquad (B.33)$$

Thus $F_n(1) = \left(1 - 2^{1-n}\right) \zeta(n)$ for $n > 1$.

We often want the Fermi-Dirac integrals for large values of z, which corresponds to low temperature.[2] (For Bose-Einstein functions, low temperature corresponds to $z \to 1$.) In this case, it's convenient to introduce a variable $y \equiv \ln z$ (thus, $y = \beta\mu$), by which

$$F_n(y) = \frac{1}{\Gamma(n)} \int_0^{\infty} \frac{x^{n-1}}{e^{x-y} + 1} dx . \qquad (B.34)$$

For large y, the integrand in Eq. (B.34) is controlled by the factor $(e^{x-y} + 1)^{-1}$, which varies from approximately unity $((1 + e^{-y})^{-1})$ for $x \to 0$ and zero for $x \to \infty$. Its *departure* from these asymptotic values occurs in the vicinity of $x \approx y$; see Fig. B.2. An approximation of Eq. (B.34) can

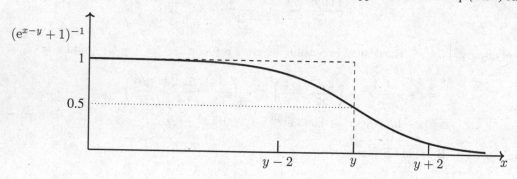

Figure B.2: Fermi-Dirac distribution for $T > 0$ (solid line). Dashed line corresponds to $T = 0$. Here $y = 7$.

be obtained by replacing $(e^{x-y} + 1)^{-1}$ by its values for $T = 0$, shown as a dashed line in Fig. B.2,

$$F_n(y) \approx \frac{1}{\Gamma(n)} \int_0^y x^{n-1} dx = \frac{1}{n\Gamma(n)} x^n \Big|_0^y = \frac{y^n}{\Gamma(n+1)} . \qquad (B.35)$$

An improved approximation is found by adding back what's left out of Eq. (B.35):

$$\Gamma(n) F_n(y) = \int_0^y x^{n-1} \left[1 - \frac{1}{1 + e^{-x+y}} \right] dx + \int_y^{\infty} \frac{x^{n-1}}{e^{x-y} + 1} dx , \qquad (B.36)$$

---

[2]Learned discussion. The interval of convergence of the series in Eq. (B.32) is (like those for the Bose-Einstein functions) $0 < z \leq 1$. Unlike the expansions of the Bose-Einstein functions, Eq. (B.21), the terms of the series in Eq. (B.32) alternate in sign, and hence the series is not bounded from above as $z \to 1^-$. It's clear from the integral representation, Eq. (B.29), that the denominator of the integrand never vanishes for $z > 0$, and hence the function exists for all $0 < z < \infty$. For $z > 1$, we require a different representation of the Fermi-Dirac functions than power-series expansions.

where the factor in square brackets comes from the identity

$$\frac{1}{e^{x-y}+1} = 1 - \frac{1}{e^{-x+y}+1}.$$

In the integrand in square brackets in Eq. (B.36), change variables to $u = y - x$, and in the second integral change variables to $t = x - y$. Then,

$$\Gamma(n)F_n(y) = \frac{y^n}{n} + \int_y^0 \frac{(y-u)^{n-1}}{e^u+1}du + \int_0^\infty \frac{(t+y)^{n-1}}{e^t+1}dt \qquad \text{(B.37)}$$

For large y, the contributions to the first integral in Eq. (B.37) are from small values of u; we can therefore extend the lower limit to ∞. Doing so, and combining the integrals, we have

$$\Gamma(n)F_n(y) = \frac{y^n}{n} + \int_0^\infty \frac{(y+u)^{n-1}-(y-u)^{n-1}}{e^u+1}du.$$

Apply the binomial theorem, Eq. (3.7), to the terms in the numerator:

$$(y+u)^{n-1}-(y-u)^{n-1} = \sum_{k=0}^{n-1}\binom{n-1}{k}y^{n-1-k}u^k\left(1-(-1)^k\right) = 2\sum_{k=1,3,5,\cdots}^{n-1}\binom{n-1}{k}y^{n-1-k}u^k.$$

Then,

$$\Gamma(n)F_n(y) = \frac{y^n}{n} + 2\sum_{k=1,3,5,\cdots}^{n-1}\binom{n-1}{k}y^{n-1-k}\int_0^\infty \frac{u^k}{e^u+1}du. \qquad \text{(B.38)}$$

Using Eqs. (B.29) and (B.32),

$$\int_0^\infty \frac{u^k}{e^u+1}du = \Gamma(k+1)F_{k+1}(1) = \Gamma(k+1)\left(1 - \frac{1}{2^{k+1}} + \frac{1}{3^{k+1}} - \cdots\right)$$

Using Eq. (B.33),

$$1 - \frac{1}{2^{k+1}} + \frac{1}{3^{k+1}} - \cdots = \left(1 - \frac{1}{2^k}\right)\zeta(k+1).$$

We therefore have the result that

$$\int_0^\infty \frac{u^k}{e^u+1}du = \left(1 - \frac{1}{2^k}\right)\Gamma(k+1)\zeta(k+1). \qquad \text{(B.39)}$$

Equation (B.39) doesn't apply for $k = 0$, but that term is excluded from the series in Eq. (B.38). Combining Eq. (B.39) with Eq. (B.38),

$$\Gamma(n)F_n(y) = \frac{y^n}{n} + 2\sum_{k=1,3,5,\cdots}^{n-1}\binom{n-1}{k}y^{n-(k+1)}\left(1-\frac{1}{2^k}\right)\Gamma(k+1)\zeta(k+1)$$

$$= y^n\left[\frac{1}{n} + 2\sum_{k=1,3,5,\cdots}^{n-1}\binom{n-1}{k}\frac{1}{y^{k+1}}\left(1-\frac{1}{2^k}\right)\Gamma(k+1)\zeta(k+1)\right].$$

Let $m \equiv k+1$ be a new summation index. Then,

$$\Gamma(n)F_n(y) = y^n\left[\frac{1}{n} + 2\sum_{m=2,4,6,\cdots}^{n}\binom{n-1}{m-1}\frac{1}{y^m}\left(1-\frac{1}{2^{m-1}}\right)\Gamma(m)\zeta(m)\right].$$

This expression can be simplified:

$$F_n(y) = \frac{y^n}{\Gamma(n+1)}\left[1 + 2\sum_{m=2,4,6,\cdots}^{n} n(n-1)\cdots(n-m+1)\left(1 - \frac{1}{2^{m-1}}\right)\frac{\zeta(m)}{y^m}\right]$$

Writing out the first two terms in the series, we have for large y and $n > 0$:

$$F_n(y) \sim \frac{y^n}{\Gamma(n+1)}\left[1 + n(n-1)\frac{\pi^2}{6}\frac{1}{y^2} + n(n-1)(n-2)(n-3)\frac{7\pi^4}{360}\frac{1}{y^4} + \cdots\right], \qquad \text{(B.40)}$$

where we've used the values of the zeta function shown in Table B.1. The reader is urged not to confuse $y = \ln z = \beta\mu$ in Eq. (B.40) with the variable z that occurs in formulas such as Eq. (B.31).

B.8 THE JOYCE-DIXON APPROXIMATION

For the important case of $F_{3/2}$ (related to the density of fermions; Eq. (5.69)), we're often required to invert $\phi = F_{3/2}(y)$ to obtain y given the value of ϕ, which we can write as $y = F_{3/2}^{-1}(\phi)$. For $y \gg 1$, we could use the leading term in Eq. (B.40) to estimate $y \approx (\Gamma(5/2)\phi)^{2/3}$ (such an approximation is accurate for $\phi \gtrsim 5$). Is there a systematic way to invert $F_{3/2}(y)$? The *Joyce-Dixon approximation*[161] is an accurate analytic expression for $F_{3/2}^{-1}$ for values of y required in practical calculations involving semiconductor lasers and doped semiconductor devices.

The Joyce-Dixon approximation for $y = F_{3/2}^{-1}(\phi)$ for $0.1 \lesssim \phi \lesssim 10$ is:

$$y = \ln\phi + \sum_{m=1}^{\infty} A_m\phi^m, \qquad \text{(B.41)}$$

where the first few coefficients are

$$A_1 = \frac{\sqrt{2}}{4} \approx 0.3536$$

$$A_2 = \frac{3}{16} - \frac{\sqrt{3}}{9} \approx -4.9501 \times 10^{-3}$$

$$A_3 = \frac{1}{8} + \frac{5\sqrt{2}}{48} \approx 1.4839 \times 10^{-4}$$

$$A_4 = \frac{1585}{6912} + \frac{5\sqrt{2}}{32} - \frac{5\sqrt{3}}{24} - \frac{\sqrt{5}}{25} \approx -4.4256 \times 10^{-6}.$$

The rapid decrease in the magnitudes of the coefficients in Eq. (B.41) results in a rapid convergence of the series. Keeping only the first two terms in the series, $y \approx \ln\phi + A_1\phi$, is acceptably accurate for many applications of the free fermion model.

B.9 THE SOMMERFELD EXPANSION

We often encounter integrals more general than Eq. (B.34), where we integrate the product of a function $\phi(x)$ and the Fermi-Dirac function, i.e., integrals in the form

$$I \equiv \int_0^\infty \phi(x)f(x)\,dx$$

where

$$f(x) \equiv \frac{1}{e^{x-y}+1}.$$

In physical applications, $\phi(x) \to 0$ as $x \to -\infty$ and diverges no more rapidly than some power[3] of x as $x \to \infty$. In these cases, we can extend the lower limit of integration to $-\infty$; note that $f(x) \to 1$ as $x \to -\infty$ for any y. Thus, we develop an approximation for large y of integrals in the form

$$I = \int_{-\infty}^{\infty} \phi(x) f(x) dx . \tag{B.42}$$

Let

$$\psi(x) \equiv \int_{-\infty}^{x} \phi(x') dx' ,$$

so that

$$d\psi(x) = \phi(x) dx .$$

Integrate by parts the integral[4] in Eq. (B.42),

$$I = \int_{-\infty}^{\infty} \phi(x) f(x) dx = - \int_{-\infty}^{\infty} \psi(x) \frac{\partial f}{\partial x} dx . \tag{B.43}$$

The derivative of the Fermi-Dirac function is zero for $x \ll y$ and for $x \gg y$ (see Fig. B.3), and thus

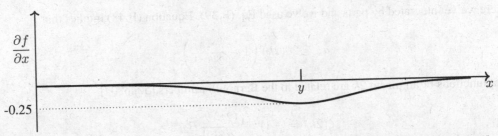

Figure B.3: Derivative of the Fermi-Dirac function, $\partial f / \partial x$, where $f(x, y) \equiv (e^{x-y} + 1)^{-1}$.

$\partial f / \partial x$ is appreciable only for $x \approx y$. Provided that $\phi(x)$ is sufficiently smooth and not too rapidly varying a function for $x \approx y$, we can develop a Taylor series expansion for $\psi(x)$ about $x = y$:

$$\psi(x) = \psi(y) + \sum_{k=1}^{\infty} \frac{(x-y)^k}{k!} \frac{d^k \psi(x)}{dx^k} \bigg|_{x=y} . \tag{B.44}$$

Substituting Eq. (B.44) into Eq. (B.43), we have

$$I = \psi(y) + \sum_{k=1}^{\infty} \frac{1}{k!} \left(\frac{d^k \psi}{dx^k} \right) \bigg|_{x=y} \int_{-\infty}^{\infty} (x-y)^k \left(-\frac{\partial f}{\partial x} \right) dx . \tag{B.45}$$

The leading term is $\psi(y)$ because

$$- \int_{-\infty}^{\infty} \frac{\partial f}{\partial x} dx = - (f(\infty) - f(-\infty)) = 1 .$$

---

[3]The free-particle density of states function, for example (see Section 2.1.5), diverges as a power law as $x \to \infty$, but is identically zero for $x \le 0$.

[4]The integrated part vanishes as $x \to \infty$ because the Fermi-Dirac function vanishes more rapidly with x than $\psi(x)$ diverges, and as $x \to -\infty$ because the Fermi-Dirac function approaches unity in this limit while $\psi(x)$ approaches zero.

Moreover, because $\partial f/\partial x$ is an even function of $(x-y)$, only terms even in k will contribute to the integral in Eq. (B.45). Equation (B.45) can therefore be written

$$I = \int_{-\infty}^{x} \phi(x')\mathrm{d}x' + \sum_{k=1}^{\infty} a_k \frac{\mathrm{d}^{2k-1}\phi}{\mathrm{d}x^{2k-1}}\bigg|_{x=y}, \tag{B.46}$$

where

$$a_k \equiv \frac{1}{(2k)!} \int_{-\infty}^{\infty} (x-y)^{2k} \left(-\frac{\partial f}{\partial x}\right) \mathrm{d}x = \frac{1}{(2k)!} \int_{-\infty}^{\infty} (x-y)^{2k} \frac{e^{x-y}}{(e^{x-y}+1)^2} \mathrm{d}x. \tag{B.47}$$

Change variables in Eq. (B.47) with $u = x - y$; Eq. (B.47) can be written

$$(2k)!a_k = \int_{-\infty}^{\infty} u^{2k} \frac{e^u}{(e^u+1)^2} \mathrm{d}u = 2\int_{0}^{\infty} u^{2k} \frac{e^u}{(e^u+1)^2} \mathrm{d}u = -2\int_{0}^{\infty} u^{2k} \frac{\mathrm{d}}{\mathrm{d}u}\left(\frac{1}{e^u+1}\right) \mathrm{d}u \tag{B.48}$$

$$= 4k \int_{0}^{\infty} \frac{u^{2k-1}}{e^u+1} \mathrm{d}u = 4k\left(1 - \frac{1}{2^{2k-1}}\right)\Gamma(2k)\zeta(2k),$$

where we've integrated by parts and we've used Eq. (B.39). Equation (B.48) implies that

$$a_k = 2\zeta(2k)\left(1 - \frac{1}{2^{2k-1}}\right).$$

Zeta functions of argument $2k$ are related to the Bernoulli numbers,[50, p807]

$$\zeta(2k) = (-1)^{k+1}\frac{(2\pi)^{2k}}{2(2k)!}B_{2k},$$

where the first few Bernoulli numbers are

$$B_2 = \frac{1}{6} \qquad B_4 = -\frac{1}{30} \qquad B_6 = \frac{1}{42} \qquad B_8 = -\frac{1}{30} \qquad B_{10} = \frac{5}{66}.$$

Thus,

$$a_k = (-1)^{k+1}\left(1 - \frac{1}{2^{2k-1}}\right)\frac{(2\pi)^{2k}}{(2k)!}B_{2k}.$$

The first few values of a_k are:

$$a_1 = \frac{\pi^2}{6} \qquad a_2 = \frac{7\pi^4}{360} \qquad a_3 = \frac{31\pi^6}{15120}.$$

We therefore have from Eq. (B.46) the first few terms of the *Sommerfeld expansion*

$$\int_{-\infty}^{\infty} \frac{\phi(x)}{e^{x-y}+1}\mathrm{d}x = \int_{-\infty}^{y} \phi(x)\mathrm{d}x + \frac{\pi^2}{6}\left(\frac{\mathrm{d}\phi}{\mathrm{d}x}\right)_{x=y} + \frac{7\pi^4}{360}\left(\frac{\mathrm{d}^3\phi}{\mathrm{d}x^3}\right)_{x=y} + \frac{31\pi^6}{15120}\left(\frac{\mathrm{d}^5\phi}{\mathrm{d}x^5}\right)_{x=y} + \cdots.$$

$$\tag{B.49}$$

Equation (B.49) is more general than Eq. (B.40); it reduces to Eq. (B.40) for the case of

$$\phi(x) = \begin{cases} x^{n-1} & x \geq 0 \\ 0 & x < 0. \end{cases}$$

Classical mechanics

$\mathbf{W}$ E review topics from classical mechanics that are used in this book.

C.1   THE LEGENDRE TRANSFORMATION

A function $f(x)$ is *convex* if the line segment between any two points on the graph of the function lies above or on the graph, such as in the left part of Fig. C.1. Tangents to convex functions lie below

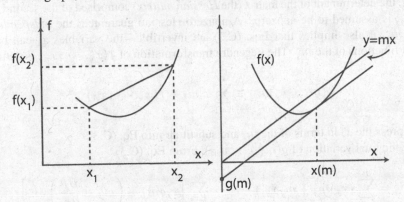

Figure C.1: Convex function $f(x)$ (left) and its Legendre transformation $g(m)$ (right)

the graph. A function is convex if $f''(x) \geq 0$; a function is *strictly convex* if $f''(x) > 0$. Strictly convex functions have no more than one minimum. A *concave* function $f(x)$ is such that $-f(x)$ is convex. Entropy $S(U, V, N)$ is a concave function—the second derivatives of S with respect to (U, V, N) are negative [3, Section 3.10], while internal energy $U(S, V, N)$ is a convex function—the second derivatives of U with respect to (S, V, N) are positive.[3, Exercise 3.5]

The *Legendre transformation* of a convex function $f(x)$ is a new convex function $g(m)$ of a new variable, m. It's not an integral transform, it's a geometric construction.[1] For a given number m, draw the line $y = mx$ (see right part of Fig. C.1). Let $D(x, m) \equiv mx - f(x) \geq 0$ denote the distance between $y = mx$ and the function $f(x)$ (if an m meeting the criterion exists). Define $x(m)$ as the value of x where $D(x, m)$ is maximized, which occurs where $f'(x) = m$. At $x(m)$, the tangent to the curve is parallel to $y = mx$. The Legendre transformation is defined as

$$g(m) \equiv D(x(m), m) = mx(m) - f(x(m)) .$$
(C.1)

---

[1] The Legendre transformation (the subject of this appendix) is not a Legendre *transform*, which is an integral transform $\widetilde{f}(n)$ of a function $f(x)$ defined on $[-1, 1]$, $\widetilde{f}(n) \equiv \int_{-1}^{1} P_n(x) f(x) \mathrm{d}x$, where $P_n(x)$ is a Legendre polynomial.

Geometrically, $g(m)$ is the number such that $f(x(m)) = -g(m) + mx$. It provides a *duality between points and tangents as equivalent ways of characterizing functions*: Instead of specifying a function pointwise, the values of f are constructed from the *slope* of the function at that point, plus the "offset," the y-intercept, which is the Legendre transformation.

Example. Find the Legendre transformation of the convex function $f(x) = x^2$.
Form the construct $D(x, m) = mx - x^2$. That quantity is maximized at $x(m) = m/2$. Then,

$$g(m) = D(x(m), m) = m \cdot \frac{m}{2} - \left(\frac{m}{2}\right)^2 = \frac{m^2}{4},$$

which is another convex function. Thus, the Legendre transformation of the kinetic energy $\frac{1}{2}mv^2$ (m is mass) is $p^2/(2m)$, where p is the transformation variable, with $v(p) = p/m$.

For multivariable convex functions $f(x_1, \cdots, x_n)$, introduce a *set* of new variables that are the slopes of f associated with each of the variables x_i:

$$m_i \equiv \frac{\partial f}{\partial x_i}. \qquad (i = 1, \cdots, n) \tag{C.2}$$

The *Hessian*, the determinant of the matrix (the *Hessian matrix*) comprised of the second derivatives $\partial^2 f/\partial x_i \partial x_j$, is assumed to be nonzero.[2] A nonzero Hessian guarantees the *independence* of the n variables m_i; it also implies that Eqs. (C.2) are invertible—the variables x_i can be uniquely expressed as functions of the m_j. The Legendre transformation of $f(x_1, \cdots, x_n)$ is defined as

$$g(m_1, \cdots, m_n) \equiv \sum_{i=1}^{n} x_i m_i - f(x_1, \cdots, x_n), \tag{C.3}$$

where we express the x_i in terms of the m_i and substitute into Eq. (C.3).
Consider the total variation of $g(m_1, \cdots, m_n)$. From Eq. (C.3),

$$\delta g = \sum_{j=1}^{n} (x_j \delta m_j + m_j \delta x_j) - \delta f = \sum_{j=1}^{n} \left[x_j \delta m_j + \left(m_j - \frac{\partial f}{\partial x_j} \right) \delta x_j \right]. \tag{C.4}$$

Using Eq. (C.2) in Eq. (C.4), we conclude that $g(m_1, \cdots, m_n)$ is a function of the independent variables m_j, and moreover that

$$x_j = \frac{\partial g}{\partial m_j}. \qquad (j = 1, \cdots, n) \tag{C.5}$$

There is a *duality* embodied in the Legendre transformation—the "new" variables are the partial derivatives of the "old" function with respect to the old variables, and the old variables are the partial derivatives of the new function with respect to the new variables—the Legendre transformation of the Legendre transformation is therefore the original function.[3] See Table C.1.

The class of functions amenable to Legendre transformation can be enlarged. Assume that $f = f(x_1, \cdots, x_n; y_1, \cdots, y_p)$ is a function of *two* sets of variables (where $p \neq n$): $\{x_i\}$, of which f is a convex function, and another set $\{y_i\}$. The variables y_i, which are independent of the x_i, are *parameters* for the purposes of the Legendre transformation, and do not participate in the transformation, which is still defined by Eq. (C.3). The transformed function will then be a function

---

[2]The Hessian is in essence the Jacobian determinant of the matrix of derivatives, $\partial m_i/\partial x_j$.
[3]Mathematicians say that the transformation is *involutive*—the Legendre transformation is its own inverse.

Table C.1: Dual nature of the Legendre transformation

Transformation of convex function	$f(x_1, \cdots, x_n)$	$g(m_1, \cdots, m_n)$
Transform variables	$m_i = \dfrac{\partial f}{\partial x_i}$	$x_i = \dfrac{\partial g}{\partial m_i}$
Legendre transformation	$g = \sum_{i=1}^n m_i x_i - f$	$f = \sum_{i=1}^n m_i x_i - g$
New convex function	$g = g(m_1, \cdots, m_n)$	$f = f(x_1, \cdots, x_n)$

of the extra variables, $g = g(m_1, \cdots, m_n; y_1, \cdots, y_p)$. Repeating the derivation in Eq. (C.4) for the variation of g in all variables, Eq. (C.5) still holds, but we have the additional relations:

$$\frac{\partial g}{\partial y_i} = -\frac{\partial f}{\partial y_i} . \qquad (i = 1, \cdots, p) \qquad (C.6)$$

The Legendre transformation is used in physics in two different ways. First, in thermodynamics it's used to shift emphasis away from variables that may be difficult to control experimentally, such as entropy, in favor of other variables that are more amenable to experimental control, such as temperature. (The thermodynamic potentials (see Section 1.4) are each a Legendre transformation of the internal energy function.) Second, in classical mechanics it's used to derive Hamiltonian mechanics from Lagrangian mechanics; Section C.5.

C.2 THE EULER-LAGRANGE EQUATION

The *calculus of variations* is concerned with finding the extremum of *functionals*, mappings between *functions* and real numbers.[4] A functional $J[y]$ (J acts on function $y(x)$ to produce a number) is often given in the form of a definite integral, $J[y] \equiv \int_{x_1}^{x_2} F(y(x)) \mathrm{d}x$, where F and the endpoints x_1, x_2 are known. The task is to find the *extremal function* $y(x)$ that makes $J[y]$ attain an extremum. Often F involves not just $y(x)$ but also $y' = \mathrm{d}y/\mathrm{d}x$, in which case F is denoted $F(y, y')$. We show that $J[y]$ attains an extremum when $y(x)$ satisfies the Euler-Lagrange equation, Eq. (C.18).

First, however, an example of how the type of problem posed by the calculus of variations could arise. Consider, for fixed points (a, b) in the Euclidean plane, the path of shortest distance between them. The question can be formulated as which function $y(x)$ *extremizes* the integral

$$J = \int_a^b \mathrm{d}s = \int_a^b \sqrt{(\mathrm{d}x)^2 + (\mathrm{d}y)^2} = \int_{x_a}^{x_b} \mathrm{d}x \sqrt{1 + (\mathrm{d}y/\mathrm{d}x)^2} = \int_{x_a}^{x_b} \mathrm{d}x \sqrt{1 + (y')^2} . \qquad (C.7)$$

Thus, $F(y') = \sqrt{1 + (y')^2}$. Pick a function $y(x)$, find its derivative y', substitute into Eq. (C.7), and do the integral. Which function results in the smallest value of J?

The problem sounds fantastically general, too general perhaps to make progress. Plan on success, however: *Assume* an extremal function $y(x)$ exists. Determine $y(x)$ through the requirement that small *functional variations* away from the extremal function produce even smaller changes in the value of J. To do this, introduce a class of varied functions,

$$Y(x, \alpha) \equiv y(x) + \alpha \eta(x) , \qquad (C.8)$$

---

[4]In quantum mechanics one encounters *linear* functionals—*bras* in Dirac notation. Functionals J considered in the calculus of variations are usually not linear.

where α is a number and $\eta(x)$ is any function that vanishes at the endpoints, $\eta(x_1) = \eta(x_2) = 0$, but is otherwise arbitrary. The functional can then be parameterized,

$$J(\alpha) = \int_{x_1}^{x_2} F\left(Y(x, \alpha), Y'(x, \alpha)\right) \mathrm{d}x , \tag{C.9}$$

where by assumption $J(\alpha = 0)$ is the extremum value. We expect that $J(\alpha)$ is a smooth function of α and is such that

$$\left.\frac{\partial J}{\partial \alpha}\right|_{\alpha=0} = 0 , \tag{C.10}$$

i.e., *to first order in small values of α there are only second-order changes in $J(\alpha)$*. When Eq. (C.10) is satisfied, J is said to be *stationary* with respect to small values of α; Eq. (C.10) is the *stationarity condition*. We show that Eq. (C.10) can be achieved for arbitrary $\eta(x)$.

Differentiate Eq. (C.9) (the limits of integration are independent of α):

$$\frac{\partial J}{\partial \alpha} = \int_{x_1}^{x_2} \frac{\partial}{\partial \alpha} F\left(Y(\alpha, x), Y'(\alpha, x)\right) \mathrm{d}x = \int_{x_1}^{x_2} \left(\frac{\partial F}{\partial y}\frac{\partial Y}{\partial \alpha} + \frac{\partial F}{\partial y'}\frac{\partial Y'}{\partial \alpha}\right) \mathrm{d}x . \tag{C.11}$$

The *infinitesimal variation*, $\delta y(x)$, is the variation in functional form *at a point* x,

$$\delta y(x) \equiv Y(x, \mathrm{d}\alpha) - y(x) = \frac{\partial Y}{\partial \alpha}\mathrm{d}\alpha = \eta(x)\mathrm{d}\alpha , \tag{C.12}$$

where we've used Eq. (C.8). The infinitesimal variation of $y'(x)$ is similarly defined

$$\delta y'(x) = \delta\left(\frac{\mathrm{d}y}{\mathrm{d}x}\right) \equiv Y'(x, \mathrm{d}\alpha) - y'(x) = \frac{\partial Y'}{\partial \alpha}\mathrm{d}\alpha = \eta'(x)\mathrm{d}\alpha = \frac{\mathrm{d}}{\mathrm{d}x}(\delta y) . \tag{C.13}$$

The variation of the derivative equals the derivative of the variation. The change in J under the variation is $\delta J \equiv (\partial J/\partial \alpha)\mathrm{d}\alpha$, and thus the stationarity condition is expressed by writing $\delta J = 0$. Multiplying Eq. (C.11) by $\mathrm{d}\alpha$ we have, using Eqs. (C.12) and (C.13):

$$\delta J = \int_{x_1}^{x_2} \left(\frac{\partial F}{\partial y}\delta y(x) + \frac{\partial F}{\partial y'}\delta y'(x)\right) \mathrm{d}x . \tag{C.14}$$

Equation (C.13) allows the second term in Eq. (C.14) to be integrated by parts:

$$\int_{x_1}^{x_2} \frac{\partial F}{\partial y'}\delta y' \mathrm{d}x = \int_{x_1}^{x_2} \frac{\partial F}{\partial y'}\frac{\mathrm{d}}{\mathrm{d}x}(\delta y)\mathrm{d}x = \frac{\partial F}{\partial y'}\delta y(x)\Big|_{x_1}^{x_2} - \int_{x_1}^{x_2} \frac{\mathrm{d}}{\mathrm{d}x}\left(\frac{\partial F}{\partial y'}\right)\delta y(x)\mathrm{d}x$$

$$= -\int_{x_1}^{x_2} \frac{\mathrm{d}}{\mathrm{d}x}\left(\frac{\partial F}{\partial y'}\right)\delta y(x)\mathrm{d}x , \tag{C.15}$$

where the integrated part vanishes because $\delta y(x)$ vanishes at the endpoints. Combining Eq. (C.15) with Eq. (C.14), the change of J under the variation $\delta y(x)$ is

$$\delta J = \int_{x_1}^{x_2} \left[\frac{\partial F}{\partial y} - \frac{\mathrm{d}}{\mathrm{d}x}\left(\frac{\partial F}{\partial y'}\right)\right]\delta y(x)\mathrm{d}x \equiv \int_{x_1}^{x_2} \frac{\delta F}{\delta y}\delta y(x)\mathrm{d}x . \tag{C.16}$$

The terms in square brackets in Eq. (C.16) define the *functional derivative* of $F(y, y')$,

$$\frac{\delta F}{\delta y(x)} \equiv \frac{\partial F}{\partial y} - \frac{\mathrm{d}}{\mathrm{d}x}\left(\frac{\partial F}{\partial y'}\right) . \tag{C.17}$$

We want Eq. (C.16) to vanish for *any* variation $\delta y(x)$. The only way that can happen is if the functional derivative of F vanishes identically,

$$\frac{\delta F}{\delta y} = \frac{\partial F}{\partial y} - \frac{\mathrm{d}}{\mathrm{d}x}\left(\frac{\partial F}{\partial y'}\right) = 0 . \tag{C.18}$$

Equation (C.18), the *Euler-Lagrange equation*, solves the problem of the calculus of variations by producing a differential equation to be satisfied by the extremal function, $y(x)$.

C.3 LAGRANGIAN MECHANICS

The problem posed by Newtonian dynamics (find the trajectory of particle given the forces acting on it) can be formulated as a problem in the calculus of variations. We seek a function $F = F(T, V)$, where T, V are the kinetic and potential energy functions of the particle, such that the condition

$$\delta \int_{t_1}^{t_2} F(T(\dot{x}), V(x))\mathrm{d}t = 0 \tag{C.19}$$

is equivalent to the Newtonian equation of motion. Here we've chosen a "clean" separation[5] between $x(t)$ and $\dot{x}(t)$, with $T = T(\dot{x})$ and $V = V(x)$. Substituting $F(T, V)$ in Eq. (C.18):

$$\frac{\partial F}{\partial V}\frac{\partial V}{\partial x} - \frac{\mathrm{d}}{\mathrm{d}t}\left(\frac{\partial F}{\partial T}\frac{\partial T}{\partial \dot{x}}\right) = 0. \tag{C.20}$$

Equation (C.20) reproduces Newton's second law, $\dot{p} = -\partial V/\partial x$, if we identify $p \equiv \partial T/\partial \dot{x}$ and if F is such that $\partial F/\partial T = -\partial F/\partial V = $ a nonzero constant. Take $F = T - V \equiv L$, where F has been renamed L, the *Lagrangian function*.

The Lagrangian is a function of position and speed: $L(x, \dot{x}, t)$ is a function on a three-dimensional space concatenated out of all possible positions, speeds, and times,[6] denoted $\mathbb{R} \times \mathbb{R} \times \mathbb{R}$. For trajectories $r(t)$, L is a function on a seven-dimensional space (r, v, t), $\mathbb{R}^3 \times \mathbb{R}^3 \times \mathbb{R}$. In Lagrangian mechanics *velocity is treated as an independent variable*. That seems not to make sense, however. Velocity $v = \mathrm{d}r/\mathrm{d}t$ is *determined* by the time derivative of the position vector and can't be treated as an independent variable. Such reasoning is valid when $r(t)$ denotes the *actual* trajectory of a particle, but not when all possible paths are considered. From basic uniqueness theorems, only one solution of a second-order differential equation can pass through a given point with a given slope, implying there can be a multitude of *possible* trajectories through a given point having different slopes. For every possible path described by seven numbers, $(x(t), y(t), z(t), \dot{x}(t), \dot{y}(t), \dot{z}(t), t)$, a path $r(t)$ is selected by the Euler-Lagrange equation such that $v(t) \equiv \dot{r}(t)$.

The Lagrangian can be extended to point masses having n *independent degrees of freedom* described by *generalized coordinates* $q_i, i = 1, \cdots, n$. The space spanned by these coordinates is termed *configuration space*. The distance between nearby points in configuration space is[7] $(\mathrm{d}s)^2 = \sum_{i=1}^{n}(\mathrm{d}q_i)^2$. With time as a parameter,[8] the kinetic energy $T = \frac{1}{2}m(\mathrm{d}s/\mathrm{d}t)^2 = \frac{1}{2}m\sum_{i=1}^{n}(\dot{q}_i)^2$. The Lagrangian is then

$$L(q_k, \dot{q}_k) = \frac{1}{2}m\sum_{i=1}^{n}(\dot{q}_i)^2 - V(q_j). \tag{C.21}$$

The time derivatives $\dot{q}_i$ are called *generalized velocities*.

Hamilton's principle is that of all the paths $q_i(t)$ a point mass *could* take through configuration space between fixed endpoints, the actual path is the one which extremizes the *action integral S*,

$$\delta S \equiv \delta \int_{t_1}^{t_2} L(q_k, \dot{q}_k, t)\mathrm{d}t = 0. \tag{C.22}$$

The time development of each generalized coordinate is found its own Euler-Lagrange equation:

$$\frac{\delta L}{\delta q_k} = \frac{\partial L}{\partial q_k} - \frac{\mathrm{d}}{\mathrm{d}t}\left(\frac{\partial L}{\partial \dot{q}_k}\right) = 0. \qquad k = 1, \cdots, n \tag{C.23}$$

Along the path of stationary action, the Lagrangian is *variationally constant* ($\delta L/\delta q_k = 0$), invariant under small functional variations δq_k *taken for the motion as a whole* between fixed endpoints.

---

[5]More generally, one can have $T = T(x, \dot{x}) = (m/2)\sum_{ij}g_{ij}(x)\dot{x}_i\dot{x}_j$ where $g_{ij}(x)$ is the metric tensor field, or $V = V(x, \dot{x})$ with a charged particle in a magnetic field.

[6]The Lagrangian could have an explicit time dependence.

[7]More generally, we could have a metric tensor $(\mathrm{d}s)^2 = \sum_{i=1}^{n}\sum_{j=1}^{n}g_{ij}\mathrm{d}q_i\mathrm{d}q_j$.

[8]In nonrelativistic mechanics all paths in configuration space are parameterized by the same time.

C.4 CHARGED PARTICLES

A charged particle coupled to electric and magnetic fields E and B is an important yet atypical case in that, even though the Lorentz force is not conservative,[9] it can nevertheless be derived from a scalar function $V(q_k, \dot{q}_k)$ such that (as we'll show)

$$F_k = q(E + v \times B)_k = -\frac{\partial V}{\partial q_k} + \frac{d}{dt}\left(\frac{\partial V}{\partial \dot{q}_k}\right) \equiv -\frac{\delta V}{\delta q_k}, \tag{C.24}$$

where q is the charge and F_k is the kth component of F in the system of generalized coordinates. Lagrange's equations have the structure of Newton's equation, that force is balanced by acceleration:

$$\frac{\delta L}{\delta q_k} = \frac{\delta T}{\delta q_k} - \frac{\delta V}{\delta q_k} = 0 \implies F_k = -\frac{\delta T}{\delta q_k},$$

i.e., the acceleration term $-\delta T/\delta q_k = (d/dt)(\partial T/\partial \dot{q}_k) - \partial T/\partial q_k$.

The homogeneous Maxwell equations are solved by introducing scalar and vector potential functions ϕ and A such that $B = \nabla \times A$ and $E = -\nabla\phi - \partial A/\partial t$. The Lorentz force can therefore be written

$$F = q\left(-\nabla\phi - \frac{\partial A}{\partial t} + v \times \nabla \times A\right). \tag{C.25}$$

Let's massage the term in Eq. (C.25) involving v:

$$(v \times \nabla \times A)_i = \sum_{jk} \epsilon_{ijk} v_j (\nabla \times A)_k = \sum_{jklm} \epsilon_{ijk}\epsilon_{klm} v_j \partial_l A_m = \sum_{jlm} (\delta_{il}\delta_{mj} - \delta_{jl}\delta_{im}) v_j \partial_l A_m \tag{C.26}$$

$$= \sum_m v_m \partial_i A_m - \sum_l v_l \partial_l A_i \equiv \sum_m v_m \partial_i A_m - (v \cdot \nabla) A_i,$$

where we've used the properties of the Levi-Civita symbol, ϵ_{ijk}. The vector potential is a function of space and time, $A = A(r, t)$, and thus its total time derivative can be written

$$\frac{dA}{dt} = \frac{\partial A}{\partial t} + (v \cdot \nabla) A. \tag{C.27}$$

Combining Eqs. (C.26) and (C.27), we have the (not particularly pretty) identity

$$\left(v \times \nabla \times A - \frac{\partial A}{\partial t}\right)_i = \sum_m v_m \partial_i A_m - \frac{dA_i}{dt}.$$

An equivalent way of writing the components of the Lorentz force is therefore

$$q\left(-\nabla\phi - \frac{\partial A}{\partial t} + v \times \nabla \times A\right)_k = q\left(-\partial_k\phi + \sum_m v_m \partial_k A_m - \frac{dA_k}{dt}\right). \tag{C.28}$$

The virtue of this (somewhat tortured) manipulation is that Eq. (C.28) can be written in the form of Eq. (C.24) if we take the potential energy function to be $V \equiv q\phi - qv \cdot A$ (show this). The Lagrangian for a particle coupled to the electromagnetic field is thus

$$L = T - q\phi + qv \cdot A. \tag{C.29}$$

---

[9] Is it true that $\nabla \times (E + v \times B) = 0$? One has to take into account the field momentum, a digression for another day.

C.5   HAMILTONIAN MECHANICS

The Lagrangian $L(q, \dot{q})$ is a convex function of generalized velocities. Its *slope* $\partial L / \partial \dot{q}_k$ is defined as the *canonical momentum* p_k associated with generalized coordinate q_k,

$$p_k \equiv \frac{\partial L}{\partial \dot{q}_k} = \frac{\partial T}{\partial \dot{q}_k} - \frac{\partial V}{\partial \dot{q}_k}. \tag{C.30}$$

For conservative forces there is no difference between canonical and *kinetic momentum*, $\partial T / \partial \dot{q}_k$. The distinction is necessary for magnetic forces; from Eq. (C.29), $p = mv + qA$. The *Hamiltonian function* is the Legendre transformation of the Lagrangian with respect to the generalized velocities. Using Eq. (C.3),

$$H(p_j, q_j, t) \equiv \sum_{k=1}^{n} p_k \dot{q}_k - L(q_j, \dot{q}_j, t) . \tag{C.31}$$

Combining Eqs. (C.31), (C.30) (show this):

$$H(p_k, q_k) = \frac{1}{2m} \sum_{k=1}^{n} (p_k)^2 + V(q_j) . \tag{C.32}$$

Note the simplification achieved by the Hamiltonian: It does not contain any derivatives of p_j or q_j with respect to t. We note that, through the Legendre transformation, p_k is *uniquely* determined in terms of $\dot{q}_k$, and thus the canonical momenta are independent of the generalized coordinates.

From the differential of Eq. (C.31):[10]

$$dH = \sum_{k=1}^{n} \left[p_k d\dot{q}_k + \dot{q}_k dp_k - \frac{\partial L}{\partial q_k} dq_k - \frac{\partial L}{\partial \dot{q}_k} d\dot{q}_k \right] = \sum_{k=1}^{n} [\dot{q}_k dp_k - \dot{p}_k dq_k] , \tag{C.33}$$

where we've used Eqs. (C.23) and (C.30) to equate $\partial L / \partial q_k = \dot{p}_k$. From Eq. (C.33) we identify[11]

$$\frac{\partial H}{\partial p_k} = \dot{q}_k \qquad \frac{\partial H}{\partial q_k} = -\dot{p}_k . \qquad k = 1, \cdots, n \tag{C.34}$$

Hamilton's equations of motion, an equivalent formulation of classical mechanics. Rather than n second-order differential equations, Hamilton's equations are $2n$ first-order differential equations.

C.6   DYNAMIC REVERSIBILITY

Under time inversion, the canonical momentum becomes its negative, even in a magnetic field—if currents reverse, so does the vector potential $A(r) \approx \int d^3 r' J(r') / (|r - r'|)$. Consider the transformation $t \to t' = -t$, and $p_i \to p_i' = -p_i$, $q_i \to q_i' = q_i$. The Hamiltonian is invariant under time inversion because it's an even function of the momenta, $H' = H(p', q') = H(-p, q) = H(p, q)$. Hamilton's equations of motion are therefore *time-reversal symmetric*:

$$\frac{dq_i'}{dt'} = -\frac{dq_i}{dt} = -\frac{\partial H}{\partial p_i} = \frac{\partial H'}{\partial p_i'} \qquad \frac{dp_i'}{dt'} = \frac{dp_i}{dt} = -\frac{\partial H}{\partial q_i} = -\frac{\partial H'}{\partial q_i'} .$$

C.7   COVARIANCE UNDER POINT TRANSFORMATIONS

The methods of analytical mechanics afford considerable flexibility in choosing coordinates. Indeed, there are an *unlimited* number of possible systems of generalized coordinates (with the same number

---

[10] We're ignoring the possibility that L is an explicit function of time.
[11] Because p_k, q_k are independent of each other.

of degrees of freedom). As we now show, *the form of the Lagrange and Hamilton equations is invariant under invertible coordinate transformations in configuration space.*

Any set $\{Q_k\}_{k=1}^n$ of independent functions[12] of the q_i provides another set of coordinates to describe the system, a new set of numbers to attach to the same points in configuration space,

$$Q_k = Q_k(q_1, \cdots, q_n) . \qquad\qquad k = 1, \cdots, n \qquad\qquad (C.35)$$

We restrict ourselves to coordinate transformations not explicitly involving the time. The transformation equations (C.35) are invertible (independent $\Leftrightarrow$ non-vanishing Jacobian determinant), with $q_j = q_j(Q_1, \cdots, Q_n)$, $j = 1, \cdots, n$. The generalized velocities then transform as

$$\dot{q}_j = \sum_{k=1}^n \frac{\partial q_j}{\partial Q_k} \dot{Q}_k , \qquad\qquad (C.36)$$

i.e., the generalized velocities in one coordinate system, $\dot{q}_j$, are linear combinations of those in another, $\dot{Q}_k$, multiplied by functions solely of the Q_i, $\partial q_j / \partial Q_k$, which we can indicate by writing $\dot{q}_j = \dot{q}_j(Q, \dot{Q})$. Equation (C.36) implies the relation[13]

$$\frac{\partial \dot{q}_j}{\partial \dot{Q}_k} = \frac{\partial q_j}{\partial Q_k} . \qquad\qquad (C.37)$$

The Lagrangian function in the "old" coordinates, $L(q_k, \dot{q}_k)$, *defines* through Eq. (C.36) and the inverse of Eq. (C.35) the Lagrangian $\widetilde{L}(Q_j, \dot{Q}_j)$ in the new coordinate system:

$$L(q_k, \dot{q}_k) \equiv \widetilde{L}(Q_j, \dot{Q}_j) . \qquad\qquad (C.38)$$

Equation (C.38) indicates that the *value* of a scalar field is invariant under coordinate transformations, but the *form* of the function of the transformed coordinates may change.[14] From Eq. (C.23),

$$0 = \frac{\delta L}{\delta q_k} = \sum_{j=1}^n \frac{\partial Q_j}{\partial q_k} \frac{\delta \widetilde{L}}{\delta Q_j} , \qquad\qquad k = 1, \cdots n \qquad\qquad (C.39)$$

where we've used Eq. (C.38). We therefore have (because the terms $\partial Q_j / \partial q_k$ are arbitrary)

$$\frac{\delta \widetilde{L}}{\delta Q_j} \equiv \frac{\partial \widetilde{L}}{\partial Q_j} - \frac{\mathrm{d}}{\mathrm{d}t} \left(\frac{\partial \widetilde{L}}{\partial \dot{Q}_j} \right) = 0 . \qquad j = 1, \cdots, n \qquad\qquad (C.40)$$

The form of the Lagrange equations is independent of the choice of coordinates. Equation (C.23) (or Eq. (C.40)) is a necessary and sufficient condition for the action integral to be stationary. Finding the extremum of the action integral is thus independent of coordinate system; we can adopt whatever coordinates are best suited to the problem. Because the Lagrangian is invariant—Eq. (C.38)—we arrive at the important conclusion: *The value of the action integral S is invariant.*

The canonical momentum in the new coordinates is defined as in Eq. (C.30):

$$P_k \equiv \frac{\partial \widetilde{L}}{\partial \dot{Q}_k} = \sum_{j=1}^n \frac{\partial \dot{q}_j}{\partial \dot{Q}_k} \frac{\partial L}{\partial \dot{q}_j} = \sum_{j=1}^n \frac{\partial q_j}{\partial Q_k} p_j , \qquad\qquad (C.41)$$

---

[12] A set of functions is *independent* if the Jacobian determinant—the determinant of the matrix of partial derivatives $\partial Q_j / \partial q_k$ (the Jacobian matrix) does not vanish identically.

[13] The reciprocal to Eq. (C.36) is $\dot{Q}_k = \sum_{j=1}^n (\partial Q_k / \partial q_j) \dot{q}_j$, with $\partial \dot{Q}_k / \partial \dot{q}_j = \partial Q_k / \partial q_j$.

[14] The temperature at a point in a room doesn't care what coordinates you use to locate points in the room.

where we've used Eq. (C.38) and the chain rule in the second equality, and Eq. (C.37) in the last. The canonical momenta P_k therefore transform just as do the velocities, Eq. (C.36), and we can write that $P_k = P_k(Q, p) = P_k(q, p)$. Note the invariance of the scalar product:

$$\sum_{j=1}^{n} p_j \dot{q}_j = \sum_{k=1}^{n} \sum_{j=1}^{n} \sum_{l=1}^{n} P_k \frac{\partial Q_k}{\partial q_j} \frac{\partial q_j}{\partial Q_l} \dot{Q}_l = \sum_{k=1}^{n} \sum_{l=1}^{n} P_k \dot{Q}_l \delta_{kl} = \sum_{k=1}^{n} P_k \dot{Q}_k , \qquad (C.42)$$

where we've used the inverse of Eq. (C.41) and Eq. (C.36).

The "old" Hamiltonian defines, through these transformation equations, its form for the new coordinates and momenta:

$$H(p_k, q_k) \equiv \tilde{H}(P_j, Q_j) . \qquad (C.43)$$

(Equivalently, Eq. (C.43) follows from the Legendre transform, $\tilde{H} = \sum_{k=1}^{n} P_k \dot{Q}_k - \tilde{L}(Q, \dot{Q})$, together with Eq. (C.38) and Eq. (C.42).) The form invariance of Hamilton's equations then follows using the steps by which Eq. (C.34) was derived:

$$\frac{\partial \tilde{H}}{\partial P_k} = \dot{Q}_k \qquad \frac{\partial \tilde{H}}{\partial Q_k} = -\dot{P}_k .$$

Form invariance under coordinate transformations is referred to as *covariance*. The Lagrange and Hamiltonian equations of motion are covariant under invertible coordinate transformations in configuration space.

C.8 CANONICAL TRANSFORMATIONS

The canonical momenta $\{p_k\}$ are independent of the generalized coordinates, $\{q_i\}$. The p_i are, through the Legendre transformation, uniquely associated with the generalized velocities $\dot{q}_i$, and velocities are treated as independent variables in the Lagrangian function. The Lagrangian is the Legendre transformation of the Hamiltonian, and hence Hamilton's principle can be expressed by combining Eqs. (C.22) and (C.31):

$$\delta \int_{t_1}^{t_2} \left(\sum_{k=1}^{n} p_k \dot{q}_k - H(p_j, q_j) \right) dt = 0 . \qquad (C.44)$$

Performing the variations indicated in Eq. (C.44),

$$\sum_{k=1}^{n} \int_{t_1}^{t_2} \left(p_k \delta \dot{q}_k + \dot{q}_k \delta p_k - \frac{\partial H}{\partial p_k} \delta p_k - \frac{\partial H}{\partial q_k} \delta q_k \right) dt = 0 . \qquad (C.45)$$

The first term in Eq. (C.45) can be integrated by parts using $\delta \dot{q}_k = \mathrm{d}(\delta q_k)/\mathrm{d}t$—Eq. (C.13)—to yield ($\delta q_k$ vanishes at t_1 and t_2):

$$\sum_{k=1}^{n} \int_{t_1}^{t_2} \left[\delta p_k \left(\dot{q}_k - \frac{\partial H}{\partial p_k} \right) - \delta q_k \left(\dot{p}_k + \frac{\partial H}{\partial q_k} \right) \right] dt = 0 . \qquad (C.46)$$

Invoking Hamilton's equations, Eq. (C.34), we have in Eq. (C.46) that the stationarity of the action integral is independent of variations of the q_i *and* the p_i.

The extra set of independent variables implies the possibility of a wider class of transformations in *phase space*, the $2n$-dimensional space spanned by coordinates and momenta. We consider as the generalization of Eq. (C.35), time-independent, invertible transformations

$$\left. \begin{array}{l} Q_i = Q_i(q_1, \cdots, q_n; p_1, \cdots, p_n) \\ P_i = P_i(q_1, \cdots, q_n; p_1, \cdots, p_n) . \end{array} \right\} \qquad i = 1, \cdots, n \qquad (C.47)$$

We're not interested in *arbitrary* phase-space transformations, just those that preserve the form of Hamilton's equations: *canonical transformations*. Under canonical transformations, the new coordinates are functions of *both* the old coordinates and momenta, blurring the distinction between coordinates and momenta. The variables Q_i and P_i are called *canonically conjugate*.

One might think that transformations in the form of Eq. (C.47) would be canonical if the integrand of the action integral is preserved, i.e., if (from Eq. (C.44))

$$\sum_{k=1}^{n} P_k \dot{Q}_k - \tilde{H}(P_j, Q_j) \stackrel{?}{=} \sum_{k=1}^{n} p_k \dot{q}_k - H(p_j, q_j) . \tag{C.48}$$

The equality indicated in Eq. (C.48) *would* hold under coordinate transformations in configuration space, Eq. (C.35) (use Eqs. (C.42) and (C.43)), but not necessarily under phase-space transformations, Eq. (C.47). We can "cancel" the Hamiltonians in Eq. (C.48): It's still true that $H(p, q) \equiv \tilde{H}(P, Q)$ through Eq. (C.47). It would seem sufficient then to seek transformations that preserve Eq. (C.42), $\sum_k p_k \dot{q}_k = \sum_k P_k \dot{Q}_k$. That requirement, however, is too restrictive. The two integrands (of the action integral, shown in Eq. (C.48)) can differ by the total time derivative of a function and not affect the stationarity of the action integral between definite limits.[15] We require of canonical transformations the generalization of Eq. (C.42):

$$\sum_k P_k \dot{Q}_k + \frac{dF}{dt} = \sum_k p_k \dot{q}_k , \tag{C.49}$$

where F, the *generating function* of the transformation, is a function of the new and old phase-space variables.[16] There are four possibilities: $F_1(q, Q)$, $F_2(q, P)$, $F_3(p, Q)$, and $F_4(p, P)$. Note that Eq. (C.49) can be written

$$dF = \sum_k (p_k dq_k - P_k dQ_k) . \tag{C.50}$$

Equation (C.50) is another way to characterize canonical transformations: If the differential forms on the right side of Eq. (C.50) are equal to a total differential, they are related by a canonical transformation. Note that if $F = 0$, Eq. (C.50) reduces to Eq. (C.42).

Consider a generating function of the type $F_1(q, Q)$. From Eq. (C.49):

$$\sum_j \left[P_j \dot{Q}_j + \frac{\partial F_1}{\partial q_j} \dot{q}_j + \frac{\partial F_1}{\partial Q_j} \dot{Q}_j \right] = \sum_k p_k \dot{q}_k . \tag{C.51}$$

Equation (C.51) can be satisfied by requiring

$$P_k = -\frac{\partial F_1(q, Q)}{\partial Q_k} \qquad p_k = \frac{\partial F_1(q, Q)}{\partial q_k} . \tag{C.52}$$

For a given function $F_1(q, Q)$, the second equality in Eq. (C.52) implies $Q_k = Q_k(q, p)$, which combined with the first equality implies $P_k = P_k(q, p)$, and thus a transformation of the form of Eq. (C.47) is determined. Moreover, P_k and Q_k are *canonical variables*—they satisfy Hamilton's equations with $\tilde{H}$ as the Hamiltonian.

---

[15]The variation of $\int_{t_1}^{t_2} (dF/dt) dt = F(t_2) - F(t_1)$ vanishes because the variation vanishes at the end points.

[16]The generating function must link the old and new phase-space coordinates.

A generating function of the type $F_2(q, P)$, can be obtained from the Legendre transformation[17] of $F_1(q, Q)$: $F_2(q, P) \equiv F_1(q, Q) + \sum_k Q_k P_k$. Thus,

$$\frac{dF_2}{dt} \equiv \sum_k \left(\frac{\partial F_2}{\partial q_k} \dot{q}_k + \frac{\partial F_2}{\partial P_k} \dot{P}_k \right) = \sum_k \left(\frac{\partial F_1}{\partial q_k} \dot{q}_k + \frac{\partial F_1}{\partial Q_k} \dot{Q}_k + Q_k \dot{P}_k + \dot{Q}_k P_k \right)$$

$$= \sum_k \left(Q_k \dot{P}_k + p_k \dot{q}_k \right) ,$$

where we've used Eq. (C.52). We then have the relations (because $F_2 = F_2(q, P)$):

$$p_k = \frac{\partial F_2(q, P)}{\partial q_k} \qquad\qquad Q_k = \frac{\partial F_2(q, P)}{\partial P_k} . \qquad\qquad \text{(C.53)}$$

Similar transformation relations can be developed for the other generating functions [162, p241].

Example. Consider the generating function $F_1(q, Q) = \sum_k q_k Q_k$. From Eq. (C.52),

$$P_j = -\frac{\partial}{\partial Q_j} \sum_k q_k Q_k = -q_j \qquad\qquad p_j = \frac{\partial}{\partial q_j} \sum_k q_k Q_k = Q_j .$$

The *exchange transformation* $q_j \to -P_j$ and $p_j \to Q_j$ is therefore a canonical transformation in which the roles of coordinates and momenta are reversed. One can show directly that Hamiltons equations are preserved under the exchange transformation. Consider the generating function $F_2(q, P) = \sum_k q_k P_k$. From Eq. (C.53),

$$p_j = \frac{\partial}{\partial q_j} \sum_k q_k P_k = P_j \qquad\qquad Q_j = \frac{\partial}{\partial P_j} \sum_k q_k P_k = q_j .$$

The *identity transformation* $p_j \to P_j$ and $q_j \to Q_j$, where the momenta and coordinates have simply been given new names, is a canonical transformation.

C.9   JACOBIAN OF CANONICAL TRANSFORMATIONS

C.9.1   Jacobian determinants

Many calculations can be simplified using the properties of Jacobian determinants, or simply *Jacobians*. Consider n functions $u, v, \cdots, w$, that each depend on n variables $x, y, \cdots, z$. The Jacobian is the determinant of the matrix of partial derivatives. There's a highly practical notation for Jacobians that we'll employ. Let the Jacobian be denoted by what resembles a partial derivative:

$$\frac{\partial(u, v, \cdots, w)}{\partial(x, y, \cdots, z)} \equiv \begin{vmatrix} \dfrac{\partial u}{\partial x} & \dfrac{\partial u}{\partial y} & \cdots & \dfrac{\partial u}{\partial z} \\ \dfrac{\partial v}{\partial x} & \dfrac{\partial v}{\partial y} & \cdots & \dfrac{\partial v}{\partial z} \\ \vdots & \vdots & \ddots & \vdots \\ \dfrac{\partial w}{\partial x} & \dfrac{\partial w}{\partial y} & \cdots & \dfrac{\partial w}{\partial z} \end{vmatrix} .$$

---

[17] Our definition of canonical transformation, Eq. (C.50), implies that the generating function is naturally of the type $F(q, Q)$. Legendre transformations are a way to shift the emphasis onto new sets of variables.

Jacobians have the following three properties. *First*, a single partial derivative can be expressed as a Jacobian:

$$
\left(\frac{\partial u}{\partial x}\right)_{y,\cdots,z} = \frac{\partial(u, y, \cdots, z)}{\partial(x, y, \cdots, z)} = \begin{vmatrix} \dfrac{\partial u}{\partial x} & \dfrac{\partial u}{\partial y} & \cdots & \dfrac{\partial u}{\partial z} \\ 0 & 1 & \cdots & 0 \\ \vdots & & \ddots & \vdots \\ 0 & \cdots & \cdots & 1 \end{vmatrix}
\tag{C.54}
$$

(after the first row in Eq. (C.54), there are only ones on the diagonal and zeros everywhere else). *Symbols common to the numerator and denominator in the notation for the Jacobian "cancel." Second*, Jacobians obey a product rule that resembles the ordinary chain rule from calculus,[18]

$$
\frac{\partial(u, v, \cdots, w)}{\partial(x, y, \cdots, z)} = \frac{\partial(u, v, \cdots, w)}{\partial(r, s, \cdots, t)} \frac{\partial(r, s, \cdots, t)}{\partial(x, y, \cdots, z)}.
\tag{C.55}
$$

Equation (C.55) *is* in fact the chain rule, combined with the properties of determinants. Consider that one has a set of functions $\{f_i\}$ that each depend on a set of variables $\{x_j\}$, and that the $\{x_j\}$ are functions of another set of variables $\{y_k\}$. Using the chain rule,

$$
\frac{\partial f_i}{\partial x_j} = \sum_m \frac{\partial f_i}{\partial y_m} \frac{\partial y_m}{\partial x_j}.
$$

The chain rule is thus in the form of matrix multiplication of the Jacobian matrices, the determinants of which are embodied in Eq. (C.55). Equation (C.55) is a result of the property of determinants, that if matrices A, B, C are connected by matrix multiplication, $A = BC$, the determinant of A equals the product of the determinants of B and C, $|A| = |B| \cdot |C|$. *Third*, it follows from Eq. (C.55) that

$$
\frac{\partial(u, v, \cdots, w)}{\partial(x, y, \cdots, z)} = \left[\frac{\partial(x, y, \cdots, z)}{\partial(u, v, \cdots, w)}\right]^{-1}.
$$

C.9.2 Canonical transformations

The Jacobian of canonical transformations is unity, a fundamental result of classical mechanics that we now demonstrate, which is ostensibly the content of *Liouville's theorem* (see Section 2.5). Referring to Eq. (C.47) and the notation for Jacobian determinants,

$$
J \equiv \frac{\partial(Q, P)}{\partial(q, p)} \equiv \frac{\partial(Q_1, \ldots, Q_n; P_1, \ldots, P_n)}{\partial(q_1, \ldots, q_n; p_1, \ldots, p_n)}
$$

$$
= \begin{vmatrix}
\dfrac{\partial Q_1}{\partial q_1} & \cdots & \dfrac{\partial Q_1}{\partial q_n} & \dfrac{\partial Q_1}{\partial p_1} & \cdots & \dfrac{\partial Q_1}{\partial p_n} \\
\vdots & & \vdots & \vdots & & \vdots \\
\dfrac{\partial Q_n}{\partial q_1} & \cdots & \dfrac{\partial Q_n}{\partial q_n} & \dfrac{\partial Q_n}{\partial p_1} & \cdots & \dfrac{\partial Q_n}{\partial p_n} \\
\dfrac{\partial P_1}{\partial q_1} & \cdots & \dfrac{\partial P_1}{\partial q_n} & \dfrac{\partial P_1}{\partial p_1} & \cdots & \dfrac{\partial P_1}{\partial p_n} \\
\vdots & & \vdots & \vdots & & \vdots \\
\dfrac{\partial P_n}{\partial q_1} & \cdots & \dfrac{\partial P_n}{\partial q_n} & \dfrac{\partial P_n}{\partial p_1} & \cdots & \dfrac{\partial P_n}{\partial p_n}
\end{vmatrix}.
\tag{C.56}
$$

[18]An explicit demonstration of the chain-rule property of Jacobians is shown in Mazenko.[86, p570] Jacobians are also discussed in Landau and Lifshitz,[99, p51].

Equation (C.56) can be simplified using the properties of Jacobians:

$$J = \frac{\partial(Q,P)}{\partial(q,p)} = \frac{\partial(Q,P)}{\partial(q,P)} \frac{\partial(q,P)}{\partial(q,p)} = \frac{\frac{\partial(Q,P)}{\partial(q,P)}}{\frac{\partial(q,p)}{\partial(q,P)}} \equiv \frac{D_n}{D_d}. \tag{C.57}$$

The determinant of a $2n \times 2n$ matrix has been reduced to the ratio of determinants of $n \times n$ matrices,

$$D_n = \frac{\partial(Q_1, \cdots, Q_n)}{\partial(q_1, \cdots, q_n)} \equiv \left| \frac{\partial Q_i}{\partial q_j} \right| \qquad D_d = \frac{\partial(p_1, \cdots, p_n)}{\partial(P_1, \cdots, P_n)} \equiv \left| \frac{\partial p_i}{\partial P_j} \right|.$$

We have yet to use that Q and P are obtained from a canonical transformation. Using Eq. (C.53),

$$\frac{\partial Q_i}{\partial q_j} = \frac{\partial^2 F_2}{\partial q_j \partial P_i} \qquad \frac{\partial p_i}{\partial P_j} = \frac{\partial^2 F_2}{\partial P_j \partial q_i},$$

and thus

$$\frac{\partial p_i}{\partial P_j} = \frac{\partial^2 F_2}{\partial q_i \partial P_j} = \frac{\partial Q_j}{\partial q_i}.$$

Therefore, D_n is the determinant of the transpose of the matrix of which D_d is the determinant of. But the determinant of the transpose of a matrix is the same as the determinant of the matrix, $\det A^T = \det A$. Thus, $D_n = D_d$ and by Eq. (C.57), $J = 1$.

Example. The Jacobian of the identity transformation $F = \sum_k q_k P_k$ (see page 331) is the determinant of $I_{2n \times 2n}$, the $2n \times 2n$ identity matrix (show this), which is clearly unity. The Jacobian of the exchange transformation $F = \sum_k q_k Q_k$ is the determinant

$$J = \begin{vmatrix} [0]_{n \times n} & [I]_{n \times n} \\ -[I]_{n \times n} & [0]_{n \times n} \end{vmatrix},$$

where $[0]_{n \times n}$ is the $n \times n$ matrix with every element zero, and $[I]_{n \times n}$ is the $n \times n$ identity matrix. Such a determinant has the value unity. (The determinant of a 2×2 "block matrix" $\begin{pmatrix} A & B \\ C & D \end{pmatrix}$ is $\det A \det D - \det C \det B$.)

C.9.3 Geometric significance

Under a change of variables, the volume element of a multiple integral transforms with the Jacobian:

$$\int_\Omega dq_1 \cdots dq_n dp_1 \cdots dp_n = \int_{\Omega'} \frac{\partial(q_1, \cdots, q_n, p_1, \cdots, p_n)}{\partial(Q_1, \cdots, Q_n, P_1, \cdots, P_n)} dQ_1 \cdots dQ_n dP_1 \cdots dP_n,$$

where Ω denotes a volume in phase space. If the transformation $(q, p) \to (Q, P)$ is canonical, then, by what we've just shown:

$$\int_\Omega dq_1 \cdots dq_n dp_1 \cdots dp_n = \int_{\Omega'} dQ_1 \cdots dQ_n dP_1 \cdots dP_n. \tag{C.58}$$

Volume in phase space is preserved under canonical transformations.

C.10 INFINITESIMAL CANONICAL TRANSFORMATIONS

Let the generating function for a canonical transformation consist of the identity transformation (page 331) plus an infinitesimal quantity:

$$F_2(q, P) = \sum_k q_k P_k + \epsilon G(q, P),$$

where ϵ is an infinitesimal and the function $G(q, P)$ is the generator of *infinitesimal canonical transformations*. From Eq. (C.53), the new coordinates and momenta are such that

$$p_k = \frac{\partial F_2(q, P)}{\partial q_k} = P_k + \epsilon \frac{\partial G(q, P)}{\partial q_k}$$

$$Q_k = \frac{\partial F_2(q, P)}{\partial P_k} = q_k + \epsilon \frac{\partial G(q, P)}{\partial P_k}.$$

Write this transformation as

$$Q_k = q_k + \delta q_k \implies \delta q_k = \epsilon \frac{\partial G(q, P)}{\partial P_k}$$

$$P_k = p_k + \delta p_k \implies \delta p_k = -\epsilon \frac{\partial G(q, P)}{\partial q_k}. \tag{C.59}$$

Because P_k differs from p_k by an infinitesimal quantity, we can, to first order in small quantities, replace P_k on the right side of Eq. (C.59) by p_k:

$$\delta q_k = \epsilon \frac{\partial G(q, p)}{\partial p_k}$$

$$\delta p_k = -\epsilon \frac{\partial G(q, p)}{\partial q_k}. \tag{C.60}$$

A useful result emerges from Eq. (C.60) if $G(q, p) = H(q, p)$ and $\epsilon = dt$:

$$\delta q_k = dt \frac{\partial H(q, p)}{\partial p_k} = dt \dot{q}_k = dq_k$$

$$\delta p_k = -dt \frac{\partial H(q, p)}{\partial q_k} = dt \dot{p}_k = dp_k,$$

where we've used Hamilton's equations, (C.34). Thus, *an infinitesimal canonical transformation is generated by the Hamiltonian*, under which canonical coordinates at time $t + dt$ are those that evolve from their values at time t according to the time evolution specified by Hamilton's equations.[19]

That statement, however, is not restricted to infinitesimal time differences. *Finite* canonical transformations can be realized by compounding many successive infinitesimal canonical transformations. Canonical transformations have the property that a canonical transformation followed by a canonical transformation is itself a canonical transformation. Let canonical transformation C_1 be such that $(q, p) \overset{C_1}{\to} (q', p')$, and let C_2 be such that $(q', p') \overset{C_2}{\to} (q'', p'')$. Then there is a canonical transformation C_3 such that $(q, p) \overset{C_3}{\to} (q'', p'')$. *Proof*: Canonical transformations have the form shown in Eq. (C.50). Thus, C_1 is associated with the differential of a function f_1, where $df_1 = \sum_k (p_k dq_k - p'_k dq'_k)$, and C_2 is associated with the differential of a function f_2, where $df_2 = \sum_k (p'_k dq'_k - p''_k dq''_k)$. Add these equations: $d(f_1 + f_2) = \sum_k (p_k dq_k - p''_k dq''_k)$. Thus, C_3 is a canonical transformation. The time evolution of canonical coordinates corresponds to the continuous unfolding of a canonical transformation generated by the Hamiltonian.

---

[19]Depending on one's outlook, this point is either highly significant or trivial. One might take for granted that a set of variables that satisfy Hamilton's equations at time t evolve into variables that satisfy Hamilton's equations at time $t + dt$. We've shown that canonical coordinates stay canonical through their evolution under Hamilton's equations of motion.

C.11 SELF-ADJOINTNESS OF THE LIOUVILLE OPERATOR

The Liouville operator, L, defined in Eq. (2.5), includes a factor of i (unit imaginary number) to make it self-adjoint. We show here that the Liouville operator is self-adjoint. To discuss self-adjointness, an inner product must be specified for phase space. We require that a phase-space function $A(p, q)$ (which can be considered a vector in a $2n$-dimensional space) have finite *norm*:

$$\langle A|A \rangle \equiv \int A(p,q)^* A(p,q) \mathrm{d}p_1 \cdots \mathrm{d}p_n \mathrm{d}q_1 \cdots \mathrm{d}q_n < \infty .$$

An operator L is self-adjoint[20] if $\langle u|Lv \rangle = \langle Lu|v \rangle$ for any functions u, v. By definition,[21]

$$\langle u|Lv \rangle = \mathrm{i} \int u^* \sum_k \left(\frac{\partial H}{\partial q_k} \frac{\partial v}{\partial p_k} - \frac{\partial H}{\partial p_k} \frac{\partial v}{\partial q_k} \right) \mathrm{d}p\mathrm{d}q . \qquad (C.61)$$

Integrate by parts for each value of k. Make use of the identities

$$u^* \frac{\partial v}{\partial p_k} = \frac{\partial}{\partial p_k}(u^* v) - v \frac{\partial u^*}{\partial p_k} \qquad u^* \frac{\partial v}{\partial q_k} = \frac{\partial}{\partial q_k}(u^* v) - v \frac{\partial u^*}{\partial q_k} .$$

The integrated parts vanish because the norm is finite. After integrating by parts, Eq. (C.61) becomes

$$\langle u|Lv \rangle = -\mathrm{i} \int v \sum_k \left(\frac{\partial H}{\partial q_k} \frac{\partial u^*}{\partial p_k} - \frac{\partial H}{\partial p_k} \frac{\partial u^*}{\partial q_k} \right) \mathrm{d}p\mathrm{d}q . \qquad (C.62)$$

Assuming that the Hamiltonian function is pure real,[22] Eq. (C.62) is equivalent to:

$$\langle u|Lv \rangle = \int v \left(Lu \right)^* \mathrm{d}p\mathrm{d}q \equiv \langle Lu|v \rangle .$$

C.12 LAGRANGIAN DENSITY

We now consider the dynamics of a *continuous system*. To do so, we examine a system of discrete masses connected by springs and show how it transitions to a continuous elastic medium in the limit as the separation between masses vanishes. Figure C.2 shows a segment of a one-dimensional

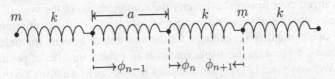

Figure C.2: System of coupled masses and springs

array of identical masses m connected by identical springs k at equilibrium positions $x = na$, where n is an integer and a is the equilibrium spring length. The figure shows a configuration of the instantaneous *displacements* ϕ_n of the masses away from their equilibrium positions. The Lagrangian for the system of masses and springs is

$$L = \frac{1}{2} \sum_n \left[m\dot{\phi}_n^2 - k \left(\phi_n - \phi_{n-1} \right)^2 \right] . \qquad (C.63)$$

---

[20] The adjoint operator $L^\dagger$ is a new operator obtained from L through the requirement $\langle u|L^\dagger v \rangle = \langle Lu|v \rangle$.
[21] To save writing, we use the convention $\mathrm{d}p\mathrm{d}q \equiv \mathrm{d}p_1 \cdots \mathrm{d}p_n \mathrm{d}q_1 \cdots \mathrm{d}q_n$.
[22] There is no reason in general to expect that H is complex valued.

The equations of motion for the displacements $\{\phi_n\}$ are, from Eq. (C.23), a coupled set of differential-difference equations

$$m\ddot{\phi}_n = -k\left(2\phi_n - \phi_{n-1} - \phi_{n+1}\right) . \tag{C.64}$$

At this point, one could Fourier transform Eq. (C.64) to get the normal-mode frequencies. But, as they say, that's not important now.

Rewrite Eq. (C.63),

$$L = \frac{1}{2}\sum_n a\left[\frac{m}{a}\dot{\phi}_n^2 - ka\left(\frac{\phi_n - \phi_{n-1}}{a}\right)^2\right] \equiv \sum_n aL_n . \tag{C.65}$$

Now let $a \to 0$ (while the number of masses $N \to \infty$) such that there is a uniform mass per length $m/a \to \rho$ and a string tension $ka \to T$. We also generalize the notion of displacement, away from its definition at discrete positions $x = na$, to a continuous *function*, $\phi(x = na, t) \to \phi(x, t)$ as $a \to 0$. The *displacement field* $\phi(x, t)$ is the elastic distortion of the medium at position x at time t. We can also think of the process $a \to 0$ on the right side of Eq. (C.65) as replacing a with dx, and converting the sum to an integral,

$$L \to \frac{1}{2}\int dx\left[\rho\left(\frac{\partial\phi}{\partial t}\right)^2 - T\left(\frac{\partial\phi}{\partial x}\right)^2\right] \equiv \int \mathscr{L}dx , \tag{C.66}$$

where we've replaced $(\phi_n - \phi_{n-1})/a$ with $\partial\phi/\partial x$ as $a \to 0$ and we've defined the *Lagrangian density* for a one-dimensional elastic medium,

$$\mathscr{L} = \frac{1}{2}\left[\rho\left(\frac{\partial\phi}{\partial t}\right)^2 - T\left(\frac{\partial\phi}{\partial x}\right)^2\right] . \tag{C.67}$$

Whenever we have a continuous system we work with a Lagrangian density, $L = \int \mathscr{L}dx$. In three dimensions, $L = \int \mathscr{L}dxdydz$. The dimension of $\mathscr{L}$ in d dimensions is energy/(length)^d.

What are the generalized coordinates for a continuous system? It's *not* the position coordinate x and it's useful to keep in mind the process by which we arrived at the Lagrangian density. The discrete index n "becomes" the position coordinate x in the transition from a discrete to a continuous system. The generalized coordinates that specify the *configuration* of the system, the displacements $\{\phi_n\}$ in the discrete case, become the displacement *field* $\phi(x)$ for the continuous system. It might seem there is only one generalized coordinate for the continuous system, the field $\phi(x, t)$, but that's misleading: The field ϕ represents the displacement at an *infinite* number of positions, those labeled by the variable x. We won't use Lagrangian densities in this book; we bring up the topic to illustrate the passage from a system of discrete dynamical variables to a field as a dynamical quantity.

Quantum mechanics of identical particles

A tenet of quantum mechanics is that identical particles (same mass, charge, spin, etc.) *are indistinguishable*. Classically, particles are unambiguously located by their spatial positions. In quantum mechanics, particles are not characterized by their positions, but rather by their *wavefunctions* $\psi(x)$ specifying the probability of finding a particle at a given position. Consider two identical waves on a rope propagating towards each other. As they collide, do they pass through each other or do they bounce off each other? Can one say, by means of measurements, which "particle" (wave) is which? Quantum mechanics asserts not: What we can know about a system is what we can measure. We have the non-classical idea that *identical particles cannot be distinguished by measurement*, i.e., identical particles are indistinguishable. The impact of this *one piece of physics* is that states in which identical particles might conceivably be distinguished *do not occur*, which affects our ability to count configurations of particles, a basic task of statistical mechanics.[1] As an example, a description of two identical coins, one in state "heads" and the other in state "tails," do not occur in quantum mechanics because it implies that the particles (coins) could be distinguished. Rather, a characteristic quantum description would be (heads-tails + tails-heads); the quantum description is such that all we know is there is one coin in state heads and one in state tails, but we cannot say which coin is in which state. In this appendix we review the quantum mechanics of identical particles.

D.1 WAVEFUNCTIONS OF IDENTICAL-PARTICLE SYSTEMS HAVE DEFINITE SYMMETRY UNDER PERMUTATIONS OF PARTICLES

Measurable quantities of stationary quantum systems are represented by expectation values of Hermitian operators $\hat{A}$, $\langle A \rangle \equiv \langle \psi | \hat{A} | \psi \rangle$. The quantum state of an N-particle system is characterized by a complex-valued, square-integrable multivariable function, $\psi(x_1, \cdots, x_N)$. Consider (for a system of identical particles), whether there are observable consequences of interchanging the positions of two particles. Imagine a "snapshot" (photograph) of a system of identical particles. With each particle instantaneously frozen in place, mentally swap two particles. Can there be any physical consequences of such an interchange? We assert not.[2] The mathematical expression of the indistinguishability of identical particles is that expectation values are unaltered by the interchange of particle coordinates in the many-body wavefunction. If we *could* find expectation values not invariant under particle exchange, we would have found measurable quantities that distinguish identical particles.

---

[1]Gibbs noted, as early as 1901, that if two particles are regarded as indistinguishable, "it seems in accordance with the spirit of the statistical method" to regard them as identical; see Section 2.3

[2]This conclusion applies whether or not the particles are interacting.

Consider the (normalized) N-particle wavefunction $\psi(x_1, \cdots, x_N)$. Just as in single-particle quantum mechanics, $|\psi(x_1, \cdots, x_N)|^2 \, d^3x_1 \cdots d^3x_N$ is the probability of finding particle 1 in volume d^3x_1, particle 2 in volume d^3x_2, up to the probability of finding particle N in d^3x_N. One might question that if the particle positions $x_1, \cdots, x_N$ cannot be distinguished, why base the theory on ill-defined quantities? In the following sections we "bootstrap" ourselves away from a wavefunction based on positions to the *occupation number formalism*, which is devoid of coordinate labels.

We require, for any possible state of the system ψ, for all observables $\hat{A}$, that under the interchange of any two particles (j, k),

$$\int d^N x \, \psi^*(x_1, \cdots, x_j, \cdots, x_k, \cdots, x_N) \hat{A} \psi(x_1, \cdots, x_j, \cdots, x_k, \cdots, x_N)$$
$$= \int d^N x \, \psi^*(x_1, \cdots, x_k, \cdots, x_j, \cdots, x_N) \hat{A} \psi(x_1, \cdots, x_k, \cdots, x_j, \cdots, x_N), \qquad (D.1)$$

where $\int d^N x = \int dx_1 \int dx_2 \cdots \int dx_N$, with $\int dx \equiv \sum_s \int d^3 r$, and where $\sum_s$ indicates a sum over spin states. Thus, we're attributing to identical particles spatial degrees of freedom and spin degrees of freedom.[3] Equation (D.1) is the mathematical expression of a physical idea; it's remarkable how much of the mathematical structure of the quantum theory of identical particles emerges from this one physical requirement.

We need specify that Eq. (D.1) holds only under the interchange of *two* particles, because an arbitrary permutation can be realized from a sequence of two-particle interchanges.[4] If we define an interchange operator

$$\hat{P}_{jk} \psi(x_1, \cdots, x_j, \cdots, x_k, \cdots, x_N) \equiv \psi(x_1, \cdots, x_k, \cdots, x_j, \cdots, x_N),$$

then an arbitrary permutation can be expressed as a succession of interchanges, $\hat{P} = \prod_{jk} \hat{P}_{jk}$. (Clearly, $\hat{P}_{jk} = \hat{P}_{kj}$.) Example: $\hat{P}_{jk} = \hat{P}_{1j} \hat{P}_{2k} \hat{P}_{12} \hat{P}_{2k} \hat{P}_{1j}$ (try it!). Applying the operator $\hat{P}_{jk}$ twice restores the original order, so that

$$\hat{P}_{jk} \hat{P}_{jk} = \hat{I} \implies \hat{P}_{jk}^{-1} = \hat{P}_{jk}, \qquad (D.2)$$

where $\hat{I}$ is the identity operator.

Writing Eq. (D.1) in Dirac notation,

$$\langle \psi | \hat{A} | \psi \rangle = \langle \hat{P}_{jk} \psi | \hat{A} | \hat{P}_{jk} \psi \rangle \equiv \langle \psi | \hat{P}_{jk}^\dagger \hat{A} \hat{P}_{jk} | \psi \rangle, \qquad \text{for all } (j, k) \qquad (D.3)$$

which defines the adjoint operator $\hat{P}_{jk}^\dagger$. The adjoint of any operator $\hat{A}$ (not necessarily Hermitian) is a new operator, denoted $\hat{A}^\dagger$, so that $\langle \hat{A}^\dagger \phi | \psi \rangle \doteq \langle \phi | \hat{A} \psi \rangle$ holds for any functions ϕ and ψ (that are elements of the Hilbert space of square-integrable functions). Equation (D.3) is not quite in the form required to define $\hat{P}_{jk}^\dagger$. For any vectors ϕ and ψ, the *polarization identity* holds (show this)

$$4 \langle \phi | \hat{A} | \psi \rangle = \langle \phi + \psi | \hat{A} | \phi + \psi \rangle - \langle \phi - \psi | \hat{A} | \phi - \psi \rangle - i \langle \phi + i\psi | \hat{A} | \phi + i\psi \rangle + i \langle \phi - i\psi | \hat{A} | \phi - i\psi \rangle. \quad (D.4)$$

Applying Eq. (D.3) to each of the terms in Eq. (D.4) we conclude that

$$\langle \phi | \hat{A} | \psi \rangle = \langle \phi | \hat{P}_{jk}^\dagger \hat{A} \hat{P}_{jk} | \psi \rangle \qquad (D.5)$$

for all (j, k) and for arbitrary functions ϕ and ψ. Equation (D.5) implies the operator identity

$$\hat{A} = \hat{P}_{jk}^\dagger \hat{A} \hat{P}_{jk}. \qquad (D.6)$$

---

[3] We argue in Section 5.5 that in discussing the quantum mechanics of identical particles, there is no need to include internal degrees of freedom such as rotation or vibration.

[4] See, for example, Halmos[163, p44].

In particular, for $\hat{A} = \hat{I}$, Eq. (D.6) implies $\hat{I} = \hat{P}_{jk}^{\dagger}\hat{P}_{jk}$, so that $\hat{P}_{jk}^{-1} = \hat{P}_{jk}^{\dagger}$. Thus, from Eq. (D.2),

$$\hat{P}_{jk}^{-1} = \hat{P}_{jk} = \hat{P}_{jk}^{\dagger} . \tag{D.7}$$

The operator $\hat{P}_{jk}$ is therefore Hermitian and unitary. For an arbitrary permutation $\hat{P} = \prod_{jk} \hat{P}_{jk}$, it follows that $\hat{P}^{\dagger} = \hat{P}^{-1}$ because $\hat{P}^{\dagger}$ is the product of the $\hat{P}_{jk}^{\dagger} = \hat{P}_{jk}$ applied in the reverse order.

If we let $\hat{P}_{jk}$ operate on Eq. (D.6) from the left and use Eq. (D.7), we find that $\hat{P}_{jk}$ commutes with operators representing physical observables, $\hat{P}_{jk}\hat{A} = \hat{A}\hat{P}_{jk}$, or

$$\left[\hat{A}, \hat{P}_{jk}\right] = 0 . \tag{D.8}$$

Because $\hat{A}$ commutes with any of the $\hat{P}_{jk}$, it commutes with an arbitrary permutation, $[\hat{A}, \hat{P}] = 0$. In particular, Eq. (D.8) holds for $\hat{A} = \hat{H}$, and thus $\hat{P}_{jk}$ commutes with the Hamiltonian,

$$\left[\hat{H}, \hat{P}_{jk}\right] = 0 . \tag{D.9}$$

Commuting operators have common sets of eigenfunctions,[5] and thus solutions of the Schrödinger equation are eigenfunctions of $\hat{P}_{jk}$. This is an extremely important point. *Every solution of the many-particle Schrödinger equation (for identical particles) is a state of definite permutation symmetry.* Because the particles are identical, the eigenvalues of *all* the $\hat{P}_{jk}$ must be the same (for a given set of particles). Denote the eigenvalues of $\hat{P}_{jk}$ as λ_{jk}: $\hat{P}_{jk}\psi = \lambda_{jk}\psi$. It follows that $\psi = \hat{P}_{jk}^2\psi = \lambda_{jk}^2\psi$, implying[6] $\lambda_{jk} = \pm 1$. Using the identity $\hat{P}_{jk} = \hat{P}_{1j}\hat{P}_{2k}\hat{P}_{12}\hat{P}_{2k}\hat{P}_{1j}$,

$$\hat{P}_{jk}\psi = \hat{P}_{1j}\hat{P}_{2k}\hat{P}_{12}\hat{P}_{2k}\hat{P}_{1j}\psi = \lambda_{1j}^2\lambda_{2k}^2\lambda_{12}\psi = \lambda_{12}\psi .$$

Thus, if ψ is an eigenfunction of $\hat{P}_{jk}$, its eigenvalues $\lambda_{jk} = \lambda_{12}$ for all (j, k), i.e., the eigenvalues of all operators $\hat{P}_{jk}$ are the same and depend on the nature of the wavefunction. If under one exchange $\lambda_{12} = +1(-1)$, then $\lambda_{jk} = +1(-1)$ for *all* exchanges. This must be the case for *identical* particles! We're led to the following definitions:

- If $\hat{P}_{jk}\psi = \psi$ for all (j, k), ψ is called *symmetric* and is denoted ψ_S.

- If $\hat{P}_{jk}\psi = -\psi$ for all (j, k), ψ is called *antisymmetric* and is denoted ψ_A.

Wavefunctions of systems of identical particles are therefore a special kind of function: *They have definite symmetry under the permutation of particle labels.* In the *Heisenberg representation* of quantum mechanics,[5] Eq. (D.9) indicates that $\hat{P}_{jk}$ *is a constant of the motion*—that the symmetry properties of the state vector of the system, if initially symmetric (antisymmetric), remains symmetric (antisymmetric) as the system evolves. We now come to an important point: *Symmetric and antisymmetric wave functions are orthogonal,*

$$\langle\psi_A|\psi_S\rangle = \langle\psi_A|\hat{P}_{jk}\psi_S\rangle = \langle\psi_A|\hat{P}_{jk}^{\dagger}\psi_S\rangle = \langle\hat{P}_{jk}\psi_A|\psi_S\rangle = -\langle\psi_A|\psi_S\rangle \implies \langle\psi_A|\psi_S\rangle = 0 .$$

The Hilbert space of state vectors for a given system of identical particles can contain only symmetric or antisymmetric wavefunctions.[7] Which type of space the state vector resides in depends on the nature of the particles involved. The *spin-statistics theorem*[8] classifies particles as *bosons* (symmetric wavefunctions), which have integer spin (integer multiples of $\hbar$), and *fermions* (antisymmetric wavefunctions) have half-integer spin (in units of $\hbar$).

---

[5] See most any text on quantum mechanics.

[6] $\hat{P}_{jk}$ is Hermitian and unitary and thus it has real eigenvalues of unit magnitude.

[7] A more careful statement is that the Hilbert space of all *possible* many-particle quantum states (of identical particles) partitions into orthogonal subspaces of symmetric and antisymmetric states.

[8] A proof of the spin-statistics theorem, which requires relativistic quantum mechanics and which is clearly outside the scope of this book, is given in Schwabl[164, p296].

D.2 SYMMETRIZATION AND ANTISYMMETRIZATION OPERATORS

Define the *symmetrization operator*, $\hat{S}$, and the *antisymmetrization operator*, $\hat{A}$,

$$\hat{S} \equiv \sum_{P \in S_N} \hat{P} \qquad \hat{A} \equiv \sum_{P \in S_N} \operatorname{sgn}(P)\hat{P},$$

where S_N denotes the *permutation group*. A *group* is a collection (set) of objects that are closed under an operation that combines two elements of the set to produce another element of the set. A group must have an identity element, and every operation defined on the group must have an inverse operation that's also a member of the set. The permutation group S_N is the set of $N!$ permutations of N objects. Example: the group S_2 of permutations of two objects has $(1, 2)$ as the identity permutation, and the permutation $(2, 1)$, with the inverse permutation applying the swap of $(2, 1)$ into $(1, 2)$, the identity. Below, we make use of the group property of S_N that a permutation followed by a permutation is itself a permutation. The quantity $\operatorname{sgn}(P)$ denotes the sign (*signum*) of the permutation, defined as $\operatorname{sgn}(P) = 1\,(-1)$ if $\hat{P}$ contains an even (odd) number of two-particle exchanges. A given permutation can be realized in many ways through a succession of two-particle interchanges, but the *parity* of a permutation is an invariant—an even or an odd number, irrespective of how the permutation is created.[9]

Let $f(x_1, \cdots, x_N)$ be an arbitrary function of N variables. We can use $\hat{S}$ and $\hat{A}$ to construct new functions that are totally symmetric (antisymmetric) functions of their arguments,

$$\psi_S(x_1, \cdots, x_N) \equiv \hat{S}f(x_1, \cdots, x_N) \qquad \psi_A(x_1, \cdots, x_N) \equiv \hat{A}f(x_1, \cdots, x_N). \quad \text{(D.10)}$$

Example. Starting with a function of two variables, $f(x_1, x_2)$, we can construct symmetric and antisymmetric functions using $\hat{S}$ and $\hat{A}$:

$$\psi_S(x_1, x_2) = \hat{S}f(x_1, x_2) = f(x_1, x_2) + f(x_2, x_1)$$
$$\psi_A(x_1, x_2) = \hat{A}f(x_1, x_2) = f(x_1, x_2) - f(x_2, x_1).$$

Let's verify that $\hat{S}$ and $\hat{A}$ do their intended jobs. For any $\hat{P}_{jk}$,

$$\hat{P}_{jk}\hat{S} = \sum_{P \in S_N} \hat{P}_{jk}\hat{P} = \sum_{P' \in S_N} \hat{P}' = \hat{S},$$

and thus $\hat{P}_{jk}\psi_S = \psi_S$. We've used the group property that $\hat{P}_{jk}\hat{P} = \hat{P}'$, where $\hat{P}'$ is another element of S_N. Because we're summing over *all* permutations, whatever permutation is produced by $\hat{P}_{jk}\hat{P}$ is included in the sum. The antisymmetrization operator is trickier. We have

$$\hat{P}_{jk}\hat{A} = \sum_{P \in S_N} \operatorname{sgn}(P)(\hat{P}_{jk}\hat{P}) = \sum_{P \in S_N} \left[-\operatorname{sgn}(\hat{P}_{jk}\hat{P}) \right] \hat{P}_{jk}\hat{P} = -\sum_{P' \in S_N} \operatorname{sgn}(\hat{P}')\hat{P}' = -\hat{A},$$

so that $\hat{P}_{jk}\psi_A = -\psi_A$. We've invoked the property that for each $P \in S_N$ it's true that $\operatorname{sgn}(\hat{P}_{jk}\hat{P}) = -\operatorname{sgn}(\hat{P})$, that whatever is $\operatorname{sgn}(P)$, adding one more permutation reverses the sign. Thus we have a recipe to *construct* symmetrized and antisymmetrized functions $\psi_S = \hat{S}f$ and $\psi_A = \hat{A}f$ such that

$$\hat{P}_{jk}\psi_S = \psi_S \qquad \hat{P}_{jk}\psi_A = -\psi_A. \quad \text{(D.11)}$$

---

[9]That is, no matter how many different ways of realizing a given permutation through a succession of two-particle interchanges, all such permutations have an even or an odd number of interchanges. See for example [163, p46].

For an arbitrary permutation $\hat{P} = \prod_{jk} \hat{P}_{jk}$,

$$\hat{P}\psi_S = \psi_S \qquad\qquad \hat{P}\psi_A = \text{sgn}(P)\psi_A . \qquad\qquad\text{(D.12)}$$

By this construction, ψ_S and ψ_A are symmetrized and antisymmetrized functions, but they're not necessarily solutions of the many particle Schrödinger equation. We want symmetrized and anti-symmetrized solutions of the Schrödinger equation, however, the topic of Section D.4.

Because $[\hat{H}, \hat{P}_{jk}] = 0$, Eq. (D.9), any solution of the Schrödinger equation (eigenfunction of $\hat{H}$) is also an eigenfunction of the permutation operators $\hat{P}_{jk}$, for any (j, k). Permutations, however, do not commute (example: $\hat{P}_{12}\hat{P}_{23} \neq \hat{P}_{23}\hat{P}_{12}$). It would seem then that we're not guaranteed there are functions that are simultaneously eigenfunctions of all $\hat{P}_{jk}$. Yet, we've just shown how to construct symmetric and antisymmetric functions having the requisite symmetry under all conceivable permutations. A further discussion of this point requires a new mathematical tool.

D.3 MATHEMATICAL DIGRESSION: THE TENSOR PRODUCT

The state of a many-particle quantum system is built up out of information from the individual quantum systems of its parts. This is implicit, for example, in the many-particle wavefunction with $|\psi(x_1, \cdots, x_N)|^2 \, d^3x_1 \cdots d^3x_N$ the joint probability of finding particle 1 in $d^3x_1, \cdots$, and particle N in d^3x_N. States of quantum systems are represented as vectors in a Hilbert space, $\mathcal{H}$. We need a mathematical tool that allows us to combine the Hilbert spaces for each subsystem, $\mathcal{H}_i$, into a Hilbert space for the composite system, $\mathcal{H}$. The *tensor product* is just such a vehicle.

Let V_1 and V_2 be vector spaces, with basis vectors $\psi_i \in V_1$, $\phi_i \in V_2$. Consider sums of the form

$$\chi \equiv \sum_{ij} A_{ij} \psi_i \phi_j , \qquad\qquad\text{(D.13)}$$

where A_{ij} are constants. Such quantities form a vector space: For $\chi^{(1)} \equiv \sum_{ij} A_{ij}^{(1)} \psi_i \phi_j$ and $\chi^{(2)} \equiv \sum_{ij} A_{ij}^{(2)} \psi_i \phi_j$ the sum is of the same form, $\chi^{(1)} + \chi^{(2)} = \sum_{ij} D_{ij} \psi_i \phi_j$ where $D_{ij} \equiv A_{ij}^{(1)} + A_{ij}^{(2)}$. The space of all sums of the form of Eq. (D.13)—the *tensor product space*—is denoted $V_1 \otimes V_2$. Basis vectors for $V_1 \otimes V_2$ are denoted $\psi_i \otimes \phi_j$. The dimension of the tensor product space is therefore the product of the dimensions of its constituent spaces. Tensor products can be likened to the menu in a restaurant where one constructs a meal from by ordering one item from column A and another from column B: Choose one item (vector) from column A (space V_1), one from column B (space V_2). The set of all combinations of items (vectors) from columns A and B (spaces V_1 and V_2) comprises the space of all possible meals (tensor products). While we will be mainly interested in identical subsystems, the spaces V_1 and V_2 making up the tensor product need not be associated with identical systems. Consider a molecule where the combined quantum system consists of rotational and vibrational states. If, however, V_1 and V_2 *are* copies of the same vector space, a common notation for $V \otimes V$ is $V^{\otimes 2}$. More generally, the tensor product of N copies of the same space is denoted $V \otimes \cdots \otimes V \equiv V^{\otimes N}$.

Products of wavefunctions arise naturally when we combine quantum systems. To take a simple example, suppose the Hamiltonian, $\hat{H}$, of the composite system that results from combining subsystems 1 and 2 can be written in separable form, $\hat{H} = \hat{H}_1 + \hat{H}_2$, where solutions to the Schrödinger equations for each subsystem are known, with $\hat{H}_1\psi = E_1\psi$ and $\hat{H}_2\phi = E_2\phi$. Then the wavefunction for the combined system is $\psi\phi$, with $\hat{H}\psi\phi = E\psi\phi$, with $E = E_1 + E_2$ (show this).

If $\hat{A}_1$ ($\hat{A}_2$) is a linear operator on V_1 (V_2) then operators $\hat{C} \equiv \hat{A}_1 \otimes \hat{I}$ and $\hat{B} \equiv \hat{I} \otimes \hat{A}_2$ are defined on $V_1 \otimes V_2$ as follows: (where χ is defined in Eq. (D.13))

$$\hat{C}\chi \equiv \sum_{ij} A_{ij}(\hat{A}_1\psi_i) \otimes \phi_j \qquad\qquad \hat{B}\chi \equiv \sum_{ij} A_{ij}\psi_i \otimes (\hat{A}_2\phi_j) .$$

The action of a linear operator $\hat{A} \equiv \hat{A}_1 \otimes \hat{A}_2$ on $V_1 \otimes V_2$ is defined by

$$\hat{A}(\psi_i \otimes \phi_j) = (\hat{A}_1 \otimes \hat{A}_2)(\psi_i \otimes \phi_j) \equiv (\hat{A}_1 \psi_i) \otimes (\hat{A}_2 \phi_j) \,.$$

Basically the tensor product is a bookkeeping device that keeps track of which operator acts on the vectors from which space. If $\hat{A}_1, \hat{B}_1$ ($\hat{A}_2, \hat{B}_2$) are operators on V_1 (V_2), then

$$\hat{A}_1 \hat{B}_1 \otimes \hat{A}_2 \hat{B}_2 = (\hat{A}_1 \otimes \hat{A}_2)(\hat{B}_1 \otimes \hat{B}_2) \,.$$

Finally, we have the important result that inner products in $V_1 \otimes V_2$ factor among the inner products in the respective spaces,

$$\langle \chi^{(1)} | \chi^{(2)} \rangle = \sum_{ij} \sum_{lm} \langle A_{ij} \psi_i \phi_j | B_{lm} \psi_l \phi_m \rangle \equiv \sum_{ijlm} A_{ij}^* B_{lm} \langle \psi_i | \psi_l \rangle_1 \langle \phi_j | \phi_m \rangle_2 \,.$$

Between basis vectors of $V_1 \otimes V_2$ the inner product is then given by $\langle \psi_i \otimes \phi_j | \psi_l \otimes \phi_m \rangle = \langle \psi_i | \psi_l \rangle_1 \langle \phi_j | \phi_m \rangle_2$.

D.4 THE PAULI PRINCIPLE

The indistinguishability of identical particles implies the many-body wavefunction is either symmetric or antisymmetric under exchange of coordinate labels (Section D.1). While we *know* that wavefunctions of identical-particle systems must be states of definite permutation symmetry, can we *calculate* many-body wavefunctions? Therein lies the rub. With very few exceptions we cannot obtain exact solutions of the Schrödinger equation for interacting particles. We must rely on perturbation theory to include the effects of inter-particle interactions. *Even at zeroth order*, however (that is, for noninteracting particles), the wavefunctions must exhibit the proper permutation symmetry.[10]

Consider the many-body Hamiltonian operator

$$\hat{H} = \sum_{i=1}^{N} \left(-\frac{\hbar^2}{2m} \nabla_i^2 + u(r_i) \right) + \frac{1}{2} \sum_{\substack{i,j=1 \\ (i \neq j)}}^{N} v(r_i, r_j) \equiv \hat{H}_0 + \hat{H}' \equiv \sum_{i=1}^{N} \hat{h}_i + \hat{H}' \,, \qquad \text{(D.14)}$$

where $u(r)$ is a *one-body potential* that couples to all particles (such as that derived from an applied electric field), and $v(r_1, r_2)$ is a *two-body potential*, such as the Coulomb interaction between charges. The Hamiltonian $\hat{H}_0$, for noninteracting identical particles, is a sum of identical single-particle Hamiltonians, $\hat{H}_0 \equiv \sum_{i=1}^{N} \hat{h}_i$. We assume that the single-particle quantum problem has been completely solved

$$\hat{h}\phi_\nu(x) = \epsilon_\nu \phi_\nu(x) \,, \qquad \text{(D.15)}$$

where ν denotes a *set* of quantum numbers that depend on the nature of the problem—the functions $\phi_\nu(x)$ could be plane waves for point particles, in which case ν would denote the allowed wavevectors k (see Section 2.1.5), or they could be hydrogenic wavefunctions for atoms, in which case ν would denote the set of quantum numbers (n, l, m). Note there is no index on $\hat{h}$ in Eq. (D.15)—it's the same operator for all particles. The set of eigenfunctions $\{\phi_\nu(x)\}$, being eigenfunctions of a Hermitian operator, are a complete, orthonormal set: $\langle \phi_\nu | \phi_{\nu'} \rangle = \delta_{\nu,\nu'}$ that one can use to express any square-integrable function, $f(x) = \sum_\nu c_\nu \phi_\nu(x)$, where the expansion coefficients $c_\nu = \langle \phi_\nu | f \rangle$.

Let's consider what the wavefunction of N noninteracting particles looks like before we impose the symmetrization requirement. The eigenfunction of the separable Hamiltonian $\hat{H}_0$ is the product of N single-particle wavefunctions

$$\psi_{\nu_1 \cdots \nu_N}(x_1, \cdots, x_N) = \phi_{\nu_1}(x_1) \otimes \phi_{\nu_2}(x_2) \otimes \cdots \otimes \phi_{\nu_N}(x_N) \,. \qquad \text{(D.16)}$$

---

[10]Let's say that again: Even noninteracting particles must exhibit permutation symmetry, and thus the permutation properties must be related to an intrinsic property of particles, which the spin-statistics theorem shows is the spin.

Just to be clear, the quantum numbers ν_i represent *any* of the quantum numbers available to a single particle. The quantity ν_1 for example does not necessarily represent the ground state; ν_1 could label the 27th energy state. The energy of the state with wavefunction Eq. (D.16) is found from

$$\hat{H}_0 \psi_{\nu_1 \cdots \nu_N}(x_1, \cdots, x_N) = \sum_i \hat{h}_i \left(\phi_{\nu_1}(x_1) \otimes \cdots \otimes \phi_{\nu_i}(x_i) \otimes \cdots \otimes \phi_{\nu_N}(x_N) \right)$$

$$= \left(\sum_i \epsilon_{\nu_i} \right) \psi_{\nu_1 \cdots \nu_N}(x_1, \cdots, x_N) .$$

To be exceedingly correct we should write $\hat{H}_0 = \sum_i \hat{I} \otimes \cdots \otimes \hat{h}_i \otimes \cdots \otimes \hat{I}$ as a sum of operators that act on the ith Hilbert space, $\mathcal{H}_i$, but act trivially on all others. Unless there is a possibility for confusion we will use $\hat{h}_i$ to denote the single-particle operator that acts on the space associated with particle i and leaves the other spaces alone. Likewise we could rig up some notation (but we won't) for the interaction Hamiltonian $\hat{H}'$ as an operator that acts on the spaces associated with particles (i, j), but acts trivially on the other spaces. The normalization of the wavefunction Eq. (D.16) is simple to compute:

$$\langle \psi_{\nu_1 \cdots \nu_N}(x_1, \cdots, x_N) | \psi_{\nu_1 \cdots \nu_N}(x_1, \cdots, x_N) \rangle = \langle \phi_{\nu_1}(x_1) \otimes \cdots \otimes \phi_{\nu_N}(x_N) | \phi_{\nu_1}(x_1) \otimes \cdots \otimes \phi_{\nu_N}(x_N)$$

$$= \langle \phi_{\nu_1} | \phi_{\nu_1} \rangle_1 \cdots \langle \phi_{\nu_N} | \phi_{\nu_N} \rangle_N = 1 .$$

This is a special case of the inner product between wavefunctions of the form of Eq. (D.16),

$$\langle \phi_{\mu_1}(x_1) \otimes \cdots \otimes \phi_{\mu_N}(x_N) | \phi_{\nu_1}(x_1) \otimes \cdots \otimes \phi_{\nu_N}(x_N) \rangle = \langle \phi_{\mu_1} | \phi_{\nu_1} \rangle_1 \cdots \langle \phi_{\mu_N} | \phi_{\nu_N} \rangle_N$$

$$= \delta_{\nu_1 \mu_1} \cdots \delta_{\nu_N \mu_N} .$$

To build in the required permutation symmetry, we apply the symmetrization and antisymmetrization operators $\hat{S}$ and $\hat{A}$ to wavefunctions of the form of Eq. (D.16). We now consider each in turn.

D.4.1 Symmetric wavefunctions of noninteracting particles

A totally symmetric wavefunction for noninteracting identical particles is obtained by *constructing it* through an application of the symmetrization operator $\hat{S}$ to a product of single-particle wavefunctions:

$$\Phi^{(S)}_{\nu_1 \cdots \nu_N} \equiv \frac{1}{\sqrt{N!}} \frac{1}{\sqrt{\prod_{k=1}^{N} n_k!}} \sum_{P \in S_N} \hat{P} \left(\phi_{\nu_1}(x_1) \otimes \phi_{\nu_2}(x_2) \otimes \cdots \otimes \phi_{\nu_N}(x_N) \right) . \tag{D.17}$$

The prefactors in Eq. (D.17) are such that $\Phi^{(S)}$ is normalized to unity, $\langle \Phi^{(S)}_{\nu_1 \cdots \nu_N} | \Phi^{(S)}_{\nu_1 \cdots \nu_N} \rangle = 1$. This can be seen as follows. There are $N!$ terms represented by the sum in Eq. (D.17) and thus $\langle \Phi^{(S)} | \Phi^{(S)} \rangle$ entails $(N!)^2$ integrals to evaluate. Fortunately most of them vanish. If all quantum numbers $\nu_1, \cdots, \nu_N$ are distinct, there are $N!$ cases where the quantum numbers "line up" between bra and ket, $\langle \phi_{\nu_1} \otimes \cdots \otimes \phi_{\nu_N} | \phi_{\nu_1} \otimes \cdots \otimes \phi_{\nu_N} \rangle = \langle \phi_{\nu_1} | \phi_{\nu_1} \rangle_1 \cdots \langle \phi_{\nu_N} | \phi_{\nu_N} \rangle_N = 1$; each such integral contributes a factor of unity to the normalization calculation. Integrals where the quantum numbers don't all line up, vanish. This explains the factor of $\sqrt{N!}$ in Eq. (D.17) . However, if the quantum numbers are *not* all distinct, with say the quantum number ν_k occuring n_k times in the list $\nu_1, \cdots, \nu_N$, there are $n_k!$ permutations of ν_k in *each* of the $N!$ non-vanishing integrals. As an example, suppose ν_i occurs twice in the list of quantum numbers, i.e., $n_i = 2$. Then, we have integrals of the form

$$\langle \phi_{\nu_1} \otimes \cdots \otimes \phi_{\nu_i} \otimes \cdots \otimes \phi_{\nu_i} \otimes \cdots \phi_{\nu_N} | \phi_{\nu_1} \otimes \cdots \otimes \phi_{\nu_i} \otimes \cdots \otimes \phi_{\nu_i} \otimes \cdots \phi_{\nu_N} \rangle .$$

In the sum over all permutations in Eq. (D.17) this integral will occur twice. In general there will be $n_k!$ non-vanishing terms contributing to the normalization calculation for each of the $N!$ terms where the quantum numbers line up on both sides of $\langle|\rangle$. This explains the factor of $\sqrt{\prod_{k=1}^{N} n_k!}$ in Eq. (D.17).

Example. Consider the case where quantum number ν_4 occurs twice in a symmetric wavefunction of $N = 3$ particles. From Eq. (D.17),

$$\Phi_{\nu_1\nu_4\nu_4}^{(S)} = \frac{1}{\sqrt{12}} \sum_{P \in S_3} \hat{P}\left(\phi_{\nu_1}(x_1)\phi_{\nu_4}(x_2)\phi_{\nu_4}(x_3)\right)$$

$$= \frac{1}{\sqrt{3}} \left[\phi_{\nu_1}(x_1)\phi_{\nu_4}(x_2)\phi_{\nu_4}(x_3) + \phi_{\nu_4}(x_1)\phi_{\nu_1}(x_2)\phi_{\nu_4}(x_3) + \phi_{\nu_4}(x_1)\phi_{\nu_4}(x_2)\phi_{\nu_1}(x_3)\right].$$

It's readily verified that this result is symmetric under $x_1 \leftrightarrow x_2$, $x_2 \leftrightarrow x_3$, and $x_1 \leftrightarrow x_3$, and moreover that $\langle \Phi_{\nu_1\nu_4\nu_4}^{(S)} | \Phi_{\nu_1\nu_4\nu_4}^{(S)} \rangle = 1$.

Note that we've not written $\Phi^{(S)}$ in the aforementioned example as a function of coordinates. This is because the identity of the particles is lost. All we can say is that there are particles in states ν_1, $\cdots$, ν_N, but we cannot say *which* particle is in a particular single-particle energy state. We might as well denote $\Phi_{\nu_1\nu_4\nu_4}$ as $|\nu_1\nu_4\nu_4\rangle$, which we're going to do in short order.

D.4.2   Antisymmetric wavefunctions of noninteracting particles

A totally antisymmetric wavefunction is constructed using the antisymmetrization operator $\hat{A}$ on products of single-particle wavefunctions,

$$\Phi_{\nu_1\cdots\nu_N}^{(A)} \equiv \frac{1}{\sqrt{N!}} \sum_{P \in S_N} \text{sgn}(P)\hat{P}\left(\phi_{\nu_1}(x_1) \otimes \phi_{\nu_2}(x_2) \otimes \cdots \otimes \phi_{\nu_N}(x_N)\right)$$

$$= \frac{1}{\sqrt{N!}} \begin{vmatrix} \phi_{\nu_1}(x_1) & \cdots & \phi_{\nu_1}(x_N) \\ \vdots & & \vdots \\ \phi_{\nu_N}(x_1) & \cdots & \phi_{\nu_N}(x_N) \end{vmatrix}. \tag{D.18}$$

The antisymmetrized wavefunction $\Phi^{(A)}$ a *Slater determinant*.[11] From the rules of determinants, $\Phi^{(A)} = 0$ if two quantum numbers are the same, e.g., if $\nu_i = \nu_j$ with $j \neq i$, because two rows of the determinant are the same. Likewise if $x_i = x_j$ with $j \neq i$ then $\Phi^{(A)} = 0$ because two columns of the determinant are the same. This brings us to the *Pauli principle* that: 1) It's impossible to have two fermions in the same state; one state can be occupied by at most one fermion, and 2) It's impossible to bring two fermions with the same spin projection to the same spatial point. This is implied because the spin quantum number is included in the set of quantum numbers denoted by ν_j. There is no such principle for bosons: An unlimited number of bosons can occupy the same state, and this is at the root of collective quantum phenomena such as Bose-Einstein condensation (see Section 5.9). The normalization factor in Eq. (D.18) is the same as that in Eq. (D.17) because for fermions, $n_k! = 1$.

---

[11]Determinants can be defined in several ways. One way to define the determinant of a square matrix A (with elements a_{ij}) is as the sum of all signed permutations of the matrix elements, $\det(A) \equiv \sum_{(i_1 i_2 \cdots i_n)} \pm a_{1 i_1} a_{2 i_2} \cdots a_{n i_n}$, where $(i_1 \cdots i_n)$ is a permutation of the reference sequence $i_1 \cdots i_n$ and $\pm$ is the sign of the permutation. The antisymmetrization of the product of single-particle wavefunctions is therefore a functional determinant with $a_{p i_p} = \phi_{i_p}(x_p)$. We don't have to make the "leap" (as is sometimes taught) that the antisymmetrized product of wavefunctions in Eq. (D.18) behaves *like* a determinant; it *is* a determinant. The reference sequence is chosen such that the quantum numbers occur in ascending order, $i_1 < i_2 < \cdots < i_n$.

The definition of $\Phi^{(A)}$ in Eq. (D.18) entails an ambiguity in *sign* because the order in which we write the rows (or columns) of the determinant is not specified. Consider the two wavefunctions:

$$\Phi^{(A_1)} = \frac{1}{\sqrt{2}} \begin{vmatrix} \phi_{\nu_1}(x_1) & \phi_{\nu_1}(x_2) \\ \phi_{\nu_4}(x_1) & \phi_{\nu_4}(x_2) \end{vmatrix} \qquad \Phi^{(A_2)} = \frac{1}{\sqrt{2}} \begin{vmatrix} \phi_{\nu_4}(x_1) & \phi_{\nu_4}(x_2) \\ \phi_{\nu_1}(x_1) & \phi_{\nu_1}(x_2) \end{vmatrix} .$$

Both are properly antisymmetric, yet $\Phi^{(A_1)} = -\Phi^{(A_2)}$. This sign ambiguity can be eliminated by agreeing to write the quantum numbers in *ascending order*, with $\nu_1 < \nu_2 < \cdots < \nu_N$. Such a convention is arbitrary of course, but once chosen if consistently employed leads to an unambiguous sign of Slater determinants. It will be convenient to symbolize the entire set of quantum numbers with a single label, e.g., $c \equiv (c_1, c_2, \cdots, c_N)$. With this notation, the antisymmetric wavefunction is given by

$$\Phi_c^{(A)} \equiv \frac{1}{\sqrt{N!}} \sum_{P \in S_N} \text{sgn}(P) \hat{P} \left(\phi_{c_1}(x_1) \otimes \cdots \otimes \phi_{c_N}(x_N) \right) . \qquad (c_1 < \cdots < c_N)$$

We now derive a useful result that holds for symmetric and antisymmetric wavefunctions, that for any operator $\hat{A}$

$$\langle \Phi_b^{(A/S)} | \hat{A} | \Phi_c^{(A/S)} \rangle = \frac{\sqrt{N!}}{\sqrt{\prod_k n_k^{(c)}!}} \langle \Phi_b^{(A/S)} | \hat{A} \left(\phi_{c_1}(x_1) \otimes \cdots \otimes \phi_{c_N}(x_N) \right) \rangle . \qquad (D.19)$$

We show Eq. (D.19) for the antisymmetric case explicitly. We have, using Eq. (D.18),

$$\langle \Phi_b | \hat{A} | \Phi_c \rangle = \frac{1}{\sqrt{N!}\sqrt{\prod_k n_k^{(c)}!}} \sum_{P \in S_N} \text{sgn}(P) \langle \Phi_b | \hat{A}\hat{P} \left(\phi_{c_1} \otimes \cdots \otimes \phi_{c_N} \right) \rangle . \qquad (D.20)$$

Because $\hat{A}$ commutes with any $\hat{P}$, and because $\hat{P}^\dagger = \hat{P}^{-1}$,

$$\langle \Phi_b | \hat{A}\hat{P} \left(\phi_{c_1} \otimes \cdots \otimes \phi_{c_N} \right) \rangle = \langle \Phi_b | \hat{P}\hat{A} \left(\phi_{c_1} \otimes \cdots \otimes \phi_{c_N} \right) \rangle = \langle \hat{P}^{-1}\Phi_b | \hat{A} \left(\phi_{c_1} \otimes \cdots \otimes \phi_{c_N} \right) \rangle$$

$$= \text{sgn}(P^{-1}) \langle \Phi_b | \hat{A} \left(\phi_{c_1} \otimes \cdots \otimes \phi_{c_N} \right) \rangle = \text{sgn}(P) \langle \Phi_b | \hat{A} \left(\phi_{c_1} \otimes \cdots \otimes \phi_{c_N} \right) \rangle ,$$

where we've used Eq. (D.12). Combining with Eq. (D.20),

$$\langle \Phi_b | \hat{A} | \Phi_c \rangle = \frac{1}{\sqrt{N!}\sqrt{\prod_k n_k^{(c)}!}} \sum_{P \in S_N} \left(\text{sgn}(P) \right)^2 \langle \Phi_b | \hat{A} \left(\phi_{c_1} \otimes \cdots \otimes \phi_{c_N} \right) \rangle$$

$$= \frac{1}{\sqrt{N!}\sqrt{\prod_k n_k^{(c)}!}} \sum_{P \in S_N} \langle \Phi_b | \hat{A} \left(\phi_{c_1} \otimes \cdots \otimes \phi_{c_N} \right) \rangle = \frac{N!}{\sqrt{N!}\sqrt{\prod_k n_k^{(c)}!}} \langle \Phi_b | \hat{A} \left(\phi_{c_1} \otimes \cdots \otimes \phi_{c_N} \right) \rangle ,$$

from which Eq. (D.19) follows.

D.5 COMPLETENESS

We now prove a result that is crucial for the edifice of the theory: Symmetric and antisymmetric wavefunctions of noninteracting particles form a *basis set* with which to express symmetric and antisymmetric wavefunctions of *interacting* particles. That is, arbitrary wavefunctions of interacting identical particles (showing the proper permutation symmetries) can be expanded in terms of the wavefunctions for noninteracting particles $\Phi^{(A/S)}$ developed in Section D.4.

To show this, start with an arbitrary function of N coordinates, $\psi(x_1, \cdots, x_N)$. If we hold $N-1$ coordinates fixed, with $x_j = x_j^0$, $j = 2, \cdots, N$, $\psi(x_1, x_2^0, \cdots, x_N^0)$ defines a function of a single variable that can be expanded in terms of the single-particle eigenstates $\{\phi_\nu(x)\}$,

$$\psi(x_1, x_2^0, \cdots, x_N^0) = \sum_{\nu_1} a_{\nu_1} \phi_{\nu_1}(x_1) , \qquad (D.21)$$

where the quantities $\{a_{\nu_1}\}$ are expansion coefficients. Such an expansion is guaranteed to exist because of the completeness of the functions $\{\phi_\nu(x)\}$. Using the orthonormality of the $\phi_\nu(x)$, the expansion coefficient in Eq. (D.21) is a function of the other $N-1$ coordinates,

$$a_{\nu_i} = \langle \phi_{\nu_i} | \psi(x_1, x_2^0, \cdots, x_N^0) \rangle_1 \equiv a_{\nu_i}(x_2^0, \cdots, x_N^0) \,.$$

We can then write Eq. (D.21) as

$$\psi(x_1, x_2^0, \cdots, x_N^0) = \sum_{\nu_1} a_{\nu_1}(x_2^0, \cdots, x_N^0) \phi_{\nu_1}(x_1) \,.$$

Now, erase the superscript on x_2^0 holding the other $N-2$ coordinates fixed, $x_j = x_j^0, j = 3, \cdots, N$, and expand again in the single-particle eigenstates (because $a_{\nu_1}(x_2, x_3^0, \cdots, x_N^0)$ is a function of a single variable)

$$a_{\nu_1}(x_2, x_3^0, \cdots, x_N^0) = \sum_{\nu_2} a_{\nu_1 \nu_2} \phi_{\nu_2}(x_2) \,,$$

where by the same reasoning $a_{\nu_1 \nu_2} = a_{\nu_1 \nu_2}(x_3^0, \cdots, x_N^0)$. At this point Eq. (D.21) can be written

$$\psi(x_1, x_2, x_3^0 \cdots, x_N^0) = \sum_{\nu_1} \sum_{\nu_2} a_{\nu_1 \nu_2}(x_3^0, \cdots, x_N^0) \phi_{\nu_1}(x_1) \phi_{\nu_2}(x_2) \,.$$

The process can be repeated until we have the expansion of an arbitrary function $\psi(x_1, \cdots, x_N)$ in terms of products of single-particle wavefunctions,

$$\psi(x_1, \cdots, x_N) = \sum_{\nu_1 \cdots \nu_N} a_{\nu_1 \cdots \nu_N} \phi_{\nu_1}(x_1) \cdots \phi_{\nu_N}(x_N) \,, \tag{D.22}$$

where the $a_{\nu_1 \cdots \nu_N} = \langle \phi_{\nu_1}(x_1) \cdots \phi_{\nu_N}(x_N) | \psi(x_1, \cdots, x_N) \rangle$ are constants.

The functions $\psi(x_1, \cdots, x_N)$ that we want are not *completely* arbitrary, however: The physics requires that wavefunctions of systems of identical particles have the proper symmetry under exchange of spatial coordinates. That can be achieved by imposing an additional structure on the *expansion coefficients* in Eq. (D.22). We require that $a_{\nu_1 \cdots \nu_N}$ has the desired symmetries of the wavefunction under permutation of *quantum numbers*. For, if we require that

$$\hat{P}_{jk} a_{\nu_1 \cdots \nu_j \cdots \nu_k \cdots \nu_N} = a_{\nu_1 \cdots \nu_k \cdots \nu_j \cdots \nu_N} = \pm a_{\nu_1 \cdots \nu_j \cdots \nu_k \cdots \nu_N} \,, \tag{D.23}$$

then the wavefunctions have the proper symmetry under two-particle exchanges:

$$\hat{P}_{jk} \psi(x_1 \cdots x_j \cdots x_k \cdots x_N) = \psi(x_1 \cdots x_k \cdots x_j \cdots x_N)$$
$$= \sum_{\nu_1 \cdots \nu_N} a_{\nu_1 \cdots \nu_j \cdots \nu_k \cdots \nu_N} \phi_{\nu_1}(x_1) \cdots \phi_{\nu_j}(x_k) \cdots \phi_{\nu_k}(x_j) \cdots \phi_{\nu_N}(x_N)$$
$$= \sum_{\nu_1 \cdots \nu_N} a_{\nu_1 \cdots \nu_k \cdots \nu_j \cdots \nu_N} \phi_{\nu_1}(x_1) \cdots \phi_{\nu_k}(x_k) \cdots \phi_{\nu_j}(x_j) \cdots \phi_{\nu_N}(x_N)$$
$$= \pm \sum_{\nu_1 \cdots \nu_N} a_{\nu_1 \cdots \nu_j \cdots \nu_k \cdots \nu_N} \phi_{\nu_1}(x_1) \cdots \phi_{\nu_j}(x_j) \cdots \phi_{\nu_k}(x_k) \cdots \phi_{\nu_N}(x_N) = \pm \psi(x_1 \cdots x_j \cdots x_k \cdots x_N) \,,$$

where in the second line we've used Eq. (D.22), in third line we've changed the dummy indices $\nu_j \leftrightarrow \nu_k$, and in the fourth line we've used Eq. (D.23). Under an arbitrary permutation of quantum numbers, $P(\nu_1 \cdots \nu_N)$, we require that

$$a_{P(\nu_1 \cdots \nu_N)} = \binom{+1}{\text{sgn}(P)} a_{\nu_1 \cdots \nu_N} \,, \tag{D.24}$$

where $+1$ is for bosons and $\text{sgn}(P)$ is for fermions.

The infinite sum in Eq. (D.22) is over all values of the quantum numbers $(\nu_1, \cdots, \nu_N)$. We can rewrite the sum in Eq. (D.22) as an infinite sum over *ordered* sets of quantum numbers $c \equiv (\nu_1, \cdots, \nu_N)$, combined with a finite sum over all permutations of the particular set of N quantum numbers $P(c) = P(\nu_1, \cdots, \nu_N)$,

$$\psi(x_1, \cdots, x_N) = \sum_c \sum_{P(c)} P\left(a_{\nu_1 \cdots \nu_N} \phi_{\nu_1}(x_1) \cdots \phi_{\nu_N}(x_N)\right).$$

Using Eq. (D.24) we have

$$\psi(x_1, \cdots, x_N) = \sum_c a_c \sum_{P \in S_N} \binom{+1}{\text{sgn}(P)} P\left(\phi_{\nu_1}(x_1) \cdots \phi_{\nu_N}(x_N)\right) \equiv \sum_c f_c \Phi_c^{(S/A)}(x_1, \cdots, x_N),$$

where the coefficient f_c accounts for the normalization factors in $\Phi^{(S/A)}$. The normalization of ψ requires that $\sum_c |f_c|^2 = 1$. We can conclude, finally, that because the single-particle eigenstates ϕ_ν are a complete orthonormal set of functions, *so too are the wavefunctions of noninteracting identical particles* $\Phi^{(S/A)}$, *and can serve as basis functions in the Hilbert space of many-body wavefunctions*.

D.6 FURTHER MATHEMATICAL DIGRESSION: FOCK SPACE

There is an additional technical nicety on combining vector spaces that we require, beyond the tensor product (Section D.3). We first introduce the idea of the *Cartesian product*, which for two sets C and D is a new set $C \times D$ consisting of all *ordered pairs* (c, d) where $c \in C$ and $d \in D$. Let V and W be vector spaces, which need not have the same dimensions. The Cartesian product $V \times W$ can be given a vector space structure: Add ordered pairs to produce new ordered pairs, $(v_1, w_1) + (v_2, w_2) \equiv (v_1 + v_2, w_1 + w_2)$, and define scalar multiplication of an ordered pair to produce a new ordered pair, $\alpha(v, w) \equiv (\alpha v, \alpha w)$ where $v \in V$, $w \in W$, and α is a scalar. The resulting vector space is called the *direct sum* of V and W, written $V \oplus W$. The dimension of $V \oplus W$ is equal to the sum of the dimensions of V and W. For example, the real vector space $\mathbb{R}^2$, the xy plane $\{(x, y) : x, y \in \mathbb{R}\}$ is the direct sum of the x-axis $\{(x, 0) : x \in \mathbb{R}\}$ and the y-axis $\{(0, y) : y \in \mathbb{R}\}$, $\mathbb{R}^2 = \mathbb{R} \oplus \mathbb{R}$.

The physical interpretation of the direct sum differs from that of the tensor product: Each describes a different way of combining quantum systems. The tensor product space $\mathcal{H} \otimes \mathcal{H}$ contains all possible entanglements of the states of the two individual particles. Note that we don't have to have interactions between the separate components, the quantum states of even noninteracting identical particles have new features (states of permutation symmetry) *not contained in the behavior of individual particles*; the whole *is* different than the sum of its parts. The direct sum describes superpositions of a quantum entity "exploring" separate systems. For example, let the spaces V_1 and V_2 contain the quantum states of a particle in two weakly coupled quantum wells. The most general state of the combined system $V_1 \oplus V_2$ is (v_1, v_2) with $v_1 \in V_1$ and $v_2 \in V_2$. This state can be regarded as the superposition of the state $(v_1, 0)$ (in which the particle is certainly in the first well) with the state $(0, v_2)$ (in which the particle is certainly in the second well). In such a state the particle is not definitely in either well.

These two ways of combining the vector spaces associated with quantum systems (tensor product and direct sum) can be brought together in describing a system in which the number of particles ("particles" being considered now as quantum entities) is *not a fixed number*. Such situations arise naturally in relativistic quantum mechanics, in which processes such as pair production occur. In nonrelativistic quantum mechanics, however, there are systems in which the number of particles is not fixed, such as electrons interacting with phonons, or in the grand canonical ensemble of statistical mechanics. What is the mathematical arena for such systems? If $\mathcal{H}$ is the Hilbert space of single-particle quantum states, then $\mathcal{H} \otimes \mathcal{H}$ is the vector space of all two-particle states, and $\mathcal{H} \otimes \mathcal{H} \otimes \mathcal{H}$ is the vector space of all three-particle states, etc. Suppose we have a system in which by some means we could have one, two, or three particles. The vector space containing all possible states of such a system is $\mathcal{H} \oplus (\mathcal{H} \otimes \mathcal{H}) \oplus (\mathcal{H} \otimes \mathcal{H} \otimes \mathcal{H})$. This space is the direct sum of spaces each containing states with a *definite* number of particles. We can now generalize this idea to a grand vector space capable of describing the states of systems with an *indefinite* number of particles. Define *Fock space* (after V.A. Fock who introduced the idea in 1932)

$$\mathcal{F} \equiv \mathbb{C} \oplus \mathcal{H} \oplus (\mathcal{H} \otimes \mathcal{H}) \oplus (\mathcal{H} \otimes \mathcal{H} \otimes \mathcal{H}) \oplus \cdots ,$$

where $\mathbb{C}$ is the space of all complex numbers, the Hilbert space of *zero* particles, the vacuum. Fock space can be defined using a very succinct notation, $\mathcal{F} \equiv \oplus_{n=0}^{\infty} \mathcal{H}^{\otimes n}$. Such a definition, however, does not take into account the permutation symmetry of identical particles. This is corrected by defining *two* Fock spaces as

$$\mathcal{F}_{\pm} \equiv \oplus_{n=0}^{\infty} \left(\hat{S}_{\pm} \mathcal{H}^{\otimes n} \right) = \mathbb{C} \oplus \mathcal{H} \oplus \left(\hat{S}_{\pm} (\mathcal{H} \otimes \mathcal{H}) \right) \oplus \left(\hat{S}_{\pm} (\mathcal{H} \otimes \mathcal{H} \otimes \mathcal{H}) \right) \oplus \cdots ,$$

where $\hat{S}_{\pm}$ denotes the symmetrization (antisymmetrization) operators defined in Section D.2, $\hat{S}_{+} \equiv \hat{S}$ ($\hat{S}_{-} \equiv \hat{A}$). The quantity $\mathcal{F}_{+}$ ($\mathcal{F}_{-}$) is called the symmetric (antisymmetric) Fock space. The spaces $\mathcal{F}_{\pm}$ are actually orthogonal subspaces of $\mathcal{F}$.

What is the dimension of Fock space? Best not to dwell on that question. Even in the simplest case of a single localizable particle (and thus one whose energy eigenvalue spectrum is discrete), the dimension of $\mathcal{H}$ is denumerably infinite. What then is the dimension of $\mathcal{H} \otimes \mathcal{H}$? Infinity squared? Now, what is the dimension of $\mathcal{F}$? Fock space is the "theater of all possibilities"—big enough to contain all possible states of all possible numbers of particles.

D.7 OCCUPATION-NUMBER FORMALISM FOR FERMIONS

D.7.1 Creation and annihilation operators

Starting in this section we use a simpler notation for the Hilbert space of N identical particles: Instead of $\mathcal{H}^{\otimes N}$ we use the less-imposing $\mathcal{H}(N)$. As discussed in Section D.6, we need a mathematical framework that allows us to treat systems where the number of particles is not fixed. This is accomplished with operators that map between Hilbert spaces of different particle numbers. We define operators $\hat{c}_k$ (*annihilation operator*) and $\hat{c}_k^{\dagger}$ (*creation operator*) such that

$$\hat{c}_k : \mathcal{H}(N) \to \mathcal{H}(N-1)$$
$$\hat{c}_k^{\dagger} : \mathcal{H}(N-1) \to \mathcal{H}(N) .$$

The operator $\hat{c}_k$ maps an element of $\mathcal{H}(N)$ into an element of $\mathcal{H}(N-1)$; it *removes* the single-particle state ϕ_k (when it exists) from a Slater determinant of N particles. Likewise, $\hat{c}_k^{\dagger}$ maps an element of $\mathcal{H}(N-1)$ into an element of $\mathcal{H}(N)$; it *adds* ϕ_k to a Slater determinant of $N-1$ particles (if ϕ_k is not already occupied). Each fermionic wavefunction (Slater determinant) Φ_d is labeled by a list of quantum numbers $d \equiv (\nu_1 \cdots \nu_N)$, where by definition the quantum numbers are listed in ascending order, $\nu_1 < \cdots < \nu_N$. To properly define $\hat{c}_k$ and $\hat{c}_k^{\dagger}$, this order must be respected as we add and remove rows to and from determinants.

The annihilation operator $\hat{c}_k$ is defined by the requirements that

$$\hat{c}_k \Phi_{\nu_1 \cdots \nu_N} = 0 \qquad k \notin (\nu_1 \cdots \nu_N) \tag{D.25}$$

In other words, Φ_d must contain the single-particle quantum state ϕ_k before it can be removed by $\hat{c}_k$. If, however, $k = \nu_j \in d$, the action of $\hat{c}_k$ is defined by

$$\hat{c}_k \Phi_{\nu_1 \cdots \nu_N}(x_1, \cdots, x_N) \equiv (-1)^{j-1} \Phi_{\nu_1 \cdots \nu_{j-1} \nu_{j+1} \cdots \nu_N}(x_1, \cdots, x_{N-1})$$

$$= \frac{(-1)^{j-1}}{\sqrt{(N-1)!}} \sum_{P \in S_{N-1}} \mathrm{sgn}(P)\hat{P}\left(\phi_{\nu_1}(x_1) \cdots \phi_{\nu_{j-1}}(x_{j-1})\phi_{\nu_{j+1}}(x_{j+1}) \cdots \phi_{\nu_N}(x_N)\right).$$

We have included the spatial coordinates here only for clarity. As mentioned in Section D.4, once we have symmetrized or antisymmetrized the products of single-particle wavefunctions, the notion of which particle is in which state is lost; spatial coordinates are superfluous. In this chapter we shift to the view that the many-body wavefunction Φ_d is a function solely of the quantum numbers $d = (\nu_1 \cdots \nu_N)$, without reference to spatial quantities. Thus, we can equally write the above equation as

$$\hat{c}_k \Phi_{\nu_1 \cdots \nu_N} \equiv (-1)^{j-1} \Phi_{\nu_1 \cdots \nu_{j-1} \nu_{j+1} \cdots \nu_N}, \qquad k = \nu_j \qquad \text{(D.26)}$$

Equations (D.25) and (D.26) define the action of $\hat{c}_k$. Below we define $\hat{c}_k^\dagger$ as the adjoint of $\hat{c}_k$.

What's with the sign convention in Eq. (D.26)? A determinant has an unambiguous sign only when the rows are written in a standard order, which we have taken to be the list of quantum numbers $\nu_1 \cdots \nu_N$ in ascending order. The quantum number that is eliminated, ν_j, has a definite position in that list. Suppose that ν_1 is removed from the list. By Eq. (D.26) the action of $\hat{c}_1$ is

$$\hat{c}_1 \frac{1}{\sqrt{N!}} \begin{vmatrix} \phi_{\nu_1}(x_1) & \cdots & \phi_{\nu_1}(x_N) \\ \phi_{\nu_2}(x_1) & \cdots & \phi_{\nu_2}(x_N) \\ \vdots & & \vdots \\ \phi_{\nu_N}(x_1) & \cdots & \phi_{\nu_N}(x_N) \end{vmatrix} = \frac{1}{\sqrt{(N-1)!}} \begin{vmatrix} \phi_{\nu_2}(x_1) & \cdots & \phi_{\nu_2}(x_{N-1}) \\ \vdots & & \vdots \\ \phi_{\nu_N}(x_1) & \cdots & \phi_{\nu_N}(x_{N-1}) \end{vmatrix}.$$

In this case we have effectively "crossed out" the top row of the determinant (to eliminate reference to ν_1) and the last column (to eliminate reference to x_N). If we agree that crossing out the top row and last column of a Slater determinant does not introduce a relative change in sign, then we can put any Slater determinant into that form by exchanging rows until the row containing the single-particle state to be eliminated appears in the top row.[12] Suppose that ν_j is to be eliminated. Then

$$\begin{vmatrix} \phi_{\nu_1}(x_1) & \cdots & \phi_{\nu_1}(x_N) \\ \vdots & & \vdots \\ \phi_{\nu_j}(x_1) & \cdots & \phi_{\nu_j}(x_N) \\ \vdots & & \vdots \\ \phi_{\nu_N}(x_1) & \cdots & \phi_{\nu_N}(x_N) \end{vmatrix} = (-1)^{j-1} \begin{vmatrix} \phi_{\nu_j}(x_1) & \cdots & \phi_{\nu_j}(x_N) \\ \phi_{\nu_1}(x_1) & \cdots & \phi_{\nu_1}(x_N) \\ \vdots & & \vdots \\ \phi_{\nu_{j-1}}(x_1) & \cdots & \phi_{\nu_{j-1}}(x_N) \\ \phi_{\nu_{j+1}}(x_1) & \cdots & \phi_{\nu_{j+1}}(x_N) \\ \vdots & & \vdots \\ \phi_{\nu_N}(x_1) & \cdots & \phi_{\nu_N}(x_N) \end{vmatrix}.$$

Having put the row with ϕ_{ν_j} at the top of the determinant, we can cross out the top row and last column to effect the action of $\hat{c}_k$

$$\hat{c}_k \frac{1}{\sqrt{N!}} \begin{vmatrix} \phi_{\nu_1}(x_1) & \cdots & \phi_{\nu_1}(x_N) \\ \vdots & & \vdots \\ \phi_{\nu_j}(x_1) & \cdots & \phi_{\nu_j}(x_N) \\ \vdots & & \vdots \\ \phi_{\nu_N}(x_1) & \cdots & \phi_{\nu_N}(x_N) \end{vmatrix} = \frac{(-1)^{j-1}}{\sqrt{(N-1)!}} \begin{vmatrix} \phi_{\nu_1}(x_1) & \cdots & \phi_{\nu_1}(x_{N-1}) \\ \vdots & & \vdots \\ \phi_{\nu_{j-1}}(x_1) & \cdots & \phi_{\nu_{j-1}}(x_{N-1}) \\ \phi_{\nu_{j+1}}(x_1) & \cdots & \phi_{\nu_{j+1}}(x_{N-1}) \\ \vdots & & \vdots \\ \phi_{\nu_N}(x_1) & \cdots & \phi_{\nu_N}(x_{N-1}) \end{vmatrix}.$$

---

[12] It is up to us to define the rules for mapping a Slater determinant for N particles into one for $N - 1$ particles. Merely setting the wavefunction to be eliminated to zero would make the entire determinant vanish.

The creation operator $\hat{c}_k^\dagger$ is the adjoint of $\hat{c}_k$. For $\Phi_d \in \mathcal{H}(N-1)$ and $\Phi_b \in \mathcal{H}(N)$,

$$
\langle \hat{c}_k^\dagger \Phi_d | \Phi_b \rangle = \langle \Phi_d | \hat{c}_k \Phi_b \rangle
$$

$$
= \begin{cases} (-1)^{j-1} \langle \Phi_d | \Phi_{b-k} \rangle = (-1)^{j-1} \delta_{d,b-k} & \text{if there is a } j \text{ with } k = b_j \\ 0 & \text{otherwise} \end{cases},
$$

where we have used the orthonormality between Φ_d and Φ_{b-k}, with the latter being the Slater determinant that results from $\hat{c}_k \Phi_b$. The quantity $\delta_{d,b-k}$ is equivalent to $\delta_{d+k,b}$. Remembering that d (b) is a list of $N-1$ (N) quantum numbers, define

$$
\bar{d} \equiv d + k = (d_1, \cdots, d_{j-1}, k, d_j, \cdots, d_{N-1})
$$

as a list of N quantum numbers with k such that $d_{j-1} < k < d_j$. We then have

$$
\langle \hat{c}_k^\dagger \Phi_d | \Phi_b \rangle = \begin{cases} (-1)^{j-1} \delta_{\bar{d},b} & \text{if there is a } j \text{ such that } d_{j-1} < k < d_j \\ 0 & \text{otherwise} \end{cases}.
$$

But $\delta_{\bar{d},b} = \langle \Phi_{\bar{d}} | \Phi_b \rangle$, so that

$$
\langle \hat{c}_k^\dagger \Phi_d | \Phi_b \rangle = \begin{cases} (-1)^{j-1} \langle \Phi_{\bar{d}} | \Phi_b \rangle & \text{if there is a } j \text{ such that } d_{j-1} < k < d_j \\ 0 & \text{otherwise} \end{cases}.
$$

The action of $\hat{c}_k^\dagger$ is defined as

$$
\hat{c}_k^\dagger \Phi_d \equiv \begin{cases} (-1)^{j-1} \Phi_{\bar{d}} & \text{if there is a } j \text{ such that } d_{j-1} < k < d_j \\ 0 & \text{otherwise} \end{cases}. \tag{D.27}
$$

Thus, $\hat{c}_k^\dagger$ creates a new Slater determinant with the single-particle wavefunction ϕ_k added. We can "psyche out" the factor of $(-1)^{j-1}$ by going through a two-step process of adding a new row (with ϕ_k) at the top of the determinant and a new column, and then exchanging rows until it occurs in the proper place.

$$
\hat{c}_k^\dagger \frac{1}{\sqrt{(N-1)!}} \begin{vmatrix} \phi_{d_1}(x_1) & \cdots & \phi_{d_1}(x_{N-1}) \\ \vdots & & \vdots \\ \phi_{d_{N-1}}(x_1) & \cdots & \phi_{d_{N-1}}(x_{N-1}) \end{vmatrix} = \frac{1}{\sqrt{N!}} \begin{vmatrix} \phi_k(x_1) & \cdots & \phi_k(x_N) \\ \phi_{d_1}(x_1) & \cdots & \phi_{d_1}(x_N) \\ \vdots & & \vdots \\ \phi_{d_{N-1}}(x_1) & \cdots & \phi_{d_{N-1}}(x_N) \end{vmatrix}
$$

$$
= \frac{(-1)^{j-1}}{\sqrt{N!}} \begin{vmatrix} \phi_{d_1}(x_1) & \cdots & \phi_{d_1}(x_N) \\ \vdots & & \vdots \\ \phi_{d_{j-1}}(x_1) & \cdots & \phi_{d_{j-1}}(x_N) \\ \phi_k(x_1) & \cdots & \phi_k(x_N) \\ \phi_{d_j}(x_1) & \cdots & \phi_{d_j}(x_N) \\ \vdots & & \vdots \\ \phi_{d_{N-1}}(x_1) & \cdots & \phi_{d_{N-1}}(x_N) \end{vmatrix}.
$$

D.7.2 Occupation-number representation of fermionic creation and annihilation operators

Having to write out Slater determinants is cumbersome. Fortunately there is a compact way of expressing the action of the creation and annihilation operators known as the *occupation number representation.* For a given many-fermion state with quantum numbers $d = (d_1, \cdots, d_N)$, define the *occupation number*[13]

$$n_i \equiv \begin{cases} 1 & \text{if } i \in (d_1, \cdots, d_N) \\ 0 & \text{if } i \notin (d_1, \cdots, d_N) \end{cases}. \tag{D.28}$$

Using the flexibility afforded by Dirac notation we can denote the many-particle state vector either as a list of quantum numbers, $|d\rangle$, or as a list of occupation numbers

$$\Phi_d = |d\rangle = |n_1, n_2, n_3, \cdots\rangle .$$

In writing $|n_1, n_2, n_3, \cdots\rangle$ it's understood that the ket notation subsumes the normalization factor of $\sqrt{N!}$. Note that we don't have to explicitly keep track of the fact that Φ_d (or $|d\rangle$) is an N-particle quantum state; this information is stored in the sum of the occupation numbers, $N = \sum_i n_i$.

Example. Suppose ϕ_1, ϕ_3, and ϕ_4 are occupied, so that $d = (1, 3, 4)$. We can write this as

$$|1, 3, 4\rangle = |1_1, 0_2, 1_3, 1_4, 0_5, 0_6, \cdots\rangle .$$

The action of the creation and annihilation operators can be expressed compactly in the occupation number formalism. Instead of Eqs. (D.25) and (D.26), we have

$$\begin{aligned} \hat{c}_k |n_1, \cdots, 1_k, \cdots\rangle &= \theta_k |n_1, \cdots, 0_k, \cdots\rangle \\ \hat{c}_k |n_1, \cdots, 0_k, \cdots\rangle &= 0 , \end{aligned} \tag{D.29}$$

where the factor $(-1)^{j-1}$ is written

$$\theta_k \equiv (-1)^{\left[\sum_{j<k} n_j\right]}$$

The number of occupied single-particle states with $\nu_i < \nu_j$ is simply $j - 1$. For example,

$$\hat{c}_4 |1_1, 1_2, 1_3, 1_4, \cdots\rangle = (-1)^{4-1} |1_1, 1_2, 1_3, 0_4, \cdots\rangle = (-1)^{\sum_{i=1}^{3} n_i} |1_1, 1_2, 1_3, 0_4, \cdots\rangle .$$

The occupation number $n_j = 1$ corresponds to the jth row in the Slater determinant; the factor $\sum_{j<k} n_j$ is the number of row interchanges discussed previously.

In this notation, the action of the creation operator can be written (instead of Eq. (D.27)),

$$\begin{aligned} \hat{c}_k^\dagger |n_1, \cdots, 0_k, \cdots\rangle &= \theta_k |n_1, \cdots, 1_k, \cdots\rangle \\ \hat{c}_k^\dagger |n_1, \cdots, 1_k, \cdots\rangle &= 0 . \end{aligned} \tag{D.30}$$

This can be boiled down even further: Both cases in Eq. (D.30) can be written

$$\hat{c}_k^\dagger |n_1, \cdots, n_k, \cdots\rangle = \theta_k(1 - n_k) |n_1, \cdots, 1_k, \cdots\rangle . \tag{D.31}$$

Likewise, both cases in Eq. (D.29) can be written

$$\hat{c}_k |n_1, \cdots, n_k, \cdots\rangle = \theta_k n_k |n_1, \cdots, 0_k, \cdots\rangle . \tag{D.32}$$

---

[13] The term *occupation number* is used in two ways: As in Eq. (D.28), in which an occupation number for fermions has the precise values 0 or 1, or as in statistical mechanics, where occupation number refers to a thermal average of the number of particles in a given energy state, such as the Fermi-Dirac distribution function, Eq. (5.62).

Equations (D.31) and (D.32) are succinct expressions for the action of the creation and annihilation operators in the occupation number representation; they are fully equivalent to the definitions given in Section D.7.1.

Let the system consist of a single particle in state ϕ_k. The action of $\hat{c}_k$ in this case produces a special quantum state known as the *vacuum state*, $|0\rangle$, that has no occupied states

$$\hat{c}_k |0_1, \cdots, 0_{k-1}, 1_k, 0_{k+1}, \cdots\rangle \equiv |0\rangle .$$

Another way to characterize the vacuum state is that $\hat{c}_k |0\rangle = 0$ for all k.

D.7.3 Anticommutators of fermionic creation and annihilation operators

The *anticommutator* of operators $\hat{A}$ and $\hat{B}$ is defined as $\{\hat{A}, \hat{B}\} \equiv \hat{A}\hat{B} + \hat{B}\hat{A}$, which is distinct from the commutator $[\hat{A}, \hat{B}] \equiv \hat{A}\hat{B} - \hat{B}\hat{A}$. The fundamental anticommutators of fermionic creation and annihilation operators are

$$\{\hat{c}_l, \hat{c}_k\} = 0 \qquad \{\hat{c}_l^\dagger, \hat{c}_k^\dagger\} = 0 \qquad \{\hat{c}_l^\dagger, \hat{c}_k\} = \delta_{l,k} . \tag{D.33}$$

It's straightforward to prove these relations. Using Eq. (D.32), first work out

$$\hat{c}_l \hat{c}_k |n_1, \cdots, n_k, \cdots, n_l, \cdots\rangle = n_k \theta_k \hat{c}_l |n_1, \cdots, 0_k, \cdots, n_l, \cdots\rangle \tag{D.34}$$
$$= \theta_k \theta_l n_k n_l |n_1, \cdots, 0_k, \cdots, 0_l, \cdots\rangle .$$

Then evaluate

$$\hat{c}_k \hat{c}_l |n_1, \cdots, n_k, \cdots, n_l, \cdots\rangle = \tilde{\theta}_l n_l \hat{c}_k |n_1, \cdots, n_k, \cdots, 0_l, \cdots\rangle = \tilde{\theta}_l \theta_k n_l n_k |n_1, \cdots, 0_k, \cdots, 0_l, \cdots\rangle \tag{D.35}$$
$$= -\theta_l \theta_k n_l n_k |n_1, \cdots, 0_k, \cdots, 0_l, \cdots\rangle ,$$

where $\tilde{\theta}_l = -\theta_l$ because θ_l occurs for $n_k = 0$ in Eq. (D.34), but $\tilde{\theta}_l$ occurs for $n_k = 1$ in Eq. (D.35). (If $n_k = 0$ in Eq. (D.35) the entire expression vanishes anyway.) Adding Eqs. (D.34) and (D.35), we conclude that $\{\hat{c}_l, \hat{c}_k\} = 0$. The proofs for $\{\hat{c}_l^\dagger, \hat{c}_k^\dagger\} = 0$ and $\{\hat{c}_l^\dagger, \hat{c}_k\} = 0$ ($l \neq k$) are similar.

The proof for $\{\hat{c}_k^\dagger, \hat{c}_k\}$ is more involved, however. First work out

$$\hat{c}_k^\dagger \hat{c}_k |n_1, \cdots, n_k, \cdots\rangle = \theta_k n_k \hat{c}_k^\dagger |n_1, \cdots, 0_k, \cdots\rangle = \theta_k^2 n_k |n_1, \cdots, 1_k, \cdots\rangle , \tag{D.36}$$

where of course $\theta_k^2 = 1$. Then work out

$$\hat{c}_k \hat{c}_k^\dagger |n_1, \cdots, n_k, \cdots\rangle = \theta_k^2 (1 - n_k) |n_1, \cdots, 0_k, \cdots\rangle . \tag{D.37}$$

Now examine cases. For $n_k = 1$, we have from Eq. (D.36) that $\hat{c}_k^\dagger \hat{c}_k |n_1, \cdots, 1_k, \cdots\rangle = |n_1, \cdots, 1_k, \cdots\rangle$ while from Eq. (D.37), $\hat{c}_k \hat{c}_k^\dagger |n_1, \cdots, 1_k, \cdots\rangle = 0$. In is then true that $(\hat{c}_k^\dagger \hat{c}_k + \hat{c}_k \hat{c}_k^\dagger) |n_1, \cdots, 1_k, \cdots\rangle = |n_1, \cdots, 1_k, \cdots\rangle$. For $n_k = 0$, we have from Eq. (D.36) $\hat{c}_k^\dagger \hat{c}_k |n_1, \cdots, 0_k, \cdots\rangle = 0$ and from Eq. (D.37) $\hat{c}_k \hat{c}_k^\dagger |n_1, \cdots, 0_k, \cdots\rangle = |n_1, \cdots, 0_k, \cdots\rangle$. It is then true that $(\hat{c}_k^\dagger \hat{c}_k + \hat{c}_k \hat{c}_k^\dagger) |n_1, \cdots, 0_k, \cdots\rangle = |n_1, \cdots, 0_k, \cdots\rangle$. In either case $\{\hat{c}_k^\dagger, \hat{c}_k\} = 1$. This completes the proof of the anticommutation properties, Eq. (D.33).

D.7.4 Operators for observables

Out of the creation and annihilation operators, $\hat{c}_k^\dagger$ and $\hat{c}_k$, we can construct Hermitian operators, as required (by quantum mechanics) of operators that represent observables; $\hat{c}_k^\dagger \hat{c}_k$ for example is Hermitian, as is $\hat{c}_k^\dagger + \hat{c}_k$ (show this).

D.7.4.1 Number operator

Using Eq. (D.36), $\hat{c}_k^\dagger \hat{c}_k \,|n_1, \cdots, n_k, \cdots\rangle = n_k \,|n_1, \cdots, n_k, \cdots\rangle$, from which we infer the operator statement

$$\hat{c}_k^\dagger \hat{c}_k = n_k \,. \tag{D.38}$$

We introduce the *number operator*

$$\hat{N} \equiv \sum_{k=1}^{\infty} \hat{c}_k^\dagger \hat{c}_k \,, \tag{D.39}$$

where the sum is unrestricted; no harm in adding an infinite number of zeros. For any Slater determinant

$$\hat{N} \,|n_1, n_2, \cdots\rangle = \left(\sum_{k=1}^{\infty} \hat{c}_k^\dagger \hat{c}_k \right) |n_1, n_2, \cdots\rangle = \left(\sum_{k=1}^{\infty} n_k \right) |n_1, n_2, \cdots\rangle = N \,|n_1, n_2, \cdots\rangle \,.$$

D.7.4.2 Single-particle operators

The single-particle Hamiltonian $\hat{H}_0$, Eq. (D.14), which in its spatial representation is expressed in terms of differential and multiplicative operators, can be given an equivalent representation in terms of creation and annihilation operators. In what follows, we show the equivalence

$$\hat{H}_0 = \sum_{i=1}^{N} \hat{h}(x_i) \longleftrightarrow \sum_{i,j=1}^{\infty} \langle i|\hat{h}|j\rangle \,\hat{c}_i^\dagger \hat{c}_j \,, \tag{D.40}$$

where

$$\langle i|\hat{h}|j\rangle \equiv \int \phi_i^*(x)\hat{h}(x)\phi_j(x)\mathrm{d}x \,.$$

Whereas $\hat{h}(x)$ (the *first quantization* form of the Hamiltonian) acts on the spatial coordinates of wavefunctions, the creation and annihilation operators, Eqs. (D.31) and (D.32), act on occupation numbers (the *second quantization* form of operators). To demonstrate the equivalence in Eq. (D.40) we must show that the two forms of the operator (first and second quantization) give rise to identical results when acting on Slater determinants. The term *second quantization* conjures up something new. Nothing, however, is fundamentally new here, just an equivalent way of doing quantum mechanics where the occupation numbers are emphasized instead of spatial coordinates. The advantage of the occupation number formalism is that it's well suited to systems in which the number of particles is not a fixed quantity. From Eq. (D.40), the first quantization form of the Hamiltonian involves a finite sum from 1 to N; in second quantization the sums are infinite, independent of N.

Even though these kinds of proof are tedious, we now show that the two forms of the operator produce equivalent results when acting on Slater determinants, i.e.,

$$\sum_{i=1}^{N} \hat{h}(x_i) \,|\Phi_d\rangle = \sum_{i,j=1}^{\infty} \langle i|\hat{h}|j\rangle \,\hat{c}_i^\dagger \hat{c}_j \,|\Phi_d\rangle \,.$$

Two operators are identical when their matrix elements are equal in a given basis. The strategy is to show the equality of

$$\sum_{i=1}^{N} \langle \Phi_b|\hat{h}(x_i)|\Phi_d\rangle = \sum_{i,j=1}^{\infty} \langle i|\hat{h}|j\rangle \,\langle \Phi_b|\hat{c}_i^\dagger \hat{c}_j|\Phi_d\rangle \tag{D.41}$$

for all Φ_b and Φ_d. Start with the left side of Eq. (D.41),

$$\sum_{i=1}^{N} \langle \Phi_b | \hat{h}(x_i) | \Phi_d \rangle = \sum_{i=1}^{N} \sqrt{N!} \, \langle \Phi_b | \hat{h}(x_i) \left(\phi_{d_1}(x_1) \otimes \cdots \otimes \phi_{d_N}(x_N) \right) \rangle$$

$$= \sum_{i=1}^{N} \sum_{P \in S_N} \mathrm{sgn}(P) \, \langle \phi_{b_{P(1)}} | \phi_{d_1} \rangle_1 \cdots \langle \phi_{b_{P(i)}} | \hat{h} | \phi_{d_i} \rangle_i \cdots \langle \phi_{b_{P(N)}} | \phi_{d_N} \rangle_N \, ,$$

where we've used Eq. (D.19) in the first line and Eq. (D.18) in the second. There are $N-1$ integrals here which vanish unless the indices "line up." The two sets of quantum numbers b and d must therefore have at least $N-1$ elements in common. There are two ways this can happen: $b = d$ (all N elements the same) and b and d differ by at most one quantum number ($N-1$ elements the same); in the latter case $\langle \phi_{b_{P(i)}} | \hat{h} | \phi_{d_i} \rangle_i$ does not restrict the allowed quantum numbers. We consider each case in turn.

For $b = d$ there is, for each i, one permutation that leads to a nonvanishing result, that for the reference sequence (sgn(P)=1) with $b_j = d_j, \, j = 1, \cdots, N$; thus,

$$\langle \Phi_d | \hat{H}_0 | \Phi_d \rangle = \sum_{i=1}^{N} \langle i | \hat{h} | i \rangle \, , \tag{D.42}$$

where in Dirac notation $|i\rangle \equiv |\phi_{d_i}\rangle$. When b and d differ by one quantum number, let

$$b = (b_1, \cdots, b_{k-1}, b_k, b_{k+1}, \cdots, b_{l-1}, b_l, b_{l+1}, \cdots, b_N)$$
$$d = (d_1, \cdots, d_{k-1}, d_k, d_{k+1}, \cdots, d_{l-1}, d_l, d_{l+1}, \cdots, d_N) \, . \tag{D.43}$$

where $b_k \notin d$ and $d_l \notin b$. In this case there is a single permutation (call it P_0) that leads to a nonvanishing result, that which puts the dissimilar quantum numbers together with $\hat{h}_i$, $\langle b_k | \hat{h} | d_l \rangle$; this occurs only for $i = l$. Thus, in this case we have

$$\langle \Phi_b | \hat{H}_0 | \Phi_d \rangle = \mathrm{sgn}(P_0) \, \langle b_k | \hat{h} | d_l \rangle \, . \tag{D.44}$$

We must show that we're led to the same results using the second quantization form in Eq. (D.40). Consider that

$$\sum_{i,j=1}^{\infty} \langle i | \hat{h} | j \rangle \langle \Phi_b | \hat{c}_i^\dagger \hat{c}_j | \Phi_d \rangle = \sum_{i,j=1}^{\infty} \langle i | \hat{h} | j \rangle \langle \hat{c}_i \Phi_b | \hat{c}_j \Phi_d \rangle \, . \tag{D.45}$$

But,

$$\langle \hat{c}_i \Phi_b | \hat{c}_j \Phi_d \rangle = \pm \langle \Phi_{b-i} | \Phi_{d-j} \rangle \, . \tag{D.46}$$

For this integral to be nonvanishing, the sets b and d must either be identical or they can differ by at most one quantum number, because only then can the two sets be made to agree through the action of one annihilation operator and one creation operator. Consider $b = d$. Then,

$$\langle \hat{c}_i \Phi_d | \hat{c}_j \Phi_d \rangle = \begin{cases} 0 & j \neq i \\ \langle \Phi_d | \hat{c}_i^\dagger \hat{c}_i | \Phi_d \rangle = n_i \langle \Phi_d | \Phi_d \rangle = n_i & j = i \end{cases} \, . \tag{D.47}$$

Combining Eq. (D.47) with Eq. (D.45),

$$\langle \Phi_d | \hat{H}_0 | \Phi_d \rangle = \sum_{i=1}^{\infty} \langle i | \hat{h} | i \rangle \, n_i = \sum_{j=1}^{N} \langle j | \hat{h} | j \rangle \, , \tag{D.48}$$

the same as Eq. (D.42). The infinite sum in Eq. (D.48) over all possible occupation numbers is equivalent to a finite sum over states that are actually occupied. We will use this step in what follows without further comment. For $b \neq d$, with $b_k \notin d$ and $d_l \notin b$ in the notation of Eq. (D.43), there is only one possibility that leads to a nonzero result:

$$\langle \hat{c}_i \Phi_b | \hat{c}_j \Phi_d \rangle = \begin{cases} \pm 1 & i = b_k, j = d_l \\ 0 & \text{otherwise} \end{cases}, \tag{D.49}$$

where the sign corresponds to $\text{sgn}(P_0)$ in Eq. (D.44). Combining Eq. (D.49) with Eq. (D.45),

$$\langle \Phi_b | \hat{H}_0 | \Phi_d \rangle = \text{sgn}(P_0) \langle b_k | \hat{h} | d_l \rangle ,$$

the same as Eq. (D.44). We have thus established the equivalence of the first and second quantization forms of the single-particle operator in Eq. (D.40). Note that this result holds for *any* one-particle operator; nowhere did we invoke the properties of $\hat{h}(x)$ as the single-particle Hamiltonian.

D.7.4.3 Two-particle operators

The interaction Hamiltonian in Eq. (D.14) can also be written in second quantization form,

$$\hat{H}' = \frac{1}{2} \sum_{\substack{i,j=1 \\ (i \neq j)}}^{N} v(r_i, r_j) \longleftrightarrow \frac{1}{2} \sum_{ijkl=1}^{\infty} \langle ij|v|kl \rangle \, \hat{c}_i^\dagger \hat{c}_j^\dagger \hat{c}_l \hat{c}_k , \tag{D.50}$$

where

$$\langle ij|v|kl \rangle \equiv \int \int \phi_i^*(x) \phi_j^*(x') v(x, x') \phi_k(x) \phi_l(x') \mathrm{d}x \mathrm{d}x' .$$

Note carefully the order of the indices in Eq. (D.50): $|kl\rangle$ and $\hat{c}_l \hat{c}_k$; this is not a typo. As with the single-particle operator in Eq. (D.41), we need to demonstrate the equality

$$\sum_{\substack{i,j=1 \\ (i \neq j)}}^{N} \langle \Phi_b | v(r_i, r_j) | \Phi_d \rangle = \sum_{ijkl=1}^{\infty} \langle ij|v|kl \rangle \, \langle \Phi_b | \hat{c}_i^\dagger \hat{c}_j^\dagger \hat{c}_l \hat{c}_k | \Phi_d \rangle . \tag{D.51}$$

To show Eq. (D.51) is considerably more involved than the analogous proof in the single-particle case and we omit it.

D.8 OCCUPATION-NUMBER FORMALISM FOR BOSONS

Having worked out the creation and annihilation operators for fermions, where the occupation numbers are restricted to two values ($n_i = 0, 1$), the creation and annihilation operators for bosons are relatively simple. We can use occupation numbers to denote boson wavefunctions just as readily as for fermionic wavefunctions.

To develop the creation and annihilation operators for bosons, denoted $\hat{b}_k^\dagger$ and $\hat{b}_k$, it's helpful to write Eqs. (D.31) and (D.32) in a somewhat more general form,

$$\hat{c}_k |n_1, \cdots, n_k, \cdots\rangle = \theta_k n_k |n_1, \cdots, (n_k - 1), \cdots\rangle$$

$$\hat{c}_k^\dagger |n_1, \cdots, n_k, \cdots\rangle = \begin{cases} \theta_k (n_k + 1) |n_1, \cdots, (n_k + 1), \cdots\rangle & \text{if } n_k = 0 \\ 0 & \text{if } n_k = 1 \end{cases}.$$

For bosonic wavefunctions (which are not a determinant) there is no need for the factor of θ_k, so we can let $\theta_k \to 1$. As there is no Pauli principle for bosons, we can also remove the restriction that $\hat{c}_k^\dagger \Phi = 0$ if $n_k = 1$. Bosonic creation and annihilation operators are defined by

$$
\begin{aligned}
\hat{b}_k^\dagger |n_1, \cdots, n_k, \cdots\rangle &\equiv \sqrt{n_k + 1} \, |n_1, \cdots, (n_k + 1), \cdots\rangle \\
\hat{b}_k |n_1, \cdots, n_k, \cdots\rangle &\equiv \sqrt{n_k} \, |n_1, \cdots, (n_k - 1), \cdots\rangle \; .
\end{aligned}
\tag{D.52}
$$

The square root factors are introduced so that the number operator for bosons is formally the same as that for fermions, Eq. (D.38) i.e.,

$$
\hat{b}_k^\dagger \hat{b}_k = n_k \; .
\tag{D.53}
$$

Using Eq. (D.52),

$$
\langle \cdots n_k \cdots | \hat{b}^\dagger \hat{b}_k | \cdots n_k \cdots \rangle = n_k \langle \cdots n_k \cdots | \cdots n_k \cdots \rangle = n_k \; .
\tag{D.54}
$$

The bosonic operators $\hat{b}_k^\dagger$ and $\hat{b}_k$ satisfy *commutation relations* (as opposed to the anticommutation relations for the fermionic operators, Eq. (D.33))

$$
\left[\hat{b}_l, \hat{b}_k\right] = 0 \qquad \left[\hat{b}_l^\dagger, \hat{b}_k^\dagger\right] = 0 \qquad \left[\hat{b}_l, \hat{b}_k^\dagger\right] = \delta_{lk} \; .
\tag{D.55}
$$

It is straightforward to prove the commutation relations:

$$
\begin{aligned}
\hat{b}_l \hat{b}_k |n_1, \cdots, n_l, \cdots, n_k, \cdots\rangle &= \sqrt{n_k} \hat{b}_l |n_1, \cdots, n_l, \cdots, (n_k - 1), \cdots\rangle \\
&= \sqrt{n_k n_l} |n_1, \cdots, (n_l - 1), \cdots, (n_k - 1), \cdots\rangle \\
&= \hat{b}_k \hat{b}_l |n_1, \cdots, n_l, \cdots, n_k, \cdots\rangle
\end{aligned}
\tag{D.56}
$$

The commutation relations for $[\hat{b}_l^\dagger, \hat{b}_k^\dagger] = 0$ and $[\hat{b}_l, \hat{b}_k^\dagger] = 0$ $(l \neq k)$ follow in similar fashion. The only nontrivial commutation relation is $[\hat{b}_l, \hat{b}_l^\dagger] = 1$. This is shown as follows:

$$
\begin{aligned}
\hat{b}_l \hat{b}_l^\dagger |n_1, \cdots, n_l, \cdots\rangle &= \sqrt{n_l + 1} \hat{b}_l |n_1, \cdots, (n_l + 1), \ldots\rangle \\
&= (n_l + 1) |n_1, \cdots, n_l, \cdots\rangle \\
\hat{b}_l^\dagger \hat{b}_l |n_1, \cdots, n_l, \cdots\rangle &= \sqrt{n_l} \hat{b}_l^\dagger |n_1, \cdots, (n_l - 1), \cdots\rangle \\
&= n_l |n_1, \cdots, n_l, \cdots\rangle
\end{aligned}
\tag{D.57}
$$

Subtracting these two results establishes that $[\hat{b}_l, \hat{b}_l^\dagger] = 1$.

We state without proof that the second quantization representation of one- and two-particle operators has the same form with boson operators as that for fermion operators, Eqs. (D.40) and (D.50),

$$
\hat{H}_0 = \sum_{i=1}^{N} \hat{h}(x_i) = \sum_{i,j=1}^{\infty} \langle i|\hat{h}|j\rangle \, \hat{b}_i^\dagger \hat{b}_j
$$

$$
\hat{H}' = \frac{1}{2} \sum_{\substack{i,j=1 \\ (i \neq j)}}^{N} v(x_i, x_j) = \frac{1}{2} \sum_{ijkl=1}^{\infty} \langle ij|v|kl\rangle \, \hat{b}_i^\dagger \hat{b}_j^\dagger \hat{b}_l \hat{b}_k \; .
\tag{D.58}
$$

Topics in magnetism

S OME of the most successful applications of statistical mechanics involve the magnetic properties of materials, applications which require the dipole moments μ of system constituents. Statistical mechanicians have a habit of introducing various abbreviations for μ and its interaction with magnetic fields that sweep the details under the rug. In this appendix we review the physics underlying μ. We then consider the interaction between the magnetic moments of neutrons and the magnetic moments of macroscopic systems.

E.1 DIPOLE MOMENTS: THE LANDE FACTOR

The magnetic field $B(r)$ at position r, produced by a steady current density $J(r')$ at r' is, by the law of Biot and Savart,

$$B(r) = \frac{\mu_0}{4\pi} \int \frac{J(r') \times (r - r')}{|r - r'|^3} \mathrm{d}^3 r' . \tag{E.1}$$

Far from the current source, $B(r) \approx \mu_0 \left[3(\mu \cdot \hat{r})\hat{r} - \mu \right] /(4\pi r^3)$, where $\mu \equiv \frac{1}{2} \int r' \times J(r')\mathrm{d}^3 r'$ defines the dipole moment of a current distribution. There is no factor of μ_0 in μ.

If the current is in the form of a circular loop carrying current I, the dipole moment has magnitude IA, where A is the area of the loop, with direction normal to the plane of the circle as given by the right-hand rule; see Fig. E.1. For an electron in a circular orbit of radius R at constant angular

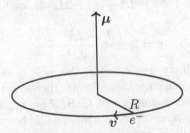

Figure E.1: Electron in a circular orbit with velocity v produces dipole moment μ.

frequency ω (so that the period is $2\pi/\omega$), the magnitude of the current is $e\omega/(2\pi)$ (where e is the magnitude of the electron charge), implying $\mu = e\omega R^2/2$. The orbital angular momentum L is opposite to the direction of μ (because of the sign convention for current), with $L = m\omega R^2$. Thus,

$$\mu = -\frac{e}{2m}L . \tag{E.2}$$

A constant applied B-field produces a torque, $\tau = \mu \times B$, so that μ precesses about the direction of B, $(\mathrm{d}/\mathrm{d}t)\mu = -(e/(2m))\mu \times B$.

Equation (E.2) pertains to the orbital angular momentum of electrons. Electrons also have an intrinsic angular momentum, its spin S. Should a "spinning" electron have a magnetic dipole? If spin was simply another type of orbital angular momentum (like Earth spins on its axis, and has orbital angular momentum about the sun), one might not think so—as far as we know, electrons are point particles. Yet, electrons do have an intrinsic magnetic moment, a purely quantum effect, one that requires the methods of quantum electrodynamics to predict. There is no reason that Eq. (E.2) should apply to spin angular momentum. We can, however, *parameterize* the dipole moment associated with spin as in Eq. (E.2):

$$\mu = -g_s \frac{e}{2m} S , \qquad (E.3)$$

where g_s is a dimensionless number, the *electron g-factor*. Experimentally (and theoretically) g_s is known with high precision, $g_s = 2.002319$. For work in magnetism, we simply take $g_s = 2$. The spin vector S also precesses about the direction of an applied field B.

We might expect the dipole moment of an electron to be a combination of Eqs. (E.2) and (E.3),

$$\mu = -\frac{e}{2m} (L + 2S) , \qquad \text{(wrong!)}$$

but it's more complicated than that. The *spin-orbit interaction* $\propto L \cdot S$ (coupling between the magnetic moment of the electron and the magnetic field of the nucleus) causes S and L to precess around the direction of the total angular momentum, $J \equiv L + S$. The component of $L + 2S$ perpendicular to J averages to zero; we want the component of $L + 2S$ along J. Thus,

$$\mu = -\frac{e\hbar}{2m} \underbrace{(L + 2S) \cdot \frac{J}{J^2}}_{g} \frac{J}{\hbar} \equiv -\mu_B g \frac{J}{\hbar} , \qquad (E.4)$$

where μ_B is the *Bohr magneton*, $\mu_B = e\hbar/(2m) = 9.27 \times 10^{-24}$ J/T (T for Tesla), and the dimensionless number g is the *Landé g factor*,

$$g = \frac{1}{J^2} (L + 2S) \cdot (L + S) = \frac{1}{J^2} (L^2 + 3L \cdot S + 2S^2) . \qquad (E.5)$$

Referring to Eq. (E.4), because J is of order $\hbar$, and the g-factor is a number of order unity, it's useful to remember that magnetic dipole moments involving electrons are of order μ_B.

Because $J^2 = (L + S) \cdot (L + S) = L^2 + S^2 + 2L \cdot S$, we have $L \cdot S = \frac{1}{2}(J^2 - L^2 - S^2)$, implying

$$g = 1 + \frac{J^2 - L^2 + S^2}{2J^2} . \qquad (E.6)$$

If the angular momentum is entirely spin in nature (because $L = 0$), then $g = 2$; if the angular momentum is entirely orbital in nature (because $S = 0$), then $g = 1$.

From quantum mechanics, the magnitudes of L, S, J are quantized: L^2 has one of the values $L'(L' + 1)\hbar^2$, S^2 has one of the values $S'(S' + 1)\hbar^2$, and J^2 has one of the values $J'(J' + 1)\hbar^2$, where L' is a positive integer, with S' and J' being either positive integer or half-integers. For an electron in a particular state with known values of L', S', and J', we have from Eq. (E.6),

$$g = 1 + \frac{J'(J' + 1) - L'(L' + 1) + S'(S' + 1)}{2J'(J' + 1)} . \qquad (E.7)$$

To find the dipole moment of a multi-electron atom, we must combine the angular momenta of all the electrons in the atom, and the quantum theory of angular momentum has its own logic for doing that. There is a procedure—*Hund's rules*—of determining the values of S', L', and J' associated with the ground state of an atom.[1] In many of the rare earth and transition metal ions,

---

[1] See most any text on quantum mechanics.

excited states are sufficiently separated in energy from the ground state that we can assume they're unoccupied. Hund's rules are as follows:

1. Maximize the total spin quantum number S' consistent with the Pauli principle. Each electron up to a half-filled shell[2] contributes $+\frac{1}{2}$ to S'. Beyond a half-filled shell, each electron contributes $-\frac{1}{2}$. Thus, a filled shell is associated with $S' = 0$.

2. Maximize the value of L' consistent with rule 1. If the shell, associated with angular momentum number l, is less than half full, then $L' = l + (l-1) + (l-2) + \cdots$, where the sum is over all electrons in the shell. A half-filled shell has $L' = 0$. For more than a half-filled shell, $L' = l + (l-1) + \cdots$, where the sum is over the electrons in excess of a half-filled shell. A filled shell therefore has $L' = 0$.

3. $J' = |L' - S'|$ if the shell is less than half full, $J' = S'$ for a half-filled shell, and $J = L' + S'$ for more than a half-filled shell. A filled shell is associated with $J' = 0$.

Examples

• Find the Landé g-factor for the ground state of the praseodymium ion Pr^{+++}, which has two electrons in an f-shell. (The praseodymium atom has the electron configuration $[Xe]4f^2 5d^1 6s^2$; it loses the three outer electrons in the solid state.) Pr^{+++} therefore has two electrons in an f-shell ($l = 3$), all other electrons are in filled shells. By Hund's rules, $S' = \frac{1}{2} + \frac{1}{2} = 1$, $L' = 3 + 2 = 5$, and $J' = 5 - 1 = 4$. From Eq. (E.7),

$$g = 1 + \frac{4 \cdot 5 - 5 \cdot 6 + 1 \cdot 2}{2 \cdot 4 \cdot 5} = 0.8 .$$

• The erbium ion Er^{+++} has 11 f-shell electrons. (The erbium atom has the electron configuration $[Xe]4f^{12} 6s^2$.) By Hund's rules, $S' = \frac{7}{2} - \frac{4}{2} = \frac{3}{2}$, $L' = 6$, $S' = 6 + \frac{3}{2} = \frac{15}{2}$. From Eq. (E.7),

$$g = 1 + \frac{\frac{15}{2} \cdot \frac{17}{2} - 6 \cdot 7 + \frac{3}{2} \cdot \frac{5}{2}}{2 \cdot \frac{15}{2} \cdot \frac{17}{2}} = 1.2 .$$

E.2   NEUTRON SCATTERING

Much of what we know about macroscopic systems comes from scattering experiments. In X-ray scattering, electromagnetic radiation scatters from the charges in the system. In neutron scattering, the magnetic moments of neutrons scatter from the magnetic moments of a system. We present a basic introduction to the interaction underlying neutron scattering.[3]

Neutrons have a magnetic moment,

$$\mu_n = -g_n \mu_N \frac{S}{\hbar} , \tag{E.8}$$

where $\mu_N \equiv e\hbar/(2m_p) = 5.051 \times 10^{-27}$ J/T is the *nuclear magneton*, with m_p the proton mass, and $g_n = 3.826$ is the *neutron g-factor*. Magnetically, neutrons act as if they're negatively charged particles, yet they're electrically neutral, indicating that they're not elementary particles.

Neutrons incident upon a macroscopic system interact with the B-field produced by the magnetic moments of the system. For one neutron, the energy of that interaction is

$$\Phi(r) = -\mu_n \cdot B(r) . \tag{E.9}$$

---

[2]A full shell has $2(2l + 1)$ electrons when spin is accounted for.
[3]For more details, see for example Squires[165].

In what follows we obtain an expression for $\Phi(r)$ and then we evaluate the quantum matrix element $\langle f|\Phi|i\rangle$ that Φ induces a transition from neutron state $|i\rangle$ to state $|f\rangle$.

To find the B-field produced by many magnetic moments, we first find the vector potential $A(r)$, from which $B(r) = \nabla \times A(r)$. Instead of starting from Eq. (E.1), we do this in steps, with $A(r)$ determined by[4]

$$A(r) = \frac{\mu_0}{4\pi} \int \frac{J(r')}{|r - r'|} d^3r' . \tag{E.10}$$

Far from sources, we can approximate Eq. (E.10) as (where a dipole μ is at the origin of coordinates)

$$A(r) = \frac{\mu_0}{4\pi} \frac{\mu \times \hat{r}}{r^2} . \tag{E.11}$$

Assume we have dipoles μ_j at positions r_j, $j = 1, \cdots, N$. The net vector potential at position r is therefore

$$A(r) = \frac{\mu_0}{4\pi} \sum_j \frac{\mu_j \times (r - r_j)}{|r - r_j|^3} . \tag{E.11}$$

To find $B(r)$ we have to take the curl of $A(r)$ in Eq. (E.11), a step facilitated by the identity

$$\frac{r - r_j}{|r - r_j|^3} = -\nabla \left(\frac{1}{|r - r_j|} \right) ,$$

and thus

$$B(r) = -\frac{\mu_0}{4\pi} \sum_j \nabla \times \left[\mu_j \times \nabla \left(\frac{1}{|r - r_j|} \right) \right] . \tag{E.12}$$

The energy of interaction is therefore, combining Eqs. (E.9) and (E.12),

$$\Phi(r) = \frac{\mu_0}{4\pi} \sum_j \mu_n \cdot \nabla \times \left[\mu_j \times \nabla \left(\frac{1}{|r - r_j|} \right) \right] . \tag{E.13}$$

Turning now to quantum mechanics, consider the initial neutron state

$$|i\rangle \equiv \frac{1}{\sqrt{V}} e^{ik_i \cdot r} |S_n\rangle_i ,$$

where k_i is the incident wave vector, $|S_n\rangle_i$ is the spin state of the neutron, and V is the volume. (That is, the incident state is the product of a plane-wave factor and the spin state, normalized to one neutron in the volume of the system.) Likewise, we take the final neutron state to be

$$|f\rangle \equiv \frac{1}{\sqrt{V}} e^{ik_f \cdot r} |S_n\rangle_f ,$$

where k_f is the outgoing wave vector. The transition matrix element is therefore

$$\langle f|\Phi|i\rangle = \frac{1}{V} \langle S_n|_f \left(\int d^3r e^{-iq \cdot r} \Phi(r) \right) |S_n\rangle_i , \tag{E.14}$$

where $q \equiv k_f - k_i$. Equation (E.14) involves the Fourier transform of the interaction potential. Using Eq. (E.13),

$$\int d^3r e^{-iq \cdot r} \Phi(r) = \frac{\mu_0}{4\pi} \sum_j \int d^3r e^{-iq \cdot r} \mu_n \cdot \nabla \times \left[\mu_j \times \nabla \left(\frac{1}{|r - r_j|} \right) \right] . \tag{E.15}$$

---

[4]Consult any textbook on electromagnetism.

The integral on the right side of Eq. (E.15) involves two factors of ∇; it can be simplified by integrating by parts twice. For vector fields $\boldsymbol{f}(\boldsymbol{r})$, scalar fields $\phi(\boldsymbol{r})$, and a constant vector $\boldsymbol{a}$:

$$\int \mathrm{d}^3 r e^{-\mathrm{i}\boldsymbol{q}\cdot\boldsymbol{r}} \boldsymbol{a} \cdot (\nabla \times \boldsymbol{f}(\boldsymbol{r})) = \mathrm{i} \int \mathrm{d}^3 r e^{-\mathrm{i}\boldsymbol{q}\cdot\boldsymbol{r}} \boldsymbol{a} \cdot (\boldsymbol{q} \times \boldsymbol{f}(\boldsymbol{r}))$$

$$\int \mathrm{d}^3 r e^{-\mathrm{i}\boldsymbol{q}\cdot\boldsymbol{r}} \nabla\phi(\boldsymbol{r}) = \mathrm{i}\boldsymbol{q} \int \mathrm{d}^3 r e^{-\mathrm{i}\boldsymbol{q}\cdot\boldsymbol{r}} \phi(\boldsymbol{r}) \,, \tag{E.16}$$

where, for these equations to hold, $\boldsymbol{f}(\boldsymbol{r})$ and $\phi(\boldsymbol{r})$ must vanish at the boundaries of the system. We assume the system is large enough that this is the case. From Eq. (E.15), $\boldsymbol{f} = \boldsymbol{\mu}_j \times \nabla\left(|\boldsymbol{r} - \boldsymbol{r}_j|^{-1}\right)$, and thus, using Eq. (E.16),

$$\int e^{-\mathrm{i}\boldsymbol{q}\cdot\boldsymbol{r}} \Phi(\boldsymbol{r}) \mathrm{d}^3 r = -\frac{\mu_0}{4\pi} \boldsymbol{\mu}_n \cdot \sum_j (\boldsymbol{q} \times \boldsymbol{\mu}_j \times \boldsymbol{q}) \int \mathrm{d}^3 r \frac{e^{-\mathrm{i}\boldsymbol{q}\cdot\boldsymbol{r}}}{|\boldsymbol{r} - \boldsymbol{r}_j|} = -\mu_0 \boldsymbol{\mu}_n \cdot \sum_j (\hat{\boldsymbol{q}} \times \boldsymbol{\mu}_j \times \hat{\boldsymbol{q}}) e^{-\mathrm{i}\boldsymbol{q}\cdot\boldsymbol{r}_j} \,,$$

where $\int \mathrm{d}^3 r e^{-\mathrm{i}\boldsymbol{q}\cdot\boldsymbol{r}} |\boldsymbol{r} - \boldsymbol{r}_j|^{-1} = (4\pi/q^2) e^{-\mathrm{i}\boldsymbol{q}\cdot\boldsymbol{r}_j}$ [165, p202] and $\hat{\boldsymbol{q}} \equiv \boldsymbol{q}/q$. We conclude from the terms $\hat{\boldsymbol{q}} \times \boldsymbol{\mu}_j \times \hat{\boldsymbol{q}}$ that *the neutron couples to the component of $\boldsymbol{\mu}_j$ perpendicular to $\hat{\boldsymbol{q}}$*. Let $\boldsymbol{p}_j \equiv \hat{\boldsymbol{q}} \times \boldsymbol{\mu}_j \times \hat{\boldsymbol{q}} = \boldsymbol{\mu}_j - \hat{\boldsymbol{q}}(\hat{\boldsymbol{q}} \cdot \boldsymbol{\mu}_j)$ denote the component of $\boldsymbol{\mu}_j$ orthogonal to $\hat{\boldsymbol{q}}$. Thus,

$$\int e^{-\mathrm{i}\boldsymbol{q}\cdot\boldsymbol{r}} \Phi(\boldsymbol{r}) \mathrm{d}^3 r = -\mu_0 \boldsymbol{\mu}_n \cdot \sum_j \boldsymbol{p}_j e^{-\mathrm{i}\boldsymbol{q}\cdot\boldsymbol{r}_j} \tag{E.17}$$

is given in terms of the coupling of the magnetic moment of the incident neutron $\boldsymbol{\mu}_n$, to the Fourier transform of the magnetic moments $\boldsymbol{p}_j$ associated with $\boldsymbol{q}$.

Combining Eqs. (E.14) and (E.17), we have for the square of the matrix element

$$|\langle f|\Phi|i\rangle|^2 = \frac{\mu_0^2}{V^2} \sum_l \sum_j e^{\mathrm{i}\boldsymbol{q}\cdot(\boldsymbol{r}_l - \boldsymbol{r}_j)} \langle \boldsymbol{S}_n|_i \boldsymbol{\mu}_n \cdot \boldsymbol{p}_l|\boldsymbol{S}_n\rangle_f \langle \boldsymbol{S}_n|_f \boldsymbol{\mu}_n \cdot \boldsymbol{p}_j|\boldsymbol{S}_n\rangle_i \,. \tag{E.18}$$

We now sum over the final spin states of the neutron, which, because they're a complete set of states is easy to do, $\sum_f |\boldsymbol{S}_n\rangle_f \langle \boldsymbol{S}_n|_f = I$. From Eq. (E.18),

$$\sum_f |\langle f|\Phi|i\rangle|^2 = \frac{\mu_0^2}{V^2} \sum_l \sum_j e^{\mathrm{i}\boldsymbol{q}\cdot(\boldsymbol{r}_l - \boldsymbol{r}_j)} \langle \boldsymbol{S}_n|_i (\boldsymbol{\mu}_n \cdot \boldsymbol{p}_l)(\boldsymbol{\mu}_n \cdot \boldsymbol{p}_j)|\boldsymbol{S}_n\rangle_i \tag{E.19}$$

Neutrons are spin-$\frac{1}{2}$ particles and therefore, from Eq. (E.8),

$$\boldsymbol{\mu}_n = -g_n \mu_N \frac{\boldsymbol{S}_n}{\hbar} = -\frac{g_n \mu_N}{2} \boldsymbol{\sigma} \,,$$

where $\boldsymbol{\sigma}$ is a "vector" of Pauli matrices $(\sigma_1, \sigma_2, \sigma_3)$. For the terms in Eq. (E.19),

$$(\boldsymbol{\mu}_n \cdot \boldsymbol{p}_l)(\boldsymbol{\mu}_n \cdot \boldsymbol{p}_j) = \left(\frac{g_n \mu_N}{2}\right)^2 (\boldsymbol{\sigma} \cdot \boldsymbol{p}_l)(\boldsymbol{\sigma} \cdot \boldsymbol{p}_j) \,. \tag{E.20}$$

The Pauli matrices satisfy the property[5] for arbitrary vectors $\boldsymbol{A}, \boldsymbol{B}$,

$$(\boldsymbol{\sigma} \cdot \boldsymbol{A})(\boldsymbol{\sigma} \cdot \boldsymbol{B}) = (\boldsymbol{A} \cdot \boldsymbol{B}) I + \mathrm{i}\boldsymbol{\sigma} \cdot (\boldsymbol{A} \times \boldsymbol{B}) \,. \tag{E.21}$$

Note that the right side of Eq. (E.21) is not symmetric in $\boldsymbol{A}$ and $\boldsymbol{B}$; the order of $\boldsymbol{A}$ and $\boldsymbol{B}$ matters in general. In our case, the term linear in $\boldsymbol{\sigma}$ on the right side of Eq. (E.21) doesn't contribute when we

---

[5] See most any textbook on quantum mechanics.

average over the initial spin states of the neutron. We find, combining Eqs. (E.21) and (E.20) with Eq. (E.19),

$$\sum_i \sum_f |\langle f|\Phi|i\rangle|^2 = 2\left(\frac{\mu_0 g_n \mu_N}{2V}\right)^2 \sum_l \sum_j e^{i\boldsymbol{q}\cdot(\boldsymbol{r}_l - \boldsymbol{r}_j)} \boldsymbol{p}_l \cdot \boldsymbol{p}_j. \tag{E.22}$$

Equation (E.22) applies for one neutron. In experiments where large numbers of neutrons are scattered over a long period of time, $\boldsymbol{p}_l \cdot \boldsymbol{p}_j$ is replaced with its ensemble average $\langle \boldsymbol{p}_l \cdot \boldsymbol{p}_j \rangle$:

$$\sum_i \sum_f |\langle f|\Phi|i\rangle|^2 = 2\left(\frac{\mu_0 g_n \mu_N}{2V}\right)^2 \sum_l \sum_j e^{i\boldsymbol{q}\cdot(\boldsymbol{r}_l - \boldsymbol{r}_j)} \langle \boldsymbol{p}_l \cdot \boldsymbol{p}_j \rangle. \tag{E.23}$$

At this point, if one were calculating the scattering cross section, Eq. (E.23) would be combined with Fermi's golden rule to find the rate of transitions $i \to f$. For our purposes, we can stop: We've achieved our objective in showing that the scattering probability associated with $\boldsymbol{q}$ is related to the Fourier transform of the correlation function $\langle \boldsymbol{p}_l \cdot \boldsymbol{p}_j \rangle$.

The method of the most probable distribution

I N this appendix we review a traditional method of reasoning in statistical mechanics—the method of the most probable distribution—that the configuration of a system having the maximum entropy is, for macroscopic systems, virtually the same as the entropy obtained by counting all possible configurations of the system.

F.1   DISTRIBUTING PARTICLES BETWEEN COMPARTMENTS

Consider two compartments A and B, each the same size, each containing identical, yet distinguishable particles that can pass between them. At an instant of time there are N_A (N_B) particles in A (B), where $N_A + N_B = N$ is a fixed quantity.[1] The number of ways that compartment A can have N_A particles is, from either Eqs. (3.6) or (3.12),

$$\binom{N}{N_A} = \frac{N!}{N_A!(N - N_A)!} = \frac{N!}{N_A!N_B!} . \tag{F.1}$$

The *total number of ways* W that particles can be situated in the two compartments is found by summing Eq. (F.1) over all possible values of N_A. Using Eq. (3.8),

$$W = \sum_{N_A=0}^{N} \binom{N}{N_A} = 2^N . \tag{F.2}$$

Using Eq. (1.62), the "entropy" of this system is, using Eq. (F.2):[2]

$$\ln W = N \ln 2 . \tag{F.3}$$

Which arrangement of particles occurs in the *maximum number of ways*? Consider, starting from an initial distribution of N_A^0 particles in A and N_B^0 in B ($N_A^0 + N_B^0 = N$), the effect of transferring ξ particles between compartments, with $N_A = N_A^0 - \xi$ and $N_B = N_B^0 + \xi$, $-N_B^0 \leq \xi \leq N_A^0$. The number of ways to transfer ξ particles is

$$w(\xi) = \frac{N!}{(N_A^0 - \xi)!(N_B^0 + \xi)!} .$$

---

[1] In this section, N_A does not mean Avogadro's number.

[2] Equation (F.3) does not qualify as entropy because it cannot satisfy the requirements given in Eq. (1.24). Nevertheless, we'll refer to it as entropy anyways. In fact, now might be a good time to mention *information*, a generalization of entropy to systems not in thermal equilibrium, of which Eq. (F.3) is a typical result. See [3, Chapter 12].

The logarithm of this number is (using the Stirling approximation)

$$\ln w(\xi) = \ln N! - \ln(N_A^0 - \xi)! - \ln(N_B^0 + \xi)!$$
$$\approx N \ln N - (N_A^0 - \xi) \ln(N_A^0 - \xi) - (N_B^0 + \xi) \ln(N_B^0 + \xi) . \tag{F.4}$$

As can be shown, the derivative of $\ln w(\xi)$ with respect to ξ is[3]

$$\frac{\partial}{\partial \xi} \ln w(\xi) = \ln \left(\frac{N_A^0 - \xi}{N_B^0 + \xi} \right) . \tag{F.5}$$

The maximum occurs for $\xi = \xi^* \equiv \frac{1}{2} \left(N_A^0 - N_B^0 \right)$. The configuration having maximum entropy is thus for A and B to each have $N/2$ particles. The maximum entropy is found by substituting $\xi = \xi^*$ into Eq. (F.4):

$$\ln w(\xi^*) = N \ln 2 . \tag{F.6}$$

Equation (F.6), the entropy of the configuration having maximum entropy is, within the accuracy of the Stirling approximation, the *same* as Eq. (F.3), the entropy of the system obtained by including *all possible arrangements of particles.*[4] Is this an accident? Can the value of the sum in Eq. (F.2) (for W) be approximated by just *one term* in the series, the number of configurations associated with maximum entropy (what we have called $w(\xi^*)$)? In short, the answer is *yes*. We cannot give a general proof, but for N sufficiently large,[5] there is almost always a configuration of the system that occurs in such a predominantly large number of ways, that the sum over system configurations[6] can be replaced with the largest term in the series. This behavior is exemplified in the formula for $\ln w(\xi)$, Eq. (F.4). From Eq. (F.5), the second derivative evaluated at the configuration ξ^* has the value $\partial^2 \ln w(\xi)/\partial \xi^2 |_{\xi^*} = -4/N$ (show this). Thus, for system configurations ξ in the vicinity of ξ^*, we have the Taylor expansion

$$\ln w(\xi) \approx N \ln 2 - \frac{2}{N} (\xi - \xi^*)^2 + \cdots ,$$

or, equivalently,

$$w(\xi) \approx 2^N \exp \left(-\frac{2}{N} (\xi - \xi^*)^2 \right) .$$

For large N, configurations relatively close to ξ^*, those with $\xi = \xi^* \pm \sqrt{N}$, make a negligible contribution to the entropy in comparison to the entropy of the most frequently occurring configuration.

F.2   THE BOLTZMANN DISTRIBUTION

Consider a system composed of N identical, weakly interacting subsystems.[7] The energy of the ith subsystem is E_i, and the total energy of the system is the sum of the subsystem energies, $E = \sum_{i=1}^{N} E_i$. Each subsystem has the same spectrum of allowed energies, $\{\lambda_j\}$. For the purposes of this calculation, we shift emphasis from the energy of each subsystem to the *number* of subsystems n_j that have energy level λ_j, the *occupation numbers* (see Section D.7.2). The occupation numbers are constrained such that

$$\sum_j n_j = N , \tag{F.7}$$

---

[3] In evaluating the derivative in Eq. (F.5), we're treating ξ as if it's a continuous variable, whereas in actuality it's an integer. This can be justified when N is sufficiently large.

[4] The "state" of the system of two compartments with N particles distributed between them has not been specified with any further refinement; we must include in W all possible configurations consistent with the specification of the system.

[5] Such as we have for macroscopic systems.

[6] Configurations compatible with the macroscopic specification of the system.

[7] The concept of weak interaction is discussed in Section 4.1.2.

where the sum in Eq. (F.7) is over all possible energy levels of the system. The total energy of the system is then

$$\sum_j n_j \lambda_j = E . \tag{F.8}$$

The total energy can be distributed over the N subsystems in many ways. For a given set of occupation numbers $\{n_j\}$, there are $W(\{n_j\})$ permutations of the system (see Eq. (3.12)),

$$W(\{n_j\}) = \frac{N!}{\prod_k (n_k!)} . \tag{F.9}$$

Which occupation numbers maximize W? We expect (based on what we found in Section F.1) that the most probable configuration occurs in such a predominantly large number of ways that the function W in Eq. (F.9) (or $\ln W$) has an extremely sharp maximum. Ordinarily, multivariable functions $f(x_1, \cdots, x_N)$ have extrema where $\mathrm{d}f = 0$, i.e., where

$$\mathrm{d}f = \sum_{i=1}^N \frac{\partial f}{\partial x_i} \mathrm{d}x_i = 0 .$$

If the variables $\{x_i\}$ can be varied independently, extrema are located by finding the solutions of the N equations $\partial f / \partial x_j = 0$, $j = 1, \cdots, N$. In our case, however, the variables *can't* be varied independently because of the two equations of constraint, (F.7) and (F.8). The method of *Lagrange multipliers* was developed just for this situation, in which we seek the extremum of the associated function (where we've added zero to $\ln W$)

$$F[n_1, n_2, \cdots] \equiv \ln W[n_1, n_2, \cdots] + \alpha \left(N - \sum_j n_j \right) + \beta \left(E - \sum_j n_j \lambda_j \right) , \tag{F.10}$$

where α, β are presently unknown quantities. By the Lagrange method, we're now free to seek the extremum of F in Eq. (F.10), treating the occupation numbers as independent variables (show this):

$$\delta F = -\sum_j \left(\ln n_j + \alpha + \beta \lambda_j \right) \delta n_j = 0 ,$$

and thus the occupation numbers that maximize the entropy are such that

$$\ln n_j + \alpha + \beta \lambda_j = 0 . \qquad \text{(for all } j\text{)}$$

The most probable value of the occupation numbers is therefore

$$n_j = \mathrm{e}^{-(\alpha + \beta \lambda_j)} ,$$

precisely in the form assumed by Gibbs (see Section 4.1.2.7). The Lagrange multipliers can be found by enforcing the equations of constraint. Using Eq. (F.7),

$$\mathrm{e}^{-\alpha} = \frac{N}{\sum_j \mathrm{e}^{-\beta \lambda_j}} \equiv \frac{N}{Z(\beta)} ,$$

The probability of finding a subsystem (*any* of the N subsystems) in energy state λ_j is therefore

$$P_j \equiv \frac{n_j}{N} = \frac{\mathrm{e}^{-\beta \lambda_j}}{Z(\beta)} ,$$

the probability distribution for the canonical ensemble. We showed in Section 4.1.2.8 that $\beta = (kT)^{-1}$.

F.3 FERMI-DIRAC AND BOSE-EINSTEIN DISTRIBUTIONS

In Section 5.5 we derived the grand partition functions for systems of identical bosons or fermions, from which we found the Bose-Einstein and Fermi-Dirac distributions for the average occupation numbers of these particles in thermal equilibrium (see Eq. (5.60)). These distribution functions can also be derived by the method of the most probable distribution, which we demonstrate in this section. To do so, we divide the spectrum of single-particle energies into energy "cells," such as indicated in Fig. F.1. For macroscopic systems, the energy levels become closely spaced.[8] Figure F.1

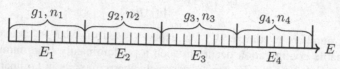

Figure F.1: Division of single-particle energy levels into energy cells, with g_i energy levels per cell, which are occupied by n_i particles.

is notional—it's not meant to imply that energy levels are equally spaced. The energies labeled E_j in Fig. F.1 are the *average* energy of each cell containing g_j energy levels, where it's presumed that $g_j \gg 1$. Assume there are n_j particles occupying the energy levels in the jth cell. The occupation numbers must satisfy the constraints $\sum_j n_j = N$ and the total energy of the system $E = \sum_j n_j E_j$.

For fermions, no single energy level can be occupied by more than one particle; thus, n_j cannot exceed g_j. The number of ways n_j particles can be distributed over the g_j energy levels is therefore the number of ways the g_j levels can be divided into two groups: one consisting of n_j levels that are filled with one particle apiece, and the other consisting of $(g_j - n_j)$ levels that are unoccupied:

$$w(j) = \frac{g_j!}{n_j!(g_j - n_j)!} .$$

The number of ways particles can be distributed over all energy levels is therefore, *for a given set of occupation numbers,*

$$W_{FD}[\{n_k\}] = \prod_i \frac{g_i!}{n_i!(g_i - n_i)!} . \tag{F.11}$$

For bosons, the number of distinct ways that n_i identical, indistinguishable particles can be distributed over g_i energy levels (where there is no restriction on the number of particles in each energy level) was shown in Section 5.3.1 to be

$$w(j) = \frac{(n_j + g_j - 1)!}{n_j!(g_j - 1)!} .$$

The number of ways particles can be distributed over all energy levels is then, *for a given set of occupation numbers,*

$$W_{BE}[\{n_k\}] = \prod_i \frac{(n_i + g_i - 1)!}{n_i!(g_i - 1)!} . \tag{F.12}$$

Which occupation numbers describe systems in thermal equilibrium? Answer: Those that maximize the entropy. The entropy is found from

$$S = k \ln \left(\sum_{\{n_i\}} W[\{n_i\}] \right) , \tag{F.13}$$

---

[8] The energy difference between successive energy levels scales with system volume as $\Delta E \sim V^{-2/3}$, Section 2.1.5.

where the sum in Eq. (F.13) is restricted to sets of occupation numbers satisfying Eq. (F.7). We expect, because of the largeness of the numbers involved, that the logarithm of the sum in Eq. (F.13) can be approximated by the logarithm of the largest term in the series, which we denote as $W[\{n_i^*\}]$. We can determine the occupation numbers associated with the largest term in the series by the same method adopted in the previous section: Form the associated function F in Eq. (F.10) and find its extremum.

From Eq. (F.11),

$$\ln W_{FD} = \sum_i \left[\ln g_i! - \ln n_i! - \ln(g_i - n_i)! \right] . \tag{F.14}$$

Assume that n_i and g_i are large enough that the Stirling approximation is accurate, Eq. (3.14). Applying Stirling's approximation to all factorials in Eq. (F.14), we have (show this)

$$\ln W_{FD} \approx \sum_i \left[n_i \ln \left(\frac{g_i}{n_i} - 1 \right) - g_i \ln \left(1 - \frac{n_i}{g_i} \right) \right] . \tag{F.15}$$

Make the same approximations for W_{BE} in Eq. (F.12) (show this):

$$\ln W_{BE} \approx \sum_i \left[n_i \ln \left(1 + \frac{g_i - 1}{n_i} \right) + (g_i - 1) \ln \left(1 + \frac{n_i}{g_i - 1} \right) \right] . \tag{F.16}$$

Because $g_i \gg 1$, we have a common expression that encompasses Eqs. (F.15) and (F.16),

$$\ln W_\theta \approx \sum_i \left[n_i \ln \left(\frac{g_i}{n_i} + \theta \right) + \theta g_i \ln \left(1 + \theta \frac{n_i}{g_i} \right) \right] , \qquad (\theta = \pm 1) \tag{F.17}$$

where $\theta = +1$ for bosons and $\theta = -1$ for fermions. As is straightforward to show from Eq. (F.17),

$$\delta \ln W_\theta = \sum_i \ln \left(\frac{g_i}{n_i} + \theta \right) \delta n_i ,$$

where each of the occupation numbers can be varied independently because we're about to use the method of Lagrange multipliers. As in Eq. (F.10), we seek the extremum of

$$F \equiv \ln W_\theta + \alpha \left(N - \sum_j n_j \right) + \beta \left(E - \sum_i n_i E_i \right) .$$

The extremum is found from $\delta F = 0$:

$$\delta F = \sum_i \left[\ln \left(\frac{g_i}{n_i} + \theta \right) - \alpha - \beta E_i \right]_{n_i = n_i^*} \delta n_i = 0 ,$$

implying that the set of occupation numbers $\{n_i^*\}$ maximizing the sum appearing in Eq. (F.13) is

$$n_i^* = \frac{g_i}{e^{\alpha + \beta E_i} - \theta} . \tag{F.18}$$

Because n_i^* is proportional to g_i (in Eq. (F.18)),

$$\frac{n_i^*}{g_i} = \frac{1}{e^{\alpha + \beta E_i} - \theta} \qquad (\theta = \pm 1) \tag{F.19}$$

can be interpreted as the most probable number of particles occupying any of the energy levels associated with energy E_i in Fig. F.1 (depending on whether they are bosons or fermions). One might question this interpretation given that the grouping of energy levels into cells in Fig. F.1 is arbitrary. We showed in Section 5.5.2 how, in the grand canonical ensemble, the thermal average of the occupation numbers of identical bosons and fermions is given by Eq. (5.60), the same as Eq. (F.19). It's only because of the agreement of Eq. (F.19) with Eq. (5.60) (the derivation of which doesn't group energy levels into cells) that the argument in this section is acceptable.

Using the occupation numbers $\{n_i^*\}$ we can calculate the entropy of the equilibrium state (in which S is a maximum),

$$S = k \ln \left(W[\{n_i^*\}] \right) .$$

Use Eq. (F.17), setting $n_i = n_i^*$,

$$\frac{S}{k} = \ln W[\{n_i^*\}] = \sum_i \left[n_i^* \ln \left(\frac{g_i}{n_i^*} + \theta \right) + \theta g_i \ln \left(1 + \theta \frac{n_i^*}{g_i} \right) \right]$$

$$= \sum_i \left[n_i^* (\alpha + \beta E_i) - \theta g_i \ln \left(1 - \theta e^{-(\alpha + \beta E_i)} \right) \right]$$

$$= \alpha N + \beta E - \theta \sum_i g_i \ln \left(1 - \theta e^{-(\alpha + \beta E_i)} \right) , \tag{F.20}$$

where in the second line we've used Eq. (F.18) and in the last line we've used the assumption that the most probable occupation numbers $\{n_i^*\}$ are so predominantly large that $N \approx \sum_j n_j^*$; likewise for the energy, $E \approx \sum_j n_j^* E_j$. From Eq. (1.53) (general thermodynamics), we know that

$$\frac{S}{k} = \frac{1}{kT} (E + PV - \mu N) . \tag{F.21}$$

and thus we can identify, comparing Eqs. (F.20) and (F.21), $\beta = 1/(kT)$ and $\alpha = -\beta \mu$. With these identifications we infer, again comparing Eqs. (F.20) and (F.21),

$$\frac{PV}{kT} = -\theta \sum_i g_i \ln \left(1 - \theta e^{-\beta(E_i - \mu)} \right) . \tag{F.22}$$

Equation (F.22) should be compared with Eq. (5.64); the two are quite similar. Equation (5.64) gives us the equation of state of an ideal gas of fermions or bosons without making the approximation (made in this appendix) that energy levels can be grouped into cells of g_i microscopic energy levels. Using Eq. (4.76), for the grand potential of the ideal gas $\Phi = -kT \ln Z_G = -PV$, we make the identification from Eq. (F.22),

$$\ln Z_G = -\theta \sum_i g_i \ln \left(1 - \theta e^{-\beta(E_i - \mu)} \right) \tag{F.23}$$

Using Eq. (4.78), we have from Eq. (F.23),

$$N = kT \frac{\partial \ln Z_G}{\partial \mu} \bigg|_{T,V} = \sum_i \frac{g_i}{e^{\beta(E_i - \mu)} - \theta} = \sum_i n_i^* ,$$

where we've used Eq. (F.19). Likewise, using Eq. (4.82), it's straightforward to show that

$$E = -\frac{\partial \ln Z_G}{\partial \beta} \bigg|_{z,V} = \sum_i n_i^* E_i ,$$

where $z \equiv e^{\beta \mu}$ is the fugacity. Thus, the approximations made in this appendix are consistent with the general relations developed in Chapters 4 and 5.

Bibliography

[1] K. De Raedt, H. Michielsen, H. De Raedt, et al. Massively parallel quantum computer simulator. *Computer Physics Communications*, 176:121–136, 2007.

[2] W. Kohn. Electronic structure of matter-wave functions and density functionals. *Reviews of Modern Physics*, 71:1253–1266, 1999.

[3] J.H. Luscombe. *Thermodynamics*. CRC Press, 2018.

[4] H. Gould and J. Tobochnik. *Statistical and Thermal Physics: With Computer Applications*. Princeton University Press, 2010.

[5] E.E. Daub. Maxwell's demon. *Studies Hist. & Phil. Sci.*, 1:213–227, 1970.

[6] A.S. Eddington. *The Nature of the Physical World*. Macmillan, 1929.

[7] G.-J. Su. Modified law of corresponding states for real gases. *Industrial and Engineering Chemistry*, 38:803–806, 1946.

[8] H.B. Callen. *Thermodynamics*. John Wiley, 1960.

[9] M. Planck. *The Theory of Heat Radiation*. Dover, 1991.

[10] M. Planck. *The Theory of Heat*. Macmillan, 1932.

[11] J.W. Gibbs. Thermodynamics. In *The Collected Works of J. Willard Gibbs*, volume I. Yale University Press, 1948.

[12] A. Einstein. *Albert Einstein: Philosopher Scientist, Vol. 1*. Open Court, 1949.

[13] J.H. Luscombe. *Core Principles of Special and General Relativity*. CRC Press, 2019.

[14] S. Putterman. *Superfluid Hydrodynamics*. North Holland Publishing, 1974.

[15] D.D. Nolte. The tangled tale of phase space. *Physics Today*, 63(4):33–38, 2010.

[16] B.H. Borden and J.H. Luscombe. *Mathematical Methods in Physics, Engineering and Chemistry*. Wiley, 2019.

[17] R.K. Pathria and R.D. Beale. *Statistical Mechanics*. Elsevier, 2011.

[18] N.W. Ashcroft and N.D. Mermin. *Solid State Physics*. Saunders College Publishing, 1976.

[19] J.W. Gibbs. Elementary Principles in Statistical Mechanics. In *The Collected Works of J. Willard Gibbs*, volume II. Yale University Press, 1948.

[20] J. Perrin. *Atoms*. D. Van Nostrand, 1916.

[21] G.D. Birkhoff. Proof of the ergodic theorem. *Proc Natl Acad Sci USA*, 17:656–660, 1931.

[22] I.E. Farquhar. *Ergodic Theory in Statistical Mechanics*. John Wiley, 1964.

[23] J.L. Lebowitz and O. Penrose. Modern ergodic theory. *Physics Today*, 26(2):23–29, 1973.

[24] N. Bohr. Faraday lecture. Chemistry and the quantum theory of atomic constitution. *Journal of the Chemical Society*, 0:349–384, 1932.

[25] L.D. Landau and E.M. Lifshitz. *Mechanics*. Pergamon Press, 1976.

[26] A. Wintner. *The Analytical Foundations of Celestial Mechanics*. Princeton University Press, 1941.

[27] V.I. Arnold. *Mathematical Methods of Classical Mechanics*. Springer-Verlag, 2nd edition, 1989.

[28] E.T. Whittaker. *A Treatise on the Analytical Dynamics of Particles and Rigid Bodies*. Cambridge University Press, fourth edition, 1937.

[29] H. Cramer. *Mathematical Methods of Statistics*. Princeton University Press, 1946.

[30] M. Kac. *Probability and Related Topics in Physical Sciences*. Interscience Publishers, 1959.

[31] R. Carnap. *Logical Foundations of Probability*. University of Chicago Press, 1962.

[32] M. Born. *Natural Philosophy of Cause and Chance*. Oxford University Press, 1949.

[33] R.C. Tolman. *The Principles of Statistical Mechanics*. Oxford University Press, 1938.

[34] Y.G. Sinai. *Probability Theory*. Springer-Verlag, 1992.

[35] A. Papoulis. *Probability, Random Variables, and Stochastic Processes*. McGraw-Hill, 1991.

[36] M.G. Kendall and A. Stuart. *The Advanced Theory of Statistics*, volume 1. Hafner, 1963.

[37] J.L. Lebowitz and E.H. Lieb. Existence of thermodynamics for real matter with Coulomb forces. *Phys. Rev. Lett.*, 22:631–634, 1969.

[38] A.I. Khinchin. *Mathematical Foundations of Statistical Mechanics*. Dover, 1949.

[39] H. Grad. Statistical mechanics, thermodynamics, and fluid dynamics of systems with an arbitrary number of integrals. *Comm. Pure Appl. Math.*, 5:455–494, 1952.

[40] E.T. Jaynes. The Gibbs Paradox. In *Maximum Entropy and Bayesian Methods*, pages 1–22. Springer, 1992.

[41] K. Denbigh. *The Principles of Chemical Equilibrium*. Cambridge University Press, 1981.

[42] D. Ruelle. *Statistical Mechanics*. W.A. Benjamin, 1969.

[43] L. Van Hove. Quelques proprieties generales de l'integrale de configuration d'un systeme de particules avec interaction. *Physica*, 15:951–961, 1949.

[44] W. Rudin. *Principles of Mathematical Analysis*. McGraw-Hill, 1964.

[45] G.H. Hardy, J.E. Littlewood, and G. Polya. *Inequalities*. Cambridge University Press, 1952.

[46] R.P. Feynman. *Statistical Mechanics*. W.A. Benjamin, 1972.

[47] H.S. Leff and A.F. Rex. *Maxwell's Demon: Entropy, Information, Computing*. Princeton University Press, 1990.

[48] W.E. Henry. Spin Paramagnetism of Cr^{+++}, Fe^{+++}, and Gd^{+++} at Liquid Helium Temperatures and in Strong Magnetic Fields. *Physical Review*, 88:559–562, 1952.

[49] P.A.M. Dirac. *The Principles of Quantum Mechanics*. Oxford University Press, 1958.

[50] M. Abramowitz and I.A. Stegun, editors. *Handbook of Mathematical Functions*. U.S. Department of Commerce, 1964.

[51] S. Chandrasekhar. *An Introduction to the Study of Stellar Structure*. University of Chicago Press, 1939.

[52] C. Bloch. Diagram expansions in quantum statistical mechanics. In J. de Boer and G.E. Uhlenbeck, editors, *Studies in Statistical Mechanics*, volume III, pages 1–118. North-Holland Publishing Company, 1965.

[53] A.L. Fetter and J.D. Walecka. *Quantum Theory of Many-Particle Systems*. McGraw-Hill, 1971.

[54] J.E. Mayer and M.G. Mayer. *Statistical Mechanics*. John Wiley, 1940.

[55] G.E. Uhlenbeck and G.W. Ford. *Lectures in Statistical Mechanics*. American Mathematical Society, 1963.

[56] G. Polya. *Mathematics and Plausible Reasoning*, volume I. Princeton University Press, 1954.

[57] D.R. Lide, editor. *CRC Handbook of Chemistry and Physics*. CRC Press, 86th edition, 2005.

[58] R.P. Feynman. The development of the space-time view of quantum electrodynamics. *Science*, 153:699–708, 1966.

[59] L. Tonks. The complete equation of state of one, two, and three-dimensional gases of hard elastic spheres. *Physical Review*, 50:955–962, 1936.

[60] H. Takahashi. Eine Einfache Methode zur Behandlung der Statistichen Mechanik Eindimensionaler Substanzen. *Proceedings of the Physico-Mathematical Society of Japan*, 24:60, 1942.

[61] E.H. Lieb and D.C. Mattis. *Mathematical Physics in One Dimension*. Academic Press, 1966.

[62] T. Trif. Multiple integrals of symmetric functions. *The American Mathematical Monthly*, 104(7):605–608, 1997.

[63] J.G. Kirkwood. Molecular distributions in liquids. *J. Chem. Phys.*, 7:919–924, 1939.

[64] B.J. Alder and T.E. Wainwright. Phase transition for a hard sphere system. *J. Chem. Phys.*, 27:1208–1209, 1957.

[65] A.P. Gast and W.B. Russel. Simple ordering in complex fluids. *Physics Today*, 51(12):24–30, 1998.

[66] S.G. Brush. History of the Lenz-Ising model. *Reviews of Modern Physics*, 39(883–893), 1967.

[67] Zhengbing Bian et al. Discrete optimization using quantum annealing on sparse Ising models. *Frontiers in Physics*, 2:56–66, 2014.

[68] P.A.M. Dirac. Quantum mechanics of many-electron systems. *Proceedings of the Royal Society A*, 123:714–733, 1929.

[69] W. Heisenberg. Zur Theorie des Ferromagnetismus. *Zeit. Phys.*, 49:619–636, 1928.

[70] M.B. Stearns. Why is iron magnetic? *Physics Today*, 31(4):34–39, 1978.

[71] H.A. Kramers and G.H. Wannier. Statistics of the two-dimensional ferromagnet. Part I. *Physical Review*, 60(252–262), 1941.

[72] H.A. Kramers and G.H. Wannier. Statistics of the two-dimensional ferromagnet. Part II. *Physical Review*, 60(263-276), 1941.

[73] E.W. Montroll. Statistical mechanics of nearest-neighbor systems. *J. Chem. Phys.*, 9(706–721), 1941.

[74] F.R. Gantmacher. *Applications of the Theory of Matrices*. Interscience Publishers, 1959.

[75] J.H. Luscombe and C.L. Frenzen. Transfer matrix redundancy and renormalization group transformations. Unpublished.

[76] J.P. Hansen and I.R. McDonald. *Theory of Simple Liquids*. Academic Press, 1976.

[77] E.C. Titchmarsh. *Introduction to the Theory of Fourier Integrals*. Oxford University Press, 1948.

[78] H.L. Frisch and J.L. Lebowitz. *The Equilibrium Theory of Classical Fluids*. W.A. Benjamin, 1964.

[79] L. Verlet. Computer "experiments" on classical fluids. II. Equilibrium correlation functions. *Physical Review*, 165:201–214, 1968.

[80] J.K. Percus and G.J. Yevick. Analysis of classical statistical mechanics by means of collective coordinates. *Physical Review*, 110:1–13, 1958.

[81] D.L. Goodstein. *States of Matter*. Prentice-Hall, 1975.

[82] A.B. Pippard. *The Elements of Classical Thermodynamics*. Cambridge University Press, 1957.

[83] E.A. Guggenheim. The principle of corresponding states. *J. Chem. Phys.*, 13:253–261, 1945.

[84] L.P. Kadanoff, W. Götze, D. Hamblen, et al. Static phenomena near critical points: Theory and experiment. *Reviews of Modern Physics*, 39:395–431, 1967.

[85] J.C. Maxwell. On the dynamical evidence of the molecular constitution of bodies. In W.D. Niven, editor, *The Scientific Papers of James Clerk Maxwell*, volume Two, pages 418–438. Cambridge University Press, 1890.

[86] G.F. Mazenko. *Equilibrium Statistical Mechanics*. John Wiley, 2000.

[87] R.O. Watts. Percus-Yevic equation applied to a Lennard-Jones fluid. *J. Chem. Phys.*, 48:50–55, 1968.

[88] M. Kac, G.E. Uhlenbeck, and P.C. Hemmer. On the van der Waals theory of the vapor-liquid equilibrium. I. Discussion of a one-dimensional model. *Journal of Mathematical Physics*, 4:216–228, 1963.

[89] J.L. Lebowitz and O. Penrose. Rigorous treatment of the van der Waals-Maxwell theory of the liquid-vapor transition. *Journal of Mathematical Physics*, 7:98–113, 1966.

[90] M.A. Anisimov et al. Critical exponents of liquids. *Sov. Phys. JETP*, 49:844–848, 1979.

[91] R. Hocken and M.R. Moldover. Ising critical exponents in real fluids: An experiment. *Phys. Rev. Lett.*, 37:29–32, 1976.

[92] F. London. *Superfluids: Macroscopic Theory of Superfluid Helium*, volume II. Dover, 1954.

[93] J.H. Luscombe. Current issues in nanoelectronic modeling. *Nanotechnology*, 4:1–20, 1992.

[94] N. Tateiwa, Y. Haga, and E. Yamamoto. Novel critical behavior of magnetization in URhSi: Similarities to the uranium ferromagnetic superconductors UGe_2 and URhGe. *Phys. Rev. B*, 99:094417, 2019.

[95] J. Mattsson, C. Djurberg, and P. Nordblad. Determination of the critical exponent β from measurements of a weak spontaneous magnetization in the 3d Ising antiferromagnet FeF_2. *Journal of Magnetism and Magnetic Materials*, 136:L23–L28, 1994.

[96] M. Itskov, D. Dargazany, and K. Hörnes. Taylor expansion of the inverse function with application to the Langevin function. *Mathematics and Mechanics of Solids*, 7:693–701, 2011.

[97] H.E. Stanley. *Introduction to Phase Transitions and Critical Phenomena*. Oxford University Press, 1971.

[98] B. Kaufman and L. Onsager. Crystal statistics. III. Short-range order in a binary Ising lattice. *Physical Review*, 76:1244–1252, 1949.

[99] L.D. Landau and E.M. Lifshitz. *Statistical Physics*. Pergamon Press, 1978.

[100] P.G. de Gennes and J. Prost. *The Physics of Liquid Crystals*. Oxford University Press, 1993.

[101] N. Goldenfeld. *Lectures on Phase Transitions and the Renormalization Group*. Addison-Wesley, 1992.

[102] L. Onsager. Crystal statistics. I. A two-dimensional model with an order-disorder transition. *Physical Review*, 65:117–149, 1944.

[103] M.E. Fisher. The theory of equilibrium critical phenomena. *Reports on Progress in Physics*, 30:615–730, 1967.

[104] R.J. Creswick, H.A. Farach, and C.P. Poole. *Introduction to Renormalization Group Methods in Physics*. John Wiley, 1992.

[105] V.L. Ginzburg. Some remarks on phase transitions of the second kind and the microscopic theory of ferroelectric materials. *Soviet Physics Solid State*, 2:1824–1834, 1961.

[106] D.J. Amit. The Ginzburg criterion rationalized. *Journal of Physics C: Solid State Physics*, 7:3369–3377, 1974.

[107] B.M. McCoy and T.T. Wu. *The Two-Dimensional Ising Model*. Harvard University Press, 1973.

[108] R.J. Baxter. *Exactly Solved Models in Statistical Mechanics*. Academic Press, 1982.

[109] C.J. Thompson. *Mathmatical Statistical Mechanics*. Macmillan, 1972.

[110] I.S. Gradshteyn and I.M. Ryzhik. *Tables of Integrals, Series, and Products*. Academic Press, 1965.

[111] C.N. Yang. The spontaneous magnetization of a two-dimensional Ising model. *Physical Review*, 85:808–816, 1952.

[112] E.W. Montroll, R.B. Potts, and J.C. Ward. Correlations and spontaneous magnetization of the two-dimensional Ising model. *Journal of Mathematical Physics*, 4:308–322, 1963.

[113] G.S. Rushbrooke. On the thermodynamics of the critical region for the Ising problem. *J. Chem. Phys.*, 39:842, 1963.

[114] D.B. Abraham. Susceptibility and fluctuations in the Ising ferromagnet. *Physics Letters A*, 43:163–164, 1973.

[115] R.B. Griffiths. Thermodynamic inequality near the critical point for ferromagnets and fluids. *Phys. Rev. Lett.*, 14:634–634, 1965.

[116] R.B. Griffiths. Ferromagnets and simple fluids near the critical point: Some thermodynamic inequalities. *J. Chem. Phys.*, 43:1958, 1965.

[117] L.E. Reichl. *A Modern Course in Statistical Physics*. John Wiley, 2nd edition, 1998.

[118] N.D. Mermin and H. Wagner. Absence of ferromagnetism or antiferromagnetism in one- or two-dimensional isotropic Heisenberg models. *Phys. Rev. Lett.*, 17:1133–1136, 1966.

[119] P.C. Hohenberg. Existence of long-range order in one and two dimensions. *Physical Review*, 158:383–386, 1967.

[120] D.R. Nelson and M.E. Fisher. Soluble renormalization groups and scaling fields for low-dimensional Ising systems. *Annals of Physics*, 91:226–274, 1975.

[121] B. Widom. Equation of state in the neighborhood of the critical point. *J. Chem. Phys.*, 11:3898–3905, 1965.

[122] J.T. Ho and J.D. Litster. Magnetic equation of state of $CrBr_3$ near the critical point. *Phys. Rev. Lett.*, 22:603–606, 1969.

[123] L.P. Kadanoff. Scaling laws for Ising models near T_c. *Physics*, 2:262–272, 1966.

[124] K.G. Wilson. Renormalization group and critical phenomena. I. Renormalization group and the Kadanoff scaling picture. *Phys. Rev. B*, 4:3174–3183, 1971.

[125] H.A. Rose and P.L. Sulem. Fully developed turbulence and statistical mechanics. *Journal de Physique*, 39:441–484, 1978.

[126] W.D. McComb. *The Physics of Fluid Turbulence*. Oxford University Press, 1990.

[127] B.B. Mandelbrot. *The Fractal Geometry of Nature*. Macmillan, 1982.

[128] A.H. Clark, J.D. Thompson, M.D. Shattuck, N.T. Ouellette, and C.S. O'Hern. Critical scaling near the yielding transition in granular media. *Physical Review E*, 97:062901, 2018.

[129] J.D. Thompson and A.H. Clark. Critical scaling for yield is independent of distance to isostaticity. *Phys. Rev. Research*, 1:012002, 2019.

[130] K.G. Wilson. The renormalization group: Critical phenomena and the Kondo problem. *Reviews of Modern Physics*, 47:773–840, 1975.

[131] J.H. Luscombe and R.C. Desai. Statistical mechanics of a fractal lattice: Renormalization-group analysis of the Sierpinski gasket. *Phys. Rev. B*, 32:1614–1627, 1985.

[132] A.J. Guttmann. Asymptotic analysis of power-series expansions. In C. Domb and J. Lebowitz, editors, *Phase Transitions and Critical Phenomena*, volume 13, pages 1–234. Academic Press, 1989.

[133] G.A. Baker and P. Graves-Morris. *Padé approximants*. Cambridge University Press, 1996.

[134] S. Ma. Introduction to the renormalization group. *Reviews of Modern Physics*, 45:589–614, 1973.

[135] K.G. Wilson and J. Kogut. The renormalization group and the ϵ expansion. *Physics Reports*, 12:75–199, 1974.

[136] M.E. Fisher. The renormalization group in the theory of critical behavior. *Reviews of Modern Physics*, 46:597–616, 1974.

[137] D.J. Wallace and R.K.P. Zia. The renormalization group approach to scaling in physics. *Reports on Progress in Physics*, 41:1–86, 1978.

[138] J. Zinn-Justin. *Quantum Fied Theory and Critical Phenomena*. Oxford University Press, 1993.

[139] T. Niemeyer and J.M.J. van Leeuwen. Wilson theory for 2-dimensional Ising spin systemss. *Physica*, 71:17–40, 1974.

[140] G.F. Mazenko, M.J. Nolan, and O.T. Valls. Real-space dynamic renormalization group. I. General formalism. *Phys. Rev. B*, 22:1263–1274, 1980.

[141] R.J. Glauber. Time-dependent statistics of the Ising model. *Journal of Mathematical Physics*, 4:294–307, 1963.

[142] I. Syozi. Transformation of Ising models. In C. Domb and M.S. Green, editors, *Phase Transitions and Critical Phenomena*, volume 1, page 269. Academic Press, 1972.

[143] A.S. Sudbo and P.C. Hemmer. Cumulant expansion in renormalization-group transformations on Ising spin systems: A third-order calculation. *Phys. Rev. B*, 13:980–982, 1976.

[144] T.W. Burkhardt and J.M.J. van Leeuwen, editors. *Real-Space Renormalization*. Springer-Verlag, 1982.

[145] G.F. Mazenko and O.T. Valls. The real space dynamic renormalization group. In T.W. Burkhardt and J.M.J. van Leeuwen, editors, *Real-Space Renormalization*, pages 87–117. Springer-Verlag, 1982.

[146] J.H. Luscombe and G.F. Mazenko. Higher-order calculations in the real space dynamic renormalization group: One dimension. *Annals of Physics*, 146:174–208, 1983.

[147] G.F. Mazenko, J.E. Hirsch, M.J. Nolan, and O.T. Valls. Real-space dynamic renormalization group. III. Calculation of correlation functions. *Phys. Rev. B*, 23:1431–1446, 1981.

[148] G.F. Mazenko and J.H. Luscombe. Application of the real space dynamic renormalization group to the one-dimensional kinetic Ising model. *Annals of Physics*, 132:121–162, 1981.

[149] K.G. Wilson. Renormalization group and critical phenomena. II. Phase-space cell analysis of critical behavior. *Phys. Rev. B*, 4:3184–3205, 1971.

[150] M. Gell-Mann and F.E. Low. Quantum electrodynamics at small distances. *Physical Review*, 95:1300–1312, 1954.

[151] N.N. Bogoliubov and D.V. Shirkov. *The Theory of Quantized Fields*. Interscience Publishers, 1959.

[152] M.E. Peskin and D.V. Schroeder. *An Introduction to Quantum Field Theory*. Westview Press, 1995.

[153] J. Schwinger, editor. *Selected Papers on Quantum Electrodynamics*. Dover, 1958.

[154] S. Weinberg. *Dreams of a Final Theory*. Hutchinson, 1993.

[155] J.R. Schrieffer. *Theory of Superconductivity*. Addison-Wesley, 1964.

[156] R.P. Feynman and A.R. Hibbs. *Quantum Mechanics and Path Integrals*. McGraw-Hill, 1965.

[157] V.N. Popov. *Functional Integrals and Collective Excitations*. Cambridge University Press, 1988.

[158] E. Artin. *The Gamma Function*. Holt, Rinehart and Winston, 1964.

[159] J.S. Blakemore. *Semiconductor Statistics*. Dover, 1987.

[160] L.D. Cloutman. Numerical evaluation of the Fermi-Dirac integrals. *The Astrophysical Journal Supplement Series*, 71:677–699, 1989.

[161] W.B. Joyce and R.W. Dixon. Analytic approximations for the Fermi energy of an ideal Fermi gas. *Applied Physics Letters*, 31:354–356, 1977.

[162] H. Goldstein. *Classical Mechanics*. Addison-Wesley, 1950.

[163] P.R. Halmos. *Finite-Dimensional Vector Spaces*. D. Van Nostrand, 1958.

[164] F. Schwabl. *Advanced Quantum Mechanics*. Springer, 2004.

[165] G.L. Squires. *Introduction to the Theory of Thermal Neutron Scattering*. Cambridge University Press, 1978.

Index

nted in the United States
Bookmasters